ACSM'S Introduction to EXERCISE SCIENCE

FOURTH EDITION

ACSM'S Introduction to EXERCISE SCIENCE

FOURTH EDITION

Jeffrey A. Potteiger, PhD, FACSM
Grand Valley State University
Grand Rapids, Michigan

Philadelphia • Baltimore • New York • London
Buenos Aires • Hong Kong • Sydney • Tokyo

Acquisitions Editor: Lindsey Porambo
Senior Development Editor: Amy Millholen
Senior Editorial Coordinator: Lindsay Ries
Marketing Manager: Phyllis Hitner
Production Project Manager: Kirstin Johnson
Design Coordinator: Stephen Druding
Manufacturing Coordinator: Margie Orzech
Prepress Vendor: S4Carlisle Publishing Services
ACSM Publications Committee Chair: Jeffrey A. Potteiger, PhD, FACSM
ACSM Chief Operating Officer: Katie Feltman
ACSM Development Editor: Angie Chastain

Fourth Edition, Revised Reprint

Copyright © 2024 American College of Sports Medicine.

3rd edition Copyright © 2018 American College of Sports Medicine. 2nd edition Copyright © 2014 American College of Sports Medicine. 1st edition Copyright © 2011 American College of Sports Medicine. All rights reserved. This book is protected by copyright. No part of this book may be reproduced or transmitted in any form or by any means, including as photocopies or scanned-in or other electronic copies, or utilized by any information storage and retrieval system without written permission from the copyright owner, except for brief quotations embodied in critical articles and reviews. Materials appearing in this book prepared by individuals as part of their official duties as U.S. government employees are not covered by the above-mentioned copyright. To request permission, please contact Wolters Kluwer at Two Commerce Square, 2001 Market Street, Philadelphia, PA 19103, via email at permissions@lww.com, or via our website at shop.lww.com (products and services).

11 10 9 8 7 6 5 4 3

Printed in Mexico

Library of Congress Cataloging-in-Publication Data

ISBN-13: 978-1-9752-1787-7

ISBN-10: 1-9752-1787-X

Cataloging in Publication data available on request from publisher.

Care has been taken to confirm the accuracy of the information present and to describe generally accepted practices. However, the authors, editors, and publisher are not responsible for errors or omissions or for any consequences from application of the information in this publication and make no warranty, expressed or implied, with respect to the currency, completeness, or accuracy of the contents of the publication. Application of this information in a particular situation remains the professional responsibility of the practitioner; the clinical treatments described and recommended may not be considered absolute and universal recommendations.

The authors, editors, and publisher have exerted every effort to ensure that drug selection and dosage set forth in this text are in accordance with the current recommendations and practice at the time of publication. However, in view of ongoing research, changes in government regulations, and the constant flow of information relating to drug therapy and drug reactions, the reader is urged to check the package insert for each drug for any change in indications and dosage and for added warnings and precautions. This is particularly important when the recommended agent is a new or infrequently employed drug. Some drugs and medical devices presented in this publication have Food and Drug Administration (FDA) clearance for limited use in restricted research settings. It is the responsibility of the health care provider to ascertain the FDA status of each drug or device planned for use in their clinical practice.

shop.lww.com

To
Ellen**, **Tate**, and **Caroline
for their never-ending support and love

Preface

Exercise science is an umbrella term used to describe the study of numerous aspects of physical activity, exercise, sport, and athletic performance that have the common characteristic of movement and the adaptations that occur as a result of physical activity, regular exercise, and training and performing in sport and athletic competition. The study of the physiological, psychological, nutritional, motor, and functional adaptations and responses to physical activity, exercise, sport, and athletic competition is key to understanding human health, disease, injury, rehabilitation, and performance. Physical activity and regular exercise play very important roles in promoting overall health and healthy living and reducing the risk for numerous lifestyle diseases. Proper physical training and preparation are crucial for maximizing individual and team performance in sports and athletic competitions, for reducing the risk of injury, and to assist in the rehabilitation and recovery from an injury. Exercise science students must, therefore, be well prepared for understanding how human movement assists individuals in their pursuit of good health and successful sport performance.

ACSM's Introduction to Exercise Science provides an overview of the components important to developing a solid foundation and appreciation of all aspects of exercise science. This book is designed for first-year students in exercise science or any of the related areas, including athletic training and sports medicine, clinical and sport biomechanics, clinical exercise physiology, exercise and sport nutrition, exercise physiology, exercise and sport psychology, and motor development, control, and learning. The 13 chapter topics and content were chosen to represent both the foundational and the broad-based professional areas and issues of exercise science.

Features

Learning Objectives at the beginning of each chapter highlight concepts of importance. Boxes throughout the chapters also highlight important **terms and their definitions**, helping students to identify immediately terminology that may be new or unfamiliar. **Interviews** with prominent exercise science professionals are featured within each chapter; these individuals provide helpful insight into proper preparation for developing a successful professional career. **Critical Thinking Questions** are placed throughout the chapters as well to help facilitate discussion and deeper application of concepts. **Review Questions** at the end of each chapter provide an opportunity for a brief review of the material covered. **Project-Based Learning** assignments help students gain knowledge by working on a topic of practical application that will reinforce understanding and comprehension of the material covered in each chapter. **Figures, tables, and illustrations** support and help explain the content information provided in the chapters.

This book is designed to provide beginning students with an overview of the foundational content within the areas of exercise science as well as options available for professional career opportunities, career development, and employment. A look into the future focuses on several key paths that exercise science may take in the next 20 years. I hope that this book provides students with valuable insight as they explore the wonderful world of the human body.

Acknowledgments

I wish to extend a heartfelt thank you to all my professional colleagues and former students who shared their thoughts and insights with me about the content and direction of this textbook. I would like to specifically acknowledge all of my ACSM colleagues, for their support and belief in me, and Ellen E. Adams, for her critical review of the material.

Jeffrey A. Potteiger, PhD, FACSM

Reviewers

R. Lee Franco, PhD, ACSM-EP
Virginia Commonwealth University
Richmond, Virginia

Diane L. Gill, PhD, FACSM
University of North Carolina
Greensboro, North Carolina

Allan H. Goldfarb, PhD, FACSM
University of North Carolina
Greensboro, North Carolina

Shannon L. Lennon, PhD
University of Delaware
Newark, Delaware

Matt Miller, PhD
Auburn University
Auburn, Alabama

Valerie Moody, PhD
University of Montana
Missoula, Montana

John C. Quindry, PhD, FACSM
University of Montana
Missoula, Montana

R. Andrew Shanely, PhD
Appalachian State University
Boone, North Carolina

Lisa Stegall, PhD
Hamline University
St. Paul, Minnesota

Erin E. Talbert, PhD
University of Iowa
Iowa City, Iowa

Herman van Werkhoven, PhD
Appalachian State University
Boone, North Carolina

Contents

Preface vi

Acknowledgments viii

Reviewers ix

CHAPTER 1 Introduction to Exercise Science 1
What Is Exercise Science 6
History of Exercise Science 9
Exercise Science and the American College of Sports Medicine 16
Academic Preparation in Exercise Science 22

CHAPTER 2 Introduction to Research 39
Research in Exercise Science 40
The Research Process 49
Evidence-Based Practice 60
Student Research in Exercise Science 62

CHAPTER 3 Exercise Science: A Systems Approach 69
Nervous System 72
Muscular System 75
Skeletal System 79
Cardiovascular System 82
Pulmonary System 84
Urinary System 87
Digestive System 90
Endocrine System 93
Immune System 97
Energy Systems 100

CHAPTER 4 Exercise Physiology 115
History of Exercise Physiology 117
The Basis of Study in Exercise Physiology 121
Areas of Study in Exercise Physiology 126
Other Areas of Study 149

CHAPTER 5 Clinical Exercise Physiology 159
History of Clinical Exercise Physiology   160
Clinical Exercise Physiologists' Duties and Responsibilities   164
Specific Disease Conditions   179
Areas of Study in Clinical Exercise Physiology   193

CHAPTER 6 Athletic Training and Sports Medicine 205
History of Athletic Training and Sports Medicine   209
Primary Responsibility Areas of Athletic Training Professionals   214
Sports Medicine   226
Areas of Study in Athletic Training and Sports Medicine   232

CHAPTER 7 Exercise and Sport Nutrition 247
History of Nutrition   250
Basic Nutrients   257
Measuring Nutritional Intake   262
Nutrition for Health   265
Areas of Study in Nutrition for Health   268
Nutrition for Sport   270
Areas of Study in Sport Nutrition   277

CHAPTER 8 Exercise and Sport Psychology 293
History of Exercise and Sport Psychology   296
Study of the Mind and Body   300
Exercise Psychology   312
Exercise Behavior   317
Areas of Study in Exercise and Sport Psychology   321

CHAPTER 9 Motor Behavior 335
History of Motor Behavior   337
Motor Development   341
Motor Learning   348
Motor Control   359
Areas of Study in Motor Behavior   365

CHAPTER 10 Clinical and Sport Biomechanics 377
History of Biomechanics   379
Study of Biomechanics   383
Basic Concepts Related to Movement   390
Complex Movement Concepts   394
Areas of Study in Biomechanics   396

CHAPTER 11 Assessment and Equipment in Exercise Science 413

Pretesting Guidelines and Procedures 414
Cardiovascular and Pulmonary Function Assessment 417
Musculoskeletal Assessment 428
Energy Balance Assessment 434
Body Composition Assessment 439
Blood Collection Equipment and Assessment 442
Rehabilitation Equipment and Assessment 446
Motor Performance Assessment 449
Behavioral and Psychological Assessment 453

CHAPTER 12 Careers and Professional Issues in Exercise Science 461

Certification, Licensure, and Registration 463
Career Employment and Professional Opportunities 468
Professional Organizations in Exercise Science 479
Professional Organizations Related to Exercise Science 486
U.S. Government Agencies with an Interest in Exercise Science 487
Additional Organizations and Agencies in Exercise Science 493

CHAPTER 13 Exercise Science in the Twenty-First Century 499

Exercise Science and Health 500
Epidemiology and Health Promotion 502
Using Past Information to Improve Future Health 509
What Will the Future Bring? 511
Exercise Science and Sport and Athletic Competition 515
What Will the Future Bring? 516

Index 535

CHAPTER 1

Introduction to Exercise Science

After completing this chapter, you will be able to:

1. Explain the importance of exercise science as it relates to enhancing our understanding of health, physical activity, exercise, sport, and athletic performance.

2. Identify the different disciplines, subdisciplines, and specialty areas of exercise science, and describe how they relate to exercise science.

3. Describe the key highlights in the historical development of exercise science.

4. Identify the key influential accomplishments of the American College of Sports Medicine for promoting exercise science.

5. Recognize the general and advanced undergraduate coursework necessary for a career in an exercise science or health care profession.

Physical activity, exercise, sport, and athletic performance are integral parts of so many societies and cultures around the world. Some of our earliest recordings of history mention the importance of physical activity and health. For example, many of the Greek philosopher Aristotle's teachings describe the significance of physical health to a good life (1). Indeed, there is considerable research evidence that supports the role of being physically active and participating in regular exercise for promoting good **health** and decreasing the risk of **morbidity** and **mortality** (2–11). Despite this information, the morbidity and mortality rates from lifestyle-related diseases remain high (12). Figure 1.1 shows the most common causes of mortality in the United States for the year 2017 (12). For many of the disease conditions contributing to early mortality, there is a lifestyle component that can be significantly affected by regular participation in physical activity and exercise.

In developed countries of the world, the changing work and living environment has for some people resulted in a decrease in physical labor and work activity, but at the same time an increase in the opportunity for participation in leisure time activities (13–15). Despite this increase in free time, many individuals in the general population do not exercise enough or are insufficiently physically active to promote and maintain good health and reduce disease risk (5). For example, Figure 1.2 shows the percentage of adults who met the Physical Activity Guidelines for aerobic physical activity, muscle-strengthening activity, and aerobic and muscle-strengthening activity for 2018 (16). While many individuals are regularly engaging in physical activity and exercise, there is still a significant number who do not participate in any type of structured physical activity. Partly as a result of the high

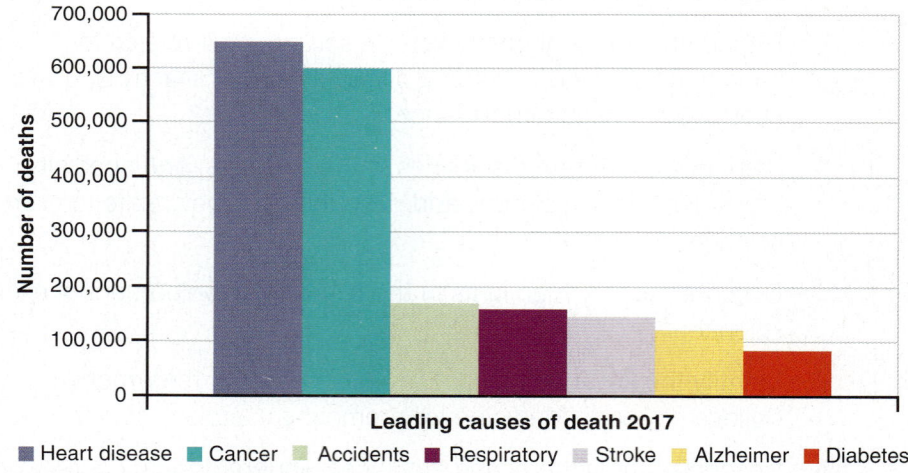

FIGURE 1.1 Common causes of mortality in the United States for the year 2017 (12).

Health A state of complete physical, mental, and social well-being; not simply the absence of disease.
Morbidity The relative incidence of a particular disease.
Mortality The rate of death in a population.

levels of physical inactivity, the United States and other countries have experienced a dramatic increase in the number of individuals who are overweight or obese, prompting health care experts to declare an obesity epidemic earlier this century (17,18). Figure 1.3 illustrates the age-adjusted prevalence of overweight and obesity among the U.S. adults aged 20 to 74 years as reported by the National Health and Nutrition Examination Surveys (NHANES) (19–23). There has also been a rise in other disease and health conditions linked to a lack of physical activity and exercise. For example, an individual's risk level for high blood pressure, Type 2 diabetes, cardiovascular disease, osteoporosis, certain forms of cancer, depression, and anxiety can be affected by the amount of physical activity and exercise in which

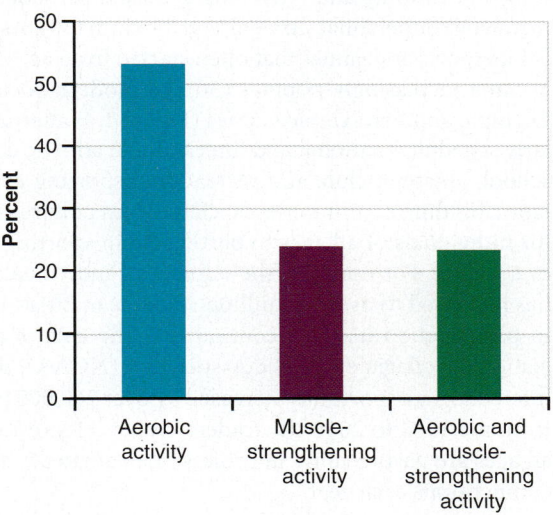

FIGURE 1.2 Percentages of U.S. adults aged 18 years and older who met the physical activity guidelines for aerobic, muscle-strengthening, and aerobic and muscle-strengthening activity for the year 2018 (16).

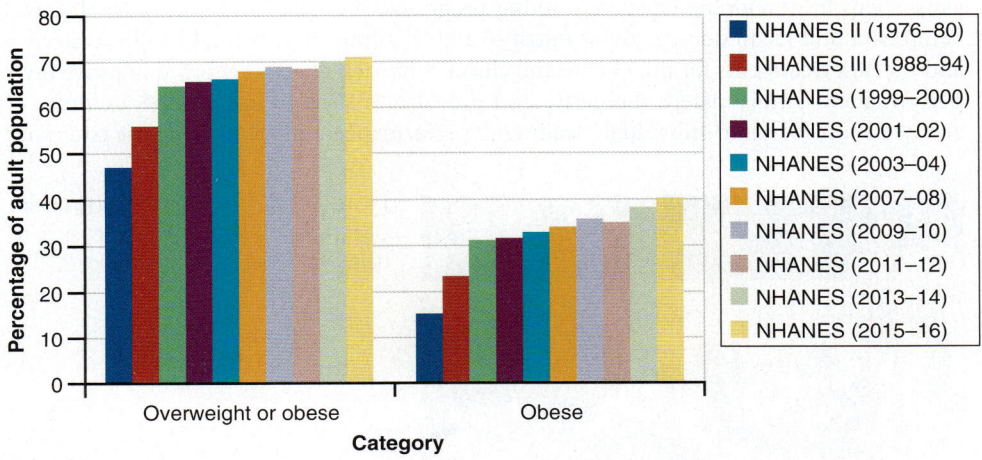

FIGURE 1.3 Age-adjusted prevalence of overweight and obesity among U.S. adults, aged 20 to 74 years as reported by the National Health and Nutrition Examination Surveys (NHANES) (19,20,21,22,23,59,60).

an individual participates (6,24–28). Therefore, exercise science professionals have an extremely important responsibility in helping individuals understand the importance of physical activity and exercise in promoting physical fitness and health, how to personally design and implement effective fitness and exercise programs, and the best practices for preventing and recovering from injury.

Throughout history, the ability to succeed in sport and athletic competition have been highly valued by people and societies. The ancient Greeks held physical prowess and successful athletes in high regard. The importance of athletic competition was so revered that regularly scheduled Olympic Games were held in Greece from 776 BC until 393 AD (29). Early sports and athletic competitions primarily involved individual events such as wrestling and running and often emerged from activities related to obtaining food (*e.g.*, javelin throwing) or personal survival (*e.g.*, boxing) (30). As societies and cultures continued to develop, sports and games that often started from activities initiated in the rural working class became increasingly popular (30). In modern society, we have the Winter and Summer Olympic and Paralympic Games (Figure 1.4) alternating on a 2-year cycle, as well as regularly scheduled national and international athletic competitions. Professional, college, high school, amateur, club, and recreational sporting events are an integral part of our social fabric binding communities, societies, and cultures together. There are more opportunities for individuals of all ages to participate in sporting events and athletic competitions than ever before. For example, the number of individuals participating in high school athletics has increased to over 7.9 million students in 2018–19 (Figure 1.5) (31). This upward trend in participation has also continued at the college and university level. According to the National Collegiate Athletic Association (NCAA), the number of students participating in intercollegiate sports has increased by over 260,000 participants, including 146,000 women, from the 1982 to 2018–19 academic year (Figure 1.6) (32). Countless other individuals of all ages are participating in professional, amateur, club, and recreational sport and athletic competitions each year.

Athletes and coaches are continuously looking for exercise science professionals to assist them in developing effective training programs, the safest and most advantageous equipment and techniques, a sound nutritional diet, appropriate mental health strategies, and the best treatments for their acute and chronic injuries: all in an effort to improve their performance. Unfortunately, the desire and pressure to succeed has created a culture of win at all cost for some individuals, leading to performance-enhancing drug use and other

FIGURE 1.4 The Summer and Winter Olympic and Paralympic Games are held every 2 years. (Shutterstock.)

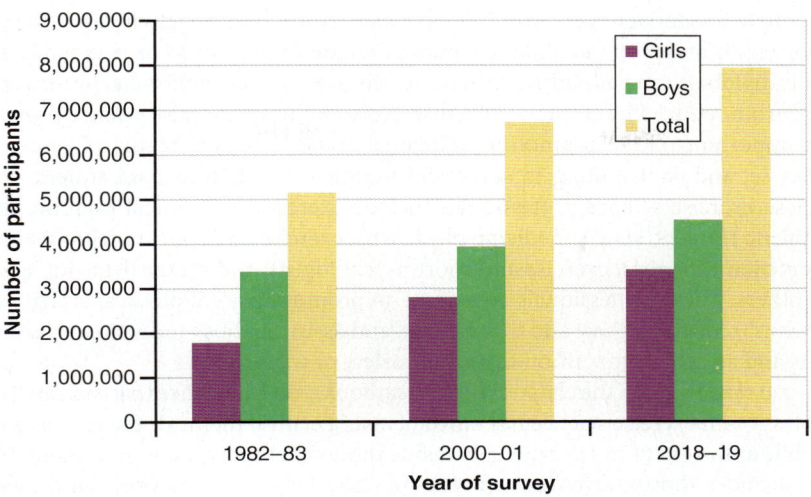

FIGURE 1.5 Number of individuals participating in high school athletics (31).

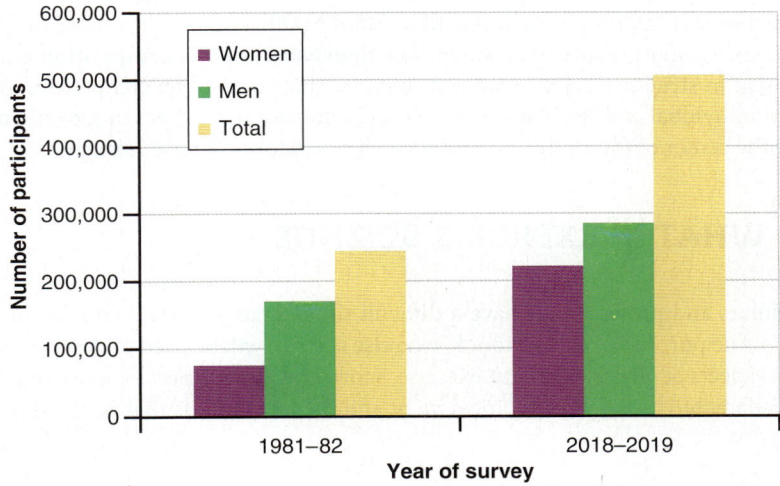

FIGURE 1.6 Number of individuals participating in NCAA-sponsored intercollegiate athletics (32).

forms of cheating during training and competition. For example, reports indicate as many as 5.4% of male and 2.9% of female adolescents have used androgenic anabolic steroids (33) and that lifetime prevalence use rates for males and females is around 6.4% and 1.6%, respectively (34). More recent evidence indicates that the prevalence of current and prior androgenic anabolic steroid users was as high as 16.9% in men and 6.5% in women (35). Each year, athletes of all ages are disqualified from participation and competition in sports because of using performance-enhancing drugs and other banned substances. Exercise science professionals can play a vital role in promoting safe and fair training methods for athletes at all ages and levels of competition.

The benefits derived from regular participation in physical activity and exercise and the involvement in sport and athletic competition are important to us personally and as a society. Individuals can gain improvements in physical, mental, and social health, and society gains from reduced levels of lifestyle diseases and illness. Exercise science professionals play an important role in promoting individual and population health, physical activity, and exercise, and contributing to successful performance in sport and athletic competition. These exercise science professionals include exercise and clinical exercise physiologists, athletic trainers, sports medicine physicians, exercise and sport nutritionists, clinical and sport biomechanists, exercise and sport psychologists, and motor behavior specialists, among others. These professionals contribute to promoting good physical activity and exercise habits, reducing injury and disease risk, and assisting those individuals participating in sports and athletic competition in a wide variety of ways.

As you read through the chapters of this textbook, you will realize that each area of study and profession in exercise science has movement as a central theme or focus. Therefore, it is worthwhile at this point to operationally define these different types of movement. **Physical activity** includes those movement activities of daily living such as work- and job-related activities, leisure time activities, and activities performed around the home. Sufficient levels of physical activity can result in improvements of fitness and health. **Exercise** is a structured movement process that individuals consciously and voluntarily engage in. Exercise includes those activities that improve or maintain fitness and health and those activities that improve performance in a sport or athletic competition. **Sport and athletic competition** are defined as movement in structured and organized activities that consist of a competitive aspect including all individual and team athletic events. Combined, these types of movement activities are at the center of the study and professional practice of exercise science.

WHAT IS EXERCISE SCIENCE

Many scholars and professionals have a difficult time agreeing on a definition of exercise science. For the purpose of this textbook, **exercise science** will be used to describe the study of various aspects of physical activity, exercise, sport, and athletic performance that have the common characteristic of physical movement and the adaptations that occur as a result of

Physical activity Movement activities of daily living, including work- and job-related activities, leisure time activities, and activities performed around the home.

Exercise A structured movement process that individuals consciously and voluntarily engage in and includes those activities that improve or maintain fitness and health.

Sport and athletic competition Movement in structured and organized activities that include a competitive aspect including all athletic events.

Exercise science An umbrella term used to describe the study of numerous aspects of physical activity, exercise, sport, and athletic performance that have the common characteristic of movement and the adaptations that occur as a result of physical activity and regular exercise.

participation in physical activity and regular exercise. The term exercise science has evolved primarily from the disciplines of physical education and exercise physiology to be much more inclusive of related areas of study and professions. Exercise science broadly includes the behavioral, functional, nutritional, physiological, psychological, and structural adaptations to movement. Over time we have seen exercise science become part of our educational and professional activities so that we now have Departments of Exercise Science in our colleges and universities, curricular programs of exercise science that can gain accreditation from national organizations, and students graduating with undergraduate and/or graduate degrees in exercise science. Kinesiology is another term commonly used to describe the study of movement. Kinesiology generally reflects a more broadly defined study of movement including the components of exercise science and the additional areas of physical education, sport history, and sport sociology. Departments or Schools of Kinesiology often include professional areas of study such as sport management, sport marketing, sport leadership, and sport journalism. Exercise science is more commonly used to reflect study, preparation, and professional practice in both basic science and applied science areas that are specifically related to health, physical activity, exercise, sport, and athletic competition.

Areas of Study in Exercise Science

An academic discipline is defined as an organized formal body of knowledge (36). Generally, in most academic disciplines the body of knowledge is limited to a specific subject matter. For example, the traditional academic areas of biology, chemistry, and mathematics are defined as disciplines because they have specific bodies of knowledge. Each of these disciplines has developed subdisciplines or specialty areas such as plant biology, nutritional biochemistry, and biostatistics. As the traditional academic disciplines have evolved, there is continuing development of subdisciplines and specialty areas. Determining whether exercise science is a specific academic discipline such as chemistry or mathematics is beyond the scope of this textbook. It is clear that related areas in exercise science such as exercise physiology and exercise and sport psychology have been recognized as disciplines (37) because each contains a distinct body of knowledge that is organized in a formal course of learning. In this textbook, exercise science is used as an umbrella term to include exercise and clinical exercise physiology, athletic training and sports medicine, exercise and sport nutrition, exercise and sport psychology, motor behavior, and clinical and sport biomechanics. Figure 1.7 illustrates the disciplines and subdisciplines covered under the exercise

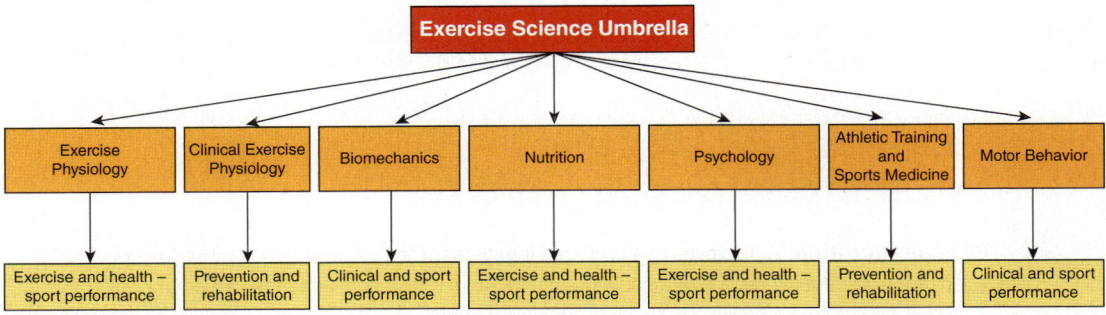

FIGURE 1.7 Relationships of the disciplines and subdisciplines of exercise science.

science umbrella. Many of the areas of study in exercise science draw on information developed by professionals in other areas, making exercise science a truly interdisciplinary field. Table 1.1 provides some examples of areas of study by exercise science students and professionals.

Exercise Science as a Field of Study

Historically, much of the study of physical activity, exercise, sport, and athletic performance originated from the academic discipline of **physical education**. Many of the early teachers and professional leaders in physical education studied and promoted the role of exercise and sport in the development of the whole person. It is clear, however, that the various areas of study in exercise science have continued to move away from physical education as the foundation academic discipline. Many of the early pioneers and leaders in the disciplines

Table 1.1 Areas of Study by Students and Professionals in Exercise Science

EXERCISE SCIENCE DISCIPLINE	AREAS OF STUDY
Exercise physiology	Physiological responses of the whole body or specific systems to physical activity, exercise, sport, and athletic competition
Clinical exercise physiology	Using movement, physical activity, and exercise in the identification, prevention, and rehabilitation of acute and chronic disease conditions
Athletic training and sports medicine	Prevention, treatment, and rehabilitation of exercise, sport, and athletic injuries
Exercise and sport nutrition	Nutritional aspects of disease prevention and health promotion and improvement of sport and athletic performance
Exercise and sport psychology	Behavioral and mental aspects of physical activity, exercise, sport, and athletic performance
Motor behavior	Development, learning, and control of body movement in healthy and diseased conditions and improvement of sport and athletic performance
Clinical and sport biomechanics	Mechanical aspects of movement in disease and injury conditions, physical activity, exercise, sport, and athletic performance

Physical education A discipline that prepares teachers to manage classroom environments, enhance student interaction and socialization, improve instruction, enhance movement skill acquisition, improve health, and increase physical activity.

and fields of study in exercise science began their careers as physical education students and teachers, but became exercise physiologists, athletic trainers, sport psychologists, or some other professional term that provided greater clarity and meaning to what they did. The areas of study in exercise science programs have moved curricular requirements away from the coursework and content knowledge of current physical education teacher licensure programs. College and university programs of study may still require physical education students to enroll in coursework in exercise physiology, biomechanics, nutrition, and motor behavior. Students of exercise science, however, are rarely required to take coursework in many of the content areas of physical education.

Over the last 60 years, professionals and leaders within exercise science and physical education have made concerted efforts to clearly define the parameters of each discipline. It is now more appropriate to define physical education as a discipline with a purpose to investigate how the teaching process can be used most effectively to acquire movement skill (38). As a discipline, physical education is primarily interested in how a body of knowledge can be developed to help individuals be better prepared as teachers, manage classroom environments, enhance student interaction and socialization, improve instruction, enhance movement skill acquisition, improve health, and increase physical activity. Conversely, exercise science is primarily interested in the study of the behavioral, functional, nutritional, physiological, and psychological functional adaptations to physical activity, exercise, sport, and athletic performance (38). As the discipline of physical education evolves to meet the expectations for the preparation of licensed teachers, there will be a further clarification of the body of knowledge required to be a successful physical education teacher. Similarly, as exercise science responds to the demands of preparing students for work in exercise and health care fields or competitive sport and athletic environments, so too will the disciplines, subdisciplines, and areas of study within exercise science continue to evolve.

Thinking Critically

Exercise science has expanded from its origins in physical education to the broad array of disciplines and subdisciplines that it encompasses today. Why do you think this trend has occurred? Why has this evolution been a good change for promoting physical activity, exercise, sport, and athletic competition?

HISTORY OF EXERCISE SCIENCE

Exploring the historic background of a specific academic discipline can provide considerable insight into the events that help shape current educational experiences and practice. This is especially true for exercise science. Answering the question "When did exercise science officially begin?" is very difficult, however, because there is no clear and definable birth of exercise science owing to its structure as a collection of disciplines and subdisciplines. Although a complete historic account of the evolution of exercise science is beyond the scope of this chapter, there are significant periods and events that deserve mentioning to allow for an understanding of how exercise science shapes our current educational programs and professional activities. Noted American College of Sports Medicine Historian Jack W. Berryman presents an excellent historic account of the important events that laid the foundation for the development of exercise science, and students are encouraged to read this account (1) and others (37,39).

Early Influences

Exercise science has evolved historically from several significant influences. Early history (starting at ~450 BC) has Greek scholars such as Hippocrates, Plato, Aristotle, and Socrates exploring physical activity in a scientific fashion and often prescribing exercise (1). Early Greek and Roman artists portrayed humans performing various feats of athletic endeavors (1). For example, one of the most well-known artistic Greek sculptures showing an athlete is the Discobolus or more commonly known as Discus Thrower (Figure 1.8). From ancient times throughout the medieval period (ca. 400 AD to 1400 AD), there was a continued interest in physical activity, exercise, and sports, primarily through studies of anatomy, physiology, and medicine (1).

During the Renaissance period (ca. fourteenth to seventeenth century), artists and scientists such as Leonardo da Vinci and Galileo Galilee had a strong analytic interest in human physical activity and health. Famous physicians such as Bombastus von Hohenheim and Thomas Linacre wrote in the areas of physiology and hygiene, respectively. In the seventeenth-century pioneers from the disciplines of biology, physics, mathematics, and chemistry were doing significant work in the area of physiology. For example, William Harvey's discovery of the circulation of the blood through arteries and veins (Figure 1.9) was instrumental in developing exciting new concepts in physiology and medicine. Harvey's most influential writings often included specific references to exercise. A significant development at this time was the use of mechanical principles to solve physiologic questions, leading to the construct of the human body as a machine. Questions about skeletal muscular contraction and pulmonary respiration interested physicians and scientists such as Robert Boyle and Robert Hooke in the 1600s. The famous English physician John Mayow wrote extensively on exercise and more specifically on muscle fibers, strength, and muscular contraction (1).

FIGURE 1.8 The Greek statue Discobolus — more commonly known as Discus Thrower. (From the Esquiline Hill; former collection Massimo-Lancellotti.)

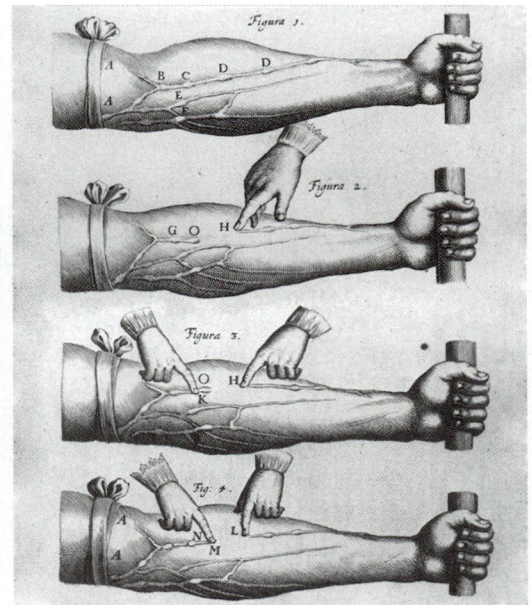

FIGURE 1.9 William Harvey described the circulation of the blood through arteries and veins. (Courtesy of the National Library of Medicine.)

The era of Enlightenment (ca. eighteenth century) saw physicians and scientists publish major works summarizing the knowledge to date on exercise, health, and life-longevity. Sir John Floyer observed and described the change in heart rate in response to moderate intensity walking, whereas London physician James Keill added to previous work on the muscular system by describing various aspects of muscle fiber size, structure, and contraction. Stephen Hales was the first to make accurate measurements of blood pressure (Figure 1.10), which ultimately contributed to Daniel Bernoulli, a Dutch mathematician, making exact calculations of the volume of blood pumped by the heart (now known as cardiac output). During this time, John Desaguliers invented the first mechanical dynamometer that was used to assess muscular force and strength. Joseph-Clement Tissot was the first to describe the effects of time, location, intensity, and duration of exercise on the physiologic processes of the body. Work by the famous French chemist Antoine-Laurent Lavoisier and French mathematician Pierre de Laplace led to the understanding of the use of oxygen to burn carbon in the body and that during physical work (*i.e.*, exercise) oxygen consumption was increased. Additional work by Lavoisier and Armand Seguin allowed for the development of the basic ideas of energy transformation and the source of heat, particularly that which occurred during exercise (1,40).

Nineteenth-Century Influences

In the early nineteenth century, the study of exercise received more attention from physicians because of its important role in the maintenance of good health and from scientists who were interested in how physical exertion and exercise affected the human body (1). During this period there was experimentation in respiratory physiology, metabolism, and nutrition, with the first experiments conducted on how dietary intake and exercise

FIGURE 1.10 Stephen Hales making the first measurement of resting blood pressure. (From the Granger Collection, New York.)

influenced the urinary system. The term physical education was introduced as a way to promote the education of individuals about their bodies, and as a result, physical exercise became more popular among the general population. The publication of several books on calisthenics and gymnastics as forms of exercise, promoted by Swedish and German physicians, occurred in the late nineteenth century (1).

In the early nineteenth century, sport and athletic competition began to be popularized in private schools and colleges and through the formation of professional teams. As a result, there was an increased interest in training the human body to improve the chance of success in sports. During the early 1800s, Sir John Sinclair of Scotland was one of the first individuals to write extensively about the role of physical training for improving performance during sport and athletic competition. Continued interest in sport and athletic performance throughout the century led to the formation of the American Alliance for Health, Physical Education, Recreation and Dance (AAHPERD), which held their first meeting in 1885 (1). AAHPERD would eventually become the Society for Health and Physical Education (SHAPE) early in the twenty-first century.

Early Twentieth-Century Influences

Exercise science continued to develop in the twentieth century, extending its roots from the discipline of physical education. Two factors played a prominent role in providing the foundation for the development of exercise science. First, colleges and universities developed specific health and physical activity–related courses in their academic curricula to promote the physical and emotional well-being of the whole person (37). The second prominent factor was the formation of specific programs of study in colleges and universities to prepare professional physical education teachers and athletic coaches (37). Early leaders in physical education, many of whom had their academic preparations in medicine, were proponents of scientifically based, systematic programs of study. Students enrolled in physical education and coaching programs were required to take coursework in anatomy, physiology, **anthropometry**, and physics. Throughout the early to mid-twentieth century, the primary focus in the field of physical education was the preparation of teachers and coaches for public schools, colleges, and universities (37). In the 1930s and 1940s, the writings of two prominent scholars, Jay Bryan Nash and Charles H. McCloy, began to set the stage for a separation of exercise science from the discipline of physical education. Specifically, Nash believed that children should be educated for lifelong leisure time pursuits, whereas McCloy believed that physical education should be used to develop the human body (37).

Other principal twentieth century figures in the development of exercise science came from a variety of academic disciplines. One of the leading pioneers in the study of physical activity was Dudley Allen Sargent. Sargent received a medical degree from Yale University in 1879 and immediately became the director of the Hemenway Gymnasium at Harvard University. Sargent pioneered an all-inclusive system for individual exercise prescriptions using information from physical examinations, muscular strength assessments, and anthropometric measurements. His work in measuring strength and power and for recording and evaluating anthropometric measurements of the human body was instrumental in the development of assessment measures of human performance (40,41). One might say that Dudley Sargent was the very first personal trainer of the twentieth century.

Sargent's advances in the discipline of physical activity strongly influenced George W. Fritz, who graduated from Harvard University with a medical degree in 1891. Fritz was an enthusiastic proponent of developing theories and beliefs regarding exercise and its effect on the human organism using scientifically based physiological research (42,43). Through this work, Fritz established the Physiology Laboratory at Harvard University, which ultimately led to the establishment of the first college degree program in physical education. Graduates of this program received a Bachelor of Science degree in anatomy, physiology, and physical training. Throughout his career, Fritz explored the relationships among physical training, anatomy, and physiology, and some expert historians consider George Fritz the "father of exercise physiology" (44).

Without question, the faculty and scholars conducting research in the Harvard Fatigue Laboratory (Figure 1.11) were the key leaders in the study of physical activity and exercise in the early twentieth century. In existence from 1927 to 1947, the laboratory was home to some of the most prominent researchers at the time, including Lawrence J. Henderson and David Bruce Dill. As the first and only director of the Harvard Fatigue Laboratory, Dill was instrumental in leading laboratory research programs. Although much of the experimental research focused on exercise and environmental physiology, many of the scientists who worked in the laboratory also conducted research in related areas such as clinical

Anthropometry The study of the physical measurements and characteristics of humans and animals.

FIGURE 1.11 Exercise and environmental experiments were conducted in the Harvard Fatigue Laboratory (ca. 1945). (Photo from Baker Library Historical Collections, Harvard Business School.)

physiology, gerontology, nutrition, and physical fitness (41,43,45). The work conducted in this laboratory provided the foundation for many of the basic theories used in the study and application of exercise science today.

Perhaps, the most significant accomplishment of the Harvard Fatigue Laboratory in the evolution of exercise science was its role in the development of other prominent laboratories for the study of physical activity and exercise. For example, Steven M. Horvath, a student of Dill, started the Institute for Environmental Stress at the University of California Santa Barbara. Some of the most well-known exercise scientists were trained at the institute under Horvath, including Jack H. Wilmore and Barbara L. Drinkwater, two influential exercise science researchers and professional leaders. The Laboratory for Physiological Hygiene at the University of Minnesota, established by Ancel Keys, produced the renowned exercise scientists Ellsworth R. Buskirk and Henry Longstreet Taylor (43). These individuals and the work from their research laboratories significantly fostered our understanding of physical activity and exercise. Many other individuals also had a profound influence on the expansion of exercise science across the United States during the early to mid-twentieth century. For example, in 1923 Author Steinhaus, founded the second laboratory devoted to the study of physiology, physical activity, and exercise at the George Williams College in Chicago. Peter V. Karpovich of Springfield College is credited with introducing physiology to physical education in the United States and was instrumental in promoting the use of weightlifting for enhancing health and human performance. Another prominent exercise scientist, Thomas K. Cureton, began his professional career at Springfield College and then later moved to the University of Illinois, where he established the Physical Fitness Research Laboratory in 1941 and an adult fitness program in 1961 (43).

Late Twentieth-Century Influences

Several landmark events following World War II resulted in the significant development of exercise science. In 1953, the poor performance of U.S. children compared to European children in the standardized Kraus–Weber physical fitness tests stimulated an increased interest in fitness assessment and the promotion of physical fitness programs in schools (Figure 1.12). The formation of the American College of Sports Medicine in the mid-1950s joined physical education–based researchers with scholars from the medical community to form an

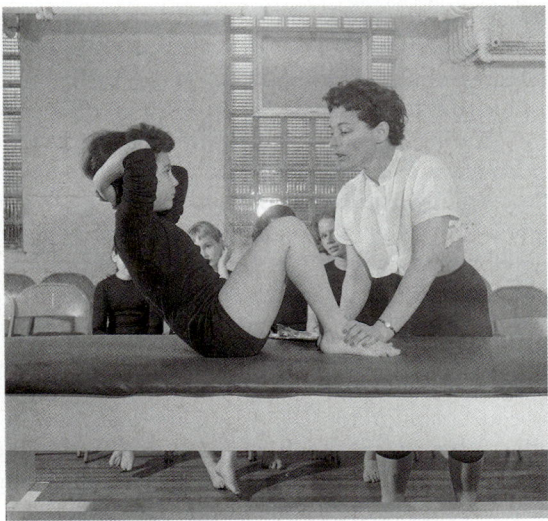

FIGURE 1.12 Administration of the Kraus–Weber physical fitness test (ca. 1954) helped promote increased interest in physical activity and exercise. (Photo by Orlando/Three Lions/Getty Images.)

organization that continues to offer significant support and promotion of exercise science and sports medicine (37). In addition, other organizations, such as the National Athletic Trainers Association (1950), the International Society of Biomechanics (1973), the National Strength and Conditioning Association (1978), and the Association for Applied Sport Psychology (1986), were created to provide support to students and professionals and to advance the mission of these professions. The decades of the 1960s and 1970s brought about the beginning of the separation of exercise science from the discipline of physical education (37) and the establishment of many of the areas of study in exercise science as stand-alone bodies of knowledge.

During the last 25 years of the twentieth century, the discipline of exercise science experienced large and profound changes. A keynote event occurred in 1970, when Kenneth H. Cooper, a physician, opened The Cooper Institute and started the Cooper Clinic. Using data from patients at the Cooper Clinic, Steven N. Blair published a landmark study in 1989 that demonstrated moderate physical activity could decrease the risk of death from all causes by 58% (46). This marked a modern shift in thought, and exercise science professionals became more actively involved in both studying and promoting the role of physical activity and exercise in health.

Additionally, during this period of time athletes at all levels became very interested in the influence of regular, systematic training on physical development and performance during sport and athletic competition. These events in the 1970s and 1980s resulted in academic departments in colleges and universities developing academic majors and minors in exercise science, as well as programs and departments of exercise science. Throughout the 1980s and 1990s, exercise science professionals continued to expand their influence in the areas of physical activity, exercise, sport, and athletic performance.

Twenty-First-Century Influences

During the first part of the twenty-first century, exercise science scholars and professionals have continued to clarify the role of physical activity and exercise in the promotion of health and the reduction of disease conditions. Over the last 20 years we have come to better understand the role of physical activity and inactivity in the development of lifestyle diseases. This knowledge has led to the further refinement of how much physical activity is sufficient to reduce the risk of cardiovascular disease (9), cancer (47), and diabetes (48).

During the last 20 years, the areas of sport and athletic performance have seen considerable attention focused on overall athlete health, and, more specifically, the impact of head trauma on health and longevity (49,50). Chronic traumatic encephalopathy (CTE) is a brain condition associated with repeated blows to the head. CTE is associated with the development of dementia and problems with thinking and memory, personality changes, and behavioral changes including aggression and depression (51). As a result of this knowledge, new and innovative assessment techniques are being developed so exercise science professionals can identify individuals who are impacted by repeated head trauma and these individuals can be treated appropriately. Of critical importance has been the development of concussion protocols (52) and alternate training methods and practice protocols in an effort to reduce the number of exposures to head trauma by athletes (53).

There continues to be an expansion of health promotion and physical activity and exercise programs by professional organizations and governmental agencies, and the participation of exercise science professionals in the success of those programs will be instrumental. For example, multidisciplinary approaches to solving individual and population-based health-related problems such as the obesity epidemic and promoting health equity are being accomplished by using the knowledge and skills of exercise physiologists, nutritionists, biomechanists, exercise behaviorists, and health care specialists working together.

In an effort to gain a greater understanding of how the disciplines and subdisciplines of exercise science have evolved, each chapter of this textbook will further examine the historic development of those areas identified in Table 1.1. This information is intended to provide you with a better understanding of how past events have affected the present and will shape the future of the exercise science disciplines.

Thinking *Critically*

In what specific ways has exercise science contributed to a broader understanding of how physical activity and exercise influence physical fitness and health?

In what specific ways has exercise science contributed to a broader understanding of how exercise and training influence sport and athletic performance?

EXERCISE SCIENCE AND THE AMERICAN COLLEGE OF SPORTS MEDICINE

There is probably no organization that has been more effective in developing and promoting exercise science than the American College of Sports Medicine (ACSM). One only has to attend a national conference of the ACSM and observe the broad array of exercise science

Health equity Absence of preventable or unfair differences among groups of people defined socially, economically, demographically, geographically, or by other methods of stratification.

and sports medicine topics presented by professionals from around the world to truly appreciate the vast impact of the ACSM. Although many of the disciplines and subdisciplines under the exercise science umbrella have their own professional organizations, none is so broad reaching in exercise science as the ACSM. Several significant events are worth discussing so that students of exercise science may have an understanding of how physical activity, exercise, sport, and athletic performance have been shaped by the ACSM since its founding. For a more detailed study of the early history and influence of the ACSM, students are directed to an additional reading (54).

Early Development of the ACSM

The first meeting of the organization that eventually became the ACSM occurred in 1954 as part of the afternoon program of the American Association for Health, Physical Education, and Recreation national meeting. The initial name chosen by the founding members "Federation of Sports Medicine" was changed to the ACSM in 1955 when the organization was officially incorporated. There were 11 professionals from the disciplines of physical education, physiology, and medicine who were instrumental in the establishment of the ACSM (54).

The ACSM developed as a result of increased interest of both the general public and professionals in health and exercise (54). Two other factors also played a prominent role in the evolution of ACSM: the interest by the U.S. military in physical fitness, physical training, and rehabilitation of military personnel in the post–World War II period and the growth of sports medicine in the international arena prompted by the Federation Internationale Medico Sportive (FIMS). At the time, FIMS was a world leader in the areas of sport and athletics. College and university team physicians and athletic trainers also contributed significantly to the formal development of the ACSM. The influence of the above factors catalyzed the founding members of the ACSM to define sports medicine as a "unique blend of physical education, medicine, and physiology" (54).

From its establishment, the ACSM and its members provided significant public outreach and worked to influence public policy. For example, in 1955, members of the ACSM provided professional guidance to U.S. President Dwight D. Eisenhower's National Conference on Physical Fitness. This commitment to public involvement has continued throughout ACSM's history and currently includes programs such as ACSM's annual International Health and Fitness Summit and the Advanced Team Physician Program, which provide opportunities for practicing professionals to receive the most current information in all areas of exercise science and sports medicine. Publications such as the *ACSM's Health and Fitness Journal*, *Fit Society Page*, and the *Sports Medicine Bulletin* allow ACSM professionals to provide meaningful health and fitness information to the general public.

The membership of the ACSM has always been actively involved in performing research and scholarship. ACSM members have provided valuable insight into the prevention and management of lifestyle-related diseases such as coronary artery disease, hypertension, diabetes, and cancer, and the rehabilitation of stroke patients and injured athletes. Much of this information is broadly available in the peer-reviewed journal *Medicine and Science in*

Peer-reviewed The formal evaluation of a professional colleague's work by experts in the field.

Sports and Exercise, published regularly by ACSM since 1969. Early in its development as an organization, the ACSM maintained an emphasis on and commitment to sound quality science and research. This commitment remains strong as the ACSM funds promising young scholars with support from the ACSM Foundation (54).

ACSM has been considered the world's leader in exercise science and sports medicine information since 1974, when it published its first **position stand** "Prevention of thermal injuries during distance running" (55). Throughout its history, ACSM has continued to publish position stands and opinion papers that have helped shape health, physical activity, exercise, sport, and athletic performance in the United States and around the world. Table 1.2 provides a list of the current position stands published by the ACSM (54). Copies of these position stands can be found on the ACSM Web page at www.acsm.org.

Table 1.2 Position Stands from the American College of Sports Medicine	
POSITION STAND	**YEAR PUBLISHED**
ADA/ACSM joint position statement: exercise and Type 2 diabetes (2)	2000
AHA/ACSM joint position statement: automated external defibrillators in health/fitness facilities (61)	2002
AHA/ACSM joint position statement: recommendations for cardiovascular screening, staffing, and emergency policies at health/fitness facilities (62)	1998
Appropriate physical activity intervention strategies for weight loss and prevention of weight regain for adults (63)	2009
Exercise and acute cardiovascular events: placing the risks into perspective (64)	2007
Exercise and fluid replacement (65)	2007
Exercise and hypertension (66)	2004
Exercise and physical activity for older adults (4)	2009
Exercise for patients with coronary artery disease (67)	1994
Exertional heat illness during training and competition (68)	2007
Nutrition and athletic performance (69)	2016
Physical activity and bone health (6)	2004
Physical activity, fitness, cognitive function, and academic achievement in children (70)	2016

Position stand An evidence-based statement on a topic of relevance to those in the fields of exercise science and sports medicine created by experts in the topic.

Table 1.2	Position Stands from the American College of Sports Medicine *(continued)*
POSITION STAND	**YEAR PUBLISHED**
Prevention of cold injuries during exercise (71)	2006
Progression models in resistance training for healthy adults (72)	2009
Quantity and quality of exercise for developing and maintaining cardiorespiratory, musculoskeletal, and neuromotor fitness in apparently healthy adults: guidance for prescribing exercise (5)	2011
The female athlete triad (73)	2007
The use of anabolic-androgenic steroids in sports (74)	1987
The use of blood doping as an ergogenic aid (75)	1996
Weight loss in wrestlers (76)	1996

The initial edition of the highly acclaimed book *Guidelines for Graded Exercise Testing and Prescription* first became available in 1974. This science and research-based book continues to be a valuable source of information for exercise science and sports medicine professionals. During the 1970s, the ACSM first awarded professional certification to individuals who completed training as exercise program directors and exercise specialists. The ACSM continues to furnish practical information through continuing education activities, periodical and nonperiodical publications, certification opportunities, and annual meetings (54).

During the early 1980s, interest in exercise and sports medicine in the United States greatly increased, and ACSM responded strongly to this interest. In 1983, a little less than 30 years after its founding, the ACSM was well over 10,000 professional and student members. During the 1980s, the professional certification program in the ACSM continued to grow and included two categories: rehabilitation and prevention. Individuals who demonstrated competencies in the rehabilitation category could become certified as program directors, exercise specialists, and exercise test technologists. By demonstrating competences in the prevention category, members became certified as a health/fitness director, health/fitness instructor, and exercise leader/aerobics (54). These programs continue today with opportunities for certification in health and fitness (ACSM Certified Personal Trainer®, ACSM Certified Exercise Physiologist, and ACSM Certified Group Exercise Instructor^{SM}), clinical (ACSM Certified Clinical Exercise Physiologist®), and Specialty credentials (Exercise is Medicine® Credential, ACSM/ACS Certified Cancer Exercise Trainer, ACSM/NCHPAD Certified Inclusive Fitness Trainer, and ACSM/NPAS Physical Activity in Public Health Specialist). During the late 1980s, the ACSM established formal relationships with the President's Council on Physical Fitness and Sports and the Office of Disease Prevention and Health Promotion within the U.S. Department of Health and Human Services (54).

Throughout the 1990s, ACSM continued to promote its research and outreach activities by making *Medicine and Science in Sports and Exercise* a monthly publication, offering the Team Physician Course, and disseminating exercise science and sports medicine position stands to ACSM members and the general public (54). These position stands continue to shape public policy and practice in the areas of health, physical activity, exercise, sport, and athletic competition. In 1994, ACSM provided consultation to the U.S. Centers for Disease Control and Prevention (CDC) and the President's Council on Physical Fitness and Sport as new recommendations for appropriate levels of physical activity were being developed. The resulting report from the U.S. Surgeon General provided guidelines for the amount of physical activity required to obtain significant health benefits and reduce the risk of developing chronic poor health conditions such as heart disease, hypertension, diabetes, and osteoporosis (56). Furthering its commitment to physical activity, the ACSM joined with several other professional organizations and government agencies in 1995 to form the National Coalition for Promoting Physical Activity (NCPPA) (54). The mission of the NCPPA is to unite the strengths of public, private, and industry organization efforts into collaborative partnerships that inspire and empower all Americans to lead more physically active lifestyles (www.ncppa.org). The ACSM has continued to develop and present key exercise science publications. A current listing of prominent ACSM publications is shown in Table 1.3.

ACSM has partnered with other professional organizations to promote improved safety and care for competitive athletes. In 1994 the ACSM, working with the American College of Cardiology, developed recommendations for determining eligibility for competition in

Table 1.3 — Key Exercise Science-Related Publications from the American College of Sports Medicine

- ACSM's Behavioral Aspects of Physical Activity and Exercise
- ACSM's Body Composition Assessment
- ACSM's Certification Review, Sixth Edition
- ACSM's Clinical Exercise Physiology
- ACSM's Complete Guide to Fitness & Health, Second Edition
- ACSM's Exercise for Older Adults
- ACSM's Exercise is Medicine
- ACSM's Exercise Management for Persons with Chronic Diseases and Disabilities, Fourth Edition
- ACSM's Exercise Testing and Prescription
- ACSM's Fitness Assessment Manual, Sixth Edition
- ACSM's Foundations of Strength Training and Conditioning, Second Edition
- ACSM's Guide to Exercise and Cancer Survivorship
- ACSM's Guidelines for Exercise Testing and Prescription, Eleventh Edition
- ACSM's Health/Fitness Facility Standards and Guidelines, Fifth Edition
- ACSM's Metabolic Calculations Handbook
- ACSM's Nutrition for Exercise Science

Table 1.3	Key Exercise Science-Related Publications from the American College of Sports Medicine (continued)
ACSM's Research Methods	
ACSM's Resources for the Exercise Physiologist, Third Edition	
ACSM's Resources for the Group Exercise Instructor	
ACSM/NCHPAD Resources for the Inclusive Fitness Trainer	
ACSM's Resources for the Personal Trainer, Sixth Edition	
ACSM's Resource Manual for Guidelines for Exercise Testing and Prescription, Seventh Edition	
ACSM's Sports Medicine: A Comprehensive Review	

athletes with cardiovascular abnormalities (57). In 1996, the ACSM partnered with the American Medical Society for Sports Medicine (AMSSM) and the American Orthopaedic Society for Sports Medicine (AOSSM) to present the first Advanced Team Physician Course. ACSM remains committed to educating team physicians through publications such as *ACSM's Primary Care Sports Medicine*, *ACSM's Exercise is Medicine*®, and *ACSM's Sports Medicine: A Comprehensive Review*. The Team Physician Consensus statement, a collective effort by members of ACSM, AOSSM, AMSSM, and the American Academy of Orthopaedic Surgeons, was first published in 2000. This statement provided physicians, school administrators, team owners, the general public, and individuals who are responsible for making decisions regarding the medical care of athletes and teams with guidelines for choosing a qualified team physician and an outline of the duties expected of a team physician (58).

Recent Role of the ACSM

The ACSM continues to increase its influence by working to shape public policy, supporting research, expanding educational opportunities for exercise science and sports medicine professionals, and disseminating knowledge to the general public. The ACSM has made commitments to influence world health through its roles with the Active Aging Partnership/National Blueprint, Musculoskeletal Partnership, and Exercise is Medicine® initiative. At the 52nd Annual Meeting in 2005, the ACSM and the President's Council on Physical Fitness and Sports announced a collaboration to benefit public health by jointly promoting physical activity, fitness, and sports. In 2006, the Federation of American Societies for Experimental Biology (FASEB) accepted ACSM into the Federation, further solidifying ACSM as a leader in the study of health, physical activity, exercise, sport, and athletic performance.

In 2007, ACSM developed the American Fitness Index™ (AFI), a signature program created in partnership with the WellPoint Foundation. The purpose was to provide an evidence-based and science-based measurement of the state of health and fitness at the community level throughout the United States. The AFI is designed to improve the health, fitness, and quality of life of Americans by promoting physical activity, using three key components: collecting, analyzing, and reporting on data related to healthy lifestyles and physical activity; serving as a resource for communities; and assisting communities interested in connecting and partnering with organizations and existing programs to collaborate on physical activity/healthy lifestyle initiatives. Additional information about the AFI can be found at www.americanfitnessindex.org.

Additionally, in 2007, the ACSM partnered with the American Medical Association (AMA) to co-launch the Exercise is Medicine® (EIM) initiative. EIM is a United States-based health initiative coordinated by the ACSM. The original purpose of EIM was to make the scientifically proven benefits of physical activity a standard component of the U.S. health care system by having health care providers promote physical activity as an important component of their patient's health. More information about EIM can be found at www.exerciseismedicine.org.

Throughout the last decade, ACSM has been at the forefront of providing leadership and creating opportunities for advancement in exercise science. The Integrative Physiology of Exercise Conference (2010), the Basic Science World Congress (2014), launching of the *Translational Journal of ACSM* (2016), and the Exercise and Cancer Prevention and Control Roundtable (2018) are just a few examples of impactful initiatives promoted and sponsored by ACSM. The Datalys Center was established in 2018 and is a collaborative venture among Biocrossroads, the NCAA, and the ACSM. This consortium conducts research and provides surveillance expertise to support the sports injury data needs of researchers, sports governing organizations, and the sports medicine community. Additionally, leadership positions have been added to the ACSM Board of Trustees in the areas of Health Equity, Diversity, and Inclusion, as well as an International Trustee position to better serve the needs of all individuals interested in the disciplines and subdisciplines of exercise science.

The ACSM continues to build upon its rich history established and fostered by its world leaders in exercise science and sports medicine as it adheres to its mission:

> To promote and integrate scientific research, education, and practical applications of sports medicine and exercise science to maintain and enhance physical performance, fitness, health, and quality of life.

Thinking Critically

In what significant ways has the ACSM contributed to the development of exercise science and the related fields of study?

The fulfillment of this mission will allow the ACSM to remain instrumental in helping exercise science professionals to create and disseminate valuable information and shape policy affecting health, physical activity, exercise, sport, and athletic performance.

ACADEMIC PREPARATION IN EXERCISE SCIENCE

As presented in this chapter and others throughout the book, you will see that exercise science professionals come from a variety of educational backgrounds and disciplines. Appropriate academic preparation is critical to developing and preparing for a successful career as an exercise science or sports medicine professional. Chapter 12 will focus on professional careers you can enter into as an exercise science major. Undergraduate students in a program of study in exercise science must pay particular attention to meeting certain prerequisites and degree requirements, especially in a program, major, or profession that requires the successful completion of a licensure or certification examination upon graduation with

FIGURE 1.13 Athletic trainers work with individuals injured during physical activity, exercise, sport, or athletic competitions. (Shutterstock.)

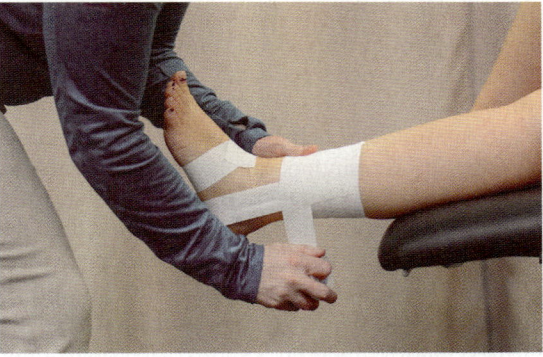

an undergraduate or graduate degree. For example, exercise science students interested in becoming certified athletic trainers (Figure 1.13) or registered dietitians (Figure 1.14) must complete the program requirements and graduate from an accredited program in athletic training or dietetics. Exercise science can also be a valuable and effective program of study for those who want to enter into postbaccalaureate study in the medical or **allied health care** fields or for completing further graduate education.

Undergraduate Coursework

Undergraduate exercise science programs and their degree majors are typically broad based and include coursework in anatomy, biology, biochemistry, chemistry, health, human development, mathematics, physics, physiology, psychology, and statistics. Information from

FIGURE 1.14 Registered dietitians counsel patients about proper nutritional intake. (Shutterstock.)

Allied health care The professional field that works to deliver patient care services for the identification, prevention, and treatment of diseases, disabilities, and disorders.

these courses helps provide a solid foundation for understanding how and why humans move. Advanced coursework provides enhanced knowledge in particular areas of study. For example, depending on the major or minor area of study, students can enroll in coursework in exercise physiology, fitness programming in health and disease, motor development, control, and behavior, nutrition, structural and functional biomechanics, exercise testing and prescription, laboratory techniques, strength and flexibility training, exercise and sport psychology, research methods, evaluation and assessment of athletic injuries, and rehabilitation modalities, among others. Additional involvement in research or scholarly projects, co-ops, internships, and clinical activities, can provide much needed preparation and experience. The foundational and advanced coursework and practical experiences are designed to provide basic and applied knowledge to prepare exercise science students for the next phase of their professional career including certification or licensure examinations, employment, graduate study, and training in professional schools. Students interested in careers in medicine or another allied health profession must also meet the major requirements for entry into that chosen field.

As you begin your course of study, it is very important that you understand the requirements of the career path you wish to take after graduation. Chapter 12 provides valuable information about professional employment and career opportunities. It is important to make sure your coursework at the undergraduate level provides you with the knowledge, skills, and abilities necessary for your chosen career path. Although the information contained in the following sections provides a general overview of the requirements for continued preparation, it is essential that you speak with a career advisor or counselor at your college or university. That individual will guide you to the most appropriate coursework and experiences for achieving your career goals. Table 1.4 provides key foundational recommendations for continued educational and professional development.

Preparation for Careers in Health Care

It is becoming increasingly common for students to major in an exercise science program as an undergraduate student and then complete postbaccalaureate work in a health care

Table 1.4 Foundational Recommendations for Continued Educational and Professional Development

CONTINUED PROFESSIONAL DEVELOPMENT	RECOMMENDATION
Graduate study	Complete the degree and prerequisite requirements for the intended graduate program of study; engage in undergraduate research if possible
Professional schools	Meet the degree requirements and prerequisites for the professional program of study; complete clinical observations and internships
Certification/licensure	Meet all eligibility requirements established by the professional certifying or licensing agency
Employment	Gain appropriate internship, clinical, or field experience in the potential field of professional employment

FIGURE 1.15 Allied health care professionals frequently come from undergraduate exercise science programs. (Shutterstock.)

field (Figure 1.15). An undergraduate degree in exercise science allows the flexibility of course selection to fulfill the requirements necessary for entry into medical, chiropractic, or dental school, a physician assistant program, and physical or occupation therapy programs. Obtaining a degree in exercise science will provide additional benefits in preparation for advanced health care education. Coursework in physical activity and exercise prescription for healthy and diseased populations, exercise testing and evaluation, nutrition for health and athletic performance, and exercise and sport psychology can provide valuable knowledge and experiences for advanced study in health care. The following sections serve as a general overview for those students intending to pursue postbaccalaureate study in a health care field or graduate school. If you have decided to pursue a career in medicine or allied health care, you are strongly encouraged to obtain information from those schools you may consider attending so that you have all necessary information regarding the schools' specific entrance requirements. You are strongly encouraged to contact your college or university premedical or health care advisor to assist you in developing your undergraduate coursework and program of study in exercise science. Additionally, almost all health care programs use an online application service, so it is important to be familiar with the requirements of the specific platform.

Preparation for Medical School

There are approximately 160 **allopathic** (MD) and 40 **osteopathic** (DO) accredited medical schools and programs in the United States and Canada. Generally, the minimum academic requirements for acceptance into these schools include one academic year of coursework in

Allopathic medicine A branch of medicine that uses medications, radiation, or surgery to treat or suppress symptoms or the ill effects of disease.

Osteopathic medicine A branch of medicine that treats diseases using a whole person approach to treatment including medications, surgery, spinal manipulation, or massage therapy, as part of the treatment.

biology, general chemistry, organic chemistry, English, and physics. Additional coursework in biochemistry, cell and molecular biology, mammalian physiology, and psychology also provides valuable preparation for medical school. Some medical schools also require a year of college mathematics, including coursework in statistics. Becoming proficient in a foreign language may also be helpful.

To enter medical school following graduation from college, you must complete all premedical requirements prior to your expected matriculation into medical school. You should take the Medical College Aptitude Test (MCAT) at least 3 to 6 months before submitting your application. Medical school admissions committees evaluate candidates largely, although not completely, on objective criteria. Therefore, a high overall grade point average, a high science course grade point average, and competitive MCAT scores are important. Other factors considered by admissions committees include well-developed interpersonal skills, evidence of leadership potential, supportive letters of recommendation, prior work or volunteer experience in a health care facility, and performance in a personal interview. Additional information for preparing for a career in medicine may be obtained from the following organizations: Association of American Medical Colleges (www.aamc.org) and American Association of Colleges of Osteopathic Medicine (www.aacom.org).

Preparation for Dental School

There are approximately 75 accredited schools or programs of dentistry in the United States and Canada. Preparation for entry into a dental program should include knowledge of the basic biological and physical sciences and proficiency in verbal communication. Minimum academic course requirements for acceptance generally include one academic year of introductory biology, inorganic chemistry, organic chemistry, English, mathematics, and physics (with each of the science courses having a laboratory component). Additional coursework in comparative anatomy, developmental biology, biochemistry, cell and molecular biology, genetics, histology, human physiology, and microbiology may also provide valuable preparation for entering a program in dentistry. Owing to the visual, mechanical, and personal nature of the activities you will participate in as a dentist, it may be advantageous to also consider taking courses in art, business, interpersonal communication, and behavioral psychology.

The Dental Admissions Test (DAT) is required by dental schools and programs and should be taken at least 3 to 6 months before submitting your application to dental schools. Your application materials should be completed and submitted approximately 1 year prior to your anticipated matriculation date. Dentistry programs consider a variety of factors in selecting students, including a high overall grade point average, good DAT scores, strong letters of recommendation (including those from a practicing dentist), and a significant number of hours of observation with a practicing dentist. Other important factors considered by admissions committees include a strong interest in dentistry, awareness of the terminology used in dentistry, demonstrated good manual dexterity, and performance during a personal interview. Additional information for preparing for a career in dentistry may be obtained from the following associations: American Dental Association (www.ada.org) and the American Academy of Pediatric Dentistry (www.aapd.org).

Preparation for Chiropractic School

There are approximately 20 fully accredited schools and programs of **chiropractic medicine** in the United States and Canada. All individuals practicing chiropractic medicine must pass the State Board of Health Examination in the state in which they intend to establish professional practice. Although all states sanction the practice of chiropractic medicine, most of the state boards of health also require that an applicant for a licensing examination be a graduate of an accredited college or university program. The primary focus of chiropractic medicine is spine manipulation, for which X-rays may be used as a diagnostic tool. Doctors of Chiropractic medicine also use nutrition and patient counseling as part of the complete health care practice. Neither drug treatment nor surgery may be performed by a Doctor of Chiropractic medicine. Most chiropractic programs require 4 years of study, although some schools may offer a 3-year program.

Schools of chiropractic medicine vary in the number of credit hours and specific courses required for admission. Students interested in applying to a program of study in chiropractic medicine should check on the specific entrance requirements. In general, a minimum of 2 years of college coursework must be completed prior to entering a chiropractic school, and several states require chiropractic physicians to have a baccalaureate degree in addition to the chiropractic degree to be licensed to practice chiropractic medicine.

Students are encouraged to check the Association of Chiropractic Colleges directory or individual school admissions requirements. The following list of minimal entrance requirements should serve as a general guide, although specific entrance requirements may vary slightly at each school. Students are required to have 1 year of the following courses: biological sciences, general chemistry, organic chemistry, and physics (with each of the courses having a laboratory component). Students should also have 1 year of English or communications, one semester of psychology, and coursework in the social sciences or humanities. Generally, multiple letters of recommendation are required with at least one from a practicing chiropractic physician. Applicants are generally evaluated on the basis of their undergraduate grade point average and coursework, the MCAT (if required by the school), clinical exposure and experience, personal attributes, involvement in extracurricular activities, service to others, and a personal interview with the admissions committee. Additional information for preparing for a career in chiropractic medicine may be obtained from the Association of Chiropractic Colleges (www.chirocolleges.org).

Preparation for Podiatric Medical School

There are approximately 10 accredited schools of **podiatric medicine** in the United States and Canada. Generally, the minimum academic requirements for acceptance into these schools include one academic year of coursework in biology, general chemistry, organic chemistry, English, communications, and physics. Podiatric medical schools may also have

Chiropractic medicine A branch of medicine that uses manual therapy and manipulations of the body to treat or suppress symptoms or the ill effects of injuries or disorders of the body.

Podiatric medicine A branch of medicine that diagnoses and treats conditions affecting the foot, ankle, and related structures of the leg.

specific requirements, so make sure to check the application requirements for all schools you may consider applying to.

To enter podiatric medical school following graduation from college, you should complete all premedical requirements at least 1 year prior to your expected matriculation into school. You should take the MCAT at least 3 to 6 months before submitting your application. Podiatric medical school admissions committees evaluate candidates on grade point average, science course grade point average, and MCAT scores. Other factors considered by admissions committees include well-developed interpersonal skills, evidence of leadership potential, letters of recommendation, and performance in a personal interview. Additional information for preparing for a career in podiatric medicine may be obtained from the American Association of Colleges of Podiatric Medicine (www.aacpm.org).

Preparation for Physician Assistant Programs

There are approximately 250 fully accredited physician assistant programs in the United States and Canada. The typical physician assistant program is 24 to 32 months in duration and requires at least 4 years of college and some health care experience prior to admission. Applicants to physician assistant programs must complete approximately 2 years of college courses in basic science and behavioral science as prerequisites to physician assistant training. This is similar to the premedical requirements for medical students. Students interested in physician assistant programs should take coursework in the basic sciences, including anatomy, biochemistry, behavioral sciences, clinical laboratory sciences, microbiology, medical ethics, pathophysiology, pharmacology, physical diagnosis, and physiology. Additional coursework in communications and psychology is encouraged. Preference in admission is usually given to candidates who have prior experience in a health care setting. The hallmark of physician assistant education is the clinical training in a variety of inpatient and outpatient settings, including family medicine, internal medicine, obstetrics and gynecology, pediatrics, general surgery, emergency medicine, and psychiatry (Figure 1.16).

The independent Accreditation Review Commission on Education for the Physician Assistant, Inc. accredits all physician assistant programs. Only graduates of accredited programs are eligible to sit for the Physician Assistant National Certifying Exam administered by the independent National Commission on Certification of Physician Assistants and developed by the National Board of Medical Examiners. All 50 states and the District of Columbia have enacted laws or regulations authorizing physician assistant practice and authorize physician assistants to prescribe controlled medicines. Once a physician assistant is certified, she or he must complete 100 hours of continuing medical education every 2 years

FIGURE 1.16 Physician assistants work in a variety of inpatient and outpatient setting. (Shutterstock.)

and take a recertification exam every 10 years to keep her/his certification current. Additional information for preparing for a career as a physician assistant may be obtained from the following associations: the American Academy of Physician Assistants (www.aapa.org) and Physician Assistant Education Association at (www.paeaonline.org).

Preparation for Physical and Occupational Therapy

There are approximately 500 fully accredited physical therapy (PT) and physical therapy assistant (PTA) programs and approximately 300 fully accredited occupational therapy (OT) and occupational therapy assistant (OTA) programs in the United States. Undergraduate degree programs in exercise science provide excellent preparation for graduate study in physical and occupational therapy because many exercise science programs offer the required coursework for direct entry into a physical or occupational therapy program (Figures 1.17 and 1.18).

The American Physical Therapy Association (APTA) has mandated that all PT programs offer a doctoral degree in physical therapy (DPT) to earn or retain certification. Therefore, students interested in PT must earn a doctoral degree from an accredited DPT program before taking the licensure examination. Each school has its own prerequisite courses for admission into a program, and therefore it is strongly recommended that you identify the programs you may wish to attend and verify the list of requirements for admission. In general, most DPT programs require the following courses for admission: one academic year

FIGURE 1.17 Physical therapists provide rehabilitation in an assortment of professional practices. (Shutterstock.)

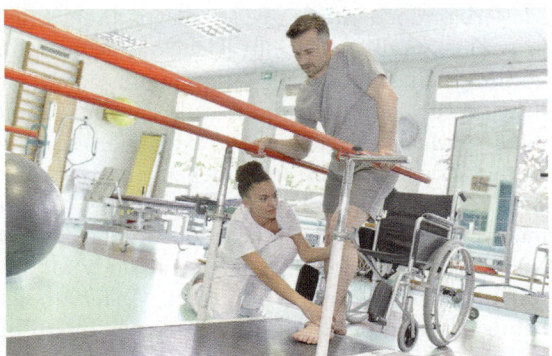

FIGURE 1.18 Occupational therapists provide functional rehabilitation to improve patients' daily routines and lives. (Shutterstock.)

of biology, chemistry, and physics (with a laboratory component for each). Additionally, most programs require at least one course in each of the following: English composition, exercise physiology, human anatomy, human physiology, biomechanics, general psychology, developmental psychology, and statistics. Additional coursework in biochemistry, organic chemistry, medical ethics, philosophy, and speech composition may also be valuable.

The American Occupational Therapy Association (AOTA) oversees the OT profession and provides support for graduate programs in OT. Students interested in OT must earn a Master's or doctoral degree from an accredited OT program and then pass a licensure examination in order to practice as an OT. Each school has its own prerequisite courses for admission into a program, and therefore it is strongly recommended that you identify the programs you may wish to attend and verify the list of requirements for admission. In general, most OT programs require the following courses for admission: human anatomy and physiology, neuroanatomy or neuropsychology or behavioral neuroscience, kinesiology, psychopathology, lifespan development, statistics, and at least one course in psychology, sociology, anthropology, public health, or related field. Additional coursework in biochemistry, organic chemistry, medical ethics, philosophy, and speech composition may also be valuable.

The majority of schools offering the Master of science in occupational therapy (MSOT), or doctorate of occupational therapy (OTD), and DPT programs either require or recommend the Graduate Record Examination (GRE) as part of the admission materials, whereas other schools may require the Allied Health Professions Aptitude Test. Physical and occupational therapy programs require that candidates have volunteered in PT or OT clinics. Many PT and OT admissions committees look for evidence of exposure to the types of patients with whom you will eventually be working and for experience in a variety of clinical settings. The number of required volunteer hours varies by school and program and must be certified by a health care professional. A general recommendation is that you complete at least 100 hours of volunteer service. Other important factors considered by admissions committees include well-developed interpersonal skills and performance during a personal interview. Additional information for preparing for a career in physical or occupational therapy may be obtained from the following associations: APTA (www.APTA.org) and the AOTA (www.aota.org).

Preparation for Graduate School

Obtaining an undergraduate degree in exercise science provides excellent preparation for continuing education in graduate school. A graduate degree is typically earned at the Master's or doctorate degree level after completing the Bachelor's degree. Graduate degree programs are different from Bachelor's degree programs in many ways, the most distinguishing being graduate degree programs do not have general education requirements and instead focus on gaining more skill and content knowledge at a higher level. Additionally, many graduate programs have extensive research components in the form of advanced methodology, statistics, and independent research.

Each of the areas of study in exercise science may be offered as graduate programs in many colleges and universities. Most graduate programs require 1 to 2 years of study for a Master's degree and 3 to 5 years of study for a doctoral degree. Many graduate degree programs require a research project, Master's thesis, or doctoral dissertation as the culminating experience, while others require an internship. Graduate programs have specific requirements for entry into the program and completion of the degree that are often unique

Table 1.5 — Guidelines for Identifying and Selecting a Program for Graduate Study

- Explore areas of interest to you, being sure to review the research interests and teaching expertise of the faculty in the graduate program
- Identify an individual who holds a position you envision yourself moving into, talk to that person about their educational journey
- Use various graduate school search engines to gain knowledge about different programs of study
- Use an admissions counselor, academic advisor, and mentor to help you learn more about which graduate programs might be the best fit for you
- Communicate with faculty from your undergraduate program about which graduate programs are appropriate for your interests

to the discipline and the particular college or university. Obtaining a graduate degree will generally result in greater marketability, a higher income, greater job responsibility, and more professional opportunities. In addition, many professional employment opportunities require an advanced degree beyond a Bachelor's degree. Furthermore, an exercise science degree can provide solid preparation for graduate degrees in areas such as athletic training, public health, epidemiology, physiology, biochemistry, psychology, and biomedical engineering.

Additional information about specific graduate programs can be obtained by visiting the home page of the college or university that you are interested in attending. Table 1.5 provides some guidance for identifying and selecting an appropriate program to continue your graduate education.

Thinking Critically

How does your choice of undergraduate coursework in exercise science best prepare you for a career in the professional fields of medicine and allied health?

Interview

Jack W. Berryman, PhD, FACSM

Professor Emeritus of Medical History at the University of Washington School of Medicine, Seattle, Washington. Jack is also ACSM Historian Emeritus and past Chair of ACSM's Office of Museum, History, and Archives

Brief Introduction – I began my college education as a Health and Physical Education major at Lock Haven University in Pennsylvania. In the early 1970s, I completed two Master's degrees at the University of Massachusetts, Amherst: one in Exercise Science and the other in History. During this time, I became involved in the early stages of the exercise science and sport studies movement. I completed a PhD at the University of Maryland in Physical Education and History, and I joined the University of Washington faculty in 1975. In 1992, I was voted a fellow in the American Academy of Kinesiology and Physical Education and in 1994, was named ACSM's official historian. I became a fellow in ACSM in 2001. Two of my books of special interest are *Sport and Exercise Science: Essays in the History of Sports Medicine* (1992) and *Out of Many, One: A History of the American College of Sports Medicine* (1995). I have been honored as ACSM's D. B. Dill Historical Lecturer an unprecedented two times, in 1994 and 2004, and received ACSM's Citation Award in 2017.

Q: What is ACSM's importance in the development of exercise science?

Because of ACSM's unique multidisciplinary membership that included physicians, physical educators, exercise scientists, and many other professions directly related to the health benefits of exercise, the organization became a leader in the field by the late 1950s. ACSM's annual meetings, along with its significant journals and books, as well as position stands and scientific roundtables, put ACSM in the forefront of health promotion and exercise science very quickly. With a strong scientific and clinical base, the ACSM has continued to be the most respected authority in the field of exercise science. The ACSM has played a leadership role in the National Coalition for Promoting Physical Activity (1995) and in the writing of *Physical Activity and Health: A Report of the Surgeon General* (1996). The College has been a dominant force in the institutionalization of exercise in the prevention, diagnosis, and treatment of cardiac and other degenerative diseases such as diabetes, osteoporosis, and obesity.

Q: What is ACSM's importance in the development of sport and athletic performance?

Although ACSM's early emphasis was on health and fitness, the ACSM began to move more in the direction of clinical sports medicine and athletic training in the 1970s. Through annual meetings and publications, those with clinical sports medicine expertise were gradually attracted to the ACSM. By the 1980s, ACSM had as professional members some of the leading sport team physicians and athletic trainers in the world. Research and publications began to focus on the prevention and treatment of sports injuries, peak athletic performance, the role of the team physician, and other important aspects of athleticism, such as the female athlete triad and proper hydration. The ACSM also offers a series of team physician courses, which are important in disseminating information to practicing physicians. In the early 2000s, ACSM began publishing *Current Sports Medicine Reports,* a clinical review journal. More recently, ACSM has been the driving force behind the movement to identify and treat concussions in a timely manner.

Q: How will ACSM contribute as a world leader in exercise science into the twenty-first century?

ACSM has in its membership the leaders in exercise science and sports medicine worldwide. This highly scientific international membership continues to position the ACSM at the forefront of any major development in the field. The ACSM also has a very proactive and highly talented professional staff at its headquarters in Indianapolis who work tirelessly to further ACSM's goals. Publications, annual meetings, special symposia, certifications, and position stands, among numerous other projects, have a constant impact on the field. Because of its stature, the ACSM has had a significant impact on policy decisions and continues to be the most dynamic force to identify exercise deficiency as a serious public health issue. The ACSM believes that exercise is, in fact, good medicine and will continue to convince federal, state, and local officials that physical activity is the key to good health and the prevention of disease through its worldwide Exercise is Medicine campaign.

Interview

Scott K. Powers, EdD, PhD, FACSM

UAA Endowed Professor and Distinguished Professor, Department of Applied Physiology & Kinesiology at the University of Florida. Scott is a past Vice President of the ACSM

Brief Introduction – I earned my BS degree in Physical Education from Carson Newman University, and after completing a Master's degree at the University of Georgia, I then earned a doctoral degree in Exercise Science (with an emphasis in Exercise Physiology) from the University of Tennessee. I desired additional training in Basic Science and Biochemistry, and subsequently completed a PhD in Physiology and Cell Biology from Louisiana State University.

Q: What are your most significant career experiences?

I began my professional career as a faculty member in the Department of Kinesiology at Louisiana State University, and I am currently a professor within the Department of Applied Physiology and Kinesiology at the University of Florida. Citing my research and teaching accomplishments, the University of Florida honored me with an endowed professorship in 2004 followed by the title of "distinguished professor" in 2005. Throughout my career, I have been active in both the American Physiological Society and the ACSM, having served both organizations via committee memberships and as an elected officer.

Q: Why did you choose to become an exercise science professional?

My personal interest in exercise physiology began during my undergraduate studies at Carson Newman University. As a distance runner on our collegiate track team, I became fascinated with the physiology of human performance and began to read all of the exercise physiology texts available at the time. I quickly realized that knowledge in exercise physiology was limited in many areas and, hence, a need for additional research existed. Fueled by natural curiosity, I decided to pursue a graduate degree in exercise physiology to learn more about this exciting field. My experience in graduate school served to motivate me to expand my knowledge about exercise physiology and to dedicate my career to research and teaching in exercise physiology.

Q: What are your top two or three professional accomplishments?

From a research perspective, my primary achievements have come in three different areas of inquiry. First, during the 1980s, my laboratory investigated the factors that control breathing and regulate pulmonary gas exchange during exercise. An important finding related to this work was that pulmonary gas exchange limits maximal oxygen uptake in many elite endurance athletes. This novel observation revealed that the lung and not the cardiovascular system limits maximal oxygen uptake in this population of elite athletes.

In 1990, my research focus shifted toward two new areas of investigation: (a) exercise-induced cardioprotection; and (b) disuse muscle atrophy. By studying how exercise changes the heart, my research group has made significant contributions to our understanding of how and why regular bouts of endurance exercise protect the heart during a heart attack. This form of cardiac safeguard is termed exercise-induced cardioprotection and provides direct evidence about the importance of exercise. In this regard, our work demonstrates that as few as 5 continuous days of endurance exercise results in cardioprotection. However, this exercise-induced protection is lost quickly (*i.e.*, within 18 days) after the cessation of exercise. Moreover, our research has revealed some of the key cellular mechanisms that are responsible for exercise-induced cardioprotection. This work is important and has led to the development of several pharmacological approaches to protect the heart against cellular injury during an ischemia-reperfusion insult (*i.e.*, heart attack).

Finally, my research team has also made some important discoveries regarding the mechanisms responsible for disuse muscle atrophy (*i.e.*, muscle loss owing to inactivity resulting from prolonged bed rest, limb immobilization, etc.). By investigating the mechanisms responsible for this type of muscle wasting, our research team hopes to develop countermeasures to prevent this type of muscle atrophy and prevent or retard this cause of muscle atrophy.

 What advice would you have for a student exploring a career in exercise science?

Successful individuals in any profession share many of the same traits. A key trait of all successful people is that they are passionate about what they do. Therefore, my advice to students exploring career options is to find your passion in life. In short, what career would make you happy so that you enjoy going to work each morning? Is it possible to enjoy going to work? YES! A job is only work if you do not like it. So, find your passion and you will never "work" a day in your life! Although locating your passion in life is the first important step in developing your career, it is critical that you receive state-of-the-art training to help you achieve your professional goals. Therefore, before choosing a major, do your homework and research the best courses that will provide you with the science background required to achieve your professional goals. Notably, choosing the "correct" faculty mentors to guide you during your graduate school education and through your postdoctoral training is vital to ensure that you receive the cutting-edge instruction that will launch your career on the right track.

SUMMARY

- Exercise science refers to the application of science to the study of movement in health, physical activity, exercise, sport, and athletic performance.
- Exercise science has evolved from the work of scholars and professionals from a variety of disciplines, subdisciplines, and specialty areas.
- Much of the educational coursework provided in a curriculum of study is valuable preparation for professional employment or continued study in graduate or professional school.
- Teachers, coaches, exercise specialists, health care providers, scholars, and researchers use the knowledge gained from exercise science in both basic and applied contexts.
- Exercise science is truly interdisciplinary with individuals from a wide array of backgrounds working together to provide the most relevant information and answer important questions related to individual health, exercise, medicine, public health, sport, and athletic competition.

FOR REVIEW

1. What is the difference between morbidity and mortality?
2. What disciplines, subdisciplines, and specialty areas constitute exercise science?

3. Describe the foundational characteristics of an academic discipline and explain how exercise science meets those characteristics?
4. What are the similarities and unique differences between physical activity, exercise, sport, and athletic competition?
5. What historical factors have contributed to the creation of exercise science as an overarching field of study?
6. In what significant ways did the Harvard Fatigue Laboratory contribute to the development of exercise science?
7. What are the primary ways the ACSM disseminates information to the professional membership and general population?
8. Why must exercise science students devote close attention to required coursework if planning on obtaining certification or licensure after graduation?
9. What are the primary requirements for entry into a professional career in a health care field after graduation?

Project-Based Learning

1. Identify the top ten graduate or professional programs you may be interested in attending after completing your undergraduate degree, review the admission criteria, and create a systematic process for selecting the program for which you want to enroll after your undergraduate education is complete.
2. Choose an ACSM position stand from Table 1.2, identify the top five take-home messages that you would share with a fellow student, and present those messages in either a written report or an oral presentation.

REFERENCES

1. Berryman JW. Ancient and early influences. In: Tipton CM, editor. *Exercise Physiology: People and Ideas*. New York (NY): Oxford University Press; 2003:1–38.
2. Albright A, Franz MS, Hornsby G, et al. Exercise and type 2 diabetes. *Med Sci Sports Exerc*. 2000;32(7):1345–60.
3. Bish CL, Blanck HM, Serdula MK, Marcus M, Kohl HW, Khan LK. Diet and physical activity behaviors among Americans trying to lose weight: 2000 behavioral risk factor surveillance system. *Obes Res*. 2005;13:596–607.
4. Chodzko-Zajio W, Proctor DN, Fiatarone-Singh MA, et al. Exercise and physical activity for older adults. *Med Sci Sports Exerc*. 2009;41(7):1510–30.
5. Garber CE, Blissmer B, Deschenes MR, et al. Quantity and quality of exercise for developing and maintaining cardiorespiratory, musculoskeletal, and neuromotor fitness in apparently healthy adults: guidance for prescribing exercise. *Med Sci Sports Exerc*. 2011;43(7):1334–59.
6. Kohrt WM, Bloomfield SA, Little KD, Nelson ME, Yingling VR. Physical activity and bone health. *Med Sci Sports Exerc*. 2004;36(11):1985–96.
7. Pate RR, Pratt M, Blair SN, et al. A recommendation from the Centers for Disease Control and Prevention and the American College of Sports Medicine. *JAMA*. 1995;273(5):402–7.
8. Al-Shaar L, Li Y, Rimm EB, et al. Physical activity and mortality among male survivors of myocardial infarction. *Med Sci Sports Exerc*. 2020;52(8):1729–36.
9. Ozemek C, Whaley MH, Finch WH, Kaminsky LA. High cardiorespiratory fitness levels slow the decline in peak heart rate with age. *Med Sci Sports Exerc*. 2016;48(1):73–81.
10. Joseph G, Marott JL, Torp-Pedersen C, et al. Dose-response association between level of physical activity and mortality in normal, elevated, and high blood pressure. *Hypertension*. 2019;74(6):1307–15.

11. Pedisic Z, Shrestha N, Kovalchik S, et al. Is running associated with a lower risk of all-cause, cardiovascular and cancer mortality, and is the more the better? A systematic review and meta-analysis. *Br J Sports Med.* 2020;54(15):898–905.
12. Centers for Disease Control and Prevention. Leading causes of death: 2017. 2020. Available from: http://www.cdc.gov/nchs/fastats/leading-causes-of-death.htm.
13. Burton NW. Occupation, hours worked, and leisure-time physical activity. *Prev Med.* 2000;31(1):673–81.
14. Centers for Disease Control and Prevention. Prevalence of no leisure-time physical activity — 35 States and the District of Columbia, 1988–2002. *MMWR Morb Mortal Wkly Rep.* 2004;53(4):82–6.
15. Porter AK, Cuthbertson CC, Evenson KR. Participation in specific leisure-time activities and mortality risk among U.S. adults. *Ann Epidemiol.* 2020:27–34.E1.
16. Centers for Disease Control and Prevention. Percent of adults 18 years of age and over who met the Physical Activity Guidelines for 2018. 2020. Available from: http://www.cdc.gov/nchs/fastats/exercise.htm.
17. Slyper AH. The pediatric obesity epidemic: causes and controversies. *J Clin Endocrinol Metab.* 2004;89(6):2540–7.
18. Apovian CM. The obesity epidemic — understanding the disease and the treatment. *N Engl J Med.* 2016;374(2):177–9.
19. Flegal KM, Carroll MD, Kuczmarski RJ, Johnson CL. Overweight and obesity in the United States: prevalence and trends, 1960–1994. *Int J Obes.* 1998;22:39–47.
20. Flegal KM, Carroll MD, Ogden CL, Curtin LR. Prevalence and trends in obesity among US Adults, 1999–2008. *JAMA.* 2010;303(3):235–41.
21. Ogden CL, Carroll MD, Curtin LR, McDowell MA, Tabak CJ, Flegal KM. Prevalence of overweight and obesity in the United States, 1999–2004. *JAMA.* 2006;295(13):1549–55.
22. Ogden CL, Carroll MD, Fryar CD, Flegal KM. Prevalence of obesity among adults and youth: United States, 2011–14. *NCHS Data Brief.* 2015;219:1–8.
23. Flegal KM, Ogden CL, Fryar C, Afful J, Klein R, Huang DT. Comparisons of self-reported and measured height and weight, BMI, and obesity prevalence from national surveys: 1999–2016. *Obesity.* 2019;27(10):1711–9.
24. Courneya KS. Physical activity and cancer survivorship: a simple framework for a complex field. *Exerc Sport Sci Rev.* 2014;42(3):102–9.
25. Kraus WE, Bittner V, Appel L, et al. The National Physical Activity Plan: a call to action from the American Heart Association: a science advisory from the American Heart Association. *Circulation.* 2015;131(21):1932–40.
26. Elhakeem A, Murray ET, Cooper R, Kuh D, Whincup P, Hardy R. Leisure-time physical activity across adulthood and biomarkers of cardiovascular disease at age 60–64: a prospective cohort study. *Atherosclerosis.* 2018;269:279–87.
27. Bridle C, Spanjers K, Patel S, Atherton NM, Lamb SE. Effect of exercise on depression severity in older people: systematic review and meta-analysis of randomised controlled trials. *Br J Psychiatry.* 2012;201(3):180–5.
28. Patterson MS, Gagnon LR, Vukelich A, Brown SE, Nelon JL, Prochnow T. Social networks, group exercise, and anxiety among college students. *J Am Coll Health.* 2019:1–9.
29. Senn AE. *Power, Politics, and the Olympic Games.* Champaign (IL): Human Kinetics; 1999.
30. Durant J. *Highlights of the Olympics: From Ancient Times to the Present.* New York (NY): Hastings House Publishers; 1973.
31. National Federation of State High School Associations. 2018–19 athletics participation summary. 2014. Available from: http://www.nfhs.org/.
32. National Collegiate Athletic Association. *NCAA® Sports Sponsorship and Participation Rates Report.* Indianapolis, IN: National Collegiate Athletic Association; 2014.
33. Castleberry T, Irvine C, Deemer SE, et al. Consecutive days of exercise decrease insulin response more than a single exercise session in healthy, inactive men. *Eur J Appl Physiol.* 2019;119(7):1591–8.
34. Sagoe D, Molde H, Andreassen CS, Torsheim T, Pallesen S. The global epidemiology of anabolic-androgenic steroid use: a meta-analysis and meta-regression analysis. *Ann Epidemiol.* 2014;24(5):383–98.
35. Pereira E, Moyses SJ, Ignácio SA, et al. Anabolic steroids among resistance training practitioners. *PLoS One.* 2019;14(10).
36. Henry FM. Physical education: an academic discipline. *J Health Phys Educ Recreat.* 1964;35:32–33, 69.
37. Swanson RA, Massengale JD. Exercise and sport science in 20th-century America. In: Massengale JD, Swanson RA, editors. *The History of Exercise and Sport Science.* Champaign (IL): Human Kinetics; 1997:1–14.

38. DeVries HA. History of exercise science. In: Housh TJ, Housh DJ, Johnson GO, editors. *Introduction to Exercise Science*. San Francisco (CA): Benjamin Cummings; 2008. p. 17–40.
39. Massengale JD, Swanson RA. *The History of Exercise and Sport Science*. Champaign (IL): Human Kinetics; 1997.
40. Tipton CM. Antiquity to the early years of the 20th century. In: Tipton CM, editor. *History of Exercise Physiology*. Champaign (IL): Human Kinetics, Inc; 2014. p. 3–32.
41. Gerber EW. *Innovators and Institutions in Physical Education*. Philadelphia (PA): Lea and Febiger; 1971.
42. Buskirk ER. From Harvard to Minnesota: keys to our history. In: Holloszy JO, editor. *Exercise and Sport Sciences Reviews*. Baltimore (MD): Williams & Wilkins; 1992. p. 1–26.
43. Tipton CM. Exercise physiology, part II: a contemporary historical perspective. In: Massengale JD, Swanson RA, editors. *The History of Exercise and Sport Science*. Champaign (IL): Human Kinetics; 1997. p. 396–438.
44. Sidentop D. *Introduction to Physical Education, Fitness, and Sport*. 2nd ed. Mountain View (CA): Mayfield Publishing Company; 1994.
45. Buskirk ER, Tipton CM. Exercise physiology. In: Massengale JD, Swanson RA, editors. *The History of Exercise and Sport Science*. Champaign (IL): Human Kinetics; 2003. p. 367–438.
46. Blair SN, Kohl HW, Paffenbarger RS, Clark DG, Cooper KH, Gibbons LW. Physical fitness and all-cause mortality. *JAMA*. 1989;262(17):2395–401.
47. Núñez C, Capelo JL, Igrejas G, Alfonso A, Botana LM, Lodeiro C. An overview of the effective combination therapies for the treatment of breast cancer. *Biomaterials*. 2016;97:34–50.
48. Aune D, Sen A, Norat T, et al. Body mass index, abdominal fatness, and heart failure incidence and mortality: a systematic review and dose-response meta-analysis of prospective studies. *Circulation*. 2016;133(7):639–49.
49. Ramkumar PN, Navarro SM, Haeberle HS, et al. Short-term outcomes of concussions in major league baseball: a historical cohort study of return to play, performance, longevity, and financial impact. *Orthop J Sports Med*. 2018;6(12).
50. Navarro SM, Sokunbi OF, Haeberle HS, et al. Short-term outcomes following concussion in the NFL: a study of player longevity, performance, and financial loss. *Orthop J Sports Med*. 2017;5(11).
51. McKee AC, Stein TD, Kiernan PT, Alvarez VE. The neuropathology of chronic traumatic encephalopathy. *Brain Pathol*. 2015;25(3):350–64.
52. Doperak J, Anderson K, Collins M, Emami K. Sport-related concussion evaluation and management. *Clin Sports Med*. 2019;38(4):16.
53. Gatorade Sport Science Institute 2021. Available from: https://www.gssiweb.org/en.
54. Berryman JW. *Out of Many, One: A History of the American College of Sports Medicine*. Champaign (IL): Human Kinetics; 1995.
55. American College of Sports Medicine. Prevention of thermal injuries during distance running. *Med Sci Sport Exerc*. 1974;19:529.
56. United States Department of Health and Human Services, Centers for Disease Control and Prevention, National Center for Chronic Disease Prevention and Health Promotion, President's Council on Physical Fitness and Sports. *Physical Activity and Health: A Report of the Surgeon General*. Atlanta (GA): United States Department of Health and Human Services; 1995.
57. Maron BJ, Isner JM, McKenna WJ. 26th Bethesda conference: recommendations for determining eligibility for competition in athletes with cardiovascular abnormalities. Task Force 3: hypertrophic cardiomyopathy, myocarditis and other myopericardial diseases and mitral valve prolapse. *J Am Coll Cardiol*. 1994;24(4):880–5.
58. Herring SA, Bergfeld JA, Boyd J, et al. Team physician consensus statement. *Med Sci Sports Exerc*. 2000;32(4):877–8.
59. Flegal KM, Kruszon-Moran D, Carroll MD, Fryar CD, Ogden CL. Trends in obesity among adults in the united states, 2005 to 2014. *JAMA*. 2016;315(21):2284–91.
60. Ogden CL, Carroll MD, Kit BK, Flegal KM. Prevalence of childhood and adult obesity in the United States, 2011–2012. *JAMA*. 2014;311(8):806–14.
61. Balady G, Chaitman BL, Foster C, Froelicher E, Gordon N, Van Camp SP. AHA/ACSM Joint Position Statement: automated external defibrillators in health/fitness facilities. *Med Sci Sport Exercise*. 2002;34(3):561–4.
62. Balady GJ, Chaitman BL, Driscoll D, et al. AHA/ACSM Joint Position Statement: recommendations for cardiovascular screening, staffing, and emergency policies at health/fitness facilities. *Med Sci Sports Exerc*. 1998;30(6):1009–18.

63. Donnelly JE, Blair SN, Jakicic JM, Manore MM, Rankin JW, Smith BK. Appropriate physical activity intervention strategies for weight loss and prevention of weight regain for adults. *Med Sci Sports Exerc.* 2009;41(2):459–71.
64. Thompson PD, Franklin BA, Balady GJ, et al. Exercise and acute cardiovascular events: placing the risks into perspective. *Med Sci Sports Exerc.* 2007;39:886–97.
65. Sawka MN, Burke LM, Eichner ER, Maughan RJ, Montain SJ, Stachenfeld NS. Exercise and fluid replacement. *Med Sci Sports Exerc.* 2007;39(2):377–90.
66. Pescatello LS, Franklin BA, Fagard R, Farquhar WB, Kelley GA, Ray CA. Exercise and hypertension. *Med Sci Sport Exerc.* 2004;36(3):533–53.
67. Van Camp SP, Cantwell JD, Fletcher GF, Smith LK, Thompson PD. Exercise for patients with coronary artery disease. *Med Sci Sports Exerc.* 1994;26(3):i–v.
68. Armstrong LE, Casa DJ, Millard-Stafford ML, et al. Exertional heat illness during training and competition. *Med Sci Sports Exerc.* 2007;39(3):556–72.
69. American College of Sports Medicine, American Dietetic Association, Dietitians of Canada. Nutrition and Athletic Performance. *Med Sci Sports Exerc.* 2016;48(3):543–68.
70. Donnelly JE, Hillman CH, Castelli D, et al. Physical activity, fitness, cognitive function, and academic achievement in children: a systematic review. *Med Sci Sports Exerc.* 2016;48(6):1197–222.
71. Castellani JW, Young AJ, Ducharme MB, et al. Prevention of cold injuries during exercise. *Med Sci Sports Exerc.* 2006;38(11):2012–29.
72. Ratamess NA, Alvar BA, Evetovich TK, et al. Progression models in resistance training for healthy adults. *Med Sci Sports Exerc.* 2009;41(3):687–708.
73. Nattiv A, Loucks AB, Manore MM, Sanborn CF, Sundgot-Borgen J, Warren MP. The female athlete triad. *Med Sci Sports Exerc.* 2007;29(5):1867–82.
74. American College of Sports Medicine. The use of anabolic-androgenic steroids in sports. *Med Sci Sport Exerc.* 1987;19(5):534–9.
75. Sawka MN, Joyner MJ, Miles DS, Robertson RJ, Spriet LL, Young AJ. The use of blood doping as an ergogenic aid. *Med Sci Sports Exerc.* 1996;28(6):i–viii.
76. Opplinger RA, Case S, Horswill CA, Landry GL, Shelter AC. Weight loss in wrestlers. *Med Sci Sports Exerc.* 1996;28(6):ix–xii.

CHAPTER 2

Introduction to Research

After completing this chapter, you will be able to:

1. Explain the importance of research as it relates to enhancing our understanding of health, physical activity, exercise, sport, and athletic performance.
2. Define the types of research commonly conducted by exercise science professionals.
3. Explain the primary components of the systematic research process.
4. Describe the dissemination and publication process for peer-reviewed research.
5. Explain evidence-based practice and how it is used to guide decision-making.
6. Explain how to become engaged in student research activities.

RESEARCH IN EXERCISE SCIENCE

Research is a carefully planned course of action designed to generate knowledge and information in an effort to expand our understanding of a specific topic, idea, or concept. Typically, the research process begins with a question that is systematically investigated using a process that involves collecting information, analyzing the information, and then reflecting on the analyzed information to formulate conclusions and possibly an answer to the research question. The investigative process to answer even a single research question can often take months, years, and sometimes decades to complete. The knowledge gained through the systematic research process is documented and disseminated for others to evaluate for its intrinsic and extrinsic value.

Conducting research and the dissemination of the knowledge gained from that research are key activities of many exercise science professionals. Much of what we know about health, physical activity, exercise, sport, and athletic performance is derived from experiments and studies conducted by exercise science professionals. Researchers work to develop or accumulate knowledge for a discipline, subdiscipline, or specialty area. As you begin your academic program of study, you will quickly learn that the knowledge that comprises the disciplines and subdisciplines of exercise science is founded in well-designed and carefully conducted research. In many of the chapters contained in this book, you will find examples and summaries of research topics and areas of study that are central to the exercise science disciplines. Research informs decision-making in all that we do as professionals. It is therefore important to provide a general overview of research so that you may have a greater appreciation for and understanding of the research work performed in the areas comprising exercise science.

Research Practices

Exercise science professionals may conduct research using a variety of different techniques, methods, designs, and models. Much of the research conducted in exercise science uses either a human (Figure 2.1) or animal model (Figure 2.2). It is also common, however, to use tissue harvested from humans or animals in addition to cell cultures (Figure 2.3), computer modeling (Figure 2.4), and advanced statistical analysis procedures to answer many of the research questions in exercise science. Examples of the different types of equipment and procedures used in exercise science research are discussed in Chapter 11. The specific processes and methods used in research are employed to assure that the conclusions reached at the end of an experiment or study are factual and stand up to rigorous analysis. These methods include the formulation of hypotheses and alternative hypotheses to test the objectivity of the experiment and experimenter (1). Researchers and scholars often disseminate the results at professional conferences and workshops and in technical reports, policy briefs, and peer-reviewed journals to allow others to review, evaluate, replicate, and often build on the work that was conducted (1).

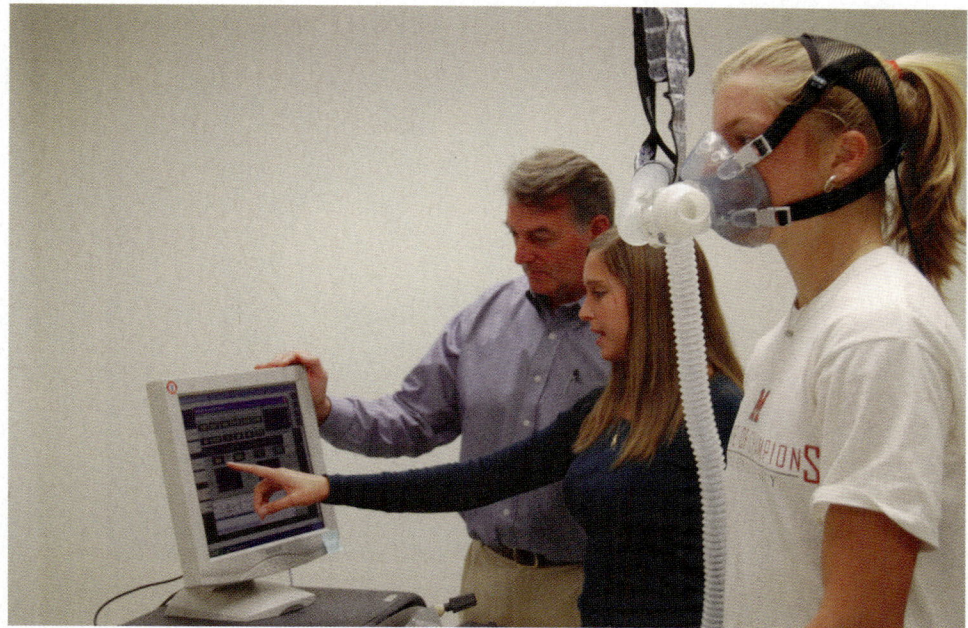

FIGURE 2.1 Much of the research in exercise science uses a human model.

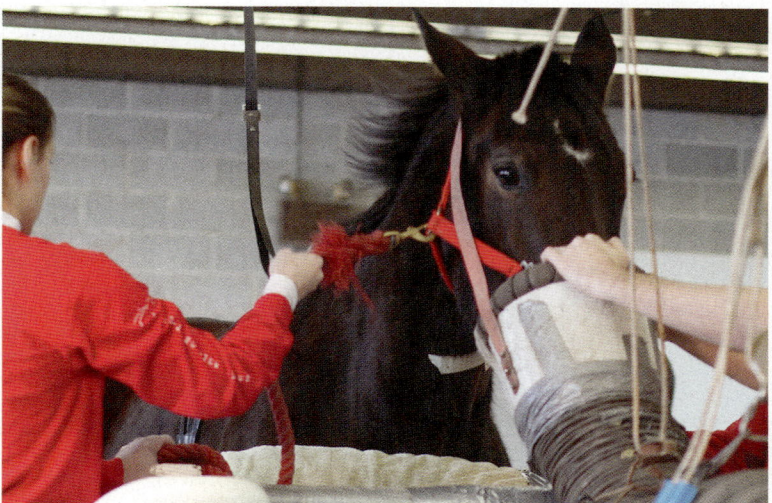

FIGURE 2.2 Animals are often used in exercise science research (provided by Equine Exercise Physiology Laboratory, Rutgers University).

FIGURE 2.3 Tissue samples and cell cultures are part of exercise science–related research. (Shutterstock.)

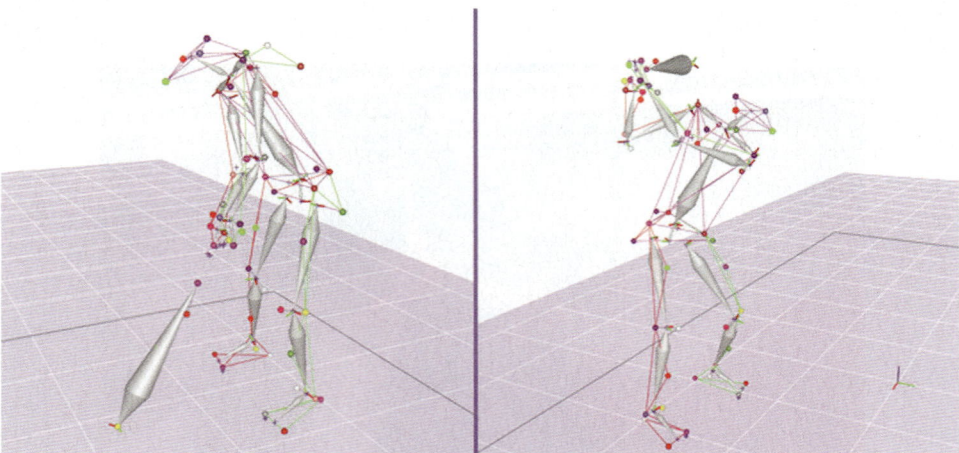

FIGURE 2.4 Computer modeling is a commonly used approach in several exercise science areas.

Investigators, scholars, and researchers use a variety of practices to solve interesting problems and answer specific questions. Individuals conducting high-quality research and scholarship acquire certain personal and intellectual characteristics that enable them to think and act in a specific manner. Some examples of these attitudes and characteristics are provided in Table 2.1 (2).

Table 2.1	Attitudes and Characteristics of Scientists and Researchers (2)
Objective and impartial, and they minimize personal bias toward the observations they make
Deal with facts and make conclusions based on the data, and do not make personal decisions without data support
Are inquisitive, wanting to seek answers from the data derived from the research
Desire to understand how things fit into an orderly system

Types of Research

Investigative work performed by researchers and scholars is typically characterized as basic, applied, or translational research. **Basic research** is sometimes referred to as pure or fundamental research and aims to expand the knowledge base by formulating, evaluating, or expanding a theory (3). Research in certain areas of exercise science can be categorized as basic research if there is, for example, a biochemical or genetic approach. In general, basic research aims to discover new and unknown knowledge often without concern for the practical application of the information that has been gained. The practical application of the information that has been generated is considered at a later time, sometimes even years in the future. The approach of basic-type research is often described as "reductionist," in that scientific methods are used that reduce the experimental method to isolate a single factor (*e.g.*, a lone gene product, protein, or other biological factor). Many of the advances seen in health, physical activity, exercise, sport, and athletic performance have arisen from practical applications of discoveries made in basic research (3).

Applied research attempts to solve practical, everyday problems, and it frequently uses the same techniques, designs, and methods as basic research. In applied research, theoretical and hypothetical concepts are tested in real world situations. Applied research attempts to find solutions to an immediate practical problem or question facing an individual or group. However, applied research is often used to make inferences beyond the specific group or situation being studied. The results of applied research are intended to be generalized and to extend to the larger target population. Applied research can be conducted in controlled laboratories, classrooms, hospitals, clinics, and field settings. Much of the research in exercise science is applied because it is concerned with testing the processes of movement and behavior in real-life conditions. Basic or applied research conducted by exercise scientists must depend on each other for the proper research process to take place

Basic research Systematic study directed toward the increase of knowledge, the primary aim being a greater knowledge or understanding of the subject under study.

Applied research Systematic research directed toward finding solutions to an immediate practical problem.

and for theories to be developed and evaluated. Some studies may use approaches that combine both basic and applied research (3).

Translational research applies findings from basic research to directly improve human health and well-being. As the name implies, translational research aims to translate findings from basic research into meaningful practices, typically in the field of health care. Translational research is often referred to as "benchtop-to-bedside" research, whereby information from laboratory experiments is used to inform **clinical trials** and ultimately patient applications. Translational research is an interface between basic research and clinical application. It uses knowledge from basic science research to generate new drugs, produce new medical devices, and create new treatment options for patients. The outcome of translational research is the production of an encouraging new treatment that can be used to improve patient health care (4). Several specialized journals are dedicated to the field of translational research, including *Translational Research* (http://www.journals.elsevier.com/translational-research-the-journal-of-laboratory-and-clinical-medicine/), the *American Journal of Translational Research* (http://www.ajtr.org/), and the *Translational Journal of the American College of Sports Medicine* (www.acsm.org). Table 2.2 provides some examples of basic research, applied research, and translational research conducted in the disciplines of exercise science.

Thinking Critically

How are basic and applied research combined to improved understanding of health, physical activity, exercise, sport, and athletic performance?

Thinking Critically

How does translational research help improve health care in patients with conditions such as cardiovascular disease or Type 2 diabetes?

Table 2.2 Examples of Basic, Applied, and Translational Research in Areas of Exercise Science

AREAS OF EXERCISE SCIENCE	BASIC RESEARCH	APPLIED RESEARCH	TRANSLATIONAL RESEARCH
Exercise physiology	The influence of muscle pH levels on lactic acid movement out of the muscle fibers	Does alteration of whole body pH levels through the use of an ergogenic aid improve distance running performance?	Does the change in muscle pH influence blood glucose control in persons with diabetes?
Clinical exercise physiology	The effect of different drugs on cardiac muscle fiber contractile strength	Do different drug therapies combined with exercise decrease the recovery time after a myocardial infarction?	Does the use of different combinations of drugs increase longevity in heart failure patients?

Translational research Applies findings from basic research into meaningful everyday practices, often in the field of health care.

Clinical trials A research study involving human volunteers that is meant to add to medical knowledge.

Table 2.2 Examples of Basic, Applied, and Translational Research in Areas of Exercise Science (continued)

AREAS OF EXERCISE SCIENCE	BASIC RESEARCH	APPLIED RESEARCH	TRANSLATIONAL RESEARCH
Athletic training and sports medicine	The effect of different durations of cold application on the change in intramuscular temperature	Does the duration of cold application improve the healing process of an injured muscle?	Does lowering muscle temperature improve functional ability in muscular dystrophy patients?
Exercise and sport nutrition	The effect of different meal composition on muscle glycogen resynthesis	Do elevated muscle glycogen levels improve endurance exercise performance?	Does changing carbohydrate intake improve blood glucose control in patients with Type 1 diabetes?
Exercise and sport psychology	The effect of different types of music on psychological arousal	Does listening to music before an athletic competition improve performance during the competition?	Does listening to upbeat music influence exercise adherence in cancer patients?
Motor control and learning	The effect of different levels of neurotransmitters in the control of movement	Does the administration of the neurotransmitter dopamine improve the quality of life in patients with Parkinson disease?	Does a drug combination designed to change neurotransmitter levels affect the ability to walk in patients with multiple sclerosis?
Clinical and sport biomechanics	The effect of different stride lengths during ambulation on balance	Does a mechanical knee brace improve the ability to freely ambulate in patients recovering from injury?	Does changing stride length influence the ability to exercise in persons with a hip replacement?

Research in exercise science can further be classified as quantitative or qualitative (3). **Quantitative research** uses a scientific approach designed for the collection and analysis of numerical data typically obtained from subjects through direct testing in a laboratory or field-based setting, or using questionnaires. Quantitative approaches are used to illustrate existing conditions or phenomena, investigate the statistical relationships between two or more variables, and explore cause and effect relationships between phenomena. The techniques, designs, and methods used in quantitative research are based on a paradigm

Quantitative research Research that uses a scientific approach designed for the collection and analysis of numerical data typically obtained from subjects through direct testing or questionnaires.

adopted from the natural sciences that believes in the assumption that reality is relatively stable, uniform, measurable, and governed by rational laws that enable generalizations to a larger population to be made (3). The quantitative research approach involves the following features:

1. clearly stated research questions,
2. rationally conceived hypotheses and alternative hypotheses,
3. fully developed and valid research procedures,
4. controlling, as much as possible, those extraneous factors that might influence the data collected,
5. using sufficiently large enough samples of participants to provide meaningful data,
6. employing data analysis techniques based upon statistical analysis procedures, and
7. depicting results in graphs, figures, and tables (5,6).

Qualitative research uses extensive observations and interviews that provide nonnumeric data obtained in natural environments in an effort to interpret concepts, experiences or opinions phenomena and discover the meaning of situations. Sometimes called naturalistic research, qualitative research is frequently conducted in natural settings and does not attempt to control the context or conditions surrounding the research setting (1,3). Unlike quantitative research, qualitative research does not subscribe to the viewpoint that the world is stable and uniform and can be explained by laws that govern phenomena. Instead, qualitative researchers use the constructionist perspective. This perspective suggests that meaning and reality are situation specific and allow for many different interpretations, none of which is necessarily more valid than any other. In qualitative research, there is usually not an attempt to generalize the results to the entire population being studied because the context of the study is population and situation specific. Specific research methods are not established prior to the beginning of the study and tend to evolve as the research is performed. In the qualitative research approach, the analysis and interpretation of the collected data is mainly interpretive and descriptive in nature. This approach results in a categorization of the data into trends and patterns and rarely uses statistical analysis procedures (1,3).

Researchers also use a combination of quantitative and qualitative methodologies in **mixed methods research** to investigate interesting research questions in many of the disciplines in exercise science. The mixed methods approach often attempts to understand the breadth and depth of research data, using the strengths of both quantitative and qualitative research. By employing a mixed methods approach, researchers are better able to expand and strengthen a research study's conclusions and interpretations.

Qualitative research Research that uses extensive observations and interviews in a natural environment to provide nonnumeric data in the form of opinions, expressions, and beliefs.
Mixed methods research Research that uses a combination of quantitative and qualitative approaches in the methodology of the study.

The use of quantitative, qualitative, and mixed methods research approaches should be based on the type of research question that is being posed. Each research method has strengths and weaknesses, which must be considered when designing a research study.

Additional Research Descriptions

A description of the different types of quantitative research and qualitative research is beneficial for understanding the strengths and weaknesses of each type and useful for determining which type of research methodology to use (1,3). Often research is categorized as either descriptive or experimental.

1. **Descriptive research**
    - describes the current state of the problem
    - does not require the manipulation of the experimental variables
    - provides no conclusions about why an effect occurs
    - offers no explanation of what happens in the response
2. **Experimental research**
    - manipulates a variable or variables to investigate the effect on some outcome
    - provides conclusions about why an effect occurs

There are also three primary forms of experimental research: longitudinal, cross-sectional, and sequential (1,3).

1. The **longitudinal research** method involves the study of change over time:
    - probably the most reliable of the three types of experimental research, because changes in societal and technological factors usually do not have a large effect on the results
    - learning or familiarization by the subjects is often a problem because repeated testing may affect the data as individuals learn how to take the test
2. The **cross-sectional research** method requires the collection of data on individuals of different characteristics or groups who represent different attributes being investigated (*e.g.*, age, gender, race, fitness levels, disease conditions):
    - allows all the data to be collected at one period in time
    - performance by certain groups might be affected for reasons other than the dependent variables being measured (*e.g.*, societal or technological factors)

Descriptive research A type of research that attempts to provide specific characteristics about a question or problem.

Experimental research A type of research that requires the manipulation of at least one variable to answer a question or problem.

Longitudinal research A type of research that involves the study of change in a variable or characteristic over time.

Cross-sectional research A type of research that requires the collection of data on individuals of different characteristics who represent different attributes being investigated.

3. The **sequential research** method combines the longitudinal and cross-sectional methods:
   - involves studying several different samples (*e.g.*, cross-sectional) over several years (*e.g.*, longitudinal)
   - allows individuals differing in some characteristic (*e.g.*, age, gender, race, fitness levels, disease conditions) to be compared at the same time to identify current differences

Additional forms of research commonly used by exercise science professionals include: systematic reviews, meta-analysis, randomized controlled trials, surveys/interviews/observations, and historical (1,3).

1. **Systematic review** involves using predetermined eligibility criteria to identify, assess, and synthesize all the empirical evidence to answer a specific research question.
   - Researchers conducting a systematic review use explicit, orderly methods that are selected with intent to minimize bias and to produce more reliable findings used to inform decision-making.
2. **Meta-analysis research** involves a reanalysis of the results from previously conducted research studies:
   - The results from a large number of research studies are systematically analyzed using special statistical procedures.
   - The intent is to reach conclusions that are supported by numerous research studies.
3. **Randomized controlled trial** involves subjects being allocated at random to receive one of several interventions:
   - Each subject has an equal chance to receive any of the interventions.
   - One of the interventions is the control condition, which may be a standard practice, a placebo, or no intervention at all.
   - It is commonly used in longitudinal research.
4. **Survey, interview, and observational research** involves documenting, describing, analyzing, and interpreting conditions that presently exist in a specific population:
   - The results show what is happening in a particular circumstance.

Sequential research A type of research that combines both cross-sectional and longitudinal research methods.
Systematic review Using predetermined eligibility criteria to identify, assess, and synthesize all the empirical evidence to answer a specific research question.
Meta-analysis research The process of statistically analyzing data from previously published research studies.
Randomized controlled trial Involves subjects being allocated at random to receive one of several interventions.
Survey, interview, and observational research The documentation, description, analysis, and interpretation of conditions that presently exist in a specific population.

5. **Historical research** The process of systematically examining and interpreting past events to provide an account of what has occurred.
   - Procedure involves exploring, recording, analyzing, and interpreting past events for the purpose of discovering generalizations.
   - Information generated is useful in understanding the past and the present, and to a limited extent, it is helpful in predicting the future (1,3).

THE RESEARCH PROCESS

While not everyone is going to conduct their own research study or experiment, it is important to have a solid understanding of what research is and how it is performed so that you may understand what information is valid and reliable. Obtaining an understanding of how good research is properly conducted and how the findings of research are disseminated to others in the field can be extremely beneficial in helping guide an exercise science professional toward evaluating information and finding answers to questions posed by patients, clients, and athletes.

Simply put, research is the process of finding answers to questions. Good-quality research involves a systematic progression of steps in an attempt to identify an answer to a question. The following sections provide a brief overview of many of the important components for conducting quality research and sharing the findings of that research with other professionals. Figure 2.5 provides an overview of the typical research process in exercise science. For a more detailed presentation of research and research methodology the student is directed to additional readings (1,7).

Designing an Experimental Research Study

The specific research question or purpose of the study will determine the design of the study. The first step in identifying the research question is to conduct a thorough literature review. This process requires reading and summarizing the literature related to the research topic. Review of **primary sources** and **secondary sources** of information is critical to understanding what research has been previously conducted on the topic of interest. A thorough literature review should assist in narrowing the previously conducted research information about the topic and identifying where there are gaps in the knowledge

Historical research The process of systematically examining and interpreting past events to give an account of what has occurred.

Primary sources Direct evidence from empirical research studies typically found in scholarly articles or research presented at professional conferences.

Secondary sources Information, reviews, or scholarly journal articles that discuss or evaluate someone else's original research.

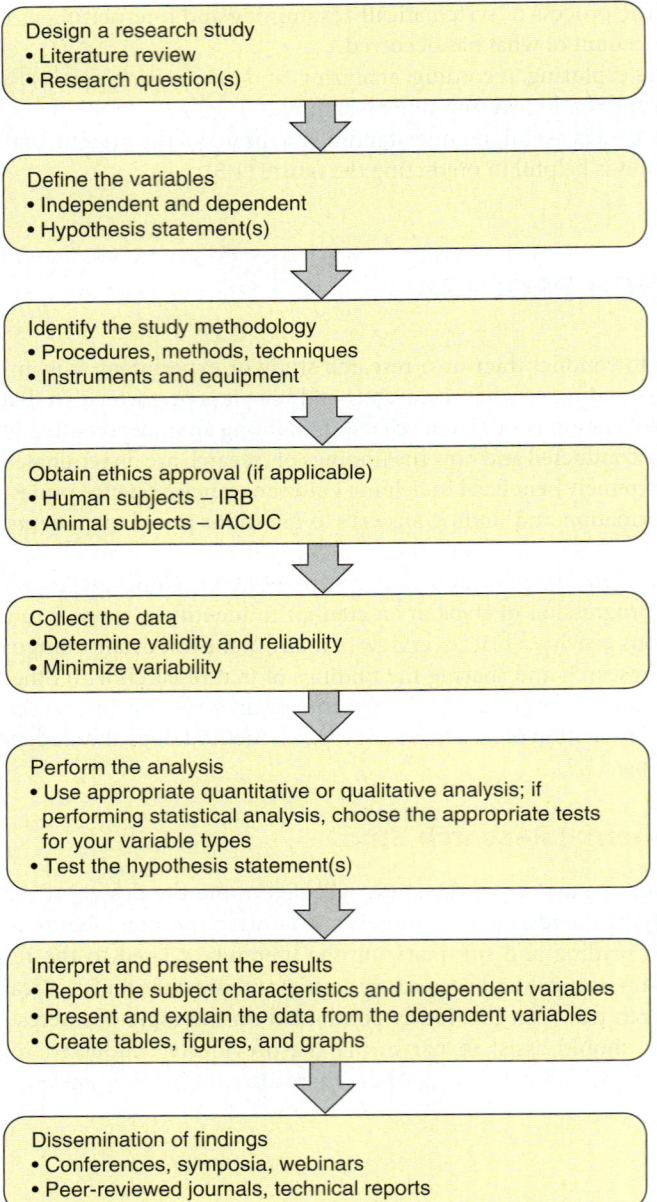

FIGURE 2.5 An overview of the typical research process in exercise science.

that need to be filled. This can help a researcher define their research question, and from there, the purpose of the study. A good research question should clearly specify the study purpose, and it should also include the subjects or population from which data will be collected.

Defining the Variables

The specific research question determines if an **independent variable** is to be controlled or manipulated during the study and what **dependent variable** is going to be measured. Research studies often include testing a single **hypothesis** or multiple hypotheses, which is generated by supporting theory and/or previous research (1). The hypothesis is often stated in a manner for which a specific outcome is expected. Researchers must attempt to identify and carefully control those independent variables (also called the experimental or treatment variable) that will have an impact or affect the outcome of the study. Additionally, the researcher should identify the most important dependent variable(s) because that information is going to be statistically analyzed in a quantitative study for supporting or rejecting the hypothesis and then used to draw conclusions from the research (1).

Study Methodology

The methodology is the specific process used to conduct the study and collect the data (*i.e.*, information) used to answer the research question. The procedures, methods, and techniques used are based on numerous factors, including the hypothesis, the study design, the subject population, the instrumentation and equipment available, and the time constraints of the researchers and the subjects. The methodology is also determined by the type of research being conducted. The methodology is set *a priori* and must be able to yield both valid and reliable data. Typically, the research procedures, methods, and techniques are not altered once the official data collection begins.

Obtaining Research Approval

Prior to beginning a research study, approval to conduct the study may be needed from an ethics board. If the research involves human subjects, then approval is sought from an Institutional Review Board, commonly known as an IRB. If animal subjects are used in the research, the approval is sought from an Institutional Animal Care and Use Committee, commonly known as an IACUC. Most colleges and universities have an IRB and an IACUC associated with their Office of Research. Both the IRB and IACUC are formally charged with making sure researchers follow the rules when engaging in human or animal research. These rules for human research are typically based on the Declaration of Helsinki or the Belmont Report, both of which maintain that basic ethical principles should underlie the conduct of biomedical and behavioral research involving human subjects and that researchers must follow guidelines to assure that such research is conducted in accordance with those principles.

> **Independent variable** The treatment or experimental variable that is manipulated by the researcher to create an effect on the dependent variable.
> **Dependent variable** A response, behavior, or outcome that a researcher wishes to predict or explain.
> **Hypothesis** A proposed explanation for the occurrence of some specified outcome tested through study and experimentation.

The rules for animal research in the United States are governed by the Animal Welfare Act, which, among other things, sets standards for animal housing and basic pain control. Those standards are enforced by the United States Department of Agriculture. Approval to conduct human or animal research usually involves the submission of an application (often called a protocol) for review by the IRB or IACUC. If the protocol does not meet the appropriate standards, then modifications to the protocol must be made. Once the protocol meets the standards set by the IRB or IACUC, then the research study can proceed. Please see the following websites for more information on the rules and standards governing human (https://www.hhs.gov/ohrp/) and animal (https://olaw.nih.gov/) research.

Identifying the Subjects

Selecting the appropriate research subjects, whether human or animal, depends on the study purpose and the research question. Characteristics such as age, gender, fitness level, health status, and disease condition are some examples of common independent variables identified by researchers when designing a study. Most research studies identify **inclusion criteria** and **exclusion criteria** that are used to guide subject selection and participation.

One of the hallmarks of human research subject selection and participation is the principle of voluntary informed consent. Each subject in a research study must be provided with information about the study, choose to participate based on their own decision, and must not be coerced into participating. Subjects are usually randomly assigned to experimental and control groups. The experimental group(s) receive a treatment (*e.g.*, drug) or condition (*e.g.*, training program), while the control group does not receive a treatment or condition and continues with their normal activities and behaviors. In some experiments, subjects may receive both an experimental and a control condition, with one of the conditions preceding the other and then subjects receiving the remaining condition after a predetermined period of time has occurred after the first intervention.

Animal subjects are also selected for participation based on specific inclusionary criteria. Because animals cannot provide voluntary consent, the university IACUC will carefully review each study protocol to ensure that animals are treated humanely and that any pain and suffering is minimized. An additional principle followed in animal research is that the least number of animals possible are used for conducting the research.

Collecting the Data

The term data is used to broadly describe the information obtained for the dependent variables, including the physical characteristics of the research participants if the research subjects are humans or animals. Data collection for the dependent variables should be

Inclusion criteria Specific characteristics that a subject must have in order to participate in a research study.

Exclusion criteria Specific characteristics that eliminate a subject from participating in a research study.

conducted with great precision and care. Only data that have **validity** and **reliability** are useful in answering the research question. The collection of the data should occur in the exact same manner for each subject, and often the same researcher or technician collects the data at each measurement time and for the same dependent variable. This minimizes the amount of variability that can occur in the research study. Collecting research data can use three different methods: observation, measurement, and questioning. The method selected is determined by the research question (3). Table 2.3 provides examples of data collection techniques for each method (3).

Table 2.3	Examples of Data Collection Techniques for Exercise Science Research (3)
OBSERVATIONAL TECHNIQUES	**DATA COLLECTION TECHNIQUES**
Direct observation	The research participants are aware of being observed and usually know why they are being observed
Indirect observation	The research participants are recorded; with the analysis of the recording occurring at a later time
Participant observation	The research observer joins the participants in the research environment and records observations
MEASUREMENT TECHNIQUES	**DATA COLLECTION TECHNIQUES**
Physical measures	Use instrumentation and equipment to record measurements
Cognitive	Use paper or computer techniques to collect information on knowledge measures
Affective	Use paper or computer techniques to collect opinions, attitude, personality, motivation, interest, and mood
QUESTIONING	**DATA COLLECTION TECHNIQUES**
Structured questionnaire	Questions can be answered by research participants with yes–no or true–false responses or by selecting an answer from a list of suggested responses
Unstructured questionnaire	Use questionnaires whereby research participants are able to answer freely in their own words
Structured interview	Use questionnaires whereby answers from the research participants are written down
Focus group interview	Information is collected from multiple participants who are led in a discussion of a particular topic by a moderator

Validity The extent to which a measuring device measures what it proposes or claims to measure.

Reliability The extent to which the same test or procedure will yield the same result either over time or with different researchers.

Performing the Analysis

The measures, scores, counts, frequencies, and other information collected during a research study are used to help answer the research question and test the hypothesis. Research studies may have multiple data sets depending on the complexity of the study and what information is collected. Researchers draw their conclusions based on an analysis and interpretation of the data. There are numerous ways to perform data analysis that are beyond the scope of this textbook. The type of data analysis depends on the nature of the information collected and the research question being asked and the hypothesis being tested (7). For further information on statistical analysis, the reader is referred to additional readings (1,3,7).

Interpreting and Presenting the Results

If the data collected during the study are qualitative in nature, the results will be reported through text using descriptive language that often provides a narrative that addresses the research question. If the data collected are quantitative, they are reported in numerical form, but the specific analysis employed and the subsequent reporting of the research findings should reflect the type of data collected during the study (7). Information on the research participants is often reported as descriptive characteristics in which the data set is described. The data from the independent variables are reported to provide a clear understanding of the factors that influenced the dependent variables. The numerical data from the dependent variable are reported as findings and used to make interpretations and conclusions about the research. Tables, figures, charts, and graphs are the most common ways to present research data.

Dissemination of the Findings

Once the data collection is complete, the analysis is performed, results and conclusions of the research study are determined, it is important for the researcher(s) to disseminate the findings to the appropriate audience. This sharing of information is critical in the advancement of knowledge in a field of study or discipline. Researchers and scholars are actively involved in the distribution of knowledge to other scientists and the general public. It is through this process that advancements in academic disciplines, treating diseases, improving health, enhancing sport and athletic performance, and developing new technologies are made.

The most commonly used ways to distribute research findings are to present the findings at a professional conference or publish the findings in a peer-reviewed journal. Additional methods of dissemination can include technical reports, blogs, preprint servers, and webinars. A key consideration when evaluating information provided by a researcher from

Peer-reviewed A process of rigorous evaluation of the research by experts working in the same area of study or discipline.

a study is whether the research presented has been peer reviewed. The peer-review process implies that an expert in the specific area of the field, with the appropriate credentials and knowledge, has evaluated the research and deemed it acceptable for dissemination. If research findings are not peer reviewed, it does not mean that the findings are inaccurate or not valid but simply that a knowledgeable professional has not certified the research as being reviewed for the appropriate level of quality and accuracy (7).

Professional Conferences

Most every professional organization in the exercise science disciplines has a conference whereby current knowledge and information and new developments about the field or discipline are presented to its members and attendees. These conferences typically occur at the international, national, regional, and state level. At many of these conferences, findings from research studies are presented, and both undergraduate and graduate students are often the presenters. The process for presenting at a professional conference begins with the submission of a **research abstract** to the presiding organization. Some professional organizations have a peer-review process whereby an individual or committee reviews the research abstract and makes a determination on the suitability of the research for presentation at the conference. Peer review often occurs with international and national conferences, where there may be a limit on the number of opportunities for research presentations. Other conferences, such as those held at the regional and state level, may allow everyone who submits a research abstract to present the research findings at the conference.

The presentations at a professional conference typically fall into one of two categories, oral and poster. Both oral and poster presentations are usually grouped together by theme or topical area. Attendees of the conference view those presentations that are of interest to them, thereby gaining knowledge and information from the dissemination of the research findings.

Oral presentations typically require the lead researcher (also called the lead author) to give the presentation to a group of individuals attending the conference. The oral presentations can be made to the entire group of conference attendees or to a small specialty subgroup within the larger conference. The presentations typically include an introduction or background information, research methods employed, results from the data analysis, and conclusions. A key component of the oral presentation is the inclusion of figures and graphs to visually display the results and findings of the study. The length of time for most oral presentations is usually 10 to 15 minutes for the presentation, with an additional 5 minutes for questions.

Poster presentations require the presenter to share information with conference attendees by placing the information from the research study onto a poster board that is generally 4 feet by 6 feet in size. The same information that is found in the oral presentation is usually also included in the poster presentation. A key component of the poster

Research abstract A short summary (~200–300 words) that describes the research study's purpose, methods, results, and conclusions.

presentation is the inclusion of figures and graphs to visually display the results and findings of the study. Poster presentations are typically available for viewing for several hours during a conference with a specific time slot during which the lead author is available to discuss the research study and findings with conference attendees.

Peer-Reviewed Journals

Peer-reviewed journals are crucial for the distribution of research and scholarly knowledge to other professionals (7). There are rating systems used to rank journals based upon the quality and impact that the published research has on the discipline. Various metrics, using objective criteria have been developed to assist in the ranking process.

As mentioned previously, the peer-review process ensures that published research is of acceptable scientific quality and importance. The peer-review process attempts to screen for research papers with serious flaws in the experimental design or the scientific methods while also ensuring that the discussion and conclusion presented by the authors are supported by the results of the data collection and analysis.

The general process of submitting the results of a research study for publication in a peer-reviewed journal is shown in Figure 2.6. The overall process begins with an important discussion among the researchers involved in the study. Agreement is reached on which researcher will lead the process (called the lead author) and which other researchers will be responsible for writing and contributing to the various sections of the manuscript (called coauthors). Many journals require the lead author and coauthors to identify what contributions each individual made to the submitted manuscript. Table 2.4 provides some examples of areas of author contribution to a manuscript as recommended by the International Committee of Medical Journal Editors (http://www.icmje.org/).

An in-depth discussion among the authors occurs to decide to which journal the manuscript should be submitted. It is important to select a journal that specializes in publishing research work from specific areas of study or disciplines in order to ensure that the research work is disseminated to the appropriate audience. There are over 70 journals that publish research conducted by professionals from exercise science disciplines. Many of these journals are published by professional organizations such as the American College of Sports Medicine (*e.g.*, *Medicine and Science in Sports and Exercise*), the National Athletic Trainers Association (*e.g.*, *Journal of Athletic Training*), and the Academy of Nutrition and Dietetics (*e.g.*, *Journal of the Academy of Nutrition and Dietetics*). Other journals are published by publishing companies who see a need for the dissemination of research and scholarship to professionals (8).

The next step in the process is preparation of the manuscript. This is done in accordance with the instructions for authors (also referred to as the author guidelines) that are found on the journal websites. For example, the instructions for authors for *Medicine and Science in Sports and Exercise* can be found at http://journals.lww.com/acsm-msse/pages/default.aspx or https://www.editorialmanager.com/msse/. Authors must closely follow the guidelines for manuscript format as failure to do so can result in significant delays in the peer-review process and publication of the research. When a draft of the manuscript is complete, it should be sent to all the coauthors for review. Often researchers will also send

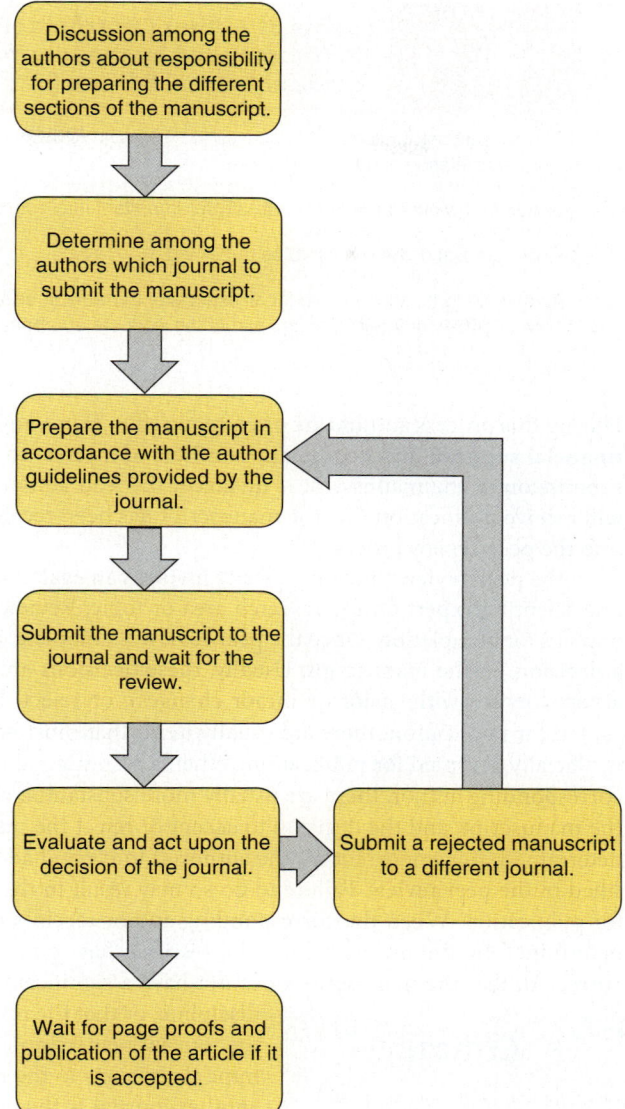

FIGURE 2.6 The general process of submitting the results of a research study for publication in a peer-reviewed journal. (Modified from (7)).

a draft of the manuscript to other professional colleagues for peer review as a professional courtesy. Once a final draft has been agreed upon by all the authors, the manuscript is submitted for review.

Submission of the manuscript for consideration of publication requires the lead author to submit the manuscript to the selected journal using the online submission process.

Table 2.4	Areas of Author Contribution to a Manuscript as Recommended by the International Committee of Medical Journal Editors (http://www.icmje.org/)
Substantial contributions to the conception or design of the work; or the acquisition, analysis, or interpretation of data for the work
Drafting the work or revising it critically for important intellectual content
Final approval of the version to be published
Agreement to be accountable for all aspects of the work in ensuring that questions related to the accuracy or integrity of any part of the work are appropriately investigated and resolved

During this process authors are often required to sign a **copyright agreement**, identify any **financial support**, and declare any **conflicts of interest** to complete submission. Once the submission of the manuscript is finalized, the lead author (or the corresponding author) will receive notification that the manuscript has been received and that it has been entered into the peer-review process (7).

The peer-review process typically involves an evaluation of the manuscript by at least one scientific expert on the research area or topic. Reviews can often take between 2 and 6 weeks for completion. Once the peer review is complete, the lead author is provided with a decision on the manuscript. Usually, these decisions are in one of the following forms: accept, revise (with major or minor changes), or reject. Even when a manuscript is accepted for publication, there are usually items that must be changed or corrected before it is officially accepted for publication. When a recommendation of revise is conveyed to the corresponding author, there are usually more substantial changes that need to be made in the manuscript, and the decision to accept or reject the manuscript has not been officially determined. It is important for the author(s) to try and address each of the concerns identified in the peer review. Failure to do so may result in the manuscript not being accepted for publication. When the corresponding author receives a rejection decision, it has been determined by the reviewers that the research has some significant flaws that cannot be corrected, that the manuscript does not have a significant impact on the area of study or discipline, or that the research does not fit the interest of the journal readership (7). The authors can then determine if revisions to the manuscript and resubmission to another journal is the appropriate course of action that may lead to publication of the manuscript.

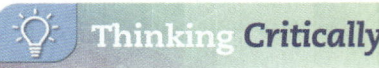

Why is the peer-review process important in the dissemination of research?

Copyright agreement An arrangement whereby the originator of the work grants to the assignee the rights to publish, reproduce, sell, or distribute the material.

Financial support Funding provided to a researcher for assistance in conducting a research study.

Conflicts of interest Occurs when an individual is involved in multiple interests, any one of which could possibly corrupt the motivation for an act in another.

Once a manuscript has been accepted for publication, the corresponding author receives additional information from the journal editorial staff. This can include formatting corrections and information on the expected publication date of the manuscript. Prior to publication, the corresponding author receives a final proof (often called a galley proof) that requires acknowledgment that the information contained in the proof is correct. Sometime thereafter the manuscript will be officially published in the journal. Publication can occur in both print and digital formats. Current technology allows many journals to make articles immediately available upon acceptance. This is referred to as being "ahead of press" or "ahead of print." This allows research to be published much faster than when all published articles have to wait until the print copy is ready for distribution to libraries and individual subscribers (8).

When a research article is published, it becomes available for viewing by anyone who has access to either the print or the digital versions. The information contained in the article is used to inform current practice by exercise science professionals and to guide the development of guidelines and position stands by professional and governmental organizations. Published articles are also carefully evaluated by other researchers, scholars, and professionals in an effort to identify future areas of study and research. Recent changes in the publication industry have allowed for articles to appear in **open access** journals. Open access journals allow individuals who are not subscribers to the journal to view the article that has been published without incurring any cost to view the article (8).

Technical Reports, Preprint Servers, and Webinars

A **technical report** is a document that describes the process, development, or results of technical or scientific research or the state of a technical or scientific research problem. It might also include recommendations and conclusions of the research, as well as complete data, design criteria, detailed procedures, extensive literature reviews, comprehensive tables, illustrations and images, and a thorough explanation of approaches that were unsuccessful. A technical report seldom goes through peer review before dissemination. A technical report is often prepared for governmental agencies and sponsors of research projects.

Preprint servers are online repositories, containing data or information associated with scholarly manuscripts that have not undergone peer review or been accepted by a published journal. Manuscripts presented on these types of repositories undergo basic screening and are checked for plagiarism. Preprint servers typically only post manuscripts that have not yet been accepted for publication by a journal. Authors can submit revised versions of their manuscripts to the preprint server at any time. Once posted, manuscripts are citable and typically cannot be removed. Preprint servers provide an opportunity for researchers to share and receive feedback on scholarly work.

Open access A process that provides free, unrestricted online access to research dissemination.
Technical report A document that describes the process, development, current state, or results of technical or scientific research or research problem.

Thinking Critically

Why is it important for researchers to disseminate the findings of their studies to other professionals?

A webinar is a presentation, lecture, workshop, or seminar that is transmitted over the internet. Using specialized software, the individuals leading the webinar can share audio, documents, and applications with webinar attendees. Additionally, attendees can often ask questions in a written format that are answered by the presenters.

EVIDENCE-BASED PRACTICE

When an exercise science professional designs a training program for a client attempting to improve their fitness, it is important that the best training program for that client be selected so that improvements in fitness can be made in the most effective and efficient way possible. The same consideration applies when designing a rehabilitation program for a patient recovering from a heart attack or when creating a training and conditioning program for an athlete working to improve their performance during competition. In these examples it is important to select or design the best program for each individual. With a wide variety of exercise and training programs available, how does a personal trainer, cardiac rehabilitation specialist, or strength and conditioning coach know which program will work the best for the client or athlete? This is where the concept of evidence-based practice plays an important role in the decision-making process.

What Is Evidence-Based Practice?

Evidence-based practice gains its historical foundation from the concept of evidence-based medicine (9). Evidence-based medicine uses the incorporation of the best research evidence with clinical expertise and patient feedback and perspective to provide the best health care plan for the patient (10). Evidence-based practice (Figure 2.7) has evolved from the foundation of evidence-based medicine to create a systematic process whereby the best available

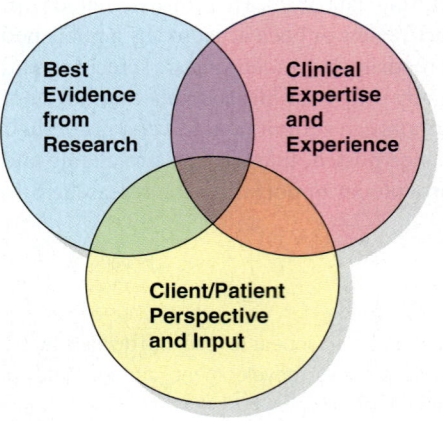

FIGURE 2.7 The evidence-based practice model.

evidence is used to inform a question, address a problem, or create a program (8). The use of the evidence-based practice process has expanded to include many of the disciplines that comprise exercise science as well as others such as nursing, social work, education, and public health.

How Evidence-Based Practice Works

There are five essential steps used in the process of creating the best evidence-based practice recommendations (10). Table 2.5 illustrates those steps involved in the process (10). While this five-step process was initially designed to inform health care providers about how to best treat patients, the steps can be readily applied to other disciplines, including those in exercise science. On a macroscale, professional organizations use the principles of evidence-based practice to create position stands that inform practicing professionals about the best way to address a disease condition, treat and manage a client, or implement a training strategy. Examples of this can be found in many of the recent position stands created by the American College of Sports Medicine (Table 1.2). On a microscale, evidence-based practice is used by professionals to help inform them as to how to best treat or train a patient or client.

While all of the five steps involved in the creation of the best evidence-based practice are critical, perhaps the most important to the process is the evaluation of published research evidence. Published information provides evidence that ranges from meta-analyses and systematic reviews to well-designed and executed scientific experiments and randomized controlled trials to expert group consensus opinion. This wide range of evidence requires careful evaluation of all publications in terms of levels of evidence (8). The thorough evaluation of published evidence results in the classification of the evidence from strong to weak (8). The evaluation process requires that the findings of each research study be evaluated on the basis of preestablished criteria. Numerous professional organizations and research consortia have developed grading systems that are intended to optimize the evidence-based practice decision-making process (8). Table 2.6 contains a listing of five of the more commonly used evidence evaluation systems (8).

Table 2.5	Steps Involved in the Process for Creating the Best Evidence-Based Practice (10)
STEP ORDER	ACTION ITEM
Step 1	Recognizing the need for information and translating that need into an answerable question(s)
Step 2	Identifying the best evidence available to answer the question(s)
Step 3	Systematically analyzing the best evidence available for its validity, impact, and applicability
Step 4	Combining the systematic analysis with clinical expertise and the patient's or client's unique circumstances into a plan of action
Step 5	Evaluating the effectiveness and efficiency of the best evidence-based practice and searching for ways for improvement

Table 2.6 Five of the More Commonly Used Evidence Evaluation Systems (8)

EVALUATION SYSTEM	WEBSITE
Strength of Recommendation Taxonomy	http://www.aafp.org/afp/2004/0201/p548.html
Health Evidence	https://www.healthevidence.org/
Grading of Recommendations, Assessment, Development, and Evaluation	http://www.gradeworkinggroup.org
United States Preventive Services Task Force	http://uspreventiveservicestaskforce.org
The Cochrane Library	http://www.cochrane.org

Thinking Critically

In what ways does evidence-based practice enhance our understanding of research?

Evidence-based practice is a commonly used process to help inform professionals about the best way to implement an intervention with an individual, team, or population. The use of evidence-based practice helps to remove the guesswork and uncertainty from determining which treatment options and training programs are the most effective.

STUDENT RESEARCH IN EXERCISE SCIENCE

Exercise science students can be actively involved in research activities while completing an undergraduate degree. The experience of being involved on a research project as a student can be very valuable and is one of the most transformative opportunities to be engaged in while in college. Undergraduate students can be actively involved in the research process in systematic investigations to create new facts, unravel new or existing problems, and explore innovative ideas and theories. Participating in laboratory, clinical, or field research helps you understand the research process firsthand and provides you with an opportunity to gain in-depth knowledge in a specific discipline or topic and develop critical thinking skills.

The best way to get involved in research at your college or university is to speak with your faculty or staff advisor about research opportunities that are available to you. Faculty members at most higher education institutions are either expected to conduct research as part of their employment responsibilities or they conduct research because they enjoy the experiences of the research process and generating answers to questions. Additionally, colleges and universities may have administrative offices that support the research activities of faculty members, staff, and students. These research offices usually have staff members who can assist students with identifying faculty members and staff who are searching for assistance with research activities. It is also common for many higher education institutions to provide special programs that help undergraduate students conduct their own research activities under the supervision of faculty and staff. Most of these student research

Table 2.7	Suggestions for Getting Involved in Research Activities as a Student

Volunteer as a research subject in a study

Observe research activities in different laboratory or field settings

Engage with faculty about their research activity

Connect with graduate students who are conducting research

Review the published research papers from faculty in your academic unit

Visit your college or university research office

Attend research fairs, seminars, and webinars

Attend a professional conference within your discipline

programs are offered during the summer semesters, but some colleges and universities have undergraduate student research support programs that are available throughout the entire year. Table 2.7 contains suggestions for finding ways to get involved in research activities as an undergraduate student.

Interview

John Thyfault, PhD

Professor, Department of Molecular and Integrative Physiology, University of Kansas Medical Center

Brief Introduction – I have been a research scientist for 21 years, including 3 as a graduate student and 3 as a postdoctoral fellow. I initially became interested in exercise physiology as an undergraduate student at Fort Hays State University. I had two interests during that time. I played football in college and needed to put on some muscle mass and get stronger to get some playing time and compete. I had already been really interested in strength training since I was in high school, but it was in college that I learned there was a whole discipline and area of research on how to get faster, stronger, and increase muscle mass. I also became very interested in the role that exercise could have on improving health. My father died from heart disease when I was very young. I think this fed my interest in how exercise could improve the structure and function of the heart and other organ systems so that heart disease could be effectively prevented and treated. After taking a personal wellness course that was mandated for all undergraduates, I discovered that I could get a degree that would allow me to learn more about exercise for both sports performance and human health. I quickly became active in different organizations within the department, involved myself in some cardiac rehabilitation experience, and immersed myself in reading strength and conditioning research. At that point, I wanted to become a strength and conditioning coach, so I obtained a certification from the National Strength and Conditioning Association. I still had a year of eligibility for football left after finishing my undergraduate degree, so I decided to start a Master's degree while I finished playing. This allowed me to serve as a strength coach intern for volleyball and football during that year, and then I stayed and finished my graduate degree in Exercise Science. I had a professor, Dr. Greg Kandt, tell me that I should consider pursuing a PhD. I had thought about it previously, but not seriously. It was his prompting that pushed me forward.

I was accepted into a PhD program at the University of Kansas in the Fall of 1999. I really had no previous research experience. It was a big change, but I quickly grew to love the idea of working on projects that created new knowledge. I also remember that I still had an interest in performance and at one point I was asked if I wanted to help the 1% of the population who were athletes or help 99% of the population that was not by focusing on the health effects of exercise. I was sold on that concept and still am today. However, I still wondered about the performance field, and after my first year of the graduate program, I was offered a Strength and Conditioning internship with the Kansas City Chiefs. My experience in the internship further solidified my interest in research on the health effects of exercise. During my time in graduate school I learned to submit my own grants, conduct my own studies, and report my results in both papers and oral presentations. At the end of my PhD work, I was not sure if I wanted to go 100% research or work at a smaller university where I could teach but do a little research. To keep both avenues open, I decided that a postdoctoral position would be necessary. I was able to secure a postdoctoral position at East Carolina University in the laboratory of Dr. Lynis Dohm. This was another important step in my career. In fact, I think my most significant career experience was getting to work as a postdoctoral fellow. At that stage in your career you are done taking classes and you really don't have any responsibilities other than thinking about science and conducting your studies. Dr. Dohm also had an open-door policy and said I could come in and discuss ideas or data with him at any time. Dr. Dohm was an excellent scientist and someone who really got excited about the process. He would always say: "We are seekers of truth" when results didn't come out the way you expected. He really helped me to understand to not be wedded to a hypothesis and that every type of outcome is an opportunity to learn something new. He also provided an example that it was good to collaborate with others from other disciplines or expertise. It was also a time when I could tell that my scientific thinking and communication skills were evolving and improving. This made me more confident in my abilities and helped me determine that research was my passion.

Q: Why did you choose to or how did you become involved with research as a student?

I have always been a very curious person. I've always wanted to know how things work, and as such I was very curious about how exercise could alter human physiology. My first love was strength and conditioning, but I always wondered if what we were doing in the weight room or on the field was actually working, and, more importantly, was it working at the cellular level the way that we wanted it to. I think this led me to a path (graduate school) where I could find out a deeper level if the programmed exercise was actually doing what it was supposed to.

Q: What are the top two or three "best things" about your job doing research?

I truly enjoy my research-focused position. I spend a significant amount of time every day thinking about new ideas and concepts that likely impact human health. When we have new results that come out in the lab, I work with my research colleagues to figure out why or how the results came out the way they did. We then design new experiments to answer further questions, creatively interpret our findings, and point us in new directions. I really think creativity and original thinking is the key to research. Conducting experiments is also fun and rewarding (although a lot of work). I also enjoy communicating our findings when I give talks or write papers. Since our work is related to exercise and obesity, we really have an advantage in terms of being able to relate our research to everyone. This makes me think that our research not only leads to new scientific knowledge, but may also actually get people to increase their daily physical activity and live a healthier lifestyle. Every now and then I receive an e-mail, or a person approaches me to say that my research made them become more active and that is just as rewarding as publishing our findings, or even more so. Finally, I really enjoy mentoring graduate students and postdoctoral fellows with their own programs and research projects. There is nothing better than helping one of your trainees achieve success.

Q: What advice would you have for a student exploring research opportunities in an exercise science profession?

Get out there and get experience. Find a laboratory and volunteer to help, and don't just help physically with their studies, but get into the research literature and determine why the researchers are doing their studies and what it really means. Always try to apply or understand the project in the bigger picture as that is when it gets most exciting.

Interview

Mary Jane De Souza, PhD, FACSM
Professor, Departments of Kinesiology and Physiology, The Pennsylvania State University

Brief Introduction – I completed my Bachelor's and Master's degrees at Springfield College, in Springfield, Massachusetts, and graduated from the doctoral program in Exercise Physiology at the University of Connecticut. I then completed a 2-year postdoctoral fellowship in Reproductive Physiology at the University of Connecticut Medical School. My first position after my postdoc was at the University of Connecticut Medical School as a Research Associate in a Menopause and Reproductive Endocrinology Research Program. I later accepted a position at the University of Toronto as an Associate Professor in the Department of Exercise Science and was there for 7 years, before my ongoing current position as Professor in the Departments of Kinesiology and Physiology at Penn State University.

Q: What are your most significant career experiences?

I have been blessed with the successes of achieving full professorship at a Tier 1 research institution and thus enjoy the opportunity of mentoring graduate students, doing research, and contributing to science. Mentoring, in my mind, is a gift that a profession in academia delivers, a profession that provides the opportunity to "mold and build" someone's career path and facilitate their successes. Becoming a researcher and contributing to science has also been very rewarding for me and something that I very much enjoy. I have also received high honor awards for a successful career contributions in research from the American College of Sports Medicine (Citation Award), the New England Chapter of the American College of Sports Medicine (Honor Award), and the College of Health and Human Development at Penn State University (Pauline Schmitt Russell Distinguished Research Career Award).

Q: How did you become involved with research as a student?

I had a very wise mentor during my Master's degree work, Dr. Mimi Murray, who advised me toward areas of research that she knew would pique my interest and provided me with avenues to consider to find an appropriate career path.

Q: What are the top two or three "best things" about your job doing research?

I thoroughly enjoy mentoring graduate students and helping them develop "passion" for their research area of interest. I have also found great enjoyment from helping the undergraduate student who is struggling and, with some investment of time, encouragement, and support, helping them blossom and find pathways to success. Lastly, I immensely enjoy conducting research, writing papers, and the wonderful collegial relationships I have developed over the years in this career.

Q: What advice would you have for a student exploring research opportunities in an exercise science profession?

I would advise students exploring research opportunities in exercise science to find a mentor that "fits" their needs, that will take the time to speak with them as often as necessary and, if possible, this same person should have a funded research program with the means to support a research pathway for student-focused research. I would also strongly advise that students consider the funding potential of the area of research they plan to engage and consider a focus area that has the potential for extramural funding. I would advise students to take every opportunity to write papers, write internal and small grants, write abstracts to present at meetings nationally and internationally, participate in various workshops, and do anything that builds their curriculum vitae. My final bit of advice is to be relentless and never give up.

SUMMARY

- Research in exercise science is important for understanding the fundamentals and mechanisms about a specific topic and informing professionals about the best practices to use when treating patients or interacting with clients.
- There are multiple procedures and systems to use when conducting research, and the methodology selected should be determined by the research question to be answered.
- Good-quality research utilizes a process that is systematic.
- The dissemination of research findings at conferences or in peer-reviewed journals is needed for professionals to understand current knowledge and application.
- Evidence-based practice is a commonly used process to help inform professionals about the best treatments or interventions to use with an individual or population.

FOR REVIEW

1. Define basic research, applied research, and translational research.
2. What are the key differences between quantitative, qualitative, and mixed methods research?
3. How does longitudinal research differ from cross-sectional research?
4. What is the difference between an independent variable and a dependent variable?

5. Define validity and reliability.
6. What are the two most common research ethics boards?
7. What is a peer-reviewed journal?
8. Define evidence-based practice.
9. What is the importance of the grading system in evidence-based practice evaluation?

Project-Based Learning

1. Identify a published research paper, in a journal that uses peer review, on a topic that involved physical activity or exercise for health promotion, and provide answers with justification to the following questions:
   a. Is this basic, applied, or translational research?
   b. Is this quantitative, qualitative, or mixed methods research?
   c. Is this longitudinal, cross-sectional, or sequential research?
   d. Is this a systematic review, meta-analysis, randomized controlled trial, survey, interview or observational study or historical research?
   e. What are the dependent and independent variables?
   f. Is this evidence-based practice research?
2. Identify a published research paper, in a journal that uses peer review, on a topic that involved sport or athletic performance enhancement, and provide answers with justification to the following questions:
   a. Is this basic, applied, or translational research?
   b. Is this quantitative, qualitative, or mixed methods research?
   c. Is this longitudinal, cross-sectional, or sequential research?
   d. Is this a systematic review, meta-analysis, randomized controlled trial, survey, interview or observational study or historical research?
   e. What are the dependent and independent variables?
   f. Is this evidence-based practice research?

REFERENCES

1. Neutens JJ, Rubinson L. *Research Techniques for the Health Sciences*. 5th ed. San Francisco (CA): Benjamin Cummings; 2014.
2. Ary D, Jacobs LC, Sorensen CK, Walker D. *Introduction to Research in Education*. 10th ed. Boston (MA): Cengage; 2021.
3. Baumgartner TA, Hensley LD. *Conducting and Reading Research in Health and Human Performance*. 4th ed. New York (NY): McGraw-Hill Publishers; 2006.
4. Zerhouni EA. Translational and clinical science — time for a new vision. *N Engl J Med*. 2005;353(15):1621–3.
5. Drew CJ, Hardman ML, Hart AW. *Designing and Conducting Research: Inquiry into Education and Social Sciences*. 2nd ed. Needham Heights (MA): Allyn & Bacon; 1996.
6. Gay LR, Airasian P. *Educational Research: Competencies for Analysis and Application*. Upper Saddle River (NJ): Prentice-Hall; 2000.
7. Armstrong LE, Kraemer WJ. *ACSM's Research Methods*. 1st ed. Baltimore (MD): Lippincott, Williams & Wilkins; 2016.

8. Armstrong LE, Kraemer WJ. Understanding research: a clinician's perspective of basic, applied, and clinical investigations. In: Armstrong LE, Kraemer WJ, editors. *ACSM's Research Methods*. 1st ed. Baltimore (MD): Lippincott, Williams & Wilkins; 2016. p. 93–119.
9. Group E-BMW. Evidence-based medicine. A new approach to teaching the practice of medicine. *JAMA*. 1992;268(17):2420–5.
10. Sackett DL, Straus SE, Richardson WS, Rosenberg W, Haynes RB. *Evidence-Based Medicine: How to Practice and Teach EBM*. 2nd ed. Edinburgh, UK: Churchill Livingstone; 2000.

CHAPTER

3

Exercise Science: A Systems Approach

After completing this chapter, you will be able to:

1. Describe the meaning and context of a systems approach to the study of exercise science.
2. Articulate the primary functions of each system of the body.
3. Provide examples of how each system of the body can influence physical activity and exercise.
4. Provide examples of how each system of the body can influence sport and athletic performance.

As identified in Chapter 1, several interrelated disciplines, subdisciplines, and specialty areas constitute exercise science. Collectively, the study of each component of exercise science is based on a core understanding of the structure (anatomy) and function (physiology) of the human body. It is expected that beginning exercise science students enroll in courses in human anatomy and physiology, often in the first year of study in college. The knowledge acquired in these courses provides the necessary foundation for advanced study in exercise science at both the undergraduate and graduate levels.

A **systems approach** to the study of exercise science allows students to understand how the various physiologic systems of the body respond in an integrated fashion to acute and chronic stimuli and conditions. Each system has specific functions that cannot be performed in the expected manner in isolation and without interaction with other systems of the body. This system integration provides for the coordinated control of the body environment and allows the body to respond to the challenges encountered every day. Appropriate responses to challenges such as physical activity, regular exercise, stress, changes in nutritional intake, and extreme environmental conditions allow us to be healthy and perform at optimal levels in daily activities as well as during sport and athletic performance.

This chapter presents a systems approach to the study of exercise science. To maintain a current and accurate knowledge base with today's rapid generation of information, exercise science students must be able to draw on their conceptual understanding of how systems work together in a healthy state as well as during physical activity, exercise, sport, and athletic performance. Athletic trainers, clinical exercise physiologists, sport biomechanists, and other exercise science professionals are better prepared to perform their job by having a solid understanding of how the various structures and systems of the body interrelate and function together. For example, a clinical exercise physiologist designing a rehabilitation program for an individual recovering from a heart attack must understand how the nervous, cardiovascular, pulmonary, endocrine, and muscular systems work together to create movement and respond appropriately to physical activity and exercise. Only with a sound understanding of the structure and functioning of the body's integrated systems can the clinical exercise physiologist design a rehabilitation program that is safe and effective in preparing the individual to return to activities of daily living and occupational activities without an elevated level of risk for an adverse medical event. Figure 3.1 illustrates the various systems of the body and their primary functions. This chapter provides a short description of the systems of the body and examples of how each system is affected by or responds to (a) physical activity and exercise and (b) sport and athletic performance. This information should provide a foundation for which exercise science students begin to understand the importance of an integrated systems approach to the study of exercise science.

Systems approach The study of how the various systems of the body respond in an integrated fashion to acute and chronic stimuli and conditions.

Nervous system
Acts through electrical signals to manage rapid responses of the body. Also responsible for higher functions including consciousness, memory, and creativity.

Endocrine system
Acts by means of hormones secreted into the blood to manage processes that require duration rather than speed (e.g., metabolic activities and water and electrolyte balance).

Respiratory system
Obtains oxygen from and eliminates carbon dioxide to the external environment. Helps regulate body pH by adjusting the rate of removal of acid-forming carbon dioxide.

Circulatory system
Transports nutrients, oxygen, carbon dioxide, waste products, electrolytes, and hormones throughout the body.

Integumentary system
Serves as protective barrier between external environment and remainder of body, also includes sweat glands. Makes adjustments in skin blood flow important to body temperature regulation.

Muscular system
Allows body movement and heat-generation through muscle contractions which is important in body temperature regulation.

Skeletal system
Supports and protects body parts and provides calcium storage in bone.

Immune system
Protects the body against foreign invaders and tumor cells, assists with tissue repair.

Energy system
Not a physically defined system but important to all life requiring processes. Provides energy through aerobic and anaerobic pathways in all cells.

Digestive system
Obtains nutrients, water, and electrolytes from the external environment and transfers them into the plasma. Eliminates undigested food residues to the external environment.

Urinary system
Important in regulating the volume, electrolyte composition, and pH of the internal environment. Removes wastes and excess water, sodium, acids, bases, and electrolytes from the plasma and excretes them in the urine.

Reproductive system
Not essential for homeostasis but essential for perpetuation of the species.

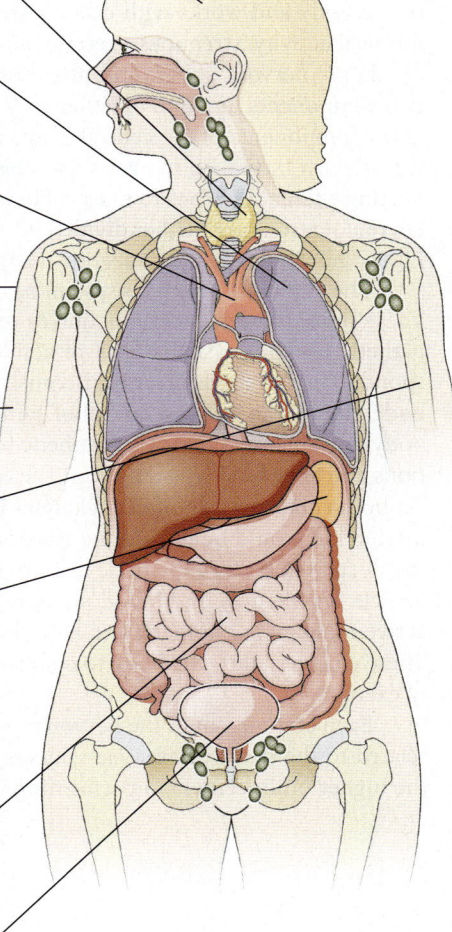

FIGURE 3.1 Systems of the body. (Adapted from (1).)

NERVOUS SYSTEM

The nervous system is one of the two primary control systems of the body (the other primary control system is the endocrine system). One advantage of the nervous system for controlling the body is that responses can occur very rapidly, typically within a fraction of a second. The nervous system controls the voluntary and involuntary actions and functions of the body and works with other systems to regulate and respond to challenges such as physical activity, strenuous exercise, or disease conditions. For ease of study, we generally divide the nervous system into the central and peripheral components, but in reality, the two components function together very closely. The brain and the spinal cord form the primary components of the central nervous system. The peripheral nervous system includes the afferent (sensory) neurons and efferent (motor) neurons, the motor end plates connecting efferent neurons to muscle fibers and the sensory receptors on sensory organs. The efferent neurons are further divided into the somatic neurons and the autonomic neurons. Figure 3.2 depicts the organizational structure of the nervous system (1).

The autonomic nervous system has two divisions: sympathetic and parasympathetic. These systems work in conjunction to regulate the various functions of the body. The sympathetic nervous system's level of activity is increased when the body is required to respond to higher levels of stress. Because physical activity and exercise act as stressors to the body, there is an increased level of sympathetic nervous system activity during increased body movement. The parasympathetic nervous system is more active during resting conditions and after food consumption. The coordinated interaction of these two systems allows for both subtle and significant changes in body function to occur. A good example of this interaction would be during the start of exercise. The increased sympathetic activity and the decreased parasympathetic activity result in an increase in heart rate, force of cardiac muscle contraction, and blood pressure, as well as a redistribution of blood flow from inactive tissues (*e.g.*, stomach and kidneys) to active tissues (*e.g.*, heart and skeletal muscle). These changes allow the body to coordinate an appropriate response to meet the demands of exercise.

Each component of the nervous system is responsible for several important functions related to the study of exercise science, with two primary interest areas being the role of the higher brain centers in performing voluntary physical activity and movement and the

Afferent neurons Nerves that carry electric impulses toward the brain and spinal cord.
Efferent neurons Nerves that carry electric impulses away from the brain and spinal cord.
Somatic Part of the nervous system that controls voluntary action.
Autonomic Part of the nervous system that regulates involuntary action.
Sympathetic Part of the autonomic nervous system that tends to act in opposition to the parasympathetic nervous system especially under conditions of stress.
Parasympathetic Part of the autonomic nervous system that tends to act in opposition to the sympathetic nervous system.

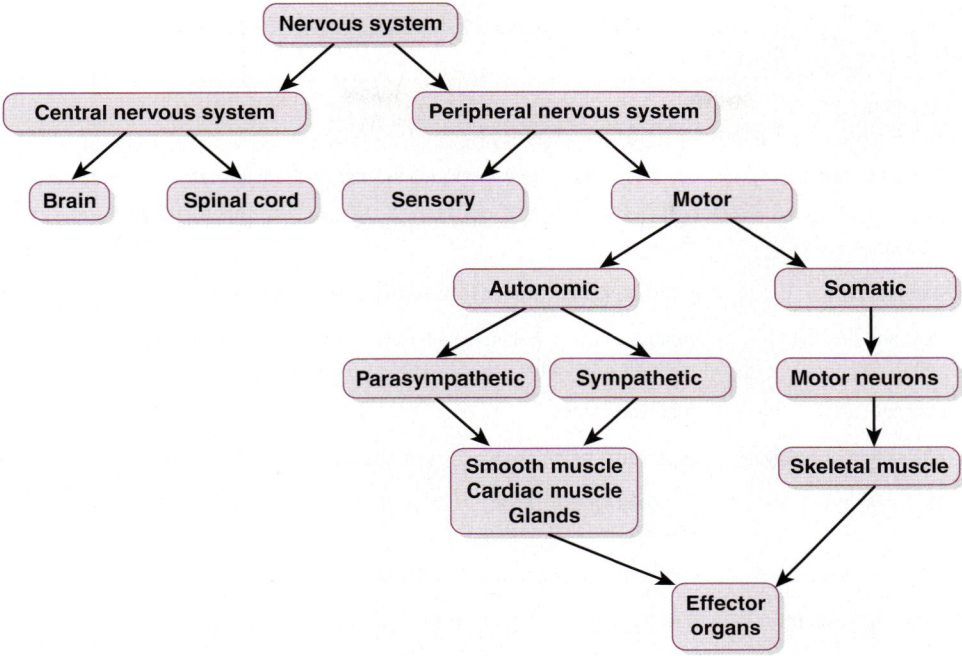

FIGURE 3.2 Organizational structure of the nervous system.

control of body movement by skeletal muscles. Chapter 8 addresses issues related to exercise behavior and sport performance, whereas Chapter 9 provides information on the neural control of movement.

Nervous System and Exercise Science

Although many neurologic disorders can affect the body's response to physical activity and exercise, many affected individuals can achieve significant health benefits with participation in regular physical activity and exercise programs. For example, cerebral palsy is a group of disorders that interfere with the normal development of areas of the brain that control muscle tone and spinal reflexes. Cerebral palsy results in limited ability to move and maintain balance and posture (2). The location and extent of the injury within the brain will influence the resulting changes in muscle tone and spinal reflex sequelae that occur (2). Medical doctors classify individuals with cerebral palsy based on functional ability, which can be beneficial for helping to identify an appropriate physical activity or exercise program (2). Individuals with cerebral palsy can benefit from participation in an exercise program focusing on the development of muscular strength, flexibility, and cardiovascular fitness (3–5). Owing to the nature of the disorder, resistance exercise may be a more suitable type of exercise for individuals with cerebral palsy to perform (5). Exercise science professionals can play a valuable role in helping individuals affected with many neurologic disorders. Table 3.1 lists the various neurologic disorders that can have improved health and fitness outcomes as a result of well-planned and appropriate physical activity and exercise program (6).

Table 3.1 — Neurologic Disorders and the Potential Benefits of Exercise (6,7)

NEUROLOGIC DISORDER	BENEFITS OF EXERCISE
Alzheimer disease	Increased fitness, physical function, cognitive function, and positive behavior
Amyotrophic lateral sclerosis	Maintain strength in healthy muscle fibers and range of motion in joints
Cerebral palsy	Improved fitness, work capacity, and sense of wellness
Deaf and hearing impaired	Improved fitness, balance, self-image, and confidence, with enhanced socialization skills
Epilepsy	Improved fitness
Intellectual disability	Improved functional capacity and strength
Mental illness	Improved fitness, mood, self-concept, and work behavior with decreased depression and anxiety
Multiple sclerosis	Improved fitness and functional performance
Muscular dystrophy	Slow down or possibly reverse the deterioration in muscle function
Parkinson disease	Enhanced functionality and movement
Polio and postpolio syndrome	Improved fitness and lower leg strength
Spinal cord injury	Improved fitness and sense of well-being
Stroke and head injury	Improved fitness and muscular strength
Visual impairment	Improved fitness, balance, self-image, and confidence, with enhanced socialization skills

For additional information about the role of exercise in treatment programs for neurologic disorders, the reader is directed to *ACSM's Exercise Management for Persons with Chronic Diseases and Disabilities* (6) and *ACSM's Clinical Exercise Physiology* (7).

Various components of the nervous system can play an important role in sport and athletic performance. For example, as an aerobic endurance athlete becomes better trained, there are changes to the autonomic nervous system that could lead to improvements in performance (8–11). Following endurance training, alterations in nervous system activity allow for a longer filling time of the heart during the period of diastole. The increased filling results in a greater **stroke volume** during each contraction of the heart, which in turn causes a higher **cardiac output**, and hence more blood being pumped throughout

Stroke volume The volume of blood pumped from the heart with each contraction.
Cardiac output The volume of blood pumped by the heart per unit of time, usually 1 minute.

Table 3.2 — Functions of the Nervous System and Their Relationship to Physical Activity, Exercise, Sport, and Athletic Performance (1)

FUNCTION	RELATIONSHIP TO PHYSICAL ACTIVITY, EXERCISE, SPORT, AND ATHLETIC PERFORMANCE
Afferent neurons provide central nervous system with sensory and visceral information	Allows for a rapid and coordinated control of body systems in response to movement
Control centers for cardiovascular, respiratory, and digestive systems	Allows for a rapid and coordinated response to movement
Controls activity of smooth muscle, cardiac muscle, and glands through the autonomic nervous system	Allows for a rapid and coordinated response of body systems to movement
Efferent neurons control movement of skeletal muscle through somatic nervous system	Allows for the body to contract skeletal muscle and create movement
Maintenance of balance	Allows for correct positioning of the body during movement
Regulation of temperature control, thirst, urine output, and food intake	Allows for the body to regulate the internal environment, remove waste products, and supply energy to the tissues in response to movement
Voluntary control of movement, thinking, memory, decision-making, creativity, self-consciousness, role in motor control	Allows for control of the body during participation in any type of movement

the body. The changes in autonomic nervous system activity also allow for an increased blood flow to active tissues (*e.g.*, skeletal muscle) during exercise (12). The higher cardiac output and enhanced blood flow to working tissues would result in a greater oxygen delivery to skeletal muscle and most likely an improvement in aerobic endurance performance (12,13). Table 3.2 provides a summary of the major functions of the nervous system and some examples of how those functions relate to physical activity, exercise, sport, and athletic performance.

MUSCULAR SYSTEM

The muscular system works in conjunction with both the nervous system and the skeletal system to create movement of the human body. In response to nervous system input, the various muscles of the body can contract and generate force. The contraction of skeletal muscle causes the bones to which they are attached to move, creating movement of the body parts. Skeletal muscle, because of its ability to generate energy and heat, also helps maintain an appropriate body temperature. Contraction of smooth muscle, found in the walls of the hollow organs and tubes in the body, regulates the movement of blood through the blood vessels, food through the digestive tract, air through the respiratory airways, and urine through the urinary tract. Contraction of cardiac muscle, found in the walls of the heart, generates the force by which the heart delivers blood to the tissues of the body (1,14).

The primary components of the muscular system are the individual muscle fibers (*i.e.*, muscle cells). Muscle fibers can generate force through the interaction of various contractile and regulatory proteins. This force allows the different types of muscle to perform their specific functions in very unique ways. Muscle fibers have distinct characteristics depending on the type of muscle (*e.g.*, skeletal, smooth, and cardiac). Skeletal muscle fibers are typically named and then classified based on certain contractile or metabolic characteristics although other classification systems exist. Table 3.3 describes the nomenclature, contractile properties, and metabolic characteristics of skeletal muscle using a traditional classification system (13).

Smooth muscle and cardiac muscle share some basic properties with skeletal muscle, yet each type also displays unique characteristics regarding the force and speed of contraction. In general, the contraction process in all three types of muscle is the same. For example, the initiation of contraction in muscle occurs through a calcium-dependent process and the generation of force occurs via the sliding protein filament theory. There are some other significant differences among the three muscle types. For example, skeletal muscle is under voluntary control, whereas smooth and cardiac muscle are controlled by the autonomic nervous system. Furthermore, the force of smooth and cardiac muscle contraction can be influenced directly by various hormones from the endocrine system, whereas skeletal muscle cannot (1,14,15).

Muscular System and Exercise Science

Previously sedentary individuals who begin an exercise program for improving their health and fitness are likely to experience delayed onset muscle soreness in the active muscles. This specific type of muscle soreness generally appears 24 to 48 hours after strenuous exercise

Table 3.3 Traditional Nomenclature, and Specific Contractile and Metabolic Characteristics of Skeletal Muscle Fibers (13)

	TYPE OF FIBER		
CHARACTERISTIC	TYPE I — SLOW OXIDATIVE	TYPE IIA — FAST OXIDATIVE GLYCOLYTIC	TYPE IIX — FAST GLYCOLYTIC
Speed of contraction	Slow	Fast	Fastest
Resistance to fatigue	High	Intermediate	Low
Myosin ATPase activity	Low	High	Highest
Oxidative energy capacity	Highest	High	Low
Nonoxidative energy capacity	Low	Intermediate	Highest

and can last for up to 72 to 96 hours. Delayed onset muscle soreness is believed to result from tissue injury caused by excessive mechanical force exerted upon the muscle fibers and connective tissue (13,16). Delayed onset muscle soreness occurs most frequently after unaccustomed exercise or activity that results in damage to cellular membranes and proteins in skeletal muscle (13). The membrane and protein damage cause the immune system to create an inflammatory response in the muscle that leads to the formation of swelling. Afferent nerves become stimulated in response to the swelling, and the individual feels soreness and pain in the muscle. **Eccentric muscle actions** appear to cause greater damage to the tissues and more substantial soreness and pain in the affected muscle (17). The phenomenon of delayed onset muscle soreness has been studied by exercise science and allied health care professionals in an effort to understand how this soreness develops and to determine if there are any preventive measures or treatment therapy that might eliminate or at least reduce the amount of muscular soreness experienced. A reduction in muscle soreness might improve an individual's compliance when starting an exercise program. Gradually increasing the intensity when beginning an exercise program, while simultaneously avoiding strenuous eccentric muscle actions, appears to be the best way to avoid developing delayed onset muscle soreness. While some types of treatment therapy have shown promising results for reducing delayed onset muscle soreness (18,19), other preventive and treatment measures such as hyperbaric oxygen therapy (20,21), prostaglandin-inhibiting drugs (22), compression garments (23), whole-body vibration (24), and acupuncture (25) seem to provide little protection or relief from delayed onset muscle soreness.

High-intensity resistance exercise training results in significant gains in muscle size and strength that can result in an improvement in sport and athletic performance. The increase in muscle size can occur as a result of either increases in the size of the individual muscle fibers (called **muscle fiber hypertrophy**) or increases in the number of individual muscle fibers (called **muscle fiber hyperplasia**). Muscle fiber hypertrophy has been demonstrated to occur in response to prolonged resistance exercise training; however, there is still some question as to whether muscle fiber hyperplasia occurs (26,27). Growth-promoting agents from the endocrine system help stimulate muscle hypertrophy in conjunction with a training program (28). Athletes who have engaged in high-intensity and high-volume resistance exercise training for many years appear to have more muscle fibers per motor unit than the average person (29). Additionally, the use of anabolic steroids and other human and synthetic growth-promoting agents may result in an increase in the number of individual fibers within a muscle (30). If hyperplasia does occur, it could result from one of two proposed mechanisms (30–32). First, the fiber number could increase, if the existing fibers hypertrophied to the extent that the individual fiber became so large that it split into more than one fiber. A second mechanism for increasing muscle fiber number could occur if existing **undifferentiated satellite cells** in the muscle are stimulated to grow into fully developed

Eccentric muscle actions When the muscle fibers lengthen when generating force.
Muscle fiber hypertrophy An increase in the muscle fiber cross-sectional size.
Muscle fiber hyperplasia An increase in the number of muscle fibers in a muscle.
Undifferentiated satellite cells An undeveloped cell that has the potential to convert to a developed cell.

muscle fibers (28). Regardless of the mechanism involved in muscle fiber hyperplasia, the increase in number appears to be small and dependent on a variety of factors including genetic profile, training history, nutritional intake, and the use of growth-promoting substances (30,32). Figure 3.3 shows a resistance-trained athlete who has experienced muscle hypertrophy. Table 3.4 provides the functions of the muscular system and examples of how those functions relate to physical activity, exercise, sport, and athletic performance (1).

> **Thinking Critically**
>
> In what ways has a systems approach to exercise science contributed to a broader understanding of the muscular and skeletal factors that influence successful sport and athletic performance?

FIGURE 3.3 A resistance-trained athlete who has experienced muscle hypertrophy (Shutterstock).

Table 3.4 Functions of the Muscular System and Their Relationship to Physical Activity, Exercise, Sport, and Athletic Performance

FUNCTION	RELATION TO PHYSICAL ACTIVITY, EXERCISE, SPORT, AND ATHLETIC PERFORMANCE
Cardiac muscle contraction propels blood through the circulatory system	Delivers nutrients and oxygen to the working tissues of the body and removes waste products
Skeletal muscle generates movement, which increases energy expenditure and heat production	Allows for body movement, responsible for the majority of daily energy expenditure
Smooth muscle contraction and dilation regulate diameter of passageways in the cardiovascular and respiratory systems	Allows for coordinated flow of blood to working tissues and air to the lungs for gas exchange

SKELETAL SYSTEM

The skeletal system serves as a structural framework of the body, protecting underlying organs and tissues of the body, providing a lever system for movement, and serving as a storage area of minerals important to the body's function. The skeletal system is also involved in blood cell formation. During body movement, the skeletal system works with the muscular system and the nervous system to create movement or respond to a stimulus. This close interaction of the muscular and skeletal systems often leads to the two systems being discussed together as the musculoskeletal system (1,33).

The minerals calcium and phosphorus and the different cells that constitute the bone marrow are the primary components of the skeletal system. The bones provide structural support, which transfers the weight of the body to the ground through the lower extremities. The bones also provide a site to which muscles can attach, subsequently acting as a lever system for the movement of body parts. Bone tissue is a rigid mass and therefore offers protection to underlying structures such as the brain, spinal cord, heart, and lungs. The skeletal system serves as an active reservoir for calcium, so when calcium levels are insufficient to meet the body's needs, calcium can be taken from bone and then replaced when calcium levels are returned to normal. Several bones of the body contain red bone marrow, which produces all types of blood cells (*e.g.*, red, white, platelets) in a process called **hematopoiesis** (1,33).

Skeletal System and Exercise Science

The interaction of physical activity, exercise, nutrition, and aging has significant implications for the health of the skeletal system. Without appropriate levels of physical activity and adequate mineral intake in the diet, the risk for developing **osteoporosis** can increase in any individual regardless of age. Osteoporosis is a disease condition characterized by low bone mineral density. Figure 3.4 illustrates the differences in bone mineral density between normal bone and osteoporotic bone. A bone that is osteoporotic has a greater risk of fracture when exposed to mechanical stress or trauma. A number of conditions, diseases, and medications have been identified that contribute to osteoporosis and increased bone fracture rate (34,35). According to the World Health Organization, an individual is clinically defined as having osteoporosis when the bone mineral density is 2.5 standard deviations or more below that of the mean level for gender-matched young adults (34). The two primary strategies for decreasing the risk of developing osteoporosis are (a) maximizing peak bone mass by age 30 years and (b) slowing the rate of bone loss over the remaining years of life (36).

Hematopoiesis The formation and development of red blood cells.

Osteoporosis A disorder in which the bones become increasingly porous, brittle, and subject to fracture, owing to loss of calcium and other mineral components.

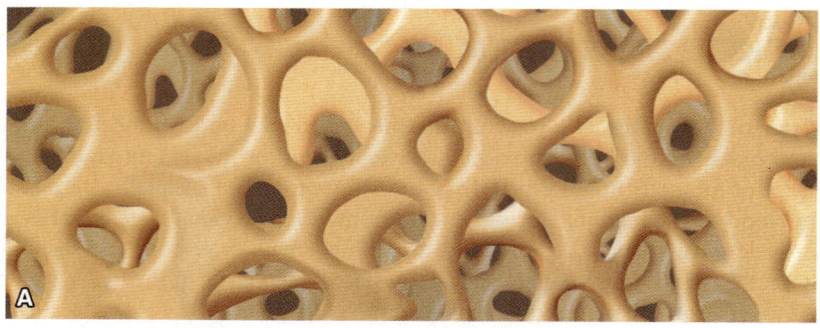
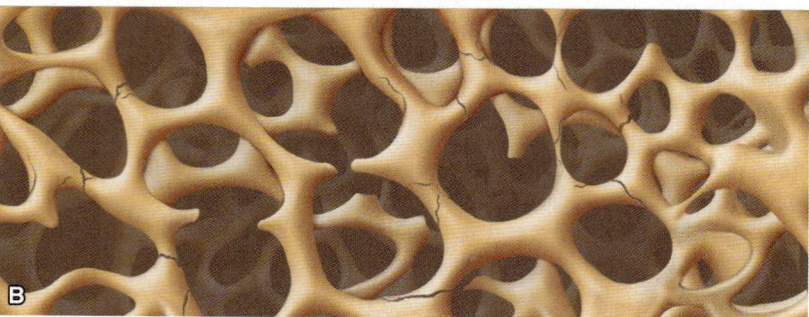

FIGURE 3.4 The differences in bone mineral density between normal bone (**A**) and osteoporotic bone (**B**).

Maximizing peak bone mass can be best accomplished by performing moderately intense exercise that requires some structural support such as walking or jogging (36–39). Consuming at least a minimal daily level of calcium also maximizes bone mass. The optimal level of calcium intake will vary for each individual and will depend on gender and age (40). Consuming an inadequate amount of calcium when young will prohibit an individual from maximizing bone mineral density and prevent older individuals from maintaining optimal bone mass (34,36,41). Exercise science professionals are instrumental in developing effective nutrition and physical activity and exercise programs for reducing the risk of osteoporosis in men and women of all ages.

The skeletal system also plays an important role in determining success in sport and athletic performance. One of the many important factors influencing success in aerobic endurance events is the ability to deliver oxygen to the working tissues of the body. Almost all of the oxygen delivery in the body occurs via red blood cells (14). The red marrow in bone generates red blood cells. The endocrine hormone **erythropoietin** controls this process (14). If the number of red blood cells can be increased in the body, oxygen delivery to

Erythropoietin A hormone that stimulates the production of red blood cells and hemoglobin.

the tissues increases as well and there are usually improvements in cardiovascular fitness and aerobic endurance performance (12). Aerobic endurance athletes can benefit from an increase in red blood cell number. **Recombinant human erythropoietin** (rEPO) has been demonstrated to increase red blood cell formation and reduce the risk of anemia as part of the therapy for a wide variety of clinically ill patients (42). If rEPO can increase red blood cell formation in well-trained athletes, there exists the potential for improving performance through an increased delivery of oxygen to the working tissues (43). In fact, rEPO has been demonstrated to improve cycling performance (44,45), and there have been numerous examples of illegal rEPO use by aerobic endurance athletes. Sports-governing associations have routinely expelled athletes from competition because the athletes test positive for rEPO during drug testing (44). Alternatively, high-altitude training may be a legal way to induce bone marrow to produce more red blood cells (46). Living at high altitudes where there is less oxygen and training at lower altitudes may be a way to stimulate bone to increase the red blood cell level, enhance oxygen delivery by the cardiovascular system, and improve aerobic endurance exercise performance (46). In response to the potential for improving aerobic endurance performance with this type of living and training environment, commercial companies have developed altitude tents that allow individuals to rest and sleep at "high altitudes" and train at low altitudes regardless of an individual's physical living location. Table 3.5 provides the functions of the skeletal system and examples of how those functions relate to physical activity, exercise, sport, and athletic performance.

Table 3.5 Functions of the Skeletal System and Their Relationship to Physical Activity, Exercise, Sport, and Athletic Performance

FUNCTION	RELATION TO PHYSICAL ACTIVITY, EXERCISE, SPORT, AND ATHLETIC PERFORMANCE
Formation of red blood cells	Carries oxygen to the tissues of the body; a key step in aerobic energy production
Lever system provides movement to the body	Allows for the contraction of the muscles to move body parts
Provides a structural framework to the body	Allows for body position and movement
Storage of minerals such as calcium and phosphorus	Appropriate levels are critical for normal physiologic function and strong bone health

Recombinant human erythropoietin The laboratory production of human erythropoietin.

CARDIOVASCULAR SYSTEM

The cardiovascular system transports blood containing oxygen, nutrients, and other substances (*e.g.*, hormones, electrolytes, and drugs) to the tissues of the body while, at the same time, facilitating the removal of carbon dioxide and other waste products from the body. The cardiovascular system also assists in body temperature regulation. Owing to the unique interaction of the cardiovascular and respiratory systems, these two systems are often referred to as one system: the cardiorespiratory or the cardiopulmonary system. Although the two systems do work in conjunction with each other, they each have other functions that allow them to be integrated with other systems of the body as well (1,14).

The primary components of the cardiovascular system include the heart, arteries, arterioles, capillaries, venules, veins, and blood. The heart is comprised of cardiac muscle and nervous tissue that generates the force that propels the blood through the body. The arteries and some of the veins are comprised of smooth muscle that helps distribute blood through the various tissues of the body. The capillaries are tubes, one cell thick, which facilitate the transfer of gases, nutrients, and waste products to and from the cells of the body. The blood is comprised of red and white blood cells, platelets, and plasma that carries the various gases, nutrients, and waste products to the body tissues (1). Figure 3.5 illustrates the various areas of the body served by the cardiovascular system. Both cardiac muscle and smooth muscle respond to input from the nervous and endocrine systems. All of these systems function together to provide a coordinated response to the challenges of physical activity and exercise.

Cardiovascular System and Exercise Science

A disease-free cardiovascular system with strength sufficient to respond to the demands of daily living, physical activity, and exercise is critical to good health. Cardiovascular disease is one of the leading causes of death in the United States (47). Years of research by public health experts and exercise science professionals have provided great insight into factors that lead to the development of cardiovascular disease (48–51). Coronary artery disease (CAD) is the primary cardiovascular disease occurring in most Americans. **Atherosclerosis**, a disease process whereby cholesterol and blood lipids build up in the arteries supplying blood to the heart, results in a reduction of blood flow to cardiac muscle (14). If blood flow to the heart is reduced to a critical level, a heart attack can result. Figure 3.6 shows how the buildup of plaque can cause a decrease in the lumen of the artery. Increased levels of physical activity and regular exercise are associated with a reduced risk of morbidity and mortality from cardiovascular disease (48,52,53). Physical activity and exercise enhance cardiac muscle function and improve blood flow to the heart and other tissues of the body. When an individual has a heart attack, that person is often referred to a cardiac

Atherosclerosis A disease process whereby cholesterol and blood lipids build up in the arteries causing a narrowing of the vessel opening.

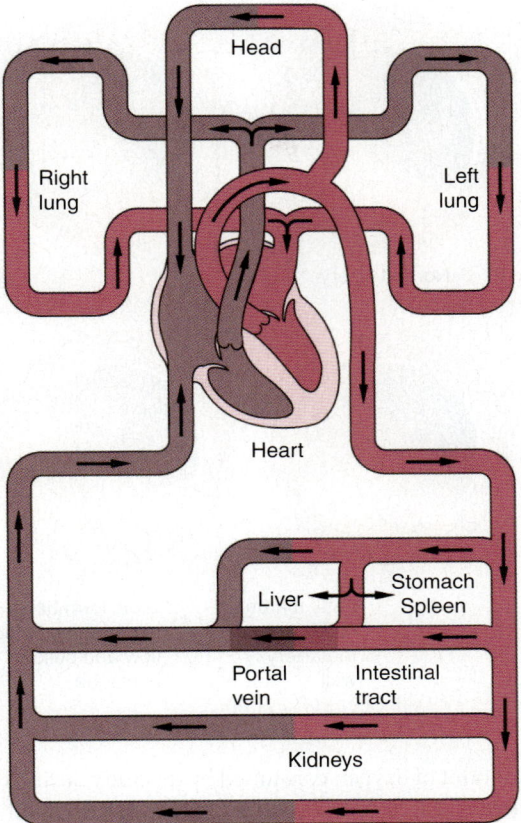

FIGURE 3.5 Areas of the body served by the cardiovascular system. (LifeART image copyright © 2010 Lippincott, Williams & Wilkins. All rights reserved.)

rehabilitation program. In these programs, clinical exercise physiologists and other allied health care professionals work to increase levels of physical fitness, improve nutritional intake, reduce stress, and modify unhealthy behaviors. Cardiac rehabilitation programs use regular exercise as an integral component for helping individuals recover from a cardiac event and to prevent events from happening in the future (54). For more information on cardiac rehabilitation, please see Chapter 5.

The cardiovascular system also plays an important role in the successful performance of various types of sport and athletic events. The delivery of oxygen and nutrients to working muscles and the removal of metabolic waste products are key in determining success during aerobic endurance events (12,55). The cardiovascular system works in conjunction with the nervous, pulmonary, and endocrine systems to accomplish this. Athletic events that last longer than approximately 3 to 5 minutes rely heavily on the adequate delivery of oxygen to tissues. **Maximal oxygen consumption** ($\dot{V}O_{2max}$) is defined as the maximal

Maximal oxygen consumption The maximum amount of oxygen the body can use during maximal effort exercise.

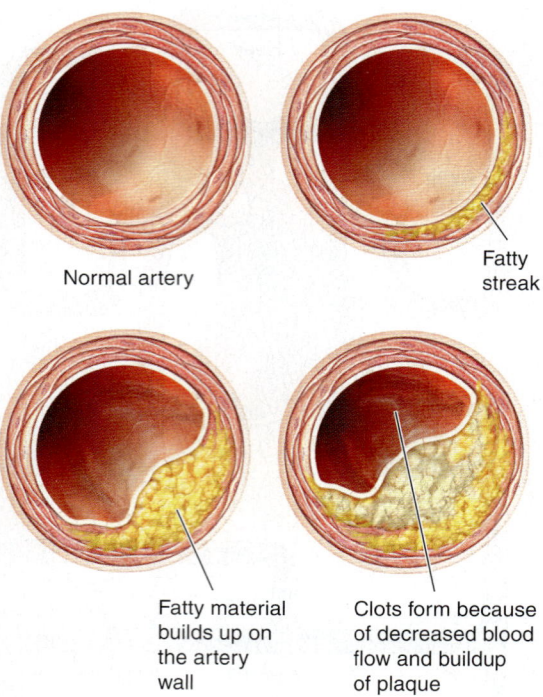

FIGURE 3.6 The buildup of plaque can cause a decrease in the lumen of the artery. (Asset provided by Anatomical Chart Co.)

amount of oxygen consumed by the body during maximal effort exercise. Much of the available research indicates that the delivery of blood and oxygen to the working tissues by the cardiovascular system is one of the limiting factors in determining an individual's $\dot{V}O_{2max}$. Because successful performance in most aerobic endurance events is determined in part by an individual's $\dot{V}O_{2max}$, this makes the contribution of the cardiovascular system critical to endurance performance success (12,55–57). Exercise science professionals are frequently involved in developing training programs, improving biomechanical movement, enhancing nutritional intake, and improving psychological factors in an effort to enhance oxygen delivery to the tissues and improving sport and athletic performance. Table 3.6 provides a summary of the functions of the cardiovascular system and examples of how those functions relate to physical activity, exercise, sport, and athletic performance.

Thinking Critically

In what ways has a systems approach to exercise science contributed to a better understanding of the role of the cardiovascular system in promoting good health by reducing the risk for heart disease and enhancing aerobic endurance performance?

PULMONARY SYSTEM

The pulmonary system brings air into the lungs, allows for oxygen to be removed from the air, and facilitates the elimination of carbon dioxide into the external environment. By regulating the carbon dioxide levels in the blood, the pulmonary system also helps maintain

Table 3.6 Functions of the Cardiovascular System and Their Relationship to Physical Activity, Exercise, Sport, and Athletic Performance

FUNCTION	RELATION TO PHYSICAL ACTIVITY, EXERCISE, SPORT, AND ATHLETIC PERFORMANCE
Assists with temperature regulation	Controls body temperature during periods of increased movement
Removes carbon dioxide and waste products	Allows for elimination of metabolic waste products of metabolism
Transports nutrients and other substances to the tissues of the body	Allows for delivery of macronutrients and substances to working tissues

the acid-base balance of the body. The lungs and associated structures create a large surface area between the blood and the external environment that allows for the fast and efficient exchange of oxygen and carbon dioxide in a healthy individual. The pulmonary system works in close conjunction with the cardiovascular system and two systems are often referred to as the cardiorespiratory or cardiopulmonary system (14).

The primary components of the pulmonary system are the respiratory muscles, the respiratory airways, and the respiratory units. The respiratory muscles (internal and external intercostals, diaphragm, and abdominals) create a pressure gradient in the chest and lungs that allow for airflow into and out of the lungs. The respiratory airways, some of which contain smooth muscle, begin with the mouth and nose, continue with the trachea, right and left bronchus, bronchiole, and end with the terminal bronchiole. Those respiratory airways that contain smooth muscle are able to dilate and constrict, in response to internal (inside the body) and external (environmental) stimuli. The bronchodilation and bronchoconstriction of the air passageways allow for increased or decreased airflow into the lungs, respectively. The respiratory unit consists of individual alveoli (where gas exchange occurs) and the pulmonary capillaries, which surround the alveoli. The majority of the exchange of oxygen and carbon dioxide occurs in the respiratory units (1,14). Figure 3.7 illustrates the primary components of the pulmonary system that facilitate air intake and gas exchange with the pulmonary capillaries.

Pulmonary System and Exercise Science

Disease conditions of the pulmonary system play a significant role in the performance of physical activity and exercise by individuals. For example, chronic obstructive pulmonary disease (COPD) is a health condition that includes chronic bronchitis and emphysema and often results in a reduced ability to perform physical activity and exercise. Often the health care management team for an individual with COPD will include physical and respiratory therapists, pulmonologists, and clinical exercise physiologists. These individuals work together to improve the functional performance of the pulmonary system and increase the individual's quality of life. Physical activity and exercise can also affect the pulmonary system and result in an asthmatic event triggered by an immune system response (58).

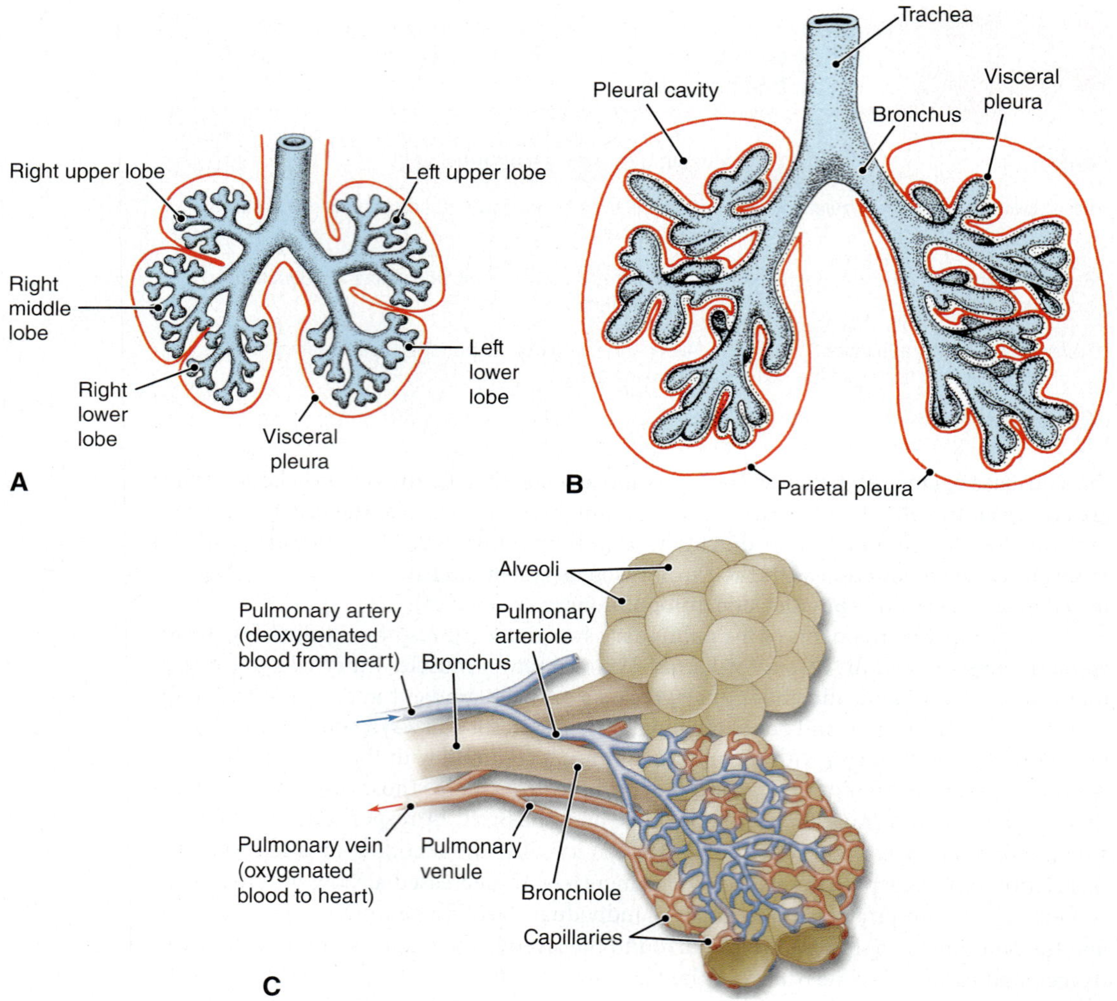

FIGURE 3.7 Primary components of the pulmonary system. **A:** Whole lung. **B:** Lung membranes and air passageways. **C:** Lung structures with blood vessels.

This asthmatic event is typically referred to as **exercise-induced asthma**. Exercise-induced asthma can result in airway constriction, shortness of breath, and wheezing similar to those experienced with asthma, but in exercise-induced asthma symptoms occur on a transient basis (59). The type, intensity, and duration of exercise can all act to trigger an asthmatic

Exercise-induced asthma A medical condition characterized by shortness of breath induced by sustained aerobic exercise.

event in susceptible individuals. Other environmental factors such as tobacco smoke, molds, dust, and cold temperatures may also play a role in triggering an exercise-induced asthma episode. An attack can result in difficult and labored exercise and reduce exercise compliance (60). The key to minimizing the occurrence of exercise-induced asthma is to avoid any predetermined factors that may trigger an episode (7,59).

The pulmonary system also plays an important role in successful sport and athletic performance. During very high to maximal intensity exercise, skeletal muscle breaks down carbohydrate and increases the production of lactic acid. The accumulation of lactic acid, which dissociates into a lactate ion and a hydrogen ion, can cause the pH (a measure of the hydrogen ion concentration) in tissues of the body to decrease and become more acidic. The increase in acidity can contribute to a decrease or cessation in sport and athletic performance. To maintain pH of the body's fluids and tissues within an acceptable range, chemical reactions in the body occur and these reactions result in an increased production of carbon dioxide (14). The primary reaction that results in the formation of carbon dioxide as a result of an increase in hydrogen ion concentration is $H^+ + HCO_3^- \rightarrow H_2CO_3 \rightarrow H_2O + CO_2$. The increased carbon dioxide is ventilated outside of the body by the lungs and this helps keep the tissues of the body from becoming too acidic as it lowers the concentration of hydrogen ions (14). Table 3.7 provides a summary of the functions of the pulmonary system and examples of how those functions relate to physical activity, exercise, sport, and athletic performance.

> **Thinking Critically**
>
> In what ways could consumption of a carbohydrate beverage improve performance during a competitive marathon or triathlon?

URINARY SYSTEM

The urinary system eliminates waste products from the body and regulates the volume, electrolyte composition, and pH of the body fluids. All tissues of the body depend on the maintenance of a stable environment of the body fluids and removal of the toxic metabolic wastes produced by the cells as they perform their normal functions. Of special importance

Table 3.7 — Functions of the Pulmonary System and Their Relationship to Physical Activity, Exercise, Sport, and Athletic Performance

FUNCTION	RELATION TO PHYSICAL ACTIVITY, EXERCISE, SPORT, AND ATHLETIC PERFORMANCE
Brings oxygen into the body	Allows for energy production during movement through aerobic metabolism
Eliminates carbon dioxide from the body	Allows for the removal of carbon dioxide, a waste product of macronutrient metabolism
Helps regulate acid-base balance	Controls pH levels in the body during periods of high-intensity movement

is the body's ability to regulate the volume and osmolarity (*i.e.*, concentration of all solutes such as electrolytes, proteins, glucose, urea, and others) of the internal fluid environment by controlling sodium and water balance (1,14).

The primary components of the urinary system are the kidneys, renal artery and vein, ureter, urinary bladder, and urethra. The kidneys play an important role in controlling body fluid levels by regulating the concentration of many of the plasma constituents, especially the electrolytes and water, and eliminating all the metabolic waste products (except for carbon dioxide). Plasma, which is the watery portion of the blood, progresses through the renal artery into the kidney where substances important to the body are retained and the undesirable or excess materials are filtered into the urine for excretion from the body. The plasma then moves back to the central circulation through the renal vein. The ureter is responsible for transporting the fluid and waste products (now called urine) to the urinary bladder for storage until the urine is eliminated from the body through the urethra (1,14). Figure 3.8 illustrates the primary components of the urinary system.

Urinary System and Exercise Science

The urinary system regulates the total fluid volume and electrolyte concentration of the body and this function can have important implications for reducing the risk for certain diseases. Hypertension affects large numbers of individuals in the United States, with approximately 33% of adults over the age of 20 years with measured high blood pressure and/or taking antihypertensive medication (61). Individuals with hypertension have an increased risk of developing cardiovascular disease (especially CAD and stroke) and kidney disease (61). Furthermore, all-cause mortality increases progressively with higher levels of both systolic blood pressure and diastolic blood pressure. The primary treatment of individuals with hypertension involves preventing morbidity and mortality associated with the high blood pressure and controlling blood pressure by the least invasive means possible (62). Individuals with Stage 1 hypertension (resting blood pressure of >130/80 mm Hg) are recommended for lifestyle modification and possible pharmacological treatment. Individuals with Stage 2 hypertension (resting blood pressure of >140/90 mm Hg) are recommended for lifestyle modification and pharmacological treatment (62,63). Changes in physical activity and exercise habits and modification of nutritional intake are at the center of effective lifestyle modification treatment for hypertension. The urinary system can play an important role in helping to manage blood pressure in hypertensive individuals by regulating fluid volume in the body. Diuretics are a group of drugs that increase the excretion of sodium and water by the kidneys (64). Diuretics block the absorption of sodium from urine by the kidney and this in turn increases the volume of fluid excreted as urine. The reduction of sodium concentration in the blood results in a reduction of total blood volume and decreases the resistance provided by the blood vessels of the body. Both of these actions have the effect of reducing blood pressure (14).

The urinary system plays an important role in sport and athletic performance in many ways. Athletes who train and play sports in hot and/or humid conditions regulate body

Osmolarity A measure of the concentration of a solution.

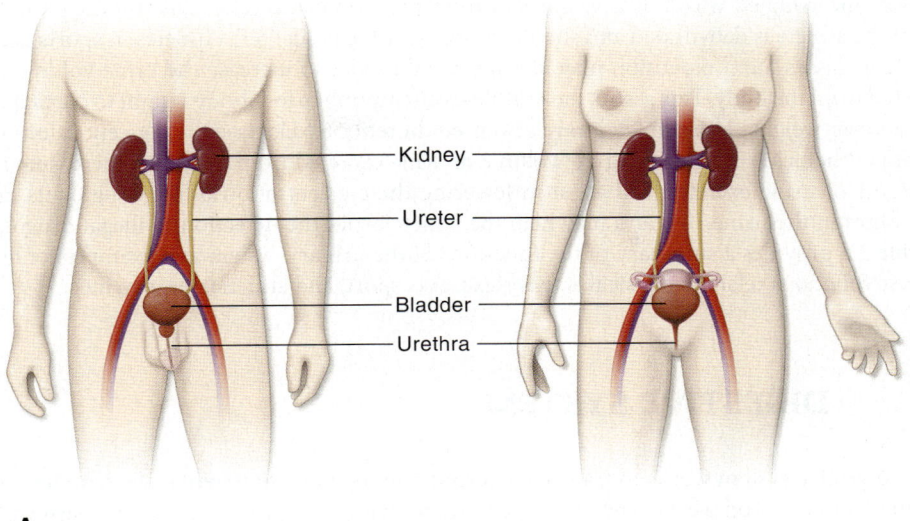

FIGURE 3.8 Primary components of the male and female urinary system. **A:** Anterior view. **B:** Lateral view. (From Premkumar K. *The Massage Connection Anatomy and Physiology*. Baltimore (MD): Lippincott, Williams & Wilkins; 2004.)

temperature primarily by sweating. As a result of sweating, the total body fluid volume decreases. In response to this decrease, the urinary system reabsorbs sodium and water from the urine in an effort to maintain an acceptable level of body water. The process of sodium and water reabsorption by the kidneys decreases the urine volume and makes the urine more concentrated. Exercise science professionals including sports nutritionists, athletic trainers, and sports medicine professionals encourage athletes to carefully examine the color of their urine as a way to monitor hydration status. When the urine is a dark concentrated color, and coupled with body weight loss from pre- to postexercise, this usually indicates that the athlete is dehydrated and should consume more fluids (65). Athletes in sports using weight classes for competition may illegally use diuretics to increase the urine volume excreted from the body. This assists those athletes attempting to lose body weight to participate in a lower weight class. Athletes have also used diuretics to decrease the concentration of a drug in the urine by increasing the volume of urine excreted from the body. Increasing the volume of urine excreted will assist in lowering the concentration of the drug or its metabolites in the urine (14) and may help the athlete avoid the detection of illegal drug use. Table 3.8 provides a summary of the functions of the urinary system and examples of how those functions relate to physical activity, exercise, sport, and athletic performance.

DIGESTIVE SYSTEM

The digestive system works to transfer **macronutrients**, **micronutrients**, **electrolytes**, and water from the food we consume into the body so that normal functions can be performed and proper health can be maintained. The macronutrients are carbohydrates, fats, and proteins, and the micronutrients are vitamins and minerals. These macro- and micronutrients contained in the food we consume represent essential sources of compounds that form the basic structures of the body and regulate the various processes of cells and tissues. Without a consistent supply of nutrients, the ability to maintain good health, perform physical

Table 3.8 Functions of the Urinary System and Their Relationship to Physical Activity, Exercise, Sport, and Athletic Performance

FUNCTION	RELATION TO PHYSICAL ACTIVITY, EXERCISE, SPORT, AND ATHLETIC PERFORMANCE
Eliminates waste products from the body	Removes waste products from increased levels of metabolism experienced during movement
Long-term regulation of acid-base balance	Helps control body pH levels that may be affected by alterations in metabolism
Regulates fluid volume and electrolyte concentrations	Controls levels of body fluids and electrolyte concentrations that are critical for efficient functioning of the body during movement

Macronutrients The foodstuffs needed in large quantities including carbohydrates, fats, and proteins that are used for numerous processes in the body.

Micronutrients The foodstuffs needed in smaller quantities including vitamins and minerals that are used for numerous processes in the body.

Electrolytes The anions and cations that are distributed in the fluid compartments of the body.

activity and exercise, and train at levels required for sport and athletic competition becomes severely compromised (1,14).

The primary components of the digestive system are the mouth, pharynx, esophagus, stomach, pancreas, liver, and small intestine and large intestine. The mouth and pharynx and the esophagus are responsible for chewing and swallowing food, respectively. The stomach mixes and promotes further digestion by using digestive juices that come from the pancreas and liver. The small intestine absorbs nutrients and most of the electrolytes and water. The large intestine is primarily responsible for absorbing salt and water and converting the remaining contents into fecal matter (1,14). Figure 3.9 illustrates the primary components of the digestive system.

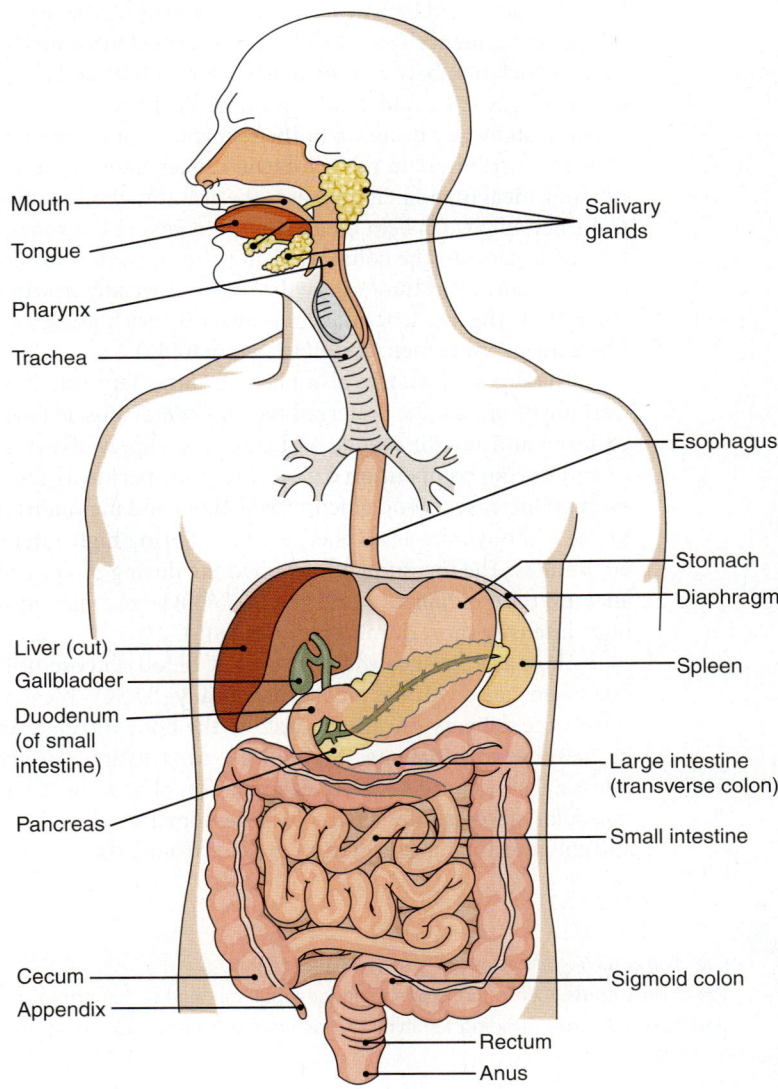

FIGURE 3.9 Primary components of the digestive system. (From Cohen BJ, Wood DL. *Memmler's the Human Body in Health and Disease.* 9th ed. Philadelphia (PA): Lippincott Williams & Wilkins; 2000.)

Digestive System and Exercise Science

Increasing physical activity and exercise and altering nutritional habits may play a role in decreasing the risk of developing certain forms of cancer, one of the leading causes of death in the United States (47). Several lines of research suggest that increasing dietary fiber intake reduces the risk of developing colorectal cancer. A protective effect of increased fiber intake may occur as a result of the dilution of fecal **carcinogens** and **procarcinogens**, reduction in the time fecal matter moves through the bowel, production of short chain fatty acids (which promote **anticarcinogenic** action), and the binding of bile acids that are carcinogenic (66). However, the results of numerous large population studies examining the role of increased fiber intake in reducing cancer risk have been inconsistent (67,68). Several investigations have demonstrated beneficial effects of increasing dietary fiber intake for reducing colorectal cancer incidence (69–71). Conversely, most prospective group studies have found little or no association between the dietary fiber intake and the risk of colorectal cancer (72–76) or adenomas (77), and randomized clinical trials of dietary fiber supplementation have failed to show any decrease in the recurrence of colorectal adenomas (78–81). One of the biggest contributors to the colorectal cancer–causing process is dietary fat. The consumption of a meal high in fat increases the amount of bile acids that are released into the digestive tract. Bile acids help break down the fats we consume, but when the large amounts of bile acids get into the colon, they may be converted to secondary bile acids, which could promote cancerous tumor growth. This is especially true of the cells that line the area of the colon (66). The incidence of colon cancer is very high in some population groups, including Black men, White men, and Black women (82).

The digestive system also plays an important role in sport and athletic performance. Carbohydrates are the preferred fuel of skeletal muscle for most high-intensity sports competitions and athletic events. An inadequate supply of dietary carbohydrate consumed prior to and during competition can lead to poor performance. Carbohydrate ingestion during exercise increases blood glucose availability and maintains the ability of the working tissues to use carbohydrate as an energy source during high-intensity exercise lasting longer than 60 minutes. The ingestion of carbohydrate during prolonged, high-intensity aerobic endurance exercise enhances performance (83,84) and may also increase performance during high-intensity short duration exercise (85).

There is a limit to the peak rate of blood glucose utilization during prolonged exercise when carbohydrates are ingested orally (83,84). The gastrointestinal system plays a key role in the delivery of carbohydrate to the body during exercise. The gastrointestinal tract has several potential sites for limiting carbohydrate use including gastric emptying and intestinal absorption (57). An examination of how the gastrointestinal, cardiovascular, and muscular systems interact can help demonstrate these limitations. Gastric emptying rates and the intestinal absorption of glucose from a 6% glucose electrolyte solution have been

Carcinogens A cancer-causing substance or agent.
Procarcinogens Compounds or substances that can lead to the formation of cancer cells.
Anticarcinogenic Tending to inhibit or prevent the activity of a carcinogen or the development of carcinoma.

measured at 1.2 and 1.3 g · min^{-1} under resting conditions and 1 g · min^{-1} during exercise (86). As the concentration of glucose in the beverage rises, the rate at which ingested glucose is supplied to the blood is lower than the rate of carbohydrate used in the muscle. This suggests either a gastrointestinal limitation to using the ingested carbohydrate as a fuel or a failure of the cardiovascular system to deliver glucose from the gastrointestinal tract into the main blood supply (87). The simultaneous consumption of both glucose and fructose, which are absorbed from the gastrointestinal tract by different mechanisms, results in the greater use of the carbohydrates for energy by muscle than the ingestion of similar amounts of glucose or fructose consumed alone (88). Furthermore, if glucose is infused into the blood, rather than consumed orally, the glucose can be used to supply energy at a faster rate (89). Collectively, this information suggests that the delivery of ingested carbohydrate from the gastrointestinal tract to the cardiovascular system might be a limiting factor in the use of ingested carbohydrate as an energy source for contracting muscle during exercise (90). Table 3.9 provides a summary of the functions of the digestive system and examples of how those functions relate to physical activity, exercise, sport, and athletic performance.

Thinking Critically

How would participation in a regular program of physical activity and exercise improve the functional ability of the pulmonary system to a disease condition of the lungs?

ENDOCRINE SYSTEM

The endocrine system helps regulate physiologic function and influence the function of other systems of the body. Along with the nervous system, the endocrine system is the other primary control system of the body. One primary advantage of the endocrine system for controlling functions is that hormones can circulate and influence cells and tissues throughout the entire body without a physical connection as the nervous system requires. Further, endocrine changes can influence function for a duration lasting from a few seconds to several hours. The endocrine system accomplishes these functions through the use of hormones secreted by the various endocrine glands of the body. Even though the endocrine glands are not connected anatomically, they function as a system in a practical sense. Many of these hormones are important for influencing both the acute responses and the chronic adaptations of the systems of the body to physical activity and exercise (1,14).

Table 3.9	Functions of the Digestive System and Their Relationship to Physical Activity, Exercise, Sport, and Athletic Performance
FUNCTION	**RELATION TO PHYSICAL ACTIVITY, EXERCISE, SPORT, AND ATHLETIC PERFORMANCE**
Delivery of macronutrients to the body	Carbohydrates, fats, and proteins are essential for the optimal function of the body during and after movement
Delivery of micronutrients to the body	Vitamins and minerals are essential for the optimal function of the body during and after movement

The primary components of the endocrine system are the glands of the body and the hormones that each gland secretes. Some endocrine glands only specialize in hormonal secretion, (*e.g.*, anterior pituitary and thyroid), whereas other components of the endocrine system consist of organs that perform other functions in addition to secreting hormones (*e.g.*, testes secrete testosterone and also produce sperm). The endocrine system, by means of the blood-borne hormones it secretes, generally regulates activities and functions at a much slower pace than the other primary control mechanism of the body, the nervous system. Most of the activities under the control of hormones are directed toward maintaining **homeostasis** (normal conditions of functioning) of the body (1,14). Table 3.10 illustrates various endocrine glands of the body of interest to exercise science and the primary hormones each secretes.

Table 3.10 Endocrine Glands and Selected Secreted Hormones (1)

ENDOCRINE GLAND	HORMONES
Hypothalamus	Releasing and inhibiting hormones
Posterior pituitary	Vasopressin
Anterior pituitary	Thyroid-stimulating hormone, adrenocorticotropic hormone, and growth hormone
Thyroid gland	Thyroxin, triiodothyronine, and calcitonin
Adrenal cortex	Aldosterone, cortisol, and androgens
Adrenal medulla	Epinephrine and norepinephrine
Pancreas	Insulin and glucagon
Parathyroid gland	Parathyroid hormone
Ovaries	Estrogen and progesterone
Testes	Testosterone
Kidneys	Renin and erythropoietin
Stomach	Gastrin
Liver	Somatomedins
Skin	Vitamin D
Heart	Atrial natriuretic peptide
Adipose tissue	Leptin and adiponectin

Homeostasis The maintenance of relatively stable internal physiologic conditions.

Endocrine System and Exercise Science

Hormones affect the systematic responses of the body in various ways and often work with other systems of the body to regulate normal functions during physical activity and exercise. For example, epinephrine and norepinephrine (also called adrenaline and noradrenaline) have been shown to increase heart rate and blood pressure in response to stress including physical activity and exercise (12). Insulin maintains blood glucose concentrations by increasing glucose uptake and utilization as an energy source in tissues of the body. The interaction of epinephrine, norepinephrine, and insulin has been associated with the development of hypertension in a disease condition called metabolic syndrome (91). Metabolic syndrome describes the clustering of several conditions of the body including elevated waist circumference, dyslipidemia (elevated triglyceride levels and decreased high-density lipoprotein [HDL] levels), impaired fasting glucose, and high blood pressure. Figure 3.10 shows the relationship among the clustering of metabolic syndrome risk factors. Central to the understanding of metabolic syndrome is the role insulin resistance plays in the development of some of the associated disease conditions. A diet high in fat and refined sugar (92) contributes significantly to the development of insulin resistance. A decreased ability of cells to absorb glucose at a given insulin concentration characterizes the condition of insulin resistance. In response to increasing insulin resistance by the tissues of the body, the pancreas secretes more insulin in an effort to promote blood glucose uptake into cells and

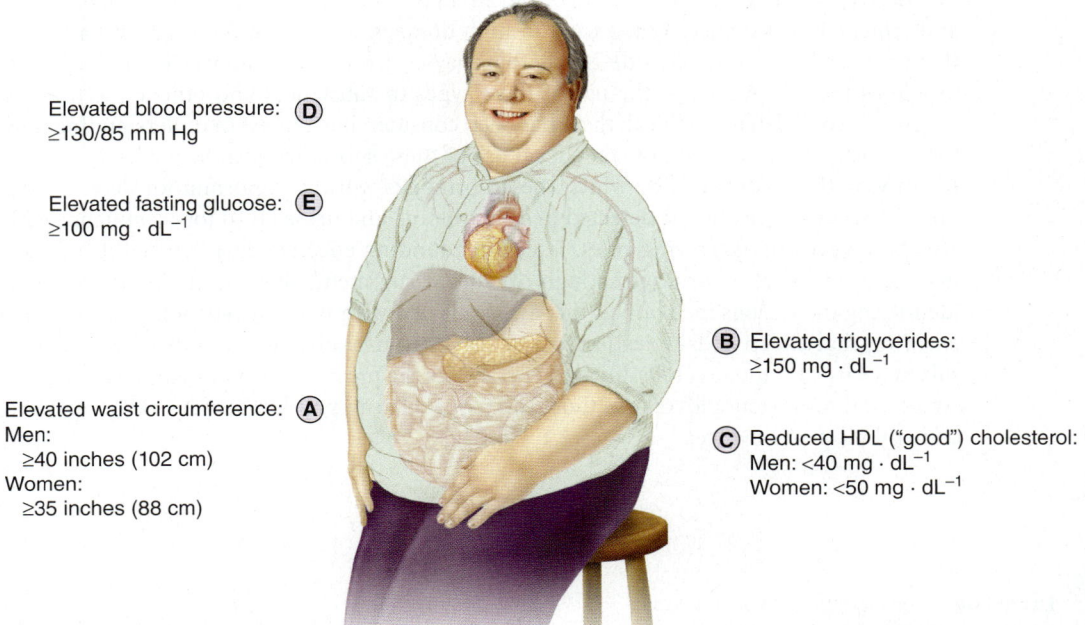

FIGURE 3.10 The relationship among the clustering of metabolic syndrome risk factors. (Asset provided by Anatomical Chart Co.)

return the blood glucose concentration to normal. If the pancreas cannot secrete sufficient amounts of insulin, the blood glucose concentration remains elevated and Type 2 diabetes results (14). When the pancreas is forced to secrete additional insulin to address the insulin resistance, the plasma insulin level becomes elevated (called hyperinsulinemia). Sympathetic nervous system activity increases in response to elevated insulin levels. This response can lead to an elevation of the hormones epinephrine and norepinephrine, which can lead to an increase in heart rate, stroke volume, and blood pressure. The elevated levels of epinephrine and norepinephrine can also interfere with insulin release from the pancreas and glucose uptake at the tissue, causing an aggravation of the problem. In this case, insulin resistance contributes significantly to the hypertension. Alternatively, given the complex interaction of the events, hypertension may in fact cause the insulin resistance. Physical activity and regular exercise can benefit individuals with insulin resistance and hypertension by improving the body's sensitivity to the hormones insulin, epinephrine, and norepinephrine (13,93).

Coaches and athletes have long been interested in the use of **exogenous** hormone supplementation to improve assorted types of sports and athletic performance. For example, the use of anabolic steroids (*e.g.*, synthetic testosterone) and human growth hormone for improving athletic performance has been common among certain groups of athletes (94–98). Anabolic steroids have both **androgenic effects** and **anabolic effects** (99). The anabolic actions cause the body to increase protein synthesis in skeletal muscles and various other tissues, which can then lead to increased nitrogen retention. The increased protein synthesis can result in an increased muscle size and strength as well as increased body weight (14). For those sports that rely on body size or the generation of power and force by the muscle, the increase in muscle mass and strength often results in improvements in sport and athletic performance. When used in high dosages, as is often the case with athletes, there are potentially serious side effects that may be irreversible and cause serious health problems (96,99). Although the use of these types of substances and others like them is illegal and considered unethical, there has been considerable interest in understanding how these substances work and how the illegal use of these substances can be detected (96,99). Identifying the mechanism by which these hormones work is important for determining how these substances can be detected in saliva, blood, and urine (100) for compliance with athletic association governing rules. Issues surrounding effective drug testing include using equipment that is sensitive enough to detect drug metabolites in the blood or urine, identifying the various metabolites that are associated with synthetic anabolic steroids, and ensuring that effective drug testing is a deterrent to anabolic steroid and other anabolic substance use by athletes (100). Table 3.11 provides the primary hormones of the endocrine system and some examples of how those functions relate to physical activity, exercise, sport, and athletic performance.

Exogenous Coming from outside the body.
Androgenic effects The development and maintenance of masculine characteristics.
Anabolic effects The development and maintenance of tissue, particularly skeletal muscle.

Table 3.11 — Primary Hormones of the Endocrine System and How Their Functions Relate to Physical Activity, Exercise, Sport, and Athletic Performance

HORMONE	FUNCTION	RELATION TO PHYSICAL ACTIVITY, EXERCISE, SPORT, AND ATHLETIC PERFORMANCE
Adiponectin	Regulates glucose and fatty acid metabolism	Plays a role in the suppression of metabolic abnormalities
Aldosterone	Increases sodium reabsorption and potassium excretion in the kidneys	Helps regulated fluid balance to prevent dehydration
Calcitonin	Decreases plasma calcium concentration	Increases calcium deposition in bone
Cortisol	Increases blood glucose concentration; contributes to stress adaptation	Helps increase blood glucose concentration to avoid hypoglycemia
Epinephrine and norepinephrine	Reinforces sympathetic nervous system activity	Assists the body when responding to the stress of movement
Erythropoietin	Stimulates red blood cell production in bone marrow	Increases oxygen delivery to working tissues
Estrogen	Responsible for development of secondary sexual characteristics	Helps regulate lean mass and skeletal mass in the body
Glucagon	Promotes maintenance of nutrient levels in blood, especially glucose	Helps regulate blood glucose levels during exercise
Growth hormone	Essential for the growth of bones and soft tissue; protein anabolism; fat mobilization	Promotes the growth of lean and skeletal tissue
Insulin	Promotes uptake of absorbed nutrients, especially insulin	Helps regulate blood glucose levels after food consumption
Leptin	Assists the brain in regulating appetite and metabolism	Helps assist the body in the regulation of an appropriate body weight
Testosterone	Responsible for development of secondary sexual characteristics	Helps regulate lean mass in the body
Vitamin D	Increases absorption of ingested calcium and phosphate in the gastrointestinal tract	Helps regulate levels of calcium in the body, especially bone

IMMUNE SYSTEM

Overall, the immune system regulates susceptibility to, severity of, and recovery from infection or injury, abnormal tissue growth, and illness. The immune system works with other body systems through a complex arrangement of structures, compounds, and cells.

The various components of the immune system are not connected anatomically, yet they do constitute a functional system in the body. The immune system allows the body to make a distinction between its normal components (*i.e.*, self) and any foreign elements (*i.e.*, non-self), with the purpose of protecting itself against those substances foreign to the body (1,14).

The primary components of the immune system are the physical, mechanical, chemical, blood, and cellular factors of the body. The **innate** (*i.e.*, natural) and **acquired** (*i.e.*, adaptive) components of the immune system function together in a complex interaction to protect the body from outside elements. The innate components of the body offer an immediate and predetermined general protection against any type of foreign challenge to the body. Conversely, the adaptive components of the immune system are extremely specific in the response to a given foreign invader (101). Table 3.12 illustrates the components of innate immunity and acquired immunity.

Table 3.12 Components of Innate and Acquired Immunity (101)

INNATE IMMUNITY	ACQUIRED IMMUNITY
Physical Barriers (to keep pathogens out)	
Skin and epithelium	
Mucus and mucus membrane structures	
Turbulent air flow	
Chemical barriers (to provide a hostile environment or neutralize pathogens inside the system)	**Chemical and soluble defenses** (to bind and neutralize pathogens)
Complement factors	Immunoglobulins (also known as antibodies)
Lysozymes	
Acute-phase proteins	
Acidic pH in stomach	
Cellular defenses (to create a toxic environment, kill bacteria or potentially infected cells, or to engulf debris; also play primary role in the inflammation process)	**Cellular defenses** (to activate immune defenses, kill infected cells, and produce antibodies)
Monocytes/macrophages[a]	T lymphocytes (T cells, multiple types)
Granulocytes, *e.g.*, neutrophils	B lymphocytes (B cells)/plasma cells
Natural killer cells	

[a]Also have a role in presenting antigens to cells of the acquired immune arm of the immune system.

Innate Immunity existing from within the body at birth.
Acquired Immunity that is derived after birth.

Immune System and Exercise Science

Over the last several decades, there has been steady growth in the study of physical activity, exercise, and immune function and how they all interact to positively and negatively affect health. Exercise may play a role in the prevention and treatment of certain illnesses such as cancer and acquired immune deficiency syndrome (AIDS) (102). Evidence from **epidemiology** studies indicates an association between regular physical activity and lower rates of certain cancers (103). Animal studies also indicate that exercise training enhances resistance to experimentally induced tumor growth (104). Furthermore, exercise may have a positive effect in stimulating the immune system during times of illness or reduced responsiveness (such as aging or AIDS). Studies over extended periods of 10 to 20 years have reported reduced incidence of cancer in physically active groups (104,105). There is also evidence that occupational physical activity is associated with a decreased risk of colorectal cancer (106) and that physical inactivity is associated with an increased risk of colorectal cancer (104). Increased levels of physical activity may also reduce the risk of cancer at other sites such as the breast (107) and reproductive organs in women (103,105). Exercise may positively affect autoimmune diseases such as rheumatoid arthritis, Graves disease, and systemic lupus erythematosus (108). In older adults, increased levels of physical activity and regular exercise may also enhance immune system function (109,110).

Some highly competitive athletes may appear to suffer high rates of certain illnesses such as mononucleosis and upper respiratory illness. The relationship between exercise and upper respiratory illness may be modeled in the form of a "J" curve as shown in Figure 3.11 (111). This model suggests that although the risk of upper respiratory illness may decrease below that of a sedentary individual when one engages in moderate exercise training, the risk may rise above average during periods of excessive amounts of high-intensity training such as that performed by some athletes (101). Frequent illness has also been observed in athletes experiencing

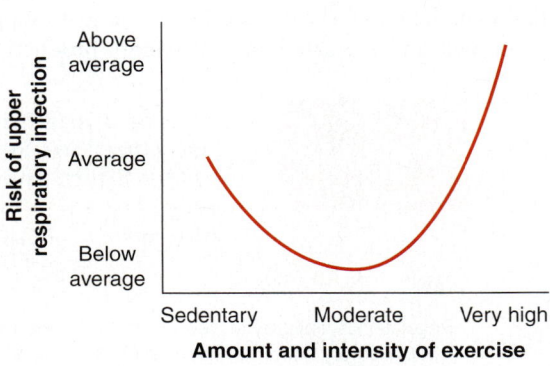

FIGURE 3.11 Relationship between exercise and upper respiratory illness may be modeled in the form of a "J" curve (111).

Epidemiology The branch of medicine dealing with the incidence and prevalence of disease in large populations.

a condition called **overtraining syndrome** (111,112). Although much of the evidence is anecdotal, the few attempts to quantify rates of illness tend to support a higher incidence of illness among certain groups of athletes (113–115). In response to this information, prospective studies have examined numerous nutritional aids in an attempt to improve immune function and reduce the incidence of illness in athletes who are training at high intensities. The effects of carbohydrate supplementation on immune function have received the majority of research work, but there are other nutritional supplements that have potential to modify immune function and reduce the risk of illness in athletes. For example, β-glucan has demonstrated potential to decrease upper respiratory tract illness and improve immune function in some (116,117) but not all studies (118). It is likely that training volume and competition exert a combined effect on the susceptibility to illness. Moderate-intensity exercise has been demonstrated to induce a much smaller change in the cellular components of the immune system. Short bouts of high-intensity exercise can cause temporary impairment of the immune response and repeated heavy training and the stress of top-level competition can have a more serious effect on immune function. Research supports the concept that heavy exertion increases an athlete's risk of upper respiratory illness because of the changes in immune function (119) and the elevation of the hormones epinephrine and cortisol (120). Table 3.13 provides the functions of the immune system and examples of how those functions relate to physical activity, exercise, sport, and athletic performance.

Thinking Critically

How does exercise training for improving health and fitness differ from exercise training for enhancing performance during sport and athletic competitions as they relate to alterations in immune system function?

ENERGY SYSTEMS

The ability to produce energy is critical for ensuring normal function in all living cells and tissues of the body. The production of energy to support the functions of the body is tightly controlled and regulated so that energy production closely matches energy utilization.

Table 3.13 Functions of the Immune System and How Those Functions Relate to Physical Activity, Exercise, Sport, and Athletic Performance

FUNCTION	RELATION TO PHYSICAL ACTIVITY, EXERCISE, SPORT, AND ATHLETIC PERFORMANCE
Regulate susceptibility to, severity of, and recovery from infection, abnormal tissue growth, and illness	Provides protective effect from internal and external illness-causing agents

Overtraining syndrome A condition whereby too much training results in the maladaptation of body responses.

Energy exchange in humans occurs when the foods (carbohydrates, fats, proteins) we eat are broken down into a usable form of energy by the body. Although energy exchange in humans is not a system in a structural sense, the various processes by which the body produces energy do constitute a system in the functional sense (1,14).

The main components of the energy system are the chemical compounds contained in the energy pathways of the cells and the macronutrients carbohydrate, fat, and protein. When the cells of the body need energy, they break down a chemical compound called adenosine triphosphate (ATP). Immediately the cells begin to resynthesize the ATP from the energy contained in the chemical bonds that hold together the macronutrients. In skeletal muscle fibers, energy to resynthesize ATP comes from one of three energy pathways: (a) **immediate sources**, which use stored energy in the form of ATP, adenosine diphosphate (ADP), and **creatine phosphate**; (b) **glycolysis** and **glycogenolysis**, which use blood glucose and stored muscle glycogen, respectively; and (c) **oxidative metabolism**, which use products of carbohydrate, fat, and protein metabolism. The immediate stores rapidly resynthesize ATP in skeletal muscle during the initiation of movement and during very high-intensity exercise. Because the immediate energy system is limited in capacity, the body quickly begins to provide energy through the other pathways. The processes of glycolysis (using glucose) and glycogenolysis (using glycogen) involve the breakdown of carbohydrates into compounds called pyruvic acid and lactic acid. As the carbohydrate is broken down, energy is used to resynthesize ATP for use by the body. The need for energy by the body, in part, determines the end product of carbohydrate metabolism. If the energy requirements of the muscle are high, then lactic acid is formed. If the energy requirements of the muscle are lower, then pyruvic acid is formed, which is then used to produce energy in oxidative metabolism. Glucose/glycogen, fats, and proteins can all be used to produce energy. Oxidative metabolism occurs in the mitochondria of cells and is an energy system that is powerful with a large capacity for producing energy (12,14). Energy production must match energy utilization during exercise or the exercise intensity must be reduced or exercise must be stopped. The energy pathway that predominates during physical activity or exercise will depend primarily on the intensity and duration of the activity. Whether carbohydrates, fats, or proteins are used to produce energy will be contingent on the pathway being used, but it can also be influenced by factors such as food consumption, training status, and the use of **ergogenic aids**. Table 3.14 illustrates the systems for energy production in the body.

Immediate energy sources Stored energy in the form of ATP and creatine phosphate.
Creatine phosphate An organic compound found in muscle and cardiac tissue and capable of storing and providing energy for muscular contraction.
Glycolysis The breakdown of glucose to produce energy.
Glycogenolysis The breakdown of glycogen to produce energy.
Oxidative metabolism The use of oxygen to break down carbohydrates, fats, and proteins to produce energy.
Ergogenic aids Any substance or device that improves physiologic or psychological performance.

Table 3.14 Energy Production Systems in the Body

ENERGY SYSTEM	SUBSTRATE USED
Immediate sources	ATP and creatine phosphate
Glycolysis and glycogenolysis	Blood glucose and muscle glycogen
Oxidative metabolism	Products of carbohydrate, fat, and protein metabolism

Energy Systems and Exercise Science

Identifying the appropriate physical activity and exercise intensity to promote the utilization of fat as an energy source is an important issue in the regulation of body weight and body composition. At the beginning of exercise, increases in energy utilization of the body are matched by increases in energy production from the immediate energy sources and the breakdown of glucose and muscle glycogen. As the exercise duration continues, energy production shifts so that under certain circumstances more energy is produced from the breakdown of fats. Several factors influence which energy pathway is used to provide energy to the body, including the nutritional status and training level of the individual, the exercise intensity and duration, and various hormonal concentrations in the body. Of those factors, exercise intensity and duration most often have the greatest influence on fat and carbohydrate utilization as energy sources (Figure 3.12). During physical activity and exercise, the percent contribution of carbohydrate and fat to energy production can be estimated by using the respiratory exchange ratio (RER), a value derived using metabolic measurement equipment that can calculate the amount of carbon dioxide produced to the amount of oxygen consumed (12). When using the RER to estimate fuel utilization during rest or exercise, the role that protein contributes to energy production is usually ignored because protein generally contributes very little to energy production during physical activity and exercise (12). The RER can be used to estimate fuel utilization because carbohydrates and fat differ in the amount of oxygen consumed and carbon dioxide produced; fat requires more oxygen than carbohydrate when used for energy production (12).

The understanding of how body weight and **body composition** affect fat utilization is of great interest for understanding the mechanisms behind the obesity epidemic (121). Increases in body weight and body fat, as well as changes in fat utilization as an energy source have significant implications for the development of several chronic disease conditions such as insulin resistance, Type 2 diabetes, and metabolic syndrome (122). Exercise science professionals are interested in determining the optimal intensity for utilizing fat during physical activity and exercise to better assist those individuals seeking to decrease body fat and reduce body weight. Recent evidence would indicate that the exercise intensity at which there

Body composition The amount of fat and nonfat tissue in the body.

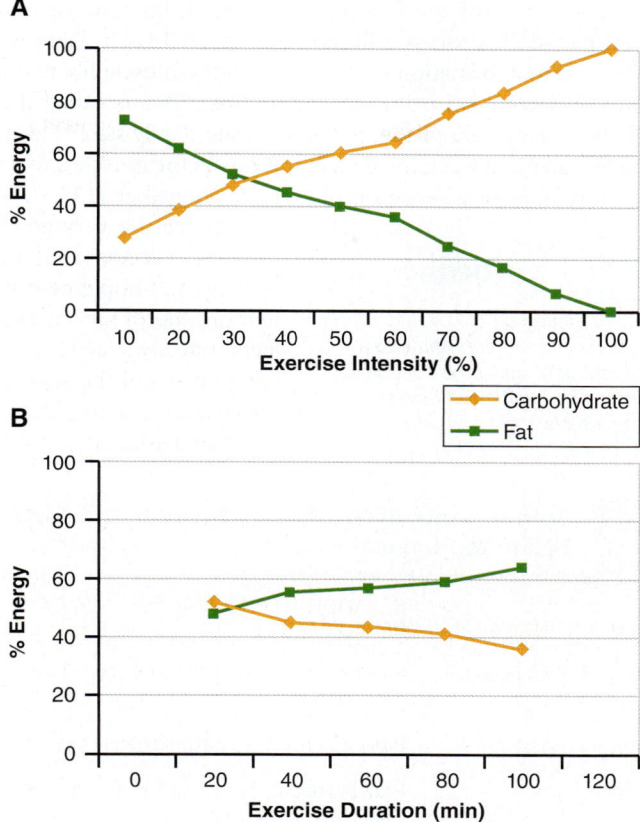

FIGURE 3.12 The relationship between exercise intensity **(A)** and duration **(B)** and the use of fat and carbohydrates as energy sources.

is maximal fat utilization is approximately 65% of an individual's $\dot{V}O_{2max}$ (1). Maximizing the rate of fat oxidation during exercise can have beneficial effects on overall health including improving insulin sensitivity (123) and managing long-term weight control (124).

Determining the optimal exercise intensity during endurance events that are longer than 10 km is critical for maximizing sport and athletic performance. Success in endurance events can be determined by examining several factors. For example, one of the strongest predictors of successful performance in distance running is the running velocity at the **maximal lactate steady state** (125). During moderate- to high-intensity exercise, the majority of energy produced to power muscular contraction comes from the breakdown of carbohydrates. During the rapid metabolism of glucose and glycogen, there is lactic acid formation, which dissociates into lactate ions and hydrogen ions. The increase in hydrogen ions can result in a decrease in the intramuscular pH. Unless the pH level is maintained, the increase in hydrogen ion concentration may contribute to fatigue and a lowering of force

Maximal lactate steady state The exercise intensity where maximal lactic acid production is matched by maximal lactic acid removal.

production in the contracting muscle (126). The removal of lactic acid occurs through several metabolic processes in the body (12,127,128). There is an exercise intensity whereby the maximal formation of lactic acid in the muscle fiber is matched by the maximal clearance of lactic acid from the muscle fiber. This is called the maximal lactate steady state (129). Exercise above the maximal lactate steady state leads to a progressive increase in the lactate and hydrogen ion concentrations in the muscle and ultimately a decrease in exercise performance or a cessation of exercise altogether (129). The maximal lactate steady state is considered a very good predictor of endurance performance success (130). Endurance coaches and athletes develop and implement training programs with the intent on improving the exercise intensity at the maximal lactate steady state (12,131,132). Table 3.15 provides the components of the energy system and examples of how those components relate to physical activity, exercise, sport, and athletic performance.

> **Thinking Critically**
>
> Why would an understanding of how the systems of the body interact contribute to a greater understanding of how physical activity and exercise influence health and how regular exercise training affects sport and athletic performance?

Table 3.15 Components of the Energy System and How Those Components Relate to Physical Activity, Exercise, Sport, and Athletic Performance

ENERGY SYSTEM	RELATION TO PHYSICAL ACTIVITY, EXERCISE, SPORT, AND ATHLETIC PERFORMANCE
Immediate sources: ATP and creatine phosphate	Provides energy during the initiation of movement and during high-intensity exercise
Glycolysis and glycogenolysis	Provides energy during moderately high-intensity exercise
Oxidative metabolism	Provides energy during resting and low- to moderately high-intensity physical activity and exercise

Interview

Nicole Keith, PhD, FACSM

Professor of Kinesiology and Associate Dean of Faculty Affairs, Indiana University-Purdue University, Indianapolis; Research Scientist, Indiana University Center for Aging Research; Investigator, Regenstrief Institute

Brief Introduction – I earned my Bachelor of Science degree in Exercise Science from Howard University in Washington, DC. I then earned a Master of Science degree in Exercise Science from the University of Rhode Island and a Doctor of Philosophy degree in Exercise Physiology from the University of

Connecticut. I completed another Master of Science degree in Clinical Research at Indiana University. I began my academic career as an instructor in the Department of Kinesiology at the University of Rhode Island and after 2 years, I was hired as an Assistant Professor at the University of Southern Indiana. Two years later, I started an Assistant Professor position at Indiana University-Purdue University, Indianapolis (IUPUI). My research focuses on addressing health disparities and achieving health equity through physical activity participation. For the last 18 years, my investigation targets have been patients served by Federally Qualified Health Centers (FQHC) in Indianapolis and adults who live in the surrounding urban communities. I have been awarded both federal and foundation funding to complete this work.

 What are your most significant career experiences?

First, establishing the Physically Active Residential Communities and Schools (PARCS) program. This program is housed in local community high schools and offers exercise opportunities for high school students and adults from the surrounding communities to exercise with IUPUI Kinesiology students at low or no cost. For the past 18 years, over 4,000 youth and adults have been served by PARCS. This program also has served as a service learning location for hundreds of Kinesiology students. Second, nearly simultaneously, my colleagues and I established Healthy Me, an FQHC-based behavioral healthy living program that has served over 7,000 patients who are overweight and obese. These two programs have made people who otherwise may not have had access to healthy living opportunities become healthy and active. Third, being appointed Associate Dean of Faculty Affairs, demonstrating that my Dean trusted me to manage the matters of over 70 faculty in the School of Health and Human Sciences and giving me the rewarding experience of supporting my peers. Fourth, as Chair of the American College of Sports Medicine's (ACSM) Committee on Diversity Action, I guided the writing of the ACSM's first diversity statement and helped lead efforts to increase diversity within the organization. I also created the ACSM Leadership and Diversity Training Program to mentor and retain underrepresented minority members. There have been 150 unique participants, 10 of whom have gone on to be elected to ACSM Regional Chapter boards or the Board of Trustees. Fifth, being elected Student Representative, Trustee, Vice President, and now President of the American College of Sports Medicine demonstrates that peers in my primary professional organization have valued my work with ACSM over the last 25 years and trust my leadership capabilities. Finally, being the first person of color to be elected to any of these offices provides me with the opportunity to show those who look like me what is possible.

 Why did you choose to become an "exercise scientist"?

My parents grew up in inner-city Chicago. Each earned undergraduate and graduate degrees and my mom earned a PhD. We left Chicago and moved to Grand Rapids, MI, when I was 4 years old. Every summer, we would visit relatives in Chicago. I noticed they ate unhealthy foods that we never ate and did very little to no physical activity — my extended family had a lifestyle that was very different from my immediate family. Over the years, I noticed my parents were aging much more slowly than their siblings. It was at that time that I realized lifestyle behaviors influenced health. I found this fascinating and wanted to explore it more. I started out pursuing Physical Education Teacher Education, but student teaching taught me this was not the right career path. I then learned about Exercise Science and discovered that I could pursue my passions through research.

Why is it important for exercise science students to understand how all of the systems of the body function?

Physical activity positively influences every system of the human body and most chronic health conditions. While clinicians and research scientists may focus on one disease or body system, nobody starts out that way. In order to discover what makes a student passionate (or disinterested), it is very important that they are exposed to all of the body's systems and functions. Having this level of knowledge will also make them more marketable in the long run. And lastly, wherever exercise scientists end up working, the other professionals around us expect us to know all things exercise. I get asked the most random questions about information I learned 30 years ago.

Q: What advice would you have for a student exploring a career in any exercise science profession?

Make decisions based on your excitement and passion. It's okay to explore different areas of focus or career options and it's very important to have a lot of different experiences so students can learn about what they love and what they don't love. Network with other students who are like you as well as those who are not, you can learn from both groups. Also, make sure that your professors get to know you before you need a letter of recommendation. And if you learn about a clinician or a scientist who has expertise in either a patient population or research area that is of interest to you, reach out to them. It's never too early to expand your professional network.

Interview

J. Timothy Lightfoot, PhD, FACSM

Registered Clinical Exercise Physiologist, Certified Exercise Specialist (ACSM). Omar Smith Endowed Chair in Kinesiology and Director, Sydney and JL Huffines Institute for Sports Medicine and Human Performance at Texas A&M University

Brief Introduction – I received my Bachelor's and Master's degrees from the University of Louisiana Monroe and my doctorate from the University of Tennessee. I completed a research consultantship with the U.S. National Aeronautics and Space Administration (NASA) at Kennedy Space Center in the Biomedical Laboratory and then a 3-year National Institutes of Health Postdoctoral Research Fellowship in the Division of Physiology at Johns Hopkins University. I have been an Assistant and Associate Professor at Florida Atlantic University and a Professor at the University of North Carolina Charlotte. I have published over 70 scientific, peer-reviewed articles on the genetics of daily physical activity and exercise endurance, as well as the physiological response to high-G exposure and hemorrhage. I have received external funding from the National Institutes of Health and Department of Defense to conduct research on the genetics of physical activity. I am a Fellow of the American College of Sports Medicine, an ACSM-certified Exercise Specialist, a Registered Clinical Exercise Physiologist, a Past President of the Southeast Regional Chapter of the American College of Sports Medicine, and a Past Member of the Board of Trustees for the American College of Sports Medicine.

Q: What are your most significant career experiences?

I have been blessed and honored to have several significant career experiences. My work at Kennedy Space Center at NASA in the mid-1980s was amazing and gave me up-close experiences as to how Exercise Physiology principles could be applied to a variety of "non-sport" venues — and I've followed that path with our work with motorsports athletes and musicians. Taking on an academic administrative role at a fairly young academic age helped me to learn how to get individuals to work together to accomplish goals. My years as a Department Chairperson also taught me many things about how "not" to get people to work together. Taking the leap and switching my primary research interest in the late 1990s to genetics was a significant experience. Suddenly, even though I was an established investigator,

I had to start over and rebuild my experience and expertise. It was so worth it because my move into genetics in 1998 gave me the foundation to receive multiple federal grants and gave my research a foundation (the biological regulation of physical activity) that can impact not only exercise science but also the health of our society.

 Why did you choose to become an "exercise scientist"?

My decision to become an exercise scientist was selfish at first: I wanted to become a better athlete. Thus, I figured the more I learned, the better I would become. However, I soon realized that I just had a passion for human performance. How humans could perform and adapt in a wide variety of environments and situations stimulated my interest and continuing efforts to understand how it all works mechanistically.

 Why is it important for exercise science students to understand how all of the systems of the body function?

As scientists, we too often work in "reductionist mode" and try to isolate everything — even down to the molecular level — so we can find "the cause." However, very few things in physiology work in isolation. So, while reductionism is critically important, the scientist always must ask how their findings work with other pieces to form the system that actually causes the body to function the way it does. Exercise Science and Exercise Physiology are terribly complex, and understanding just one small piece does not allow you to have a general knowledge about how the whole system works together.

 What advice would you have for a student exploring a career in any exercise science profession?

Find something you are passionate about: a topic that catches your imagination and won't let you go. The most successful people I know in Exercise Science are those that are passionate about the topic or the job that they are doing. So look for a topic or piece of exercise science that makes you stop and say "hey, that's cool." That's the thing you should be doing — it may be research, it may be rehabilitation, it may be coaching — but whatever it is, keep looking. You'll find it. The second piece of advice is, never be afraid to talk to people. Say "hello" and talk to them about what they are doing. Ask these people about what they are passionate about in exercise science. Chances are you'll make a great connection for your network, but you'll also get a sense for why others decide that Exercise Science is the best career path.

SUMMARY

- The various systems of the body play important roles in regulating body functions prior to, during, and following physical activity, exercise, sport, and athletic performance.
- Proper control of the body functions requires the systems to function together in a coordinated and controlled manner.
- Increased levels of physical activity and regular exercise training usually result in an improved function of the various systems, whereas physical inactivity typically leads to a decrease in functional ability and often diseased conditions in the body.
- Successful performance in sport and athletic competition depends on optimal functioning of the systems of the body.
- Various subject and content areas of exercise science, as well as the professional careers graduates enter into, will rely heavily on a sound knowledge of the systems of the body.

FOR REVIEW

1. Why should the study of the systems of the body be examined from an integrated approach?
2. What are the primary functional components of the nervous system and how do those components respond to exercise?
3. What are the three primary types of skeletal muscle fibers?
4. What are the principal functions of the skeletal system?
5. How does the cardiovascular system work to maintain challenges to homeostasis during exercise?
6. Describe the role of the pulmonary system in maintaining normal acid-base balance during rest and exercise.
7. What role does the urinary system play in the treatment of individuals with hypertension?
8. How does the gastrointestinal system influence the delivery of carbohydrate to working skeletal muscle?
9. Describe how insulin resistance influences the development of the metabolic syndrome.
10. What is the difference between innate immunity and acquired immunity?
11. Provide a list of the energy sources used in the three primary energy-producing pathways. Which pathways are used during low-, moderate-, and high-intensity exercise?

Project-Based Learning

1. Create a chart that shows the interrelationships among the 10 different systems of the body and their responses (increase, decrease, no change) for the following physical activities:
   a. Cutting the grass using a push lawn mower
   b. Walking the family dog for 45 minutes on a hot humid day
2. Create a chart that shows the interrelationships among the 10 different systems of the body and their responses (increase, decrease, no change) for the following physical activities:
   a. Playing a tennis match at 12:00 noon, on January 30th in Melbourne, Australia
   b. Running a 100-m sprint in a track and field competition

REFERENCES

1. Sherwood L. *Fundamentals of Physiology: A Human Perspective*. 3rd ed. Belmont (CA): Thompson Publishing; 2006.
2. Laskin JJ, Anderson M. Cerebral palsy. In: *ACSM's Resources for Clinical Exercise Physiology*. 2nd ed. Philadelphia (PA): Wolters Kluwer Health; 2002. p. 288–94.
3. O'Connell DG, Barnhart R, Parks L. Muscular endurance and wheelchair propulsion in children with cerebral palsy or myelomeningocele. *Arch Phys Med Rehabil*. 1992;73(8):709–11.
4. Olney S, MacPhail H, Hedden D, Boyce W. Work and power in hemiplegic cerebral palsy gait. *Phys Ther*. 1990;70:431–8.

5. Priego-Quesada JI, Lucas-Cuevas AG, Llana-Belloch S, Perez-Soriano P. Effects of exercise in people with cerebral palsy: a review. *J Phys Educ Sport*. 2014;14(1):36–41.
6. American College of Sports Medicine. *ACSM's Exercise Management for Persons with Chronic Diseases and Disabilities*. 4th ed. Champaign (IL): Human Kinetics; 2009.
7. Thompson WR. *ACSM's Clinical Exercise Physiology*. Philadelphia (PA): Wolters Kluwer Health; 2019. 792 p.
8. De Messsman RE. Respiratory sinus arrhythmia alteration following training in endurance athletes. *Eur J Appl Physiol*. 1992;64:434–6.
9. Goldsmith RL, Bigger JT, Bloomfield DM, Steinman RC. Physical fitness as a determinant of vagal modulation. *Med Sci Sports Exerc*. 1997;29:812–7.
10. Goldsmith RL, Bigger JT, Steinman RC, Fleiss JL. Comparison of 24-hour parasympathetic activity in endurance-trained and untrained young men. *J Am Coll Cardiol*. 1992;20:552–8.
11. Kowalik T, Klawe JJ, Tafil-Klawe M, et al. Multiannual, intensive strength-endurance training modulates the activity of the cardiovascular and autonomic nervous system among rowers of the International level. *BioMed Res Int*. 2019:1–6.
12. Brooks GA, Fahey TD, Baldwin KM. *Exercise Physiology: Human Bioenergetics and Its Applications*. 4th ed. Mountain View (CA): Mayfield; 2004.
13. Powers SK, Howley ET, Quindry J. *Exercise Physiology: Theory and Application to Fitness and Performance*. 11th ed. New York (NY): McGraw-Hill; 2021.
14. Guyton AC, Hall JE. *Textbook of Medical Physiology*. 13th ed. Oxford (UK): Elsevier; 2016.
15. Krans JL. The sliding filament theory of muscle contraction. *Nat Educ*. 2010;3(9):66–70.
16. Bobbert MF, Ettema G, Huijing PA. The force-length relationship of a muscle-tendon complex: experimental results and model calculations. *Eur J Appl Physiol*. 1990;61:323–9.
17. Clarkson PM. Eccentric exercise and muscle damage. *Int J Sports Med*. 1997;18(4):S314–7.
18. Nicol LM, Rowlands DS, Fazakerly R, Kellett J. Curcumin supplementation likely attenuates delayed onset muscle soreness (DOMS). *Eur J Appl Physiol*. 2015;115(8):1769–77.
19. Snyder JG, Ambeqaonkar JP, Winchester JB. Cryotherapy for treatment of delayed onset muscle soreness. *Int J Athl Ther Train*. 2011;16(4):28–32.
20. Harrison BC, Robinson D, Davison BJ, Foley B, Seda E, Byrnes WC. Treatment of exercise-induced muscle injury via hyperbaric oxygen therapy. *Med Sci Sports Exerc*. 2001;33(1):36–42.
21. Mekjavic IB, Exner JA, Tesch PA, Eiken O. Hyperbaric oxygen therapy does not affect recovery from delayed onset muscle soreness. *Med Sci Sports Exerc*. 2000;32(3):558–63.
22. Kuipers H, Keizer HA, Verstappen FT, Costill DL. Influence of a prostaglandin-inhibiting drug on muscle soreness after eccentric work. *Int J Sports Med*. 1985;6(6):336–9.
23. Heiss R, Hotfiel T, Kellermann M, et al. Effect of compression garments on the development of edema and soreness in delayed-onset muscle soreness. *J Sports Sci Med*. 2018;17:392–401.
24. Magoffin RD, Parcell AC, Hyldahl RD, Fellingham GW, Hopkins JT, Feland JB. Whole-body vibration as a warm-up before exercise-induced muscle damage on symptoms of delayed-onset muscle soreness in trained subjects. *J Strength Cond Res*. 2020;34(4):1123–32.
25. Ko GWY, Clarkson C. The effectiveness of acupuncture for pain reduction in delayed-onset muscle soreness: a systematic review. *Acupunct Med*. 2020;38(2):63–74.
26. MacDougall JD, Sale DG, Alway SE, Sutton JR. Muscle fiber number in biceps brachii of bodybuilders and control subjects. *J Appl Physiol*. 1984;57:1399–403.
27. McCall GE, Byrnes WC, Dickinson A, Pattany PM, Fleck SJ. Muscle fiber hypertrophy, hyperplasia, and capillary density in college men after resistance training. *J Appl Physiol*. 1996;81:2004–12.
28. Damas F, Libardi CA, Ugrinowitsch C. The development of skeletal muscle hypertrophy through resistance training: the role of muscle damage and muscle protein synthesis. *Eur J Appl Physiol*. 2018;118(3):485–500.
29. Larsson L, Tesch PA. Motor unit fibre density in extremely hypertrophied skeletal muscles in man: electrophysiological signs of muscle fibre hyperplasia. *Eur J Appl Physiol*. 1986;55:130–6.
30. Kadi F, Eriksson A, Holmner S, Thornell LE. Effects of anabolic steroids on the muscle cells of strength-trained athletes. *Med Sci Sports Exerc*. 1999;31(11):1528–34.
31. Adams GR, Haddad F, Baldwin KM. Time course of changes in markers of myogenesis in overloaded rat skeletal muscles. *J Appl Physiol*. 1999;87(5):1705–12.
32. Kelley G. Mechanical overload and skeletal muscle fiber hyperphasia: a meta-analysis. *J Appl Physiol*. 1996;81(4):1584–8.
33. Muscolino JE. *Kinesiology: The Skeletal System and Muscle Function*. St. Louis (MO): Mosby Elsevier; 2006.

34. Cosman F, de Beur SJ, Leboff MS, et al. Clinician's guide to prevention and treatment of osteoporosis. *Osteoporos Int*. 2014;25(10):2359–81.
35. Kanis JA, Melton LJ III, Christiansen C, Johnston CC, Khaltaev N. The diagnosis of osteoporosis. *J Bone Miner Res*. 1994;9:1137–41.
36. Kohrt WM, Bloomfield SA, Little KD, Nelson ME, Yingling VR. Physical activity and bone health. *Med Sci Sports Exerc*. 2004;36(11):1985–96.
37. French SA, Fulkerson JA, Story M. Increasing weight-bearing physical activity and calcium intake for bone mass growth in children and adolescents: a review of intervention trials. *Prev Med*. 2000;31(1):722–31.
38. Maddalozzo GF, Snow CM. High intensity resistance training: effects on bone in older men and women. *Calcif Tissue Int*. 2000;66:399–404.
39. Pagnotti GM, Styner M, Uzer G, et al. Combating osteoporosis and obesity with exercise: leveraging cell mechanosensitivity. *Nat Rev Endocrinol*. 2019;15(6):339–55.
40. Sale C, Elliott-Sale KJ. Nutrition and athlete bone health. *Gatorade Sports Sci Exch*. 2020;29(201):1–7.
41. Heaney RP. The role of calcium in prevention and treatment of osteoporosis. *Phys Sportsmed*. 1987;15(11):83–8.
42. Georgopoulos D, Matamis D, Routsi C, et al. Recombinant human erythropoietin therapy in critically ill patients: a dose-response study. *Crit Care*. 2005;9:R508–15.
43. Gledhill N, Warburton D, Jamnik V. Hemoglobin, blood volume, cardiac function, and aerobic power. *Can J Appl Physiol*. 1999;24(1):54–65.
44. Birkeland KI, Hemmersbach P. The future of doping control in athletes. *Sports Med*. 1999;28(1):25–33.
45. Heuberger JAAC, Posthuma JJ, Ziagkos D, et al. Additive effect of erythropoietin use on exercise-induced endothelial activation and hypercoagulability in athletes. *Eur J Appl Physiol*. 2020;120(8):1893–904.
46. Levine BD, Stray-Gundersen J. "Living high-training low": effect of moderate-altitude acclimatization with low-altitude training on performance. *J Appl Physiol*. 1997;83:102–12.
47. Centers for Disease Control and Prevention. Leading causes of death: 2017. 2020. Available from: http://www.cdc.gov/nchs/fastats/leading-causes-of-death.htm.
48. Blair SN, Kampert JB, Kohl HW, et al. Influences of cardiorespiratory fitness and other precursors on cardiovascular disease and all-cause mortality in men and women. *J Am Med Assoc*. 1996;276:205–10.
49. Cooper R, Cutler J, Desvigne-Nickens P, et al. Trends and disparities in coronary heart disease, stroke, and other cardiovascular diseases in the United States: findings of the national conference on cardiovascular disease prevention. *Circulation*. 2000;102:3137–47.
50. McCullough ML, Feskanich D, Rimm EB, et al. Adherence to the Dietary Guidelines for Americans and risk of major chronic disease in men. *Am J Clin Nutr*. 2000;72:1223–31.
51. McCullough ML, Feskanich D, Stampfer MJ, et al. Adherence to the Dietary Guidelines for Americans and risk of major chronic disease in women. *Am J Clin Nutr*. 2000;72:1214–22.
52. Richardson CR, Kriska AM, Lantz PM, Hayward RA. Physical activity and mortality across cardiovascular disease risk groups. *Med Sci Sports Exerc*. 2004;36(11):1923–9.
53. Sesso HD, Paffenbarger RS, Ha T, Lee IM. Physical activity and cardiovascular disease risk in middle-aged and older women. *Am J Epidemiol*. 1999;150(4):408–16.
54. Franklin BA, Brinks J. Cardiac rehabilitation: underrecognized/underutilized. *Curr Treat Options Cardiovasc Med*. 2015;17(62):1–18.
55. Jacobs RA, Rasmussen P, Siebenmann C, et al. Determinants of time trial performance and maximal incremental exercise in highly trained endurance athletes. *J Appl Physiol*. 2011;111(5):422–30.
56. Evans SL, Davy KP, Stevenson ET, Seals DR. Physiological determinants of 10-km performance in highly trained female runners of different ages. *J Appl Physiol*. 1995;78(5):1931–41.
57. Hagberg JM, Moore GE, Ferrell RE. Specific genetic markers of endurance performance and VO_2max. *Exerc Sport Sci Rev*. 2001;29(1):15–9.
58. Parsons JP, Mastronarde JG. Exercise induced asthma. *Curr Opin Pulm Med*. 2009;15(1):25–8.
59. Kovan JR, Moeller JL. Respiratory system. In: McKeag DB, Moeller JL, editors. *ACSM's Primary Care Sports Medicine*. 2nd ed. Philadelphia (PA): Lippincott, Williams, & Wilkins; 2007. p. 165–71.
60. Hough DO, Dec KL. Exercise-induced asthma and anaphylaxis. *Sports Med*. 1994;18:162–72.
61. Centers for Disease Control and Prevention. Hypertension. 2018. Available from: http://www.cdc.gov/nchs/fastats/hypertension.htm.
62. James PA, Oparil S, Carter BL, et al. 2014 Evidence-based guidelines for the management of high blood pressure in adults. *JAMA*. 2014;311(5):507–20.

63. American College of Sports Medicine. *ACSM's Guidelines for Exercise Testing and Prescription*. 11th ed. Philadelphia (PA): Lippincott, Williams, and Wilkins; 2021.
64. Krakoff LR. Diuretics for hypertension. *Circulation*. 2005;112(10):e127–9.
65. Prentice WE. *Principles of Athletic Training: A Competency-based Approach*. 15th ed. New York (NY): McGraw-Hill Companies; 2014.
66. Lipkin M, Reddy B, Newmark H, Lamprecht SA. Dietary factors in human colorectal cancer. *Annu Rev Nutr*. 1999;19:545–86.
67. Park Y, Hunter DJ, Spiegelman D, et al. Dietary fiber intake and risk of colorectal cancer: a pooled analysis of prospective cohort studies. *JAMA*. 2005;294(22):2849–57.
68. Romaneiro S, Parekb N. Dietary fiber intake and colorectal cancer risk. *Top Clin Nutr*. 2012;27(1):41–7.
69. Bingham SA, Day NE, Luben R, et al. Dietary fibre in food and protection against colorectal cancer in the European Prospective Investigation into cancer and nutrition (EPIC): an observational study. *Lancet*. 2003;361:1496–501.
70. Kaaks R, Riboli E. Colorectal cancer and intake of dietary fibre: a summary of the epidemiological evidence. *Eur J Clin Nutr*. 1995;49:S10–7.
71. Nomura AMY, Hankin JH, Henderson BE, et al. Dietary fiber and colorectal cancer risk: the multiethnic cohort study. *Cancer Causes Control*. 2007;18:753–64.
72. Fuchs CS, Giovannucci EL, Colditz GA. Dietary fiber and the risk of colorectal cancer and adenoma in women. *N Engl J Med*. 1999;340:169–76.
73. Mai V, Flood A, Peters U, Lacey N, Schairer C, Schatzkin A. Dietary fibre and risk of colorectal cancer in the Breast Cancer Detection Demonstration Project (BCDDP) follow-up cohort. *Int J Epidemiol*. 2003;14:234–9.
74. McCullough ML, Robertson AS, Chao A. A prospective study of whole grains, fruits, vegetables and colon cancer risk. *Cancer Causes Control*. 2003;14:959–70.
75. Sanjoaquin MA, Appleby PN, Thorogood M, Mann JI, Key TJ. Nutrition, lifestyle and colorectal cancer incidence: a prospective investigation of 10998 vegetarians and non-vegetarians in the United Kingdom. *Br J Cancer*. 2004;90:118–21.
76. Terry P, Giovannucci E, Michels KB. Fruit, vegetables, dietary fiber, and risk of colorectal cancer. *J Natl Cancer Inst*. 2001;93:525–33.
77. Platz EA, Giovannucci E, Rimm EB. Dietary fiber and distal colorectal adenoma in men. *Cancer Epidemiol Biomark Prev*. 1997;6:661–70.
78. Alberts DS, Martinez ME, Roe DJ. Lack of effect of a high-fiber cereal supplement on the recurrence of colorectal adenomas: Phoenix Colon Cancer Prevention Physician's Network. *N Engl J Med*. 2000;342:1156–62.
79. MacLennan R, Macrae F, Bain C. Randomized trial of intake of fat, fiber, and beta carotene to prevent colorectal adenomas: the Australian Polyp Prevention Project. *J Natl Cancer Inst*. 1995;87:1760–6.
80. McKeown-Eyssen GE, Bright-See E, Bruce WR. A randomized trial of low fat high fibre diet in the recurrence of colorectal polyps: Toronto Polyp Prevention Group. *J Cancer Epidemiol*. 1994;47:525–36.
81. Schatzkin A, Lanza E, Corle D. Lack of effect of a low-fat, high-fiber diet on the recurrence of colorectal adenomas. *N Engl J Med*. 2000;342:1149–55.
82. Centers for Disease Control and Prevention Web site [Internet]. Atlanta (GA): Centers for Disease Control and Prevention; [cited 2021]. Available from: http://www.cdc.gov.
83. Coggan AR, Coyle EF. Reversal of fatigue during prolonged exercise by carbohydrate infusion or ingestion. *J Appl Physiol*. 1987;63:2388–95.
84. Coggan AR, Coyle EF. Effect of carbohydrate feeding during high-intensity exercise. *J Appl Physiol*. 1988;65:1703–9.
85. Haub MD, Potteiger JA, Jacobsen DJ, Nau KL, Magee LM, Comeau MJ. Glycogen replenishment and repeated short duration high intensity exercise: effect of carbohydrate ingestion. *Int J Sport Nutr Exerc Metab*. 1998;9:406–15.
86. Duchman SM, Ryan AJ, Schedl HP, Summers RW, Bleiler TL, Gisolfi CV. Upper limit for intestinal absorption of a dilute glucose solution in men at rest. *Med Sci Sports Exerc*. 1997;29:482–8.
87. McConell GK, Fabris S, Proietto J, Hargreaves M. Effect of carbohydrate ingestion on glucose kinetics during exercise. *J Appl Physiol*. 1994;77:1537–41.
88. Adopo E, Peronnet F, Massicotte D, Brisson GR, Hillaire-Marcel C. Respective oxidation of exogenous glucose and fructose given in the same drink during exercise. *J Appl Physiol*. 1994;76:1014–9.
89. Coyle EF, Hamilton MT, Gonzalez-Alonso J, Montain SJ, Ivy JL. Carbohydrate metabolism during intense exercise when hyperglycemic. *J Appl Physiol*. 1991;70:834–40.

90. Hargreaves M. Metabolic responses to carbohydrate and lipid supplementation during exercise. In: Maughan RJ, Shirreffs SM, editors. *Biochemistry of Exercise IX*. 9th ed. Champaign (IL): Human Kinetics; 1996. p. 421–9.
91. Zimmet P, Boyko EJ, Collier GR, Courten M. Etiology of the metabolic syndrome: potential role of insulin resistance, leptin resistance, and other players. *Ann N Y Acad Sci*. 2000;892:25–44.
92. Barnard RJ, Roberts CK, Varon SM, Berger JJ. Diet-induced insulin resistance precedes other aspects of the metabolic syndrome. *J Appl Physiol*. 1998;84:1311–5.
93. Mark AL, Correia M, Morgan DA, Shaffer RA, Haynes WG. Obesity induced hypertension: new concepts from the emerging biology of obesity. *Hypertension*. 1999;33:537–41.
94. Buckely WE, Yesalis CE, Friedl KE, Anderson WA, Streit AL, Wright JE. Estimated prevalence of anabolic steroid use among male high school seniors. *JAMA*. 1988;260(23):3441–5.
95. DuRant R. Use of multiple drugs among adolescents who use anabolic steroids. *N Engl J Med*. 1993;328:922–6.
96. Hoffman JR, Kraemer WJ, Bhasin S, et al. Position stand on androgen and human growth hormone use. *J Strength Cond Res*. 2009;23(5):S1–59.
97. Green GA, Uryasz FD, Petr TA, Bray CD. NCAA study of substance use habits of college student athletes. *Clin J Sport Med*. 2001;11(1):51–6.
98. Windsor R, Dumitru D. Prevalence of anabolic steroid use by male and female athletes. *Med Sci Sports Exerc*. 1989;21(5):494–7.
99. Yesalis CE. *Anabolic Steroids in Sport and Exercise*. 2nd ed. Champaign (IL): Human Kinetics; 2000.
100. Cadwallader AB, Murray B. Performance-enhancing drugs I: understanding the basics of testing for banned substances. *Int J Sport Nutr Exerc Metab*. 2015;25(4):396–404.
101. Shephard RJ. *Physical Activity, Training and the Immune Response*. Carmel (IN): Cooper Publishing Group; 1997.
102. Gleeson M. Immune function in sport and exercise. *J Appl Physiol*. 2007;103:693–9.
103. Evenson KR, Stevens J, Cai J, Thomas R, Thomas O. The effect of cardiorespiratory fitness and obesity on cancer mortality in women and men. *Med Sci Sports Exerc*. 2003;35(2):270–7.
104. Hoffman-Goetz L, Husted J. Exercise, immunity, and colon cancer. In: Hoffman-Goetz L, editor. *Exercise and Immune Function*. 1st ed. Boca Raton (FL): CRC Press, Inc.; 1996. p. 179–97.
105. Dorn J, Vena J, Brasure J, Freudenheim J, Graham S. Lifetime physical activity and breast cancer risk in pre-and postmenopausal women. *Med Sci Sports Exerc*. 2003;35(2):278–85.
106. Slattery ML, Schumacher MC, Smith KR, West DW, Abd-Elghany N. Physical activity, diet, and risk of colon cancer in Utah. *Am J Epidemiol*. 1988;128(5):989–99.
107. Bernstein L. Exercise and breast cancer prevention. *Curr Oncol Rep*. 2009;11(6):490–6.
108. Ferry A. Exercise and autoimmune diseases. In: Hoffman-Goetz L, editor. *Exercise and Immune Function*. 1st ed. Boca Raton (FL): CRC Press, Inc.; 1996. p. 163–78.
109. Crist DM, Mackinnon LT, Thompson RF, Atterborn HA, Egan PA. Physical exercise increases natural cellular-mediated tumor cytotoxicity in elderly women. *Gerontology*. 1989;35:66–71.
110. Mazzeo RS. Exercise, immunity, and aging. In: Hoffman-Goetz L, editor. *Exercise and Immune Function*. 1st ed. Boca Raton (FL): CRC Press, Inc.; 1996. p. 199–241.
111. Nieman DC. Exercise, infection, and immunity. *Int J Sports Med*. 1994;15(3):S131–41.
112. Fitzgerald L. Overtraining increases the susceptibility to infection. *Int J Sports Med*. 1991;12(1):S5–8.
113. Mountjoy M, Junge A, Slysz J, Miller J. An uneven playing field: athlete injury, illness, load, and daily training environment in the year before the FINA (Aquatics) World Championships, 2017. *Clin J Sport Med*. 2019. doi:10.1097/JSM.0000000000000814.
114. Hausswirth C, Louis J, Aubry A, Bonnet G, Duffield R, Le Meur Y. Evidence of disturbed sleep and increased illness in overreached endurance athletes. *Med Sci Sports Exerc*. 2014;46(5):1036–45.
115. Nieman DC, Nehlsen-Cannarella SL, Markoff PA, et al. The effects of moderate exercise training on natural killer cells and acute upper respiratory tract infections. *Int J Sports Med*. 1990;11(6):467–73.
116. Bergendiova K, Tibenska E, Majtan J. Pleuran (beta-glucan from *Pleurotus ostreatus*) supplementation, cellular immune response and respiratory tract infections in athletes. *Eur J Appl Physiol*. 2011;111(9):2033–40.
117. Davis JM, Murphy EA, Brown AS, Carmichael MD, Ghaffar A, Mayer EP. Effects of oat beta-glucan on innate immunity and infection after exercise stress. *Med Sci Sports Exerc*. 2004;36(8):1321–7.
118. Nieman DC, Henson DA, McMahon M, et al. Beta-glucan, immune function, and upper respiratory tract infections in athletes. *Med Sci Sports Exerc*. 2008;40(8):1463–71.

119. Castell LM, Nieman DC, Bermon S, Peeling P. Exercise-induced illness and inflammation: can immunonutrition and iron help? *Int J Sport Nutr Exerc Metab*. 2019;29(2):181–8.
120. Pedersen BK, Nieman DC. Exercise immunology: integration and regulation. *Immunol Today*. 1998;19:1–3.
121. Horowitz JF. Regulation of lipid mobilization and oxidation during exercise in obesity. *Exerc Sport Sci Rev*. 2001;29(1):42–6.
122. van Baak MA. Exercise training and substrate utilization in obesity. *Int J Obes*. 1999;23(3):S11–7.
123. Robinson SL, Hattersley J, Frost GS, Chambers ES, Wallis GA. Maximal fat oxidation during exercise is positively associated with 24-hour fat oxidation and insulin sensitivity in young, healthy men. *J Appl Physiol*. 2015;118(11):1415–22.
124. Marra M, Scalfi L, Contaldo F, Pasanisi F. Fasting respiratory quotient as a predictor of long-term weight changes in non-obese women. *Ann Nutr Metab*. 2004;48(3):189–92.
125. Urhausen A, Coen B, Weiler B, Kindermann W. Individual anaerobic threshold and maximum lactate steady state. *Int J Sports Med*. 1993;14(3):134–9.
126. Westerblad H, Allen DG, Lannergren J. Muscle fatigue: lactic acid or inorganic phosphate the major cause? *News Physiol Sci*. 2002;17:17–21.
127. Bergman BC, Wolfel EE, Butterfield GE, et al. Active muscle and whole body lactate kinetics after endurance training in men. *J Appl Physiol*. 1999;87(5):1684–96.
128. Gladden LB. Lactate metabolism: a new paradigm for the third millennium. *J Physiol*. 2004;558(1):5–30.
129. Heck H, Mader A, Hess G, Mucke S, Muller R, Hollman W. Justification of the 4-mmol/lactate threshold. *Int J Sports Med*. 1985;6:117–30.
130. Bassett DR, Howley ET. Limiting factors for maximum oxygen uptake and determinants of endurance performance. *Med Sci Sports Exerc*. 2000;32(1):70–84.
131. Potteiger JA. Aerobic endurance training. In: Baechle TR, Earle RW, editors. *Essentials of Strength Training and Conditioning*. 2nd ed. Champaign (IL): Human Kinetics; 2000. p. 495–509.
132. Stepto NK, Martin DT, Fallon KE, Hawley JA. Metabolic demands of intense aerobic interval training in competitive cyclists. *Med Sci Sports Exerc*. 2001;33(2):303–10.

CHAPTER 4

Exercise Physiology

After completing this chapter, you will be able to:

1. Define exercise physiology and provide examples of the relationship between exercise physiology and exercise science.
2. Identify the important historic events in the development of exercise physiology as a scientific discipline.
3. Discuss the differences between the acute and chronic responses to exercise.
4. Discuss the basic components of a training and conditioning program.
5. Describe some of the important areas of study in the field of exercise physiology.

Exercise physiology is the study of the anatomical, physiological, and functional responses and adaptations that occur during and following movement. Understanding those responses and adaptations are important for promoting physical activity and exercise in healthy individuals and those with chronic disease. Furthermore, improving the performance of athletes of all levels of ability requires an in-depth knowledge of the foundational and advanced principles of exercise physiology. Primarily concerned with the various systems of the body (see Chapter 3), exercise physiology is the study of how the systems individually and collectively respond to acute and chronic bouts of physical activity and exercise.

The use of an integrated systems approach allows students to understand the role that physical activity and exercise play in improving health and reducing the risk of disease. There is a wealth of information that supports the role of physical activity and exercise in improving health, reducing the risk of developing a chronic disease, and being an important component of rehabilitation from an injury, disability, or disease. Physical activity and exercise are effectively used, for example, to improve bone health (1), facilitate weight loss (2), control abnormal blood pressure (3), and enhance cognition and academic performance (4). A field of study called inactivity physiology has increased in popularity and is also often studied by exercise science professionals. Inactivity physiology is concerned with responses of the systems of the body to insufficient amounts of or no physical activity or exercise (5). The lack of physical activity is associated with the loss of adaptive changes and functional responses that were developed from regular physical movement of the body (6). For example, research has demonstrated that sedentary behavior can increase the amount of heart, liver, and visceral fat in individuals at a high risk of Type 2 diabetes mellitus (7).

Athletes, coaches, and other exercise science professionals can apply the basic and advanced training and conditioning principles of exercise physiology to enhance the function of the various systems of the body. In this way, training and conditioning programs can improve sport and athletic performance and increase success during competition. The identification of those factors affecting performance and subsequent improvement as a response to appropriate training is the hallmark of using exercise physiology to improve performance. In some instances, exercise physiology is used synonymously with exercise science, and many individuals often use the terms interchangeably. However as indicated in Chapter 1, exercise science is a broad term correctly used to describe the disciplines and areas of study that involve movement as a central theme.

College and university academic programs are often structured to prepare students to function as exercise physiologists in clinical, fitness, sport, or research environments. Students who are trained to work in clinical environments, such as cardiovascular and pulmonary rehabilitation, will receive more educational training in the clinical aspects of exercise physiology (see Chapter 5). Training for a professional career in fitness and sport-oriented environments will include more experiences in understanding how untrained and trained persons will respond to the demands of physical activity, regular exercise, and training for sport and athletic competition. Individuals preparing for a career in research will be

Exercise physiology Study of the functional responses and adaptations that occur during and following physical activity and exercise.

exposed to more preparation in research design and methods, and data analysis procedures. This career track typically requires additional education in the form of a Master's or doctoral degree and may involve working with whole animal and human models, with animal and human tissue samples, or with large data sets. Regardless of the career track, preparation for a professional career in exercise physiology necessitates the study of the body's systems and how they respond to acute and chronic physical activity and exercise.

 ## HISTORY OF EXERCISE PHYSIOLOGY

Although exercise physiology is relatively new as a defined body of knowledge, individuals have been interested in the physiologic responses to physical activity and exercise since the time of the ancient Greeks (8). The early historic events in medicine, physiology, exercise, and sport that have helped define exercise physiology have also influenced the development of other disciplines in exercise science as well. Many of those important historic events are described in Chapter 1, and the reader is directed to review that information for a broad overview. The material contained in this section should be considered a supplement to Chapter 1 and will contain only information specific to the recent historical development of exercise physiology.

Early Twentieth-Century Influences

One of the most significant events for the emergence of exercise physiology as a scientific discipline was the establishment, in 1891, of the Department of Anatomy, Physiology, and Physical Training within the Lawrence Scientific School at Harvard University (9,10). Faculty and leaders in the department implemented a demanding 4-year science-based curriculum that included both theory and laboratory courses in exercise physiology (9–12). Expansion of this concept of coupling laboratory activities with course content in exercise and physical training also occurred at Springfield College in Massachusetts and George Williams College in Chicago, Illinois (9). Despite the curricular offerings in exercise physiology at these universities, there was insufficient evidence that exercise physiology had established itself as an academic discipline until the opening of the Harvard Fatigue Laboratory in 1927 (Figure 4.1) (9).

The Harvard Fatigue Laboratory's primary purpose was to study the physiological, psychological, and sociological responses of industry workers to stressful stimuli (13,14). The founders of the laboratory included fatigue in the name because it was believed the term would help attract the interest and funding of business leaders and because the concept of fatigue could be understood by many individuals without being explained or defined (13). During the years from 1941 to 1947, an additional focus of the laboratory was centered on the physical fitness of military personnel and the energy cost of performing military tasks in extreme heat and cold temperatures (15). The Harvard Fatigue Laboratory attracted high-quality scientists from around the world and produced scholars and professionals who would be instrumental in shaping exercise physiology as a scientific discipline. Paradoxically, the closure of the Harvard Fatigue Laboratory, in 1947, was instrumental in facilitating the establishment of other exercise physiology laboratories across the United

FIGURE 4.1 The Harvard Fatigue Laboratory (ca. 1946). (Photo from Baker Library Historical Collections, Harvard Business School.)

States. Many of the founders of these new laboratories were well-known individuals who had close connections to the Harvard Fatigue Laboratory and helped further expand the development of exercise physiology (10,16).

Late Twentieth-Century Influences

Beginning in the 1940s, there were a number of events that significantly shaped the discipline of exercise physiology. During this time period, several peer-reviewed journals published data from experiments involving the physiological responses to exercise (10). By the late 1940s, a sufficient body of knowledge existed in exercise physiology to merit formal instructional course offerings by colleges and universities. The further development of exercise physiology occurred in response to a number of social, political, and professional factors. In the 1950s and 1960s, poor performance by American schoolchildren on physical fitness tests and the competition between the United States and the Soviet Union to put a man in space and later land a man on the moon heightened the focus on physical fitness and performance of the American population. President Eisenhower established the President's Council on Youth Fitness, and President Kennedy further expanded the role of the Council. Other instrumental factors included the passage of funding for health-related research, facilities, and educational training programs by the National Institutes of Health; the interest of the American Physiological Society (APS) in exercise-related research; the publication of the *Journal of Applied Physiology* by the APS; and the formation of the American College of Sports Medicine (ACSM) and its peer-reviewed journal *Medicine & Science in Sports & Exercise* (9). The interest in exercise and fitness of many of the early leaders of the ACSM resulted in an increased emphasis on the study of the physiologic responses to exercise (16,17).

Throughout the 1960s and 1970s, exercise physiology continued to expand and evolve. As interest in exercise physiology continued to grow, college curricula changed from traditional programs in physical education to more science-related courses and activities. Undergraduate students enrolled in courses in biochemistry and physiology and participated in a variety of laboratory activities. This change better prepared students to enter professional careers that required a knowledge base in exercise physiology. Students now had an option to pursue a career in exercise physiology rather than be trained as physical education teachers (9). This trend continues today as many physical education programs require little formal course preparation in physiology and exercise physiology. Conversely, expanded curricula in exercise physiology include topics such as exercise testing and prescription, exercise management of chronic disease, advanced study in cellular functions and mechanisms, and genetics, though requiring little or no coursework in teacher training.

Since the establishment of the ACSM, numerous organizations have emerged to provide support for other professionals with an interest in exercise physiology. In 1997, the establishment of the American Society of Exercise Physiologists (ASEP) arose from the belief in the need for an organization to solely promote exercise physiology as a health care profession (see Chapter 12). In addition to ASEP, there are numerous other organizations such as the National Strength and Conditioning Association (NSCA) and the American Association of Cardiovascular and Pulmonary Rehabilitation (AACVPR) that have promoted the study and development of exercise physiology in health, physical activity, fitness, sport, and athletic performance.

Twenty-First Century Influences

Early in the 21st century, there was a concerted effort to create a greater understanding of the role genetics plays in health and human performance. As a result, there has been a strong emergence of molecular science and genetics to unveil and identify the underlying mechanisms of exercise-related movement and regulation. Initial work involved creating a gene map for fitness and athletic performance that resulted in a series of articles published in *Medicine & Science in Sports & Exercise* between 2001 and 2009 (18). Using observational family studies, experimental twins studies, and experimental family studies, exercise physiologists have been able to create a great understanding of the role genetics plays in health and fitness. One of the most well-known studies is the HERITAGE Family Study, which exposed samples of families to a standardized exercise training protocol and subsequent examination of the genetic influence on health, fitness, and disease (18). Additional work by exercise physiologists are helping to develop an understanding of the role genetics plays in identifying successful sport and athletic performance (19).

The last decade has seen significant expansion of areas of study within exercise physiology with the specific intent to understand how the systems of the body respond to physical activity, exercise, sport, and athletic performance. For example, the role that physical activity and a healthy lifestyle has in morbidity and mortality continues to be a focused area of research (20,21), with specific attention focused on the appropriate dose of exercise to induce positive outcomes (22) and whether physical activity and exercise can reduce the need for using medications to treat disease conditions (23). In 2018, the Physical Activity Guidelines Advisory Committee Scientific Report was presented to the United States Secretary of Health and Human Services. The report provided a detailed summary of the population benefits of a more physically active America on disease prevention and health promotion. In addition to a decreased risk of mortality, higher levels of regular physical activity of moderate-to-vigorous intensity decrease the risk of many common and costly diseases in the United States (24). Additionally, the report provided evidence that physically active individuals have improved physical function and better quality of life and that small increases in regular moderate-to-vigorous physical activity, especially if made by the least physically active individuals, would significantly decrease direct and indirect medical costs (24).

Within the last decade, there has been increased attention on the use of performance enhancing drugs by athletes. In 2012, cyclist Lance Armstrong was stripped of his seven Tour de France titles for using performance enhancing drugs, and during the 2018 Seoul Olympics the entire team of Russian athletes was barred from competition owing to systematic blood doping. Even though sport governing bodies have instituted new rules and new methods of detecting the use of blood doping agents, the athletes and their trainers and coaches have continued to devise ever-more sophisticated means of blood doping and strategies to avoid detection (25). This will continue to be an important area of physiological study in the foreseeable future.

Exercise physiology continues to expand as a discipline and organize as a profession. Students of undergraduate programs with an emphasis in exercise physiology coursework can enter professional careers as certified personal trainers, health fitness instructors, exercise specialists, strength and conditioning coaches, and wellness coaches. These professions clearly indicate the role that practical applications of exercise physiology have in healthy and diseased populations. Students can further their educational training by enrolling in Master's and doctoral degree programs to become college and university teachers and researchers in public and private settings. Students of exercise physiology are also very successful in entering into professional programs and schools of medicine, dentistry, podiatry, chiropractic, physician assistant, nursing, public health, and physical and occupational therapy. The role of exercise physiology will continue to be important in our society as we continue to identify how physical activity and exercise can improve health and reduce disease risk and as we continue to seek improvements in sport and athletic performance. Table 4.1 identifies some of the important historic events in the development of exercise physiology.

> **Thinking Critically**
>
> In what ways has exercise physiology contributed to a broader understanding of the role physical activity and exercise play in the promotion of physical fitness and health, as well as the understanding of successful sport and athletic performance?

Table 4.1 Some Significant Historic Events in the Development of Exercise Physiology

YEAR	EVENT
1891	Establishment of the Department of Anatomy, Physiology, and Physical Training at Harvard University
1927	Opening of the Harvard Fatigue Laboratory at Harvard University
1948	Publication of the *Journal of Applied Physiology* by the American Physiological Society
1954	Formation of the American College of Sports Medicine
1969	*Medicine and Science in Sports* (later titled *Medicine & Science in Sports & Exercise*®) by ACSM
1978	Establishment of the National Strength and Conditioning Association
1985	Formation of the American Association of Cardiovascular and Pulmonary Rehabilitation

Table 4.1	Some Significant Historic Events in the Development of Exercise Physiology (continued)
YEAR	EVENT
1997	Establishment of the American Society of Exercise Physiologists
2007	Acceptance of the ACSM into the Federation of American Societies for Experimental Biology
2018	Physical Activity Guidelines Committee submits report to the Secretary of Health and Human Services

THE BASIS OF STUDY IN EXERCISE PHYSIOLOGY

Two primary areas of study in exercise physiology are how the body responds to acute episodes of physical activity and exercise and how regular physical activity and exercise result in chronic adaptations of the various systems of the body. Homeostasis describes the condition of the systems of the body when the body is in a resting state. When the body is subjected to a stress, such as physical activity or exercise, the body's state of homeostasis is disrupted. When the body encounters physical work, physical activity, or exercise, numerous changes occur that affect every system of the body. Understanding homeostasis is critical to understanding the acute responses and chronic adaptations to physical activity and exercise. When homeostasis is disrupted as a result of physical activity or exercise, the various systems of the body respond by either increasing or decreasing their level of activity. This coordinated response of the systems allows the body to meet the challenges of physical activity or exercise and works to return the body to homeostasis. Repeated challenges of appropriate levels of physical activity and exercise to homeostasis result in chronic adaptations to the systems of the body. These adaptations to regular physical activity and exercise provide the foundation for improvements in health, fitness, and sport and athletic performance.

Acute Responses to Physical Activity and Exercise

The acute responses of the systems of the body are those actions that occur in response to physical activity or a single bout of exercise. When an individual engages in physical movement, there are numerous challenges to homeostasis, and almost every system of the body is involved to some degree. For example, during a brisk walk on a warm summer day, there are increases in the demand for energy production in the working muscle, a need for greater blood flow through the cardiovascular and pulmonary systems, and a rise in body

Homeostasis The maintenance of relatively stable internal physiologic conditions.
Acute responses Changes in the systems of the body that occur in response to a single bout of physical activity or exercise.

temperature. In a healthy individual, the various systems of the body respond in a coordinated fashion to meet the demands of the body during exercise and return the body to homeostasis following the completion of exercise. In this example, there would be an increase in glucose and fatty acid uptake by working skeletal muscle, an increase in heart rate, stroke volume, and cardiac output by the heart, an increase in the rate and depth of breathing by the lungs, and an increase in skin blood flow to assist with temperature regulation. Exercise science professionals must have a solid understanding of how the body systems respond to acute bouts of physical activity and exercise so that safe and effective physical activity and exercise programs can be prescribed for the individuals with which they are working. Table 4.2 illustrates how some of the major systems of the body respond to a single bout of exercise. Research and the study of the acute responses of the body during physical activity, exercise, sport, and athletic performance have allowed for a much greater understanding of how the systems of the body control the body's internal environment and response to challenges to homeostasis. Table 4.3 provides examples of areas of research in how the different systems of the body respond in an acute manner to physical activity or exercise.

Table 4.2 Acute Responses of Some Body Systems to a Single Bout of Physical Activity and Exercise

BODY SYSTEM	ACUTE RESPONSES
Cardiovascular system	Increases in heart rate, stroke volume, cardiac output, blood pressure, and a redirection of blood flow to the working tissues of the body
Pulmonary system	Increases in air movement into and out of the lungs and increased blood flow through the lungs
Muscular system	Increases in force production, utilization and production of energy, and heat production
Endocrine system	Increases in the release of epinephrine and norepinephrine

Table 4.3 Examples of Research in the Acute Responses of Some Body Systems to Physical Activity and Exercise

BODY SYSTEM	ACUTE RESPONSE
Cardiovascular system	What factors regulate the local control of blood flow to working muscle?
Pulmonary system	Does pulmonary ventilation limit maximal effort exercise?
Muscular system	What factors contribute to a loss of force production in skeletal muscle during exercise?
Endocrine system	How do different levels of carbohydrate, fat, and protein intake affect the release of the hormones insulin and glucagon?

Chronic Adaptations to Physical Activity and Exercise

The term **chronic adaptations** to physical activity and exercise refers to the adaptations that occur in the systems of the body with repeated regular physical activity and exercise. If the physical activity or exercise is performed on a regular basis and is of sufficient **intensity, duration**, and **frequency**, there will be positive adaptations of the systems of the body. These adaptations, often called training responses, occur with the primary purpose of improving the body's response to the challenges imposed by the physical activity or exercise. Table 4.4 provides some examples of the responses of the systems of the body to chronic physical activity and exercise.

In general, the chronic adaptations to physical activity and exercise improve various functions of the body while at rest and during submaximal and maximal physical work. For example, if an individual participates in a regular walking program for exercise there will be adaptations to the systems of the body that result from the chronic exercise. For a given level of exercise intensity, regular walking would result in the greater use of fat as an energy source, a lower heart rate and higher stroke volume, a more efficient exchange of oxygen and carbon dioxide in the lungs, and an improved regulation of body temperature. It is important to note that, as with the acute responses to exercise, the chronic adaptations are highly variable and strongly influenced by the genetic predisposition of the individual (26) and the type of physical activity and exercise performed (27). Table 4.5 provides some

Table 4.4	Chronic Adaptations of Some Body Systems to Physical Activity and Exercise
BODY SYSTEM	**CHRONIC ADAPTATIONS**
Cardiovascular system	Increases in stoke volume and cardiac output and decreases in heart rate at the same absolute workload
Pulmonary system	Improved air movement into and out of the lungs and increased blood flow through the lungs at the same absolute workload
Muscular system	Increased energy production from fat and decreased lactic acid formation at the same absolute workload
Endocrine system	Decreased release of epinephrine and norepinephrine at the same absolute workload

Chronic adaptations Changes in the systems of the body that occur in response to repeated regular physical activity and exercise.
Intensity How hard is the exercise, usually compared to maximal effort or maximal heart rate.
Duration How long the exercise is performed.
Frequency How often the exercise is conducted, usually measured in days per week.

Table 4.5	Examples of Research in the Study of Chronic Adaptations of Some Body Systems to Physical Activity and Exercise
BODY SYSTEM	CHRONIC ADAPTATIONS
Cardiovascular system	What factors influence the reduction in blood pressure in hypertensive individuals following different types of exercise training?
Pulmonary system	How does exercise training improve blood flow through the entire lung?
Muscular system	How does regular exercise influence the uptake of glucose by skeletal muscle in persons with diabetes?
Endocrine system	What hormones cause the shift in energy metabolism following training in athletes?

examples of areas of study in the chronic adaptations of the different systems of the body as they respond to regular physical activity and exercise. Exercise physiology is grounded in the study and understanding of the acute responses and chronic adaptations to physical activity and exercise. Much of the knowledge base in the study of exercise physiology and exercise science, in general, is the result of research and observation into the acute responses and chronic adaptations of the systems of the body to physical activity and exercise.

> **Thinking Critically**
>
> Why do athletes and coaches need to be knowledgeable about the acute responses and chronic adaptations to endurance exercise training and resistance exercise training?

Training and Conditioning Programs

Too often, **training and conditioning programs** have been designed based on what programs have been used by successful individuals, athletes, and coaches, as well as famous individuals who propose to have some expert knowledge. Effective training and conditioning programs are based on scientific principles developed from experimental research. Challenging the various systems of the body beyond what they are normally accustomed to is the fundamental principle for improving the functioning of those systems and ultimately improving health, as well as increasing physical activity, exercise, sport, and athletic performance. If a physiological system is challenged during repeated training sessions, the system will respond by improving its functional ability to meet the demands experienced by the body during rest, physical activity, and exercise. Over time, this training response allows an individual to perform more physical activity or exercise, further improving the specific systems challenged during training.

Since most physical activity, exercise, sport, or athletic performance activities involve the various physiological systems to different degrees, it is important to understand the physiological responses to acute physical activity and exercise. A well-developed training and conditioning program allocates the appropriate amount of exercise time to match the demands of the activity and the systems being challenged. By following the basic principles

Training and conditioning programs The process of regular and repeated exercise designed to increase the functioning of the body systems for improving physical performance.

of training and conditioning, changes occur in the body that allow the person to become healthier and perform better. The physiological systems improve their function only if the training and conditioning is designed to challenge the system. The principles of **overload**, **specificity**, and **reversibility** must all be considered when designing an effective training and conditioning program. The entire body or a specific system in the body must be challenged at a level beyond what it is normally accustomed to (*i.e.*, overloaded) for a training effect to occur. As overload is applied, the system(s) of the body will adapt to the physical stress and improve function. The variables that are manipulated to provide an overload and, subsequently, a training effect are intensity, duration, and frequency. How hard, how long, and how often an individual trains will impact the speed and the extent of improvement the body systems experience in response to the training and conditioning program.

There are numerous factors that can influence the individual response to a training and conditioning program. Some examples of these factors include gender, initial fitness levels, and genetic makeup. These factors must be considered by exercise science professionals when developing a training and conditioning program as not all individuals respond in a similar manner to a training and conditioning program.

The same general approach and principles of training and conditioning can be used with both males and females. However, there are specific issues that are different between genders and must be recognized when developing a training and conditioning program. Examples of some physiological issues that must be considered when developing and implementing a training and conditioning program for males and females include differences in hormonal levels, total muscle mass, body size and composition, maximum aerobic and anaerobic power, hemoglobin concentration, muscle enzyme activity, and relative use of fat and carbohydrate as fuels (28,29).

Individuals may also differ greatly in response to a training and conditioning program because of initial fitness levels. The amount of improvement is usually greater in those individuals who are less conditioned at the beginning of a training program. However, those individuals in poor physical condition will not be able to tolerate a high training intensity or a high training frequency at the initiation of the program. Less fit individuals will need to have a less strenuous training program and likely more recovery time (28).

Genetic makeup also plays an important role in how that person responds to training. An individual with a genetic predisposition for endurance training and conditioning responds differently to endurance training when compared to an individual with a markedly different genetic profile. Furthermore, some individuals may have a positive response to training and conditioning, and others do not respond at all. While training and conditioning can improve physiological function, the upper limits of adaptations and performance will be established by the individual's genetic makeup. Once the genetic ceiling is reached, then further improvement in physiological function is difficult (30).

Overload Systems must be challenged above normal activity to respond to the training stimulus.

Specificity Must train the muscles involved in the movement and the systems that support the movement.

Reversibility When training is stopped, the training effect is quickly lost.

The components of an individual training and conditioning session should include the warm-up, specific exercise(s) to be performed, and the cool-down. The warm-up serves several important objectives, including increasing cardiac output and blood flow to the muscles that are going to be used during the activity, increasing muscle temperature, and breaking up of the connective/scar tissue within the muscle, allowing for greater range of motion. The specific exercise(s) to be performed should be designed to target the muscles and physiological systems identified for improvement. The final component of the training and conditioning session is the cool-down. Following a workout, a period of low-intensity exercise or activity should be performed. The cool-down serves several important objectives, including returning "pooled" blood from skeletal muscles to central circulation and returning the body to normal levels of physiological function. The length of the warm-up and cool-down will depend on the environmental conditions, the age and fitness level of the individual, and the training intensity and duration (28).

Components of a training and conditioning session can take many different forms, intensities, durations, and exercise modes. Generally, training programs can be categorized into improvement of endurance, power, strength, and flexibility. The specific responses of the physiological systems will be dependent on the specific focus of the training and conditioning program. A presentation of the various types of training and conditioning programs is beyond the scope of this book, and as an undergraduate student you will likely receive this knowledge in coursework specifically designed for this topic. Students are recommended to read and review additional sources (28,31).

Training for the improvement of health, exercise, sport, and athletic performance requires a well-developed and scientifically based program. A training program should be developed in conjunction with periodic assessment and structured to enhance the strengths and improve the weaknesses of the individual. A combination of training practices should be used so that all physiological systems involved in successful performance are overloaded and challenged to respond with positive adaptations.

How might coursework in exercise science prepare an individual for a career as an exercise physiologist, exercise specialist, or strength and conditioning coach?

AREAS OF STUDY IN EXERCISE PHYSIOLOGY

Working as an exercise specialist, personal trainer, strength and conditioning coach, or a clinical exercise physiologist in an allied health profession requires individuals to draw upon their educational background and experiences to perform their job to the best level possible. Exercise physiology contributes to a complete understanding of health, physical activity, exercise, sport, and athletic performance through a wide variety of areas too numerous to cover them all adequately in an introductory exercise science textbook. The areas discussed in the following sections are some of the primary interest areas in exercise physiology. The areas selected are by no means meant to infer an inclusive list or indicate greater importance than a topic area not covered but are meant to provide a sampling of areas in which the knowledge of exercise physiology is generated and used. It is hoped that as you read through these sections, you will gain a better understanding of how the various systems interact to control the body functions and how various aspects of physical activity and exercise influence these responses.

Factors Controlling Substrate Metabolism

Exercise scientists and other allied health professionals have long been interested in the different factors that control energy production and utilization in the body during exercise. The energy sources, often called **substrates**, used by the body are carbohydrates, fats, and proteins. Understanding how the body tissues use these macronutrients to provide energy is critical to providing advice on how to enhance physical activity and exercise and for providing recommendations to improve sport and athletic performance. At rest and in a **postabsorptive state**, the tissues of the body rely on fat and carbohydrate as the primary sources of energy, with little energy provided from protein (32). However, the percentage of energy supplied to the body from carbohydrates, fats, and protein can be influenced by numerous factors. For example, eating meals of different macronutrient content can alter the energy sources utilized during rest and exercise (33,34). Meals high in carbohydrate content can increase glucose and glycogen utilization and enhance exercise performance, whereas meals low in carbohydrate content can decrease carbohydrate utilization and often adversely affect exercise performance (35,36). Food restriction, either through dieting or starvation and prolonged exercise can alter substrate utilization by promoting a greater contribution of energy supplied from protein (36,37).

The intensity and duration of physical activity and exercise are two primary factors affecting substrate utilization. During low-intensity physical activity and exercise, fat is the primary substrate for energy production. As physical activity or exercise intensity increases, there is a greater reliance on carbohydrate as an energy source. When exercise intensity reaches close to 100% of maximal oxygen consumption ($\dot{V}O_{2max}$), almost 100% of the energy is provided by the metabolism of carbohydrates (38,39). There is an exercise intensity that results in a point being reached when there is a shift from fat to carbohydrate as the predominant energy substrate. This point has been labeled the **crossover point** (38). During exercise, there are several factors that cause the shift in substrate utilization to occur. As exercise intensity increases, more fast glycolytic muscle fibers are recruited to provide increased power to the muscular contraction, and these fibers rely predominately on carbohydrate as the energy substrate (40). The second factor directly affecting the shift in substrate utilization is an increase in the concentration of the hormone epinephrine. As the exercise intensity increases, the sympathetic nervous system and the adrenal medulla increase the release of epinephrine into the circulation. Epinephrine increases carbohydrate metabolism by increasing muscle glycogen breakdown and inhibiting the release of fat from adipose tissue (40). Improvements in cardiorespiratory fitness as a result of increased levels of regular physical activity or exercise result in a shift in substrate utilization so that greater fat utilization occurs at higher exercise intensities (41). This results in less carbohydrate and more fat being used during physical activity or exercise (38). Figure 4.2 provides an illustration of the role of exercise intensity on substrate utilization and the crossover point.

Substrates A source of energy for the cells of the body.
Postabsorptive state The condition following the complete absorption of a meal.
Crossover point The point where the body receives more of its energy from carbohydrate rather than fat.

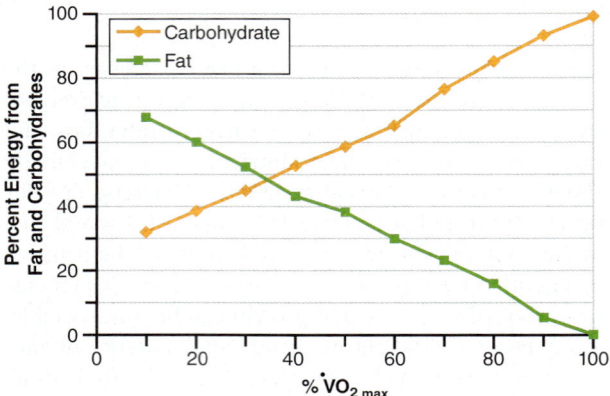

FIGURE 4.2 Effect of exercise intensity on substrate utilization and the crossover point (28).

During long-duration physical activity or exercise, there is a gradual increase in the utilization of fat as an energy substrate. This results in a decrease in the use of carbohydrates as a fuel source (40). There are several factors that cause the shift in fuel utilization during prolonged exercise to occur. The use of fat as an energy source in skeletal muscle depends in part on the release of fat from adipose tissue and delivery of fat by the circulatory system to working skeletal muscle (38). The breakdown and release of fat from adipose is controlled by enzymes called lipases. The activity level of these enzymes is influenced by the hormones epinephrine, norepinephrine, and glucagon. During long-duration exercise, the concentration of these hormones in the blood increases, resulting in an increase in fatty acid release into the circulation. The delivery of the fats to the working tissues occurs as a result of increased blood flow to the muscles that are active during physical activity and exercise (39). Fat mobilization can be inhibited by high blood levels of lactic acid and the hormone insulin. If during long-duration exercise there is an increase in the exercise intensity so that blood lactic acid levels increase, this could result in a decrease in fat mobilization and utilization as a fuel substrate (38). Insulin also inhibits fat mobilization, but generally during long-duration exercise there is a decline in blood insulin levels (38). Table 4.6 identifies some of the prominent factors affecting fuel utilization during exercise.

Table 4.6	Various Factors Affecting Fuel Utilization During Exercise
FACTOR	GENERAL RESPONSE
Increased exercise intensity	Increased use of carbohydrate
Increased exercise duration	Increased use of fat
Epinephrine and norepinephrine	Increased use of carbohydrate
Glucagon	Increased use of fat
Lactic acid	Decreased use of fat

Implications for Physical Activity and Exercise

The control of factors affecting substrate metabolism has implications for exercise science professionals designing physical activity and exercise programs. For example, if the intention is to decrease body fat and body weight, then physical activity and exercise that promote the use of fat as the substrate should be emphasized. It is often recommended that exercise intensity be kept low to promote the use of fat as a fuel substrate because there is a greater percentage of fat utilized at low intensities. However, this recommendation does not consider the total amount of energy expended during the exercise session. At higher exercise intensities, more total fat could be used as a fuel source even though the percentage of fat being utilized may be less than that which occurs at a lower intensity of exercise (28). The maximal rate of fat oxidation is defined as the highest observed use of fat as an energy source during oxidative metabolism (42). The maximal rate of fat oxidation is typically determined during a graded exercise test to exhaustion (42). Information provided from this type of testing can be used to develop effective exercise prescriptions and to monitor changes to fat oxidation rates during training. Another consideration is whether the consumption of a preexercise meal or beverage high in carbohydrate is necessary. If a meal or glucose beverage is consumed too close to the start of exercise, there will likely be an increase in the insulin concentration in the blood, which will decrease the mobilization of fat and result in a reduction in the use of fat as a fuel substrate (43,44).

Implications for Sport and Athletic Performance

Athletes who participate in sport or athletic competitions that are of prolonged duration (*e.g.*, triathlons, long-distance running, and cycling) must be careful about maintaining blood glucose and muscle glycogen concentrations during exercise. During exercise performed at moderate to high intensity, there is a significant reliance on muscle glycogen as the carbohydrate energy source. During prolonged exercise that lasts 3 to 4 hours or longer, both blood glucose and muscle glycogen contribute to energy production in an equal percentage (28). As muscle glycogen concentration decreases, the use of blood glucose increases so that when an athlete is nearing a level of muscle glycogen depletion, blood glucose may be the predominate source of carbohydrate as a fuel (45).

If blood glucose concentration is not maintained during long-duration exercise, the athlete could become hypoglycemic, which may lead to fatigue, a decrease in performance, and possible serious medical conditions. Consequently, it is important to maintain blood glucose concentrations during exercise (28,46). The timing, type, and concentration of carbohydrate consumption are two factors that are critical for helping to maintain blood glucose concentrations during long-duration exercise. It is suggested that a carbohydrate feeding of between 1 and 5 g of carbohydrate per kilogram body mass should occur between 1 and 4 hours prior to the start of exercise (47). Although this procedure results in an increased use of carbohydrate as a substrate early in the exercise bout, the additional carbohydrate consumed will help maintain blood glucose for a longer period of time during

Hypoglycemic A condition of abnormally low blood glucose levels.

FIGURE 4.3 Carbohydrate consumption during exercise can affect substrate utilization and performance. (Shutterstock.)

exercise (36,48). Carbohydrate should also be consumed during exercise (Figure 4.3), and this procedure has been shown to delay the onset of fatigue and improve long-duration exercise performance (47–49). The best type of carbohydrate to consume during exercise is glucose, sucrose, or **glucose polymer solutions** at a concentration of approximately 6% to 8% (36,50,51). Recent evidence supports the use of fructose co-ingestion to increase carbohydrate delivery to the tissues of the body (52). Exercise science professionals continue to examine ways to decrease the potential adverse effects of muscle glycogen depletion and hypoglycemia through the ingestion of glucose, glucose polymer beverages, energy gels, and bars prior to and during exercise. Maintaining blood glucose and muscle glycogen levels during exercise will likely lead to an increase in sport and athletic performance.

Muscle Control of Glucose Uptake

The control of blood glucose uptake by tissues of the body at rest and during physical activity and exercise has important implications for overall health and athletic performance and is of considerable interest to exercise science professionals. The body closely regulates energy utilization so that as skeletal muscle uses energy during physical activity or exercise there is an increase in energy production by the body tissues. As the need for energy increases, there are a variety of changes that occur in the body. For example, the use of carbohydrates as an energy source in muscle results in the breakdown and utilization of muscle glycogen and an increase in the uptake of glucose from the blood into the individual muscle cells. There is also an increase in glycogen breakdown and glucose production in the liver. The liver subsequently releases glucose into the blood in an effort to maintain blood glucose

Glucose polymer solution A beverage that contains multiple glucose molecules linked together in solution.

concentration and to provide glucose to the working tissues (29). Various hormones, including epinephrine, norepinephrine, and glucagon, can stimulate the breakdown of liver glycogen (38).

The movement of glucose from the blood into the cell predominantly depends on the interaction of the hormone insulin and a glucose transport protein. Insulin is released from the pancreas to help control blood glucose levels in the body and help glucose enter into the cells of the body. Glucose enters the cells by interacting with insulin and a glucose transport protein on the cell membrane. There are several glucose transport proteins, but the one specific to skeletal muscle is called the **glucose transport protein 4** or GLUT 4. When a muscle cell needs glucose, it increases the number of GLUT 4 proteins in the cell membrane (53). Several factors stimulate an increase in GLUT 4 proteins, including insulin, increased blood flow to skeletal muscle, increased glucose concentration, and muscle contractions (54,55). Figure 4.4 illustrates the process by which insulin facilitates the GLUT 4 transport protein to take glucose into the cell.

FIGURE 4.4 The process by which insulin and GLUT 4 facilitate glucose uptake by cells of the body.

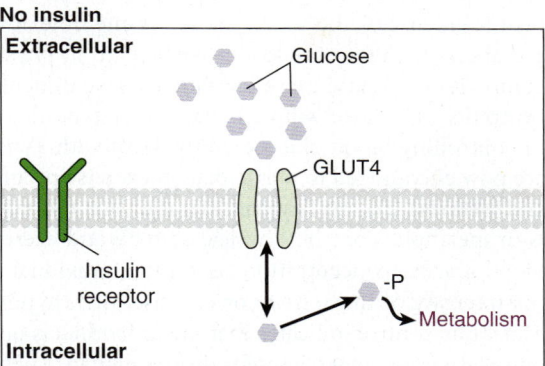

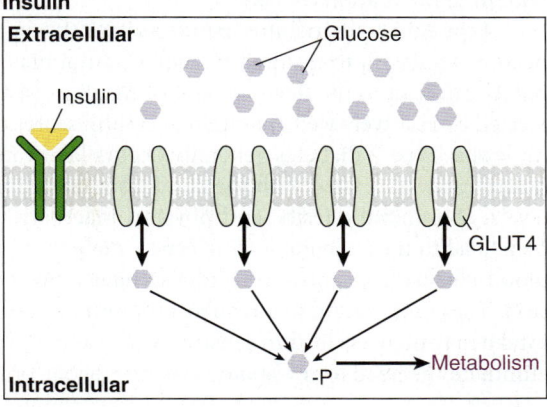

Glucose transport protein 4 A type of protein molecule that works with insulin to facilitate glucose uptake by skeletal muscle fibers.

Once glucose enters a muscle fiber, it can be stored as muscle glycogen or immediately used to produce energy. The stored muscle glycogen can be broken down later for energy in the muscle fiber. The muscle glycogen and blood glucose are metabolized through a series of chemical reactions and pathways to provide energy to support the functions of the contracting muscle fibers. As mentioned previously, the exercise intensity is one of the strongest factors influencing the use of carbohydrate as a fuel source. As exercise intensity increases, there is a need for adenosine triphosphate (ATP) to be produced at a fast rate. As a result, there is an increased reliance on carbohydrates as an energy source because the breakdown of glucose and glycogen provide ATP at a faster rate (Figure 4.2).

Implications for Physical Activity and Exercise

The role of physical activity and exercise in regulating glucose metabolism is of importance in promoting health and wellness. Diabetes mellitus is a disease condition whereby insufficient insulin is produced by the pancreas (Type 1) or the insulin does not promote the uptake of glucose by the cell (Type 2). At one time, it was believed that individuals with Type 1 diabetes mellitus should not participate in physical activity, exercise, sport, or athletic competition because these individuals have difficulty controlling their blood glucose concentration. However, with a better understanding of the role of glucose transport proteins in controlling blood glucose, individuals with Type 1 diabetes who are otherwise healthy are now encouraged to participate in exercise programs and sports to improve their health and wellness (56). A key consideration for the Type 1 diabetic is whether the individual is in metabolic control. Physical activity and exercise can have a dramatic impact on the blood glucose concentration, and if the individual is not careful about regulating diet and the intensity of the activity or exercise, then hypoglycemia and **insulin shock** can occur. Metabolic control indicates that the individual is on a regular schedule of diet, insulin, and physical activity that allows for the maintenance of blood glucose levels in the normal range with little fluctuation (57,58).

Type 2 diabetes mellitus manifests itself through a series of processes whereby the body becomes resistant to insulin. Inadequate amounts of physical activity and poor nutritional habits can lead to the development of insulin resistance (55,59). This condition is characterized by relatively well-maintained insulin secretion and normal to elevated plasma insulin levels. Type 2 diabetes generally occurs later in life, and there are usually several other health risks associated with the condition, including hypertension, high blood cholesterol levels, abdominal obesity, and physical inactivity (60). Physical activity and exercise are beneficial to the individual with Type 2 diabetes as increased levels of activity help control blood glucose concentration and also have a positive effect on the associated health risks (61). The combination of a regular program of exercise and modification to the nutritional intake may increase insulin sensitivity (62) and allow the individual with Type 2 diabetes to eliminate the need for **exogenous insulin** or the oral medications used to stimulate insulin

Insulin shock Acute hypoglycemia usually resulting from an excessive insulin and characterized by sweating, trembling, dizziness, and, if left untreated, convulsions and coma.
Exogenous insulin Insulin administered from outside the body.

secretion by the pancreas (61). Exercise science professionals involved in the prescription of physical activity and exercise must be aware of the process by which glucose is taken up and used for energy and the factors that affect that process so that effective individualized programs can be developed for healthy and diseased individuals.

Implications for Sport and Athletic Performance

Success in certain types of sport and athletic performance is dependent on the ability of the body and muscle fibers to produce energy very quickly. Those athletic events that are characterized by short duration and high-intensity periods of activity (*e.g.*, sprinting, American football, volleyball) depend heavily on the ability to rapidly use carbohydrates (primarily muscle glycogen) as an energy source. Exercise scientists, strength and conditioning coaches, sports coaches, and athletes have long been interested in improving energy production from carbohydrates and delaying the onset of fatigue that occurs from the production of lactic acid in the muscle: a by-product of carbohydrate metabolism.

The formation of lactic acid in skeletal muscle is a complex process. During rest and low-intensity exercise, most of the energy is produced by using oxygen in the **aerobic metabolism** of carbohydrates and fats. Although there is a small amount of lactic acid being produced in the muscle, it is quickly cleared by a number of processes in the body. As the exercise intensity increases and there is a need for ATP to be produced faster, there is an increase in the production of lactic acid in the contracting muscle and at the same time an increase in the removal of lactic acid from the tissues. There is, however, an exercise intensity whereby lactic acid production exceeds removal and lactic acid accumulates in the muscle and blood. As lactic acid is formed, it separates into a negatively charged lactate ion and its associated positively charged hydrogen ion. As the hydrogen ion increases in concentration, it can, along with other intramuscular metabolites, contribute to fatigue in the contracting skeletal muscle (63).

The development of various training programs designed to increase energy production from aerobic sources and decrease the fatiguing effects of lactic acid has long been a goal of coaches and athletes. In aerobic endurance events, the successful competitor among athletes with similar $\dot{V}O_{2max}$ values is usually the person who can maintain aerobic energy production at the highest percentage of his or her $\dot{V}O_{2max}$, without accumulating large amounts of lactic acid in the muscle and blood (64). Although many terms are used to describe this phenomenon, the lactate threshold is the one most commonly seen in the literature. The lactate threshold is that exercise intensity at which a specific blood lactate concentration is observed or where blood lactate concentration begins to increase above resting levels (65). Research has shown that an athlete's lactate threshold appears to be a strong predictor of aerobic endurance performance (66,67). The maximal lactate steady state is another term that also appears in the aerobic training literature. The maximal lactate steady state is the exercise intensity where maximal lactate production equals maximal lactate clearance within the body (68). Many experts consider the maximal lactate steady

Aerobic metabolism The production of energy through the use of oxygen in the cell.

state exercise intensity to be a better indicator of aerobic endurance performance than either maximal oxygen consumption or the lactate threshold (68,69). What is clear from this information is that aerobic endurance athletes must improve their ability to decrease the production of lactic acid and increase the removal of lactic acid from the muscle and blood. This requires the athlete to train at elevated levels of blood and muscle lactate to maximize training-induced improvements that decrease lactic acid production and increase lactic acid clearance by the body (70). The development and use of nutritional ergogenic aids, such as sodium bicarbonate and sodium citrate, to help minimize the effects of hydrogen ion increases during high-intensity exercise are also of interest to exercise science professionals. Both sodium bicarbonate and sodium citrate use chemical reactions in the body to help maintain the normal pH levels of the body as lactic acid concentrations increase (71).

Skeletal Muscle Physiology

Skeletal muscle has a number of important functions in the human body (see Chapter 3). As a result, exercise scientists have been studying skeletal muscle to gain a better understanding of how muscle performs various functions during exercise and sport performance. Skeletal muscle fibers develop from **embryonic myotubes** into mature muscle fibers, and this development is influenced by various growth-promoting factors (29). As the fibers develop, various regulatory and contractile proteins are arranged in a systematic pattern that allows the fiber to generate force upon stimulation from a motor neuron (29). Skeletal muscle fibers are a **heterogeneous group** of fibers with distinct contractile and metabolic characteristics (72) (see Chapter 3, Table 3.3). Each of these fiber types has characteristics that allow them to perform distinct functions and respond in different ways during muscle contraction. Skeletal muscle fibers are unique to individuals and specific muscle groups. Generally, muscle fiber types cannot be changed unless there is a dramatic change in the physical activity or exercise habits of an individual. In this instance, the muscle fiber types will take on characteristics that help the muscle meet the requirements of the physical activity or exercise.

Implications for Physical Activity and Exercise

Skeletal muscle plays an important role in promoting health and wellness for all individuals. Exercise science professionals have long been interested in identifying the most appropriate resistance exercise training program to enhance muscular strength development and improve muscular fitness, balance, coordination, and other measures of motor performance (73,74).

Through years of research, much has been learned about the various components needed for a successful resistance exercise training program. Some of the characteristics affecting strength development are shown in Figure 4.5. Resistance exercise training causes muscle fiber hypertrophy (75,76) as a result of increased protein synthesis (77). Increases

Embryonic myotubes Immature structures that can potentially convert into muscle fibers.
Heterogeneous group A collection composed of parts having dissimilar characteristics or properties.

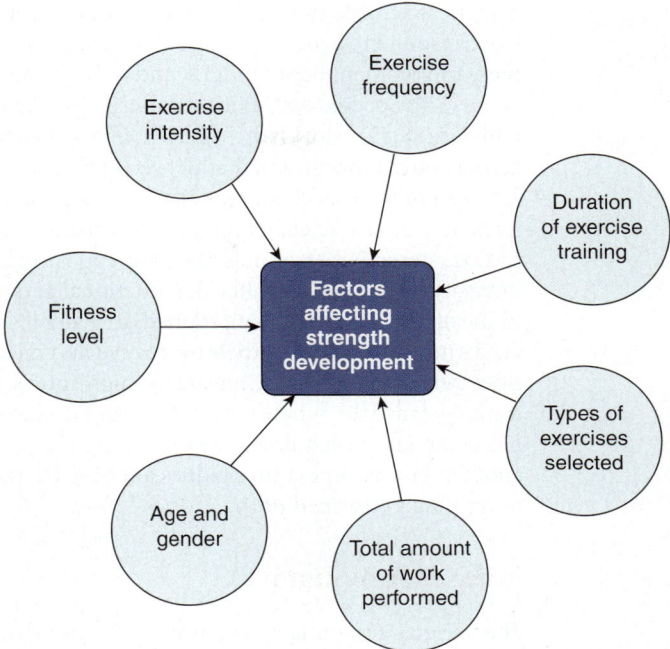

FIGURE 4.5 Predominant factors affecting strength development during a resistance exercise training program (74).

in muscular strength (78) and muscular power (79) are observed following participation in resistance training programs.

The importance of muscular strength in older adults is critical for maintaining normal functional ability and often independence (80,81). As long as training principles are met, older adults can increase muscular size and strength following resistance exercise training (82–84). Regular resistance exercise can also improve the functional status of obese overweight adults (85) and older adults with rheumatoid arthritis (86). Although much is known about how individuals respond to resistance exercise training programs, there are still many unanswered questions in the areas of gender differences in response to training, the appropriate volume of training to reduce the risk of lifestyle-related diseases, and the role of resistance exercise in treating individuals with disease conditions.

Implications for Sport and Athletic Performance

The muscle fiber type of an athlete is an important consideration in sport and athletic performance because of the unique functional and metabolic characteristics of the fibers. The use of the **muscle biopsy** (in combination with the **skinned fiber technique**) has allowed

Muscle biopsy A needle technique used to collect tissue samples from a muscle.
Skinned fiber technique A laboratory technique that removes the cell membrane and allows for more precise control of the internal environment of a muscle fiber.

exercise scientists to identify differences in fiber types between individuals and how fiber types respond to acute exercise and different types of chronic exercise training. In sedentary men and women, between 45% and 55% of the total percentage of fibers are slow twitch fibers (87). Specific distribution patterns of fiber types are evident among highly successful athletes (88,89). Slow twitch muscle fibers predominate in the active muscle of successful aerobic endurance–trained athletes, whereas fast twitch muscle fibers predominate in the active muscle of successful sprint- and resistance-trained athletes (88,89). Of particular interest to exercise science professionals, coaches, and athletes is whether muscle fiber types can be altered with training. It would appear that an appropriate duration, frequency, and intensity of training can alter the functional and metabolic characteristics of the fiber types in the muscles that are being trained. In general, low-intensity endurance training results in fast twitch fibers altering their functional and metabolic characteristics to become more like slow twitch fibers (73). Conversely, high-intensity sprint training and resistance exercise training cause the functional and metabolic characteristics of slow twitch fibers to become more like fast twitch fibers (73). Training may or may not convert fibers from one type to another, but chronic training does alter the characteristics of the fibers to support the type of training performed (90).

Bone Metabolism

The skeletal system is a very important and dynamic system of the body. In addition to providing a structural framework and protecting organs and tissues of the body, the skeletal system provides a lever system for movement, serves as a storage area of important minerals, and is also involved in blood cell formation. Though most individuals have good bone health, there are many individuals in both the general population and the athletic population that are at increased risk for poor bone health that can lead to debilitating disease conditions (1). As a result, exercise science professionals have been interested in how physical activity and exercise influence bone metabolism.

Osteoporosis is a disease condition characterized by low bone mineral density. Figure 4.6 provides examples of normal bone and osteoporotic bone. Older adults, especially females, are at an increased risk for developing osteoporosis (1). Female athletes participating in high-volume training are also at increased risk for developing low bone mineral density (osteopenia) and possibly osteoporosis (1,91,92). Of concern to those individuals with low bone mineral density and osteoporosis is the increased risk of bone fracture that is associated with repeated force or trauma (1). There are two generally accepted strategies for making the skeleton more resistant to fracture: (a) maximizing the amount of bone mineral density in the first three decades of life and (b) minimizing the decline in bone mineral density after the age of 30 years (1). The efforts of various exercise science professionals is very important for helping individuals develop physical activity and exercise programs, as well as good nutritional habits that will improve bone health.

Implications for Physical Activity and Exercise

Exercise science and allied health professionals have expanded considerably the information about the role of physical activity in maintaining good bone health. Mechanical loading is the application of force to the tissues of the body and is an important consideration

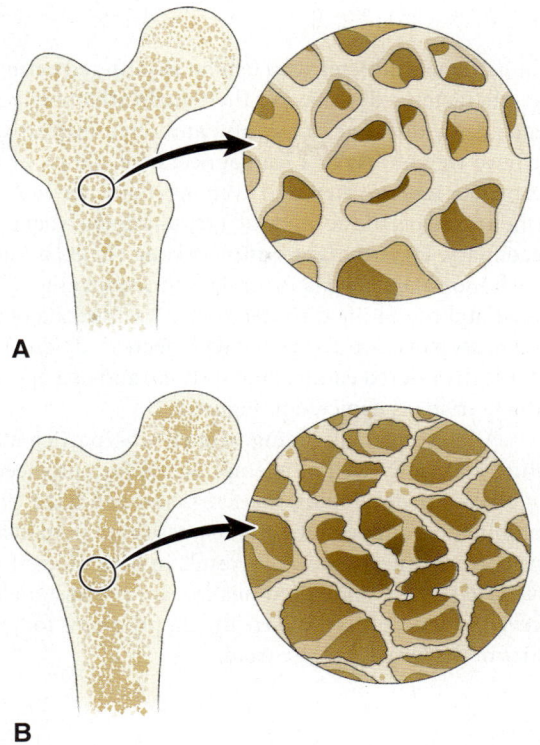

FIGURE 4.6 A comparison of normal bone and osteoporotic bone. **A:** Normal bone. **B:** Osteoporosis.

in bone health. Mechanical loading forces that are unique, variable, and dynamic in nature result in changes to the skeletal system that increase bone mineral density (1,93). It is clear that the intensity, duration, and frequency of the mechanical loading is important for maximizing bone formation, but further research is needed to determine if a specific threshold exists (1). Other factors that appear to play an important role in the development of poor bone health are calcium insufficiency (94,95) and estrogen deficiency (96). Maximizing bone mass in children, adolescents, and young adults is critical as peak bone mass is thought to be attained by the end of the third decade of life (1). Though the optimal type, frequency, intensity, and duration have yet to be conclusively determined, it is clear that both physical activity and exercise are critical for maximizing bone mineral density (1,97).

Starting at the age of about 40 years, bone mass decreases by about 0.5% or more per year, regardless of gender or ethnicity. The rate of loss varies by skeletal region and is potentially influenced by genetics, nutrition, hormonal status, and regular physical activity (1). It would appear that certain types of physical activity and exercise are more beneficial for preserving bone mass. Activities with higher intensity mechanical loading forces, such as brisk walking, stair climbing, and jogging, seem to provide a better response of the skeletal system for increasing bone mass (1,98,99). Resistance exercise training can also improve bone mineral density (97). Regular physical activity may also reduce fracture risk by increasing the bone mineral content, decreasing the bone mineral loss, or increasing muscular strength and balance and consequently reducing the risk of falling (1,100).

Implications for Sport and Athletic Performance

Young female athletes who exhibit disordered eating, **amenorrhea**, and bone mineral loss are characterized as having the female athlete triad (92). Disordered eating is characterized by a wide range of harmful and often ineffective eating behaviors used in attempts to achieve weight loss or a lean appearance. Short- and long-term morbidity, decreased performance, amenorrhea, and even mortality are outcomes of disordered eating (92,101,102). Amenorrhea is the absence of a regular menstrual cycle and can be classified as primary or secondary. **Hypothalamic amenorrhea** results in decreased ovarian hormone production, and **hypoestrogenemia** is similar to menopause. This condition is associated with both exercise and the eating disorder known as anorexia nervosa. Both hypothalamic amenorrhea and menopause are associated with decreased bone mineral density (92,103). The combination of disordered eating, amenorrhea, and osteoporosis form the foundation of the female athlete triad, as depicted in Figure 4.7.

Adolescents and young women who participate in sports or athletic competitions in which low body weight is emphasized for enhanced performance or appearance are at a high risk for developing the female athlete triad (91). If a female develops the triad, the combination of poor eating habits resulting in inadequate calcium intake and low estrogen levels from amenorrhea can result in a significant loss of bone mineral content (28,92). Exercise science and allied health care professionals continue to work to establish better prevention, detection, and treatment programs for girls and women who display characteristics of the female athlete triad.

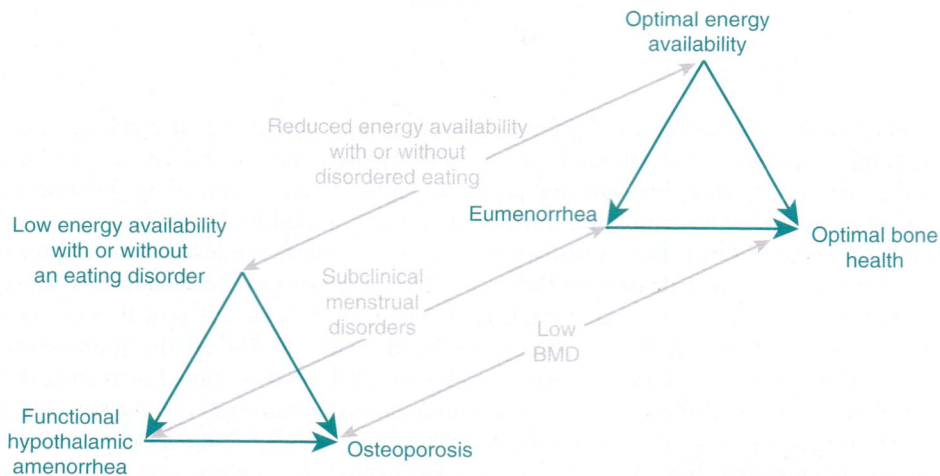

FIGURE 4.7 Interrelated factors of the female athlete triad (92). BMD, bone mineral density.

Amenorrhea Abnormal suppression or absence of menstruation.
Hypothalamic amenorrhea Condition in which there is an absence of a normal menstrual cycle that is caused by a disorder of the hypothalamus.
Hypoestrogenemia Low levels of the female hormone estrogen.

Energy Balance and Weight Control

Body weight is of critical importance for promoting overall good health and optimizing performance. Although many individuals are able to regulate and control their body weight with relative ease, there are countless others who struggle with trying to prevent excess weight gain, achieve significant weight loss, or in some instances achieve weight gain. Body weight control can be best described using the energy balance equation. This equation describes the changes that occur to body weight when there are alterations to energy intake and energy expenditure.

FACTOR		FACTOR	OUTCOME
Energy intake	=	Energy expenditure	Stable body weight
Energy intake	>	Energy expenditure	Increase in body weight
Energy intake	<	Energy expenditure	Decrease in body weight

Upon initial observation of the energy balance equation, it would appear that understanding and achieving a desired body weight is a fairly simple process. However, there are numerous genetic, environmental, socioeconomic, physiological, psychological, and behavioral factors that could potentially influence energy intake or energy expenditure, causing the equation to become unbalanced and lead to weight gain or weight loss. Factors affecting energy intake and energy expenditure of individuals are shown in Figure 4.8 (32).

It is likely that a number of these factors are interrelated and contribute to weight gain or weight loss in individuals (2). Exercise science professionals work in conjunction with other allied health professionals to identify the factors promoting weight gain and the most successful methods for losing weight and maintaining a healthy body weight.

FIGURE 4.8 Predominant factors affecting energy intake and expenditure in humans (32).

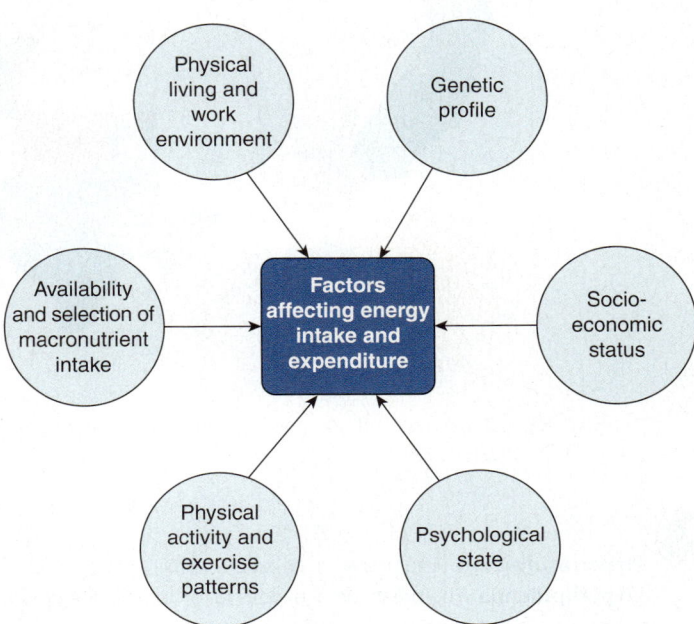

Implications for Physical Activity and Exercise

Understanding energy balance and weight control is critical for promoting good health and fitness. The United States and many other countries have seen a rapid and large increase in the number of overweight and obese individuals over the last several decades so that more individuals are overweight or obese than ever before (104–106). In Chapter 1, Figure 1.3 shows the age-adjusted prevalence of overweight and obesity among U.S. adults, age 20–74 years as reported by the National Health and Nutrition Examination Surveys. A multitude of environmental and individual factors have been suggested as contributing to this rapid weight gain (107,108). Furthermore, there is likely to be no single intervention strategy or treatment program that will assist individuals with effective weight loss and the prevention of weight gain.

Research has shown us that obesity is associated with an increased risk and prevalence of chronic diseases and health conditions such as cancer (109,110), heart disease (111), **hyperinsulinemia** (112,113), **hyperlipidemia** (114), and hypertension (115). Obesity also decreases the quality of life (116) and has a significant economic cost to our health care system (117). There is a strong connection between obesity and medical expenditures (118). Figure 4.9 illustrates, in the most recent data available, the estimates of adult

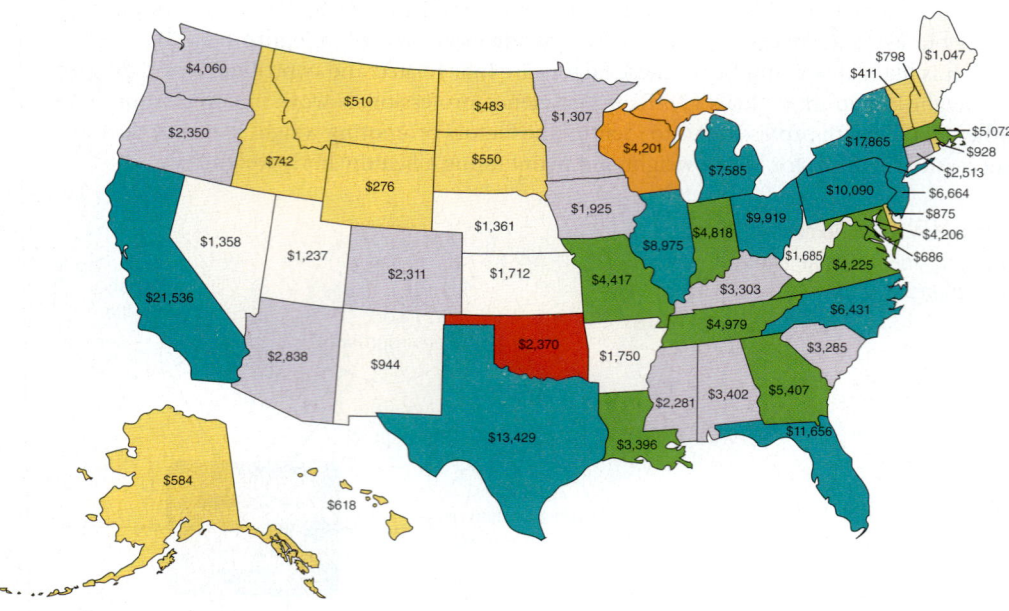

FIGURE 4.9 Estimates of adult obesity–attributable medical expenditures (in millions of dollars) by state in 2009 dollars (119).

Hyperinsulinemia The presence of excess insulin in the blood.
Hyperlipidemia The presence of excess fat or lipids in the blood.

obesity–attributable medical expenditures (in millions of dollars) by state in 2009 dollars (119). Major health initiatives, such as Healthy People 2030, are designed to reduce the number of overweight and obese individuals in the United States (www.healthypeople.gov) and other countries through the World Health Organization (www.who.int). Achieving the target goals can only be accomplished by identifying those interventions most appropriate for promoting weight loss and the prevention of weight gain (2).

There are several factors to be considered when examining the relationship of obesity to health and fitness. Body mass index (BMI) is used to quantify the amount of tissue mass based on the individual's height and weight (BMI = body weight [kg]/height [m^2]). The BMI value is used to categorize the individual as being underweight, normal weight, overweight, or obese. It has been recommended by the National Institutes of Health that weight loss is indicated in adults with a BMI greater than 25 kg·m^{-2} (120). The pattern of body fat distribution in overweight and obese individuals may also contribute to an increased health risk. Specifically, an increased level of intra-abdominal fat is positively associated with hyperinsulinemia, **hypercholesterolemia**, and hypertension (121,122). Weight loss in overweight and obese individuals does not have to be large to positively affect health. A 5% to 10% reduction of body weight can significantly improve health by decreasing blood lipids, blood pressure, and factors related to the onset of Type 2 diabetes (2,123,124); however, long-term health benefits may be maximized with sustained weight loss of greater than or equal to 10% of initial body weight (2).

Weight loss intervention programs incorporate various strategies to promote safe and effective loss of body weight. Although it remains unclear what the best short- and long-term programs for weight loss might be for individuals of various ethnic backgrounds, gender, and different levels of body weight and body fat, several characteristics of successful weight loss programs have been identified (2). These characteristics are provided in Figure 4.10.

Implications for Sport and Athletic Performance

The maintenance of energy balance and regulation of body weight for various groups of athletes is critical for ensuring success in competition. Of particular importance is the maintenance and/or increase of lean body mass in both endurance- and resistance-trained athletes. The maintenance of lean body mass is strongly influenced by the energy expenditure during training and the energy intake in the form of macronutrient consumption (32,125). It is clear that both endurance- and resistance-trained athletes have greater dietary protein needs than the current dietary reference intake (DRI) for persons over 18 years of age. The DRI for protein is currently set at 0.8 g of protein per kilogram body weight per day (126). However, the recommended requirements for dietary protein intake are higher in active individuals (127). It is recommended that endurance-trained athletes should consume between 1.2 and 1.4 g of protein per kilogram body weight per day and resistance-trained athletes should consume between 1.4 and 1.7 g of protein per kilogram body mass per day (128,129).

Hypercholesterolemia The presence of excess cholesterol in the blood.

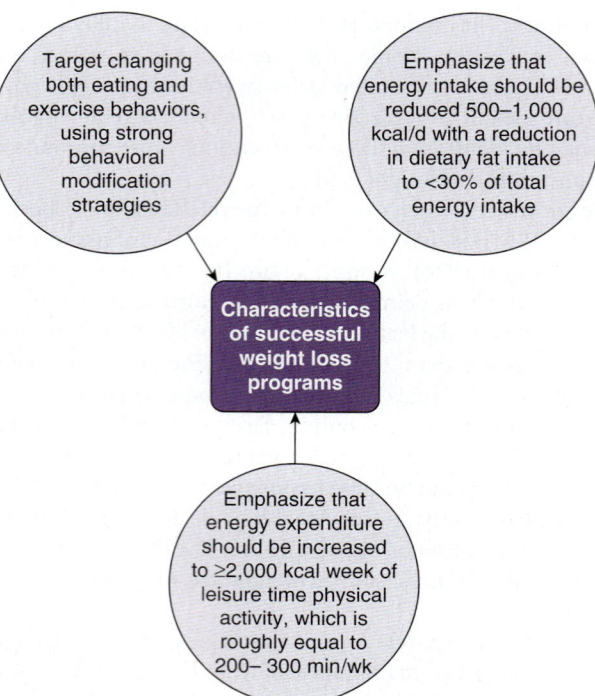

FIGURE 4.10 Characteristics of successful weight loss programs (2).

It is important for some athletes to maintain or increase lean body mass for either appearance (*e.g.*, body builders) or for force production (*e.g.*, weight lifting, sprinting, American football). To increase lean muscle mass, an appropriate intake of dietary carbohydrate and protein combined with a resistance exercise training program is essential. Considerable effort has been made by exercise science professionals to gain a better understanding of the role of muscle protein metabolism when combined with resistance exercise training. High-intensity, high-volume resistance exercise increases the rate of protein synthesis and protein breakdown in skeletal muscle for several hours after the completion of a training session (77,130). To minimize protein loss and increase protein synthesis, both protein and carbohydrate should be consumed within 30 to 60 minutes after exercise (47). During this time period, the rate of protein breakdown exceeds the rate of protein synthesis in the absence of energy intake (77,131). When food is consumed after the completion of an exercise training session, the energy consumed helps create an **anabolic** environment in the muscle that favors protein synthesis (132). The enhanced protein synthesis promotes the repair of tissues damaged during exercise and helps facilitate the development of new proteins in the muscle (133). These two factors usually lead to increases in muscle size and force production.

Thinking Critically

What information might be valuable to know when prescribing an exercise training program for an individual trying to lose weight or gain weight?

Anabolic The process where larger molecules or compounds are built from smaller molecules or compounds.

Assessment of Energy Expenditure and Physical Activity

An accurate assessment of energy expenditure during physical activity and exercise is important for promoting health and understanding factors that influence athlete performance. Exercise science professionals working with other highly skilled professionals from the disciplines of chemistry, engineering, and computer technology have developed a number of accurate and reliable assessment tools for determining energy expenditure under a variety of conditions. The assessment methods of energy expenditure and physical activity include those that are very invasive (*e.g.*, muscle biopsy) and complex (*e.g.*, doubly labeled water) to those less invasive (*e.g.*, heart rate telemetry) and simple (*e.g.*, pedometers). Figure 4.11 shows the invasive muscle biopsy procedure used to determine concentrations of intramuscular metabolites prior to and after exercise, whereas Figure 4.12 shows pedometers that can be worn on clothing for the assessment of physical activity.

In some instances, the assessment of energy expenditure can be used to answer basic research questions such as how metabolic pathways are controlled in skeletal muscle. The assessment of energy expenditure also has many practical applications. For example, knowing the amount of energy expended during a particular activity can be useful for an exercise science professional developing a nutrition program to match an exercise regimen, monitor body weight changes or identify the fitness requirement for a particular job skill.

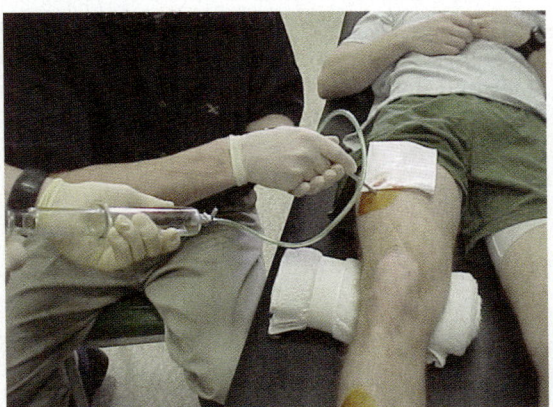

FIGURE 4.11 A muscle biopsy of the vastus lateralis can be used to assess energy expenditure in the muscle.

FIGURE 4.12 Pedometers are commonly used instruments for the assessment of physical activity.

Most assessments of energy expenditure are considered indirect. Though it is possible to directly measure energy expenditure (*e.g.*, whole room calorimeter), the process is difficult and expensive. Indirect assessments of energy expenditure generally involve a measure of some physiologic variable (*e.g.*, oxygen consumption), and then a determination of energy expenditure is derived from a mathematical prediction equation. For example, the assessment of energy expenditure by respiratory gas analysis involves the use of technical equipment for the measurement of pulmonary ventilation (*i.e.*, air flow into and out of the lungs) and the concentration of oxygen and carbon dioxide in the inspired and expired air. In this type of assessment, energy expenditure is most accurately measured if the total oxidation of carbohydrates, fats, and protein is determined. Other assessment methods of energy expenditure that require the measurement of a physiologic variable include the doubly labeled water method, which requires collection of urine or saliva and the heart rate–monitoring method that requires an accurate and continuous measurement of heart rate. These assessment methods are accurate but labor intensive and often costly (134).

An indirect assessment of physical activity for large groups of individuals can be accomplished through the use of physical activity questionnaires, pedometers, and accelerometers. Each of these methods of assessment has been validated for accuracy, usually against a method that requires the measurement of a physiologic variable (*e.g.*, respiratory gas analysis or doubly labeled water). The assessment of physical activity using a questionnaire requires an individual to accurately recall or record the activities of the day or week. The use of pedometers and accelerometers are less reliant on individual compliance and are being used more commonly to assess levels of physical activity and energy expenditure (134–136). Table 4.7 provides a summary of energy assessment methods. A more complete description of these assessment methods can be found in Chapter 11.

Implications for Physical Activity and Exercise

The use of the various methods to assess energy expenditure and physical activity has allowed for the development of a compendium of physical activities that provides an estimate

Table 4.7 Examples of Energy and Physical Activity Assessment Methods

ASSESSMENT METHOD	GENERAL COMMENTS
Muscle biopsy	Very invasive, limited to small numbers of subjects
Doubly labeled water	Very expensive, limited to small numbers of subjects, requires frequent visits to a laboratory
Respiratory gas exchange	Expensive, limited primarily to a laboratory environment, although portable units are becoming more common
Physical activity questionnaires	Inexpensive, can be used for large numbers of individuals
Pedometers and accelerometers	Inexpensive, allows for free movement of the individual

of energy expenditure during the performance of different activities (137). This compendium can be very useful for exercise science professionals when developing exercise prescriptions for promoting health and fitness. The ACSM Position Stand titled "Appropriate physical activity intervention strategies for weight loss and prevention of weight regain for adults" recommends that long-term weight loss is best accomplished if individuals perform 200 to 300 minutes of moderate intensity physical activity per week (2). Additionally, moderate intensity physical activity between 150 and 250 minutes per week is necessary to be effective at preventing weight gain (2). For exercise science professionals developing an exercise program for a client to meet this goal, it is beneficial to have a reference of physical activity energy expenditure to which they can go. Furthermore, individuals who are involved in weight loss programs may experience more success if they understand of how much energy is expended in different activities. Additionally, the accurate assessment of energy expenditure and physical activity can be beneficial to exercise science professionals and researchers testing the effectiveness and efficacy of weight loss programs.

Implications for Sport and Athletic Performance

The assessment of energy expenditure can also be useful for enhancing performance in certain sports and athletic competitions. The determination of which energy system predominates in the supply of energy to contracting muscles is important for understanding basic research questions involving the control of energy pathways and also for developing effective training and conditioning programs. For example, the development of the muscle biopsy procedure (Figure 4.11) has allowed for the accurate determination of how energy utilization and production occurs in skeletal muscle (138). Many exercise scientists have used muscle biopsies to study changes in muscle substrates, enzyme activities, and metabolite levels prior to, during, and following exercise. The use of the muscle biopsy procedure has facilitated a greater understanding of the enhancement of exercise performance following the use of supplemental nutritional aids (139). The classic muscle glycogen studies from the 1960s (138,140–142) would not have provided as much insight into the mechanisms by which altering dietary intake of carbohydrate (143) improves endurance performance without the measurement of intramuscular glycogen levels. The muscle biopsy procedure is often combined with other assessment techniques, such as blood collection and respiratory gas analysis, to help exercise science professionals gain a better understanding of energy metabolism during exercise training and athletic competition.

Thinking Critically

Why would consumption of carbohydrate prior to and during exercise have different effects on an individual exercising to improve health versus an individual competing in a prolonged athletic competition?

Environmental Exercise

Individuals often perform physical activity and exercise in a variety of environmental conditions. Ensuring safety during exercise and optimizing athletic performance in

Efficacy The power or capacity to produce a desired effect.

challenging environmental conditions have been at the forefront of study in exercise physiology for many years. For example, the 1968 Olympic Games held in Mexico City brought attention to the influence of altitude on athletic performance when several records were set in the short-duration athletic events and performance in the longer distance athletic events was adversely affected. Since that time, exercise science professionals have been interested in factors affecting short- and long-duration exercise performance at altitude. Performing physical activity and exercise in hot and cold environmental temperature conditions impacts individuals in many ways, and though we have considerable knowledge about the impact of these conditions on human responses, many individuals still suffer from dehydration, hyperthermia, and hypothermia during exercise in these environmental conditions.

Implications for Physical Activity and Exercise

Our body's core temperature (~37 °C) is closely regulated within a few degrees of a narrowly defined survival zone. Being outside that zone for too long can lead to severe injury and death from being too hot or too cold. Heat-related health problems experienced by individuals include heat syncope, heat cramps, heat exhaustion, and heat stroke (144). Figure 4.13 shows the factors that affect an individual's response to exercise in the heat (144). When an individual performs physical activity or exercise in hot and/or humid conditions, there is an increased activity in the systems of the body that control temperature regulation. The main response is the cardiovascular system's increase in blood flow to the

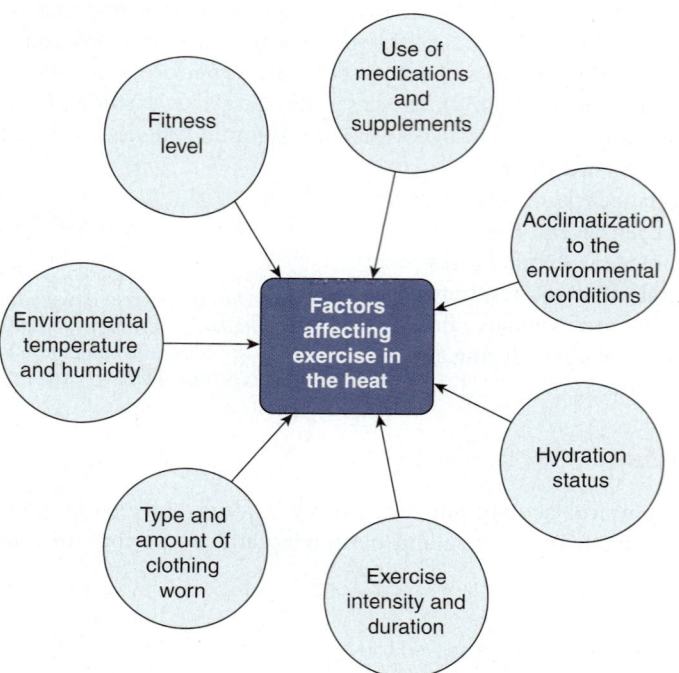

FIGURE 4.13 Various factors that affect an individual's response to exercise in the heat (144).

Table 4.8	Guidelines for Participant Safety and Reduction of Heat-Related Illness (144,149,150)
	Avoid the scheduling of athletic events and practices in extremely hot weather as much as possible
	Make sure athletes are gradually acclimatized to hot and humid environmental conditions
	Monitor athletes for signs and symptoms of heat strain
	Monitor athletes for adequate fluid replacement during and after training and competition

skin, which results in an increase in sweating. If fluids are not replaced, the increased sweat loss can lead to dehydration. When an individual becomes dehydrated, there is a reduction in physical work capacity and exercise performance, a faster time to physical exhaustion, and an increase in heat storage in the body (144–146), the latter of which increases the risk for heat-related health problems. Individuals who are physically unfit or who do not follow the guidelines for fluid replacement are at an increased risk for developing heat-related problems during physical activity or exercise in hot and/or humid conditions. Guidelines for athlete safety and the reduction of heat-related illness have been developed by a number of professional organizations, including the ACSM (144,147,148) and the National Athletic Trainers' Association (149,150), and are shown in Table 4.8. These guidelines and additional information can be found at the Web sites of the ACSM (www.acsm.org) and the National Athletic Trainers' Association (www.nata.org).

Implications for Sport and Athletic Performance

Maintaining appropriate hydration status is critical for reducing the risk of heat-related injury and optimizing athletic performance (46). Factors such as environmental conditions, gender, age, fitness level, and dietary intake affect sweat and electrolyte losses during exercise, and therefore individual fluid replacement patterns should be developed. Athletes, coaches, and exercise science professionals should ensure that fluid intake strategies include prehydration to ensure euhydration before beginning physical activity; consuming fluids during exercise to prevent excessive dehydration; and replacing fluids and electrolytes after the completion of exercise to return to euhydration. It is very important to match fluid and electrolyte intake with water loss to prevent dehydration. Conversely, overdrinking can lead to symptomatic exercise–associated **hyponatremia**, a condition that can lead to severe medical complications (46).

Hyponatremia An abnormally low concentration of sodium in the blood.

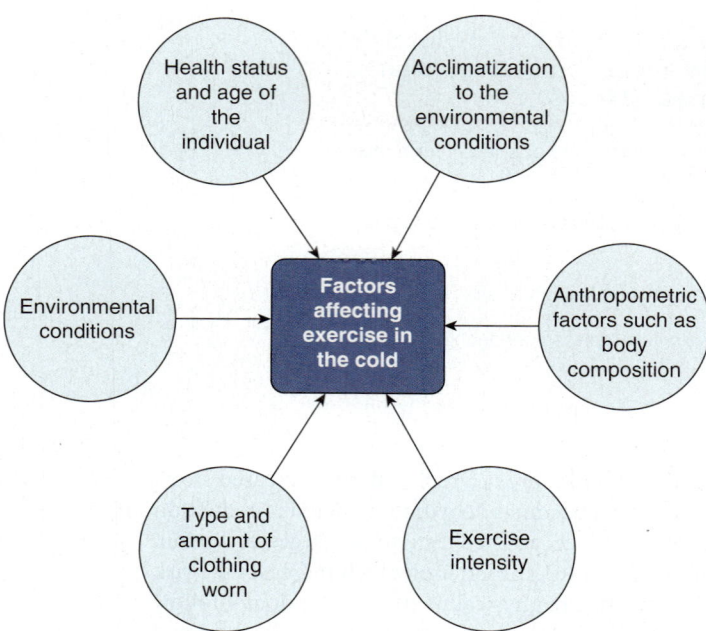

FIGURE 4.14 Various factors that affect an individual's response to exercise in the cold (153).

Physical Activity and Exercise in Cold Environmental Conditions

Physical activity and exercise in cold environmental conditions can lead to decreases in performance and an increased risk of **hypothermia** (151). There are several factors that interact during physical activity and exercise in the cold that could potentially increase the physiologic strain and injury risk beyond that associated with similar levels of activity conducted under more temperate conditions (Figure 4.14) (151).

Heat is lost from the body through four processes: **conduction**, **convection**, **radiation**, and **evaporation** (28). The two most critical environmental factors interacting with the environmental temperature to promote heat loss in cold conditions are wind and water. The wind chill index indicates how fast heat would be lost at different wind speeds and temperatures (28). Water has a thermal conductivity that is about 25 times greater than air at the same temperature (152), so more heat is lost from the body through water than air. There are several measures that can be taken to minimize the effects of cold environmental conditions on physical activity and exercise performance and to the risk of cold injury. The physiological responses to acclimatization to cold weather are very modest and depend on the severity of the exposures to the cold (151,153). To best prevent hypothermia, it is important to assess the level of cold by monitoring the environmental temperature, wind, and

Hypothermia Abnormally low body temperature.
Conduction The process of heat loss through direct transfer to a cooler object.
Convection The process of heat loss to the air surrounding the body.
Radiation The process of heat loss through the air to solid cooler objects.
Evaporation The process of heat loss through sweat evaporating from the skin surface.

solar impact, as well as the rain immersion depth and altitude if applicable (151). The risk of performing physical activity or exercise in the cold is then assessed by analyzing what is to be performed and the clothing available to wear. It is also important to identify those individuals such as children and the aged who are at higher risk of hypothermia. The physical activity and exercise intensity and duration, experience, physical condition, general health, and nutritional status of the individual should also be considered to determine if an elevated risk of cold injury exists (151). Appropriate risk management can help prevent cold injuries such as frostbite and hypothermia from occurring during physical activity and exercise in the cold (151). Exercise science professionals work to ensure that physical activity, exercise, sports, and athletic competitions can be performed safely in cold environmental conditions.

OTHER AREAS OF STUDY

The discipline of exercise physiology encompasses numerous areas of study not mentioned in this chapter. For example, strength and conditional professionals are interested in the various responses of the body to different training and conditioning methods, such as periodization, high-intensity training, plyometrics, and concurrent endurance and resistance exercise training. Exercise science professionals are also interested in the role of physical inactivity in disease risk; hormonal regulation of metabolism; the measurement of work, power, and exercise efficiency; and the role of the cardiovascular and respiratory systems in promoting improvements in health and fitness. Exercise specialists and personal trainers work to develop exercise prescriptions for health and fitness in healthy and diseased individuals as well as special populations such as children, females, the aged, and those with disabilities. Exercise science professionals are also interested in how genetic mechanisms control responses to exercise as well as regulate both aerobic fitness and physical activity levels. The interaction of nutrition and body composition on health and performance has also been studied extensively by exercise science professionals, but there is still much to be learned in these areas.

Interview

Jill A. Bush, PhD, FACSM, CSCS*D

Associate Professor in the Department of Health and Exercise Science at The College of New Jersey

Brief Introduction – A native of New Jersey, I received a BS in Exercise Science from Rutgers University, where I played varsity field hockey. I received my MS and PhD in Kinesiology from The Pennsylvania State University with an emphasis in Strength and Conditioning, Endocrinology, and Immunology.

I completed a NIH Post-Doctoral Research Fellowship at Baylor College of Medicine at The USDA/ARS Children's Nutrition Research Center in Houston, Texas. My current research examines the effects of exercise training on hormonal changes, healthy eating, and community-based programs in multiple ethnic populations of obese children and adults; and muscular and hormonal changes with the effects of vibration in strength training programs.

Q: What are your most significant career experiences?

One of my earliest significant experiences was my postdoctoral training for 2 years at The USDA/ARS Children's Nutrition Research Center at the Baylor College of Medicine in Houston, Texas, studying the effects of growth hormone and protein synthesis/degradation rates in a pig model. Having only ever worked with humans, this was a great opportunity to expand my skills and knowledge in one specific area. Upon completing my doctorate, I was immersed in research and honing my skills every day for 2 years. Not only did this allow me to work with other researchers in similar areas, it provided me with additional skills that I can use in my own research and teaching in physiology that I would have not otherwise been able to do. This experience also allowed me to spend significant time writing grants to support my research, presenting my research, and writing research publications.

Q: Why did you choose to become an "exercise scientist"?

Throughout my young life, I was an athlete in various sports and often got injured. I was very determined to gain a better understanding of how changes in the body, including those in physiology, skeletal muscle, and caloric intake, affect athletic performance. Since I am no longer an athlete, I have transitioned that excitement into studying how changes in the body across the lifespan and between genders are affected by physical activity, including competitive sport.

Q: Why is it important for exercise science students to have an understanding of exercise physiology?

Today's world we live in is very different from the world of 20 or 30 years ago. Many individuals are living at a high risk for obesity and obesity-related conditions. Most of these risk factors could be due to lifestyle choices, poor eating habits, and sedentary lifestyles. Students who have a deep understanding of exercise physiology are able to better convey the importance and benefits of regular exercise and healthy eating habits to individuals of all ages. Through this understanding, the next generation of students working in an exercise science–related field will be better equipped to interact and relay the correct and most up-to-date information regarding exercise programming, the body's adaptation to various modes of exercise, and the benefits of regular exercise.

Q: What advice would you have for a student exploring a career in any exercise science profession?

There are few women working in the strength and conditioning field, especially in research. It is important for the next generation of women to find mentors who will guide them through the career process. Being an exercise scientist combines two areas that I love: teaching and researching about the body's adaptation to exercise. Whether you want to be a personal trainer, work in a hospital as a clinical exercise physiologist, or teach and do research with the next generation, it is important as an upcoming specialist in exercise science to get involved in any way possible with those already working in an exercise science field. Not only will this allow you to observe first-hand the daily life of an exercise scientist, but provide you with important networking opportunities.

Interview

Ricardo Mora-Rodriguez, PhD

Professor in the Department of Exercise Science at the University of Castilla-La Mancha (Spain)

Brief Introduction – I grew up in Madrid (Spain) in a working class neighborhood. I started swimming when I was 12 years old, later changed to playing water polo, and ended up playing for AR Concepcion at the national Spanish division league at 18 years old. Sports were my passion, so I pursued a degree in Physical Education from the Technical University of Madrid. After working as a physical education teacher for 1 year, I earned a Fulbright scholarship to complete a degree in the United States. I arrived in the United States during the summer of 1990 and left in 1998 with an unforgettable experience and a PhD in Exercise Physiology from the Human Performance Laboratory at The University of Texas at Austin.

Q: What are your most significant career experiences?

I have been very fortunate to learn from some of the best researchers in the exercise physiology field. Among others, my mentors included Edward Coyle, Joseph Starnes, John Ivy, Jack Wilmore, and Bengt Saltin, for whom I did a short postdoctoral experience in his laboratory. The knowledge and scientific intuition of these champions of science cannot be grasped at once, so what I took from them was the passion to disentangle problems. That still serves me today in my current research. If we think with a passion that "exercise is medicine," then by training we should be able to lower (or at least prevent from increasing) drug treatment of disease conditions. Since 2012, we have been conducting experiments to address the substitution potential of exercise training in people taking medication.

Q: Why did you choose to become an "exercise scientist"?

I have liked the natural sciences since I was a kid. I still remember being amazed when in elementary school, Mr. Vicente explained to us how cones and rods worked in the retina of the eye. As an athlete, I started wondering about the training methods coaches used on us. In the early 1980s, they taught us the "revolutionary" concept that endurance performance was affected by $\dot{V}O_{2max}$, lactate threshold, and mechanical efficiency. Little did I know at the time that my future advisor and long-life friend, Dr. Ed Coyle, was one of the main proposers of the model. I guess that I decided to become an exercise scientist to decipher how to train to improve all those components and train "smarter" than anyone else did. Funny thing is that 30 years later, I have not answered how to "train smart," but I hope I have contributed to answer other questions in exercise physiology.

Q: Why is it important for exercise science students to understand exercise physiology?

Exercise physiology is at the core of the exercise sciences. Clinical exercise physiology is a great example of how exercise physiology can evolve to help us halt the growing rate of obesity-related medical conditions. Motor control, biomechanics, muscle physiology, exercise and aging, growth and maturation, body composition are all important subjects that will complete the training of an exercise scientist.

Q: What advice would you have for a student exploring a career in any exercise science profession?

When pursuing a degree in exercise science, my advice is to look for the best exercise science program. I went to study in the United States, and it was the best choice for me, but there are strong exercise science programs in other places. Learn techniques, become an expert on them, but remember that those are just a means to answer your exercise physiology questions; techniques are not what you are. We are not biochemists, we are not molecular biologists, we are not epidemiologists, we are not cardiologists, we are not mechanical engineers, but we are a solid science with more than two centuries of history. Do not get caught up in a reduced niche.

Summary

- Exercise physiology is the study of how the systems of the body respond to acute and chronic physical activity and exercise.
- Physical activity and exercise can elicit both subtle and profound challenges to the body's systems.
- Whether responding to an acute challenge or chronic exposure, the systems of the body respond in an attempt to return the body to homeostasis and ready the body for the next challenge.
- Training and conditioning programs must adhere to the principles of overload, specificity, and reversibility to be effective in causing positive changes in the physiological systems of the body.
- Exercise physiology uses a variety of approaches to answer questions and provide recommendations for improving health through physical activity and exercise.
- Areas of study in exercise physiology provide safe and effective training methods to enhance performance in sport and athletic competition.

FOR REVIEW

1. How was the Harvard Fatigue Laboratory important in the development of exercise physiology as a scientific discipline?
2. What is the difference between acute and chronic responses to exercise?
3. What are the fundamental principles that must be considered when designing a training and conditioning program?
4. What are some of the acute responses of the cardiovascular, pulmonary, muscular, and endocrine systems of the body at the start of exercise?
5. How does a regular exercise program result in chronic adaptations to the cardiovascular, pulmonary, muscular, and endocrine systems of the body?
6. What are some factors that control energy utilization during exercise?
7. Why should athletes participating in long-duration exercise be concerned about ingesting carbohydrates prior to and during exercise?
8. How do insulin and the GLUT 4 transport proteins work to increase glucose uptake by the muscle cells?
9. What is the primary factor affecting glucose uptake in an individual with Type 2 diabetes?
10. What benefits does an older adult obtain from participation in a resistance training program?

11. What are the significant health implications of the female athlete triad?
12. List some of those diseases that have an increased risk of occurrence in an obese individual.
13. What are the recommendations for protein intake for an endurance athlete and a strength athlete?
14. What are the most commonly used ways to assess physical activity in large populations?
15. Explain five factors that affect an individual's response to exercise in the heat.
16. Explain why the wind chill index is an important factor in the regulation of body temperature in a cold environment.

Project-Based Learning

1. Obtain copies of the 2018 the Physical Activity Guidelines Advisory Committee Scientific Report and the 1995 Physical Activity and Health Report of the Surgeon General. Prepare a summary report by comparing the scientific evidence for promoting physical activity between the two reports as it specifically relates to the improvement of the physiological systems of the body and how the evidence supports physical activity for reducing the leading causes of death in the United States.
2. Identify a competitive athlete to whom you can relate. Provide scientific evidence to support designing a training and conditioning program for that athlete to improve their performance during competition.

REFERENCES

1. Kohrt WM, Bloomfield SA, Little KD, Nelson ME, Yingling VR. Physical activity and bone health. *Med Sci Sports Exerc*. 2004;36(11):1985–96.
2. Donnelly JE, Blair SN, Jakicic JM, Manore MM, Rankin JW, Smith BK. Appropriate physical activity intervention strategies for weight loss and prevention of weight regain for adults. *Med Sci Sports Exerc*. 2009;41(2):459–71.
3. Pescatello LS, Franklin BA, Fagard R, Farquhar WB, Kelley GA, Ray CA. Exercise and hypertension. *Med Sci Sports Exerc*. 2004;36(3):533–53.
4. Donnelly JE, Hillman CH, Castelli D, et al. Physical activity, fitness, cognitive function, and academic achievement in children: a systematic review. *Med Sci Sports Exerc*. 2016;48(6):1197–222.
5. Hamilton MT, Healy GN, Dunstan DW, Zderic TW, Owen N. Too little exercise and too much sitting: Inactivity physiology and the need for new recommendations on sedentary behavior. *Curr Cardiovasc Risk Rep*. 2008;2(4):292–8.
6. Booth FW, Roberts CK, Thyfault JP, Ruegsegger GN, Toedebusch RG. Role of inactivity in chronic diseases: evolutionary insight and pathophysiological mechanisms. *Physiol Rev*. 2017;97(4):1351–402.
7. Henson J, Edwardson CL, Morgan B, et al. Associations of sedentary time with fat distribution in a high-risk population. *Med Sci Sports Exerc*. 2015;47(8):1727–34.
8. Berryman JW. Ancient and early influences. In: Tipton CM, editor. *Exercise Physiology: People and Ideas*. New York (NY): Oxford University Press; 2003. p. 1–38.
9. Tipton CM. Exercise Physiology, part II: a contemporary historical perspective. In: Massengale JD, Swanson RA, editors. *The History of Exercise and Sport Science*. Champaign (IL): Human Kinetics; 1997. p. 396–438.
10. Tipton CM. Contemporary exercise physiology: fifty years after the closure of the Harvard Fatigue Laboratory. In: Holloszy JO, editor. *Exercise and Sport Science Reviews*. Baltimore (MD): Williams & Wilkins; 1998. p. 315–39.

11. Kroll WP. *Graduate Study and Research in Physical Education*. Champaign (IL): Human Kinetics; 1982.
12. Park RJ. The rise and demise of Harvard's B.S. program in anatomy, physiology, and physical training: a case of conflicts of interest and scare resources. *Res Q Exerc Sport*. 1992;63:246–60.
13. Chapman CB. The long reach of Harvard's Fatigue Laboratory, 1926–1947. *Perspect Biol Med*. 1990;34:17–33.
14. Horvath SM, Horvath EC. *The Harvard Fatigue Laboratory: Its History and Contributions*. Englewood Cliffs (NJ): Prentice-Hall; 1973.
15. Folk GE. The Harvard Fatigue Laboratory: contributions to World War II. *Adv Physiol Educ*. 2010;34(3):119–27.
16. Buskirk ER, Tipton CM. Exercise physiology. In: Massengale JD, Swanson RA, editors. *The History of Exercise and Sport Science*. Champaign (IL): Human Kinetics; 2003. p. 367–438.
17. Buskirk ER. From Harvard to Minnesota: keys to our history. In: Holloszy JO, editor. *Exercise and Sport Sciences Reviews*. Baltimore (MD): Williams & Wilkins; 1992. p. 1–26.
18. Bouchard C, Leon AS, Rao DC, Skinner JS, Wilmore JH, Gagnon J. The HERITAGE family study. Aims, design, and measurement protocol. *Med Sci Sports Exerc*. 1995;27(5):721–9.
19. Chen Y, Wang D, Yan P, Yan S, Chang Q, Cheng Z. Meta-analyses of the association between the PPARGC1A Gly482Ser polymorphism and athletic performance. *Biol Sport*. 2019;36(4):301–9.
20. Journath G, Hammar N, Vikström M, et al. A Swedish primary healthcare prevention programme focusing on promotion of physical activity and a healthy lifestyle reduced cardiovascular events and mortality: 22-year follow-up of 5761 study participants and a reference group. *Br J Sports Med*. 2020;54(21):1294–9.
21. Holme I, Retterstøl K, Norum KR, Hjermann I. Lifelong benefits on myocardial infarction mortality: 40-year follow-up of the randomized Oslo diet and antismoking study. *J Intern Med*. 2016;280(2):221–7.
22. Pedisic Z, Shrestha N, Kovalchik S, et al. Is running associated with a lower risk of all-cause, cardiovascular and cancer mortality, and is the more the better? A systematic review and meta-analysis. *Br J Sports Med*. 2020;54(15):898–905.
23. Naci H, Salcher-Konrad M, Dias S, et al. How does exercise treatment compare with antihypertensive medications? A network meta-analysis of 391 randomised controlled trials assessing exercise and medication effects on systolic blood pressure. *Br J Sports Med*. 2019;53(14):859–69.
24. Physical Activity Guidelines Committee. *2018 Physical Activity Guidelines Advisory Committee Scientific Report*. Washington (DC): U.S. Department of Health and Human Services; 2018.
25. Atkinson TS, Kahn MJ. Blood doping: then and now. a narrative review of the history, science and efficacy of blood doping in elite sport. *Blood Rev*. 2020;39:1–6.
26. Wilson GC, Mavros Y, Tajouri L, Singh MF. The role of genetic profile in functional performance adaptations to exercise training or physical activity: a systematic review of the literature. *J Aging Phys Act*. 2019;27(4):594–616.
27. Harvey NR, Voisin S, Dunn PJ, et al. Genetic variants associated with exercise performance in both moderately trained and highly trained individuals. *Mol Genet Genomics*. 2020;295(2):515–23.
28. Powers SK, Howley ET, Quindry J. *Exercise Physiology: Theory and Application to Fitness and Performance*. 11th ed. New York (NY): McGraw-Hill; 2021.
29. Guyton AC, Hall JE. *Textbook of Medical Physiology*. 13th ed. Oxford (UK): Elsevier; 2016.
30. Vellers HL, Kleeberger SR, Lightfoot JT. Inter-individual variation in adaptations to endurance and resistance exercise training: genetic approaches towards understanding a complex phenotype. *Mamm Genome*. 2018;29(1):48–62.
31. McArdle WD, Katch FI, Katch VL. *Sports and Exercise Nutrition*. 4th ed. Baltimore (MD): Lippincott Williams & Wilkins; 2012.
32. Gropper SS, Smith JL. *Advanced Nutrition and Human Metabolism*. 6th ed. Belmont (CA): Wadsworth; 2013.
33. Bernardoni B, Mitchell NM, Hughes MR, Potteiger JA. Effect of different meal composition after exercise on fat and carbohydrate oxidation in women with different levels of body fat. *Appl Physiol Nutr Metab*. 2014;38:538–43.
34. Mitchell NM, Potteiger JA, Bernardoni B, Claytor RP. Effects of carbohydrate ingestion during exercise on substrate oxidation in physically active women with different body compositions. *Appl Physiol Nutr Metab*. 2013;38(3):314–9.
35. Coyle EF. Substrate utilization during exercise in active people. *Am J Clin Nutr*. 1995;61:968S–979S.
36. Sherman WM. Carbohydrate feedings before and after exercise. In: Lamb DL, Williams MR, editors. *Perspectives in Exercise Science and Sports Medicine. Volume 4: Ergogenics — Enhancement of Performance in Exercise and Sport*. New York (NY): McGraw-Hill Companies; 1991.

37. Tipton KD, Wolfe RR. Exercise-induced changes in protein metabolism. *Acta Physiol Scand.* 1998;162:377–87.
38. Brooks GA, Fahey TD, Baldwin KM. *Exercise Physiology: Human Bioenergetics and Its Applications.* 4th ed. Mountain View (CA): Mayfield; 2004.
39. Gollnick PD. Metabolism of substrates: energy substrate metabolism during exercise and as modified by training. *Fed Proc.* 1983;44(2):353–7.
40. Gollnick PD, Saltin B. Fuel for muscular exercise: role of fat. In: Horton ES, Terjung RL, editors. *Exercise, Nutrition and Energy Metabolism.* New York (NY): MacMillan; 1988. p. 72–88.
41. Astorino TA, Schubert MM, Palumbo E, Stirling D, McMillan DW. Effect of two doses of interval training on maximal fat oxidation in sedentary women. *Med Sci Sports Exerc.* 2013;45(10):1878–86.
42. Achten J, Gleeson M, Jeukendrup AE. Determination of the exercise intensity that elicits maximal fat oxidation. *Med Sci Sports Exerc.* 2002;34(1):92–7.
43. Hargreaves M. Metabolic responses to carbohydrate and lipid supplementation during exercise. In: Maughan RJ, Shirreffs SM, editors. *Biochemistry of Exercise IX.* 9th ed. Champaign (IL): Human Kinetics; 1996. p. 421–9.
44. Richter EA, Hargreaves M. Exercise, GLUT4, and skeletal muscle glucose uptake. *Physiol Rev.* 2013;93(3):993–1017.
45. Coggan AR, Coyle EF. Carbohydrate ingestion during prolonged exercise: effects on metabolism and performance. In: Holloszy JO, editor. *Exercise and Sport Sciences Reviews.* 19th ed. Philadelphia (PA): Williams & Wilkins; 1991. p. 1–40.
46. Sawka MN, Burke LM, Eichner ER, et al. Exercise and fluid replacement. *Med Sci Sports Exerc.* 2007;39(2):377–90.
47. Thomas DT, Erdman KA, Burke LM. American College of Sports Medicine Joint Position Statement. Nutrition and athletic performance. *Med Sci Sports Exerc.* 2016;48(3):543–68.
48. Coyle EF. Fluid and fuel intake during exercise. *J Sport Sci.* 2004;22:39–55.
49. Wright DA, Sherman WM, Dernbach AR. Carbohydrate feedings before, during or in combination improve cycling endurance performance. *J Appl Physiol.* 1991;71(3):1082–8.
50. American College of Sports Medicine, American Dietetic Association, Dietetic Association of Canada. Nutrition and athletic performance. *Med Sci Sports Exerc.* 2009;41(3):709–31.
51. King AJ, O'Hara JP, Arjomandkhah NC, et al. Liver and muscle glycogen oxidation and performance with dose variation of glucose–fructose ingestion during prolonged (3 h) exercise. *Eur J Appl Physiol.* 2019;119(5):1157–69.
52. Fuchs CJ, Gonzalez JT, Loon LJC. Fructose co-ingestion to increase carbohydrate availability in athletes. *J Physiol.* 2019;597(14):3549–60.
53. Cortright RN, Dohm GL. Mechanisms by which insulin and muscle contraction stimulate glucose transport. *Can J Appl Physiol.* 1997;22(6):519–30.
54. Holloszy JO. Exercise-induced increase in muscle insulin sensitivity. *J Appl Physiol.* 2005;99:338–43.
55. van Baak MA, Borghouts LB. Relationships with physical activity. *Nutr Rev.* 2000;58(3):S16-8.
56. Galassetti P, Riddell MC. Exercise and type 1 diabetes (T1DM). *Compr Physiol.* 2013;3(3):1309–36.
57. Kemmer FW, Berger M. Exercise and diabetes mellitus: physical activity as a part of daily life and its role in the treatment of diabetic patient. *Int J Sport Med.* 1983;4:77–88.
58. Richter ER, Galbo H. Diabetes, insulin, and exercise. *Sports Med.* 1986;3:275–88.
59. Barnard RJ, Roberts CK, Varon SM, Berger JJ. Diet-induced insulin resistance precedes other aspects of the metabolic syndrome. *J Appl Physiol.* 1998;84:1311–5.
60. Albright A, Franz MS, Hornsby G, et al. Exercise and type 2 diabetes. *Med Sci Sports Exerc.* 2000;32(8):1345–60.
61. American College of Sports Medicine, American Dietetic Association. Exercise and type 2 diabetes. *Med Sci Sports Exerc.* 2010;42(12):2282–303.
62. Ryan BJ, Schleh MW, Ahn C, et al. Moderate-intensity exercise and high-intensity interval training affect insulin sensitivity similarly in obese adults. *J Clin Endocrinol Metab.* 2020;105(8):1–19.
63. Enoka RM, Duchateau J. Translating fatigue to human performance. *Med Sci Sports Exerc.* 2016;48(11):2228–38.
64. Lamb DR. Basic principles for improving sport performance. *Sports Sci Exch.* 1995;8(2):1–5.
65. Wells CL, Pate RR. Training for performance of prolonged exercise. In: Lamb DL, Murray R, editors. *Perspectives in Exercise Science and Sports Medicine. 1.* Indianapolis (IN): Benchmark Press, Inc; 1995. p. 357–88.
66. Coyle EF, Coggan AR, Hopper MK, Walters TJ. Determinants of endurance in well-trained cyclists. *J Appl Physiol.* 1988;64(6):2622–30.
67. Coyle EF, Feltner ME, Kautz SA, et al. Physiological and biomechanical factors associated with endurance cycling performance. *Med Sci Sports Exerc.* 1991;23:93–107.

68. Beneke R. Anaerobic threshold, individual anaerobic threshold, and maximal lactate steady state in rowing. *Med Sci Sports Exerc*. 1995;27:863–7.
69. Foxdal P, Sjodin B, Sjodin A, Ostman B. The validity and accuracy of blood lactate measurements for the prediction of maximal endurance capacity. *Int J Sports Med*. 1994;15:89–95.
70. Potteiger JA. Aerobic endurance training. In: Baechle TR, Earle RW, editors. *Essentials of Strength Training and Conditioning*. 2nd ed. Champaign (IL): Human Kinetics; 2000. p. 495–509.
71. Potteiger JA, Webster MJ, Nickel GL, Haub MD, Palmer RJ. The effects of buffer ingestion on metabolic factors related to distance running performance. *Eur J Appl Physiol*. 1996;72:365–71.
72. Kelso TB, Hodgson DR, Visscher AR, Gollnick PD. Some properties of different skeletal muscle fiber types: comparison of reference bases. *J Appl Physiol*. 1987;62(4):1436–41.
73. Ratamess NA, Alvar BA, Evetovich TK, et al. Progression models in resistance training for healthy adults. *Med Sci Sports Exerc*. 2009;41(3):687–708.
74. Starkey DB, Pollock ML, Ishida Y, et al. Effect of resistance training volume on strength and muscle thickness. *Med Sci Sports Exerc*. 1996;28(10):1311–20.
75. Dietz WH. Does energy expenditure affect changes in body fat in children. *Am J Clin Nutr*. 1998;67:190–1.
76. McCall GE, Byrnes WC, Dickinson A, Pattany PM, Fleck SJ. Muscle fiber hypertrophy, hyperplasia, and capillary density in college men after resistance training. *J Appl Physiol*. 1996;81:2004–12.
77. Phillips SM, Tipton KD, Aarsland A, Wolf SE, Wolfe RR. Mixed muscle protein synthesis and breakdown after resistance exercise in humans. *Am J Physiol*. 1997;273:E99–E107.
78. Rhea MR, Alvar BA, Burkett LN, Ball SD. A meta-analysis to determine the dose response for strength development. *Med Sci Sports Exerc*. 2003;35(3):456–65.
79. Wilson GJ, Newton RU, Murphy AJ, Humphries BJ. The optimal training load for the development of dynamic athletic performance. *Med Sci Sports Exerc*. 1993;25(11):1279–86.
80. Chodzko-Zajio W, Proctor DN, Fiatarone-Singh MA, et al. Exercise and physical activity for older adults. *Med Sci Sports Exerc*. 2009;41(7):1510–30.
81. Haskell WL, Phillips WT. Exercise training, fitness, health, and longevity. In: Lamb DL, Gisolfi CV, editors. *Perspectives in Exercise Science and Sports Medicine: Exercise in Older Adults*. Carmel (IN): Cooper Publishing Group; 1995. p. 11–52.
82. Brown AB, McCartney N, Sale DG. Positive adaptations to weight-lifting training in the elderly. *J Appl Physiol*. 1990;69:1725–33.
83. Candow DG, Vogt E, Johannsmeyer S, Forbes SC, Farthing JP. Strategic creatine supplementation and resistance training in healthy older adults. *Appl Physiol Nutr Metab*. 2015;40(7):689–94.
84. Fiatarone MA, Marks EC, Ryan ND, Meredith CN, Lipsitz LA, Evans WJ. High-intensity strength training in nonagenarians. *JAMA*. 1990;263(22):3029–34.
85. Villareal DT, Aguirre L, Burke GA, et al. Aerobic or resistance exercise, or both, in dieting obese older adults. *N Engl J Med*. 2017;376(20):1943–55.
86. Lange E, Kucharski D, Svedlund S, et al. Effects of aerobic and resistance exercise in older adults with rheumatoid arthritis: a randomized controlled trial. *Arthritis Care Res*. 2019;71(1):61–70.
87. Edstrom L, Ekblom B. Differences in sizes of red and white muscle fibers in vastus lateralis of musculus quadriceps of normal individuals and athletes: relation to physical performance. *Scand J Clin Lab Invest*. 1972;30:175–81.
88. Costill DL, Fink WJ, Pollock ML. Muscle fiber composition and enzyme activities of elite distance runners. *Med Sci Sports*. 1976;8:96–102.
89. Tesch PA, Karlsson J. Muscle fiber types and size in trained and untrained muscles of elite athletes. *J Appl Physiol*. 1985;59(6):1716–20.
90. Kraemer WJ, Patton JF, Gordon SE, et al. Compatibility of high-intensity strength and endurance training on hormonal and skeletal muscle adaptations. *J Appl Physiol*. 1995;78(3):976–89.
91. Lambert BS, Cain MT, Heimdal T, et al. Physiological parameters of bone health in elite ballet dancers. *Med Sci Sports Exerc*. 2020;52(8):1668–78.
92. Nattiv A, Loucks AB, Manore MM, Sanborn CF, Sundgot-Borgen J, Warren MP. The female athlete triad. *Med Sci Sports Exerc*. 2007;39(10):1867–82.
93. Burr DB, Robling AG, Turner CH. Effects of biomechanical stress on bones in animals. *Bone*. 2002;30:781–6.
94. Lanyon LE, Rubin CT, Baust G. Modulation of bone loss during calcium insufficiency by controlled dynamic loading. *Calcif Tissue Int*. 1986;38:209–16.
95. Sale C, Elliott-Sale KJ. Nutrition and athlete bone health. *Gatorade Sports Sci Exch*. 2020;29(201):1–7.

96. Zaman G, Cheng MZ, Jessop HL, White R, Lanyon LE. Mechanical strain activates estrogen response elements in bone cells. *Bone*. 2000;27:233–9.
97. Chilibeck PD, Candow DG, Landeryou T, Kaviani M, Paus-Jenssen L. Effects of creatine and resistance training on bone health in postmenopausal women. *Med Sci Sports Exerc*. 2015;47(8):1587–95.
98. Chow R, Harrison JE, Notarius C. Effect of two randomized exercise programmes on bone mass of healthy post-menopausal women. *Br Med J*. 1987;295:1441–4.
99. Dalsky GP, Stocke KS, Ehsani AA, Slatopolsky E, Lee WC, Birge SJ. Weight-bearing exercise training and lumbar BMC in postmenopausal women. *Ann Intern Med*. 1988;108:824–8.
100. American College of Sports Medicine. Osteoporosis and exercise. *Med Sci Sports Exerc*. 1995;27(4):i–vii.
101. Harris RT. Bulimarexia. *Ann Intern Med*. 1983;99:800–7.
102. Herzog DB, Copeland PM. Eating disorders. *N Engl J Med*. 1985;313:295–303.
103. Cann CE, Martin MC, Genant HK, Jaffe RB. Decreased spinal mineral content in amenorrheic women. *JAMA*. 1984;251:626–9.
104. Wang Y, Beydoun MA, Min J, Xue H, Kaminsky LA, Cheskin LJ. Has the prevalence of overweight, obesity and central obesity levelled off in the United States? Trends, patterns, disparities, and future projections for the obesity epidemic. *Int J Epidemiol*. 2020;49(3):810–23.
105. Flegal KM, Carroll MD, Kit BK, Ogden CL. Prevalence of obesity and trends in the distribution of body mass index among us adults, 1999–2010. *JAMA*. 2012;307(5):491–7.
106. Flegal KM, Carroll MD, Ogden CL, Curtin LR. Prevalence and trends in obesity among US adults, 1999–2008. *JAMA*. 2010;303(3):235–41.
107. Hill JO, Peters JC. Environmental contributions to the obesity epidemic. *Science*. 2002;280:1371–4.
108. Hill JO. Genetic and environmental contributions to obesity. *Am J Clin Nutr*. 1998;68:991–2.
109. Garfinkel L. Overweight and mortality. *Cancer*. 1986;58:1826–9.
110. Giovannucci E, Ascherio A, Rimm EB, Colditz GA, Stampfer MJ, Willett WC. Physical activity, obesity, and risk for colon cancer and adenoma in men. *Ann Intern Med*. 1995;122:327–34.
111. Katzmarzyk PT, Gagnon J, Leon AS, et al. Fitness, fatness, and estimated coronary heart disease risk: the HERITAGE family study. *Med Sci Sports Exerc*. 2001;33(4):585–90.
112. Haffner SM, Mitchell BD, Hazuda HP, Stern MP. Greater influence of central distribution of adipose tissue on incidence of non-insulin-dependent diabetes in women than men. *Am J Clin Nutr*. 1991;53:1312–7.
113. Lew EA, Garfinkel L. Variations in mortality by weight among 750,000 men and women. *J Clin Epidemiol*. 1979;32:563–76.
114. Ashley FW, Kannel WB. Relation of weight change to changes in atherogenic traits: the Framingham Study. *J Chronic Dis*. 1974;27:103–14.
115. Flegal KM, Carroll MD, Kuczmarski RJ, Johnson CL. Overweight and obesity in the United States: prevalence and trends, 1960–1994. *Int J Obes Relat Metab Disord*. 1998;22:39–47.
116. Engel TJ, Crosby RD, Kolotkin RL, et al. Impact of weight loss and regain on quality of life: mirror image or differential effect? *Obes Res*. 2003;11(10):1207–13.
117. Allison DB, Zannolli R, Narayan KM. The direct health care costs of obesity in the United States. *Am J Public Health*. 1999;89(8):1194–9.
118. Finkelstein EA, Trogdon JG, Cohen JW, Dietz WH. Annual medical spending attributable to obesity: payer-and service-specific estimates. *Health Aff (Millwood)*. 2009;28(5):w822–w831.
119. Trogdon JG, Finkelstein EA, Feagan CW, Cohen JW. State-and payer-specific estimates of annual medical expenditures attributable to obesity. *Obesity*. 2012;20(1):214–20.
120. National Institutes of Health, National Heart Lung and Blood Institute. *Clinical Guidelines on the Identification, Evaluation, and Treatment of Overweight and Obesity: The Evidence Report*. Washington (DC): U.S. Department of Health and Human Services; 1998.
121. Després JP, Couillard C, Gagnon J, et al. Race, visceral adipose tissue, plasma lipids, and lipoprotein lipase activity in men and women: the health, risk factors, exercise training, and genetics (HERITAGE) family study. *Arterioscler Thromb Vasc Biol*. 2000;20:1932–8.
122. Després JP, Moorjani S, Lupien PJ, Tremblay A, Nadeau A, Bouchard C. Regional distribution of body fat, plasma lipoproteins, and cardiovascular disease. *Arteriosclerosis*. 1990;10:497–511.
123. Goldstein DJ. Beneficial health effects of modest weight loss. *Int J Obes*. 1992;16:397–415.
124. Wing RR. Physical activity in the treatment of the adulthood overweight and obesity: current evidence and research issues. *Med Sci Sports Exerc*. 1999;31(11):S547–S552.
125. Lemon PWR. Beyond the zone: protein needs of active individuals. *J Am Coll Nutr*. 2000;19(5):513S–521S.

126. National Institutes of Health Office of Dietary Supplements Web site [Internet]. Nutrient recommendations: Dietary Reference Intakes (DRI). 2021 [cited 2021]. Available from: https://ods.od.nih.gov/HealthInformation/Dietary_Reference_Intakes.aspx
127. Cortright RL, Chandler MP, Lemon PWR, DiCarlo SE. Daily exercise reduces fat, protein and body mass in male but not female rats. *Physiol Behav*. 1997;62(1):105–11.
128. Lemon PWR. Do athletes need more dietary protein and amino acids? *Int J Sport Nutr*. 1995;5:S39–S61.
129. Lemon PWR. Effects of exercise on dietary protein requirements. *Int J Sport Nutr*. 1998;8:426–447.
130. Deutser PA, Zelazowska EB, Singh AK, Sternberg EM. Expression of lymphocyte subsets after exercise and dexamethasone in high and low stress responders. *Med Sci Sports Exerc*. 1999;31(12):1799–1806.
131. Biolo G, Maggi SP, Williams BD, Tipton KD, Wolfe RR. Increased rates of muscle protein turnover and amino acid transport after resistance exercise in humans. *Am J Physiol*. 1995;268:E514–E520.
132. Miller SL, Tipton KD, Chinkes DL, Wolf SE, Wolfe RR. Independent and combined effects of amino acids and glucose after resistance exercise. *Med Sci Sports Exerc*. 2003;35(3):449–55.
133. Wolfe RR. Protein supplements and exercise. *Am J Clin Nutr*. 2000;72:551S–557S.
134. Melby CL, Ho RC, Hill JO. Assessment of human energy expenditure. In: Bouchard C, editor. *Physical Activity and Obesity*. Champaign (IL): Human Kinetics; 2000. p. 103–31.
135. Allor KM, Pivarnik JM. Stability and convergent validity of three physical activity assessments. *Med Sci Sports Exerc*. 2001;33(4):671–6.
136. Haskell WL. Assessment of physical activity. *Med Sci Sports Exerc*. 1993;25(1):60–70.
137. Ainsworth BE, Haskell WL, Whitt MC, et al. Compendium of physical activities: an update of activity codes and MET intensities. *Med Sci Sports Exerc*. 2000;32(9):S498–S516.
138. Bergstrom J. Muscle electrolytes in man. Determination by neutron activation analysis on needle biopsy specimens. A study on normal subjects, kidney patients and patients with chronic diarrhoea. *Scand J Clin Lab Invest*. 1962;14(Suppl 68):1–110.
139. Haub MD, Potteiger JA, Jacobsen DJ, Nau KL, Magee LM, Comeau MJ. Glycogen replenishment and repeated short duration high intensity exercise: effect of carbohydrate ingestion. *Int J Sport Nutr Exerc Metab*. 1998;9:406–15.
140. Ahlborg B, Bergstrom J, Ekelund LG, Hultman E. Muscle glycogen and muscle electrolytes during prolonged physical exercise. *Acta Physiol Scand*. 1967;70:129–42.
141. Bergstrom J, Hermansen L, Hultman E, Saltin B. Diet, muscle glycogen and physical performance. *Acta Physiol Scand*. 1967;71:140–50.
142. Bergstrom J, Hultman E. Muscle glycogen synthesis after exercise: an enhancing factor localized to the muscle cells in man. *Nature*. 1966;210:309–11.
143. Ivy JL, Katz A, Cutler CL, Sherman WM, Coyle EF. Muscle glycogen synthesis after exercise: effect of time of carbohydrate ingestion. *J Appl Physiol*. 1988;64(4):1480–5.
144. Armstrong LE, Casa DJ, Millard-Stafford ML, Moran DS, Pyne SW, Roberts WO. Exertional heat illness during training and competition. *Med Sci Sports Exerc*. 2007;39(3):556–72.
145. Armstrong LE, Costill DL, Fink WJ. Influence of diuretic-induced dehydration on competitive running performance. *Med Sci Sports Exerc*. 1985;17(4):456–61.
146. Cheuvront SN, Carter RC, Sawka MN. Fluid balance and endurance exercise performance. *Curr Sports Med Rep*. 2003;2:202–8.
147. Armstrong LE, Epstein Y, Greenleaf JE, et al. Heat and cold illness during distance running. *Med Sci Sports Exerc*. 1996;28(12):i–x.
148. Convertinoe VA, Armstrong LE, Coyle EF, et al. Exercise and fluid replacement. *Med Sci Sports Exerc*. 1996;28(1):1–7.
149. Casa DJ, Armstrong LE, Hillman S, et al. National Athletic Trainer's Association position statement: fluid replacement for athletes. *J Athl Train*. 2000;35(2):212–24.
150. Casa DJ, Csillan D. Preseason heat-acclimatization guidelines for secondary school athletics. *J Athl Train*. 2009;44(3):332–3.
151. Castellani JW, Young AJ, Ducharme MB, Giesbrecht GG, Glickman E, Sallis RE. Prevention of cold injuries during exercise. *Med Sci Sports Exerc*. 2006;38(11):2012–29.
152. Horvath SM. Exercise in a cold environment. In: Miller DI, editor. *Exercise and Sport Sciences Review*. 9th ed. Salt Lake City (UT): Franklin Institute; 1981. p. 221–63.
153. Young AJ. Homeostatic responses to prolonged cold exposure: human cold acclimatization. In: Fregly MJ, Blatteis CM, editors. *Handbook of Physiology: Environmental Physiology*. Bethesda (MD): American Physiological Society; 1996. p. 419–38.

CHAPTER 5

Clinical Exercise Physiology

After completing this chapter, you will be able to:

1. Describe the historical development of clinical exercise physiology.
2. Explain the duties and responsibilities of a clinical exercise physiologist.
3. List the physiologic data collected during a graded exercise test.
4. List the physiologic data collected during pulmonary function testing.
5. Describe the components of health-related physical fitness testing.
6. Identify the primary cardiovascular, respiratory, metabolic, and neuromuscular diseases.

Physical activity and exercise play an essential role in the prevention of, treatment of, and recovery from a variety of disease conditions and physical disabilities. Clinical exercise physiology involves the use of physical activity and exercise to prevent or delay the onset of chronic disease in healthy individuals or provide therapeutic or functional benefits to individuals with disease conditions or physical disabilities. Individuals trained in exercise science and possessing certification to work in clinical environments are called clinical exercise physiologists. Assisting other health care providers and specialists in diagnostic and functional capacity testing; prescribing individual exercise based on client needs, desires, and abilities; and instructing, supervising, and monitoring exercise programs in clinical settings are the primary responsibilities of clinical exercise physiologists (1). Employment as a clinical exercise physiologist can occur in a variety of medical and clinical settings such as hospitals, rehabilitation centers, outpatient clinics, weight management clinics, in community, corporate, commercial, and university fitness and wellness centers, nursing homes, and retirement communities. The range of practice includes apparently healthy individuals with no known medical problems and individuals with diagnosed cardiovascular, pulmonary, metabolic, rheumatologic, cancer, behavioral and mental health, orthopedic, and/or neuromuscular diseases and conditions (1,2).

Clinical exercise physiologists must have a solid educational background in exercise physiology, including the understanding of how the body responds to acute and chronic physical activity and exercise in both a healthy and diseased condition. Advanced knowledge of the pathophysiology of chronic diseases, pharmacology of drugs and medicines, medical terminology, medical record keeping and charting, electrocardiographic interpretation, exercise testing for special populations, and nutrition is also valuable to the clinical exercise physiologist (3). Undergraduate students interested in clinical exercise physiology should complete a clinical internship, which allows work with patients in a medical setting, interactions with a variety of health care professionals, and opportunities to prescribe and monitor exercise in a variety of locations (2,4). Individuals desiring to become clinical exercise physiologists should obtain a certification from a professional organization such as the American College of Sports Medicine (ACSM) (see Chapter 12).

HISTORY OF CLINICAL EXERCISE PHYSIOLOGY

Although clinical exercise physiology is relatively new as a defined body of knowledge, individuals have been interested in how physical activity and exercise influence health and recovery from illness and diseases since the time of the early Greeks (5). Early historic events in medicine, physiology, and exercise that helped define many of the disciplines in exercise science have also influenced the development of clinical exercise physiology. Many of those

Therapeutic Of or relating to the treatment of disease or disorders by remedial agents or methods.
Functional Performing or being able to perform a regular function.
Pathophysiology Functional changes that accompany a disease condition.

important historic events are described in Chapters 1 and 4. The material contained in this section should be considered a supplement to these chapters and contains only information specific to the development of clinical exercise physiology.

Early Influences

The use of physical activity and exercise in cardiovascular disease recovery can be traced to the eighteenth and nineteenth centuries (6). William Heberden (1710–1801) was the first to describe the condition of **angina pectoris** (chest pain) during physical exertion (7) and the use of physical activity with patients who experienced angina pectoris (8). Heberden and William Stokes (1804–1878) are credited with being the first physicians to recommend the use of physical activity and exercise to promote the recovery from heart disease (6,9). By the early 1900s, many physicians accepted that a fairly high level of physical activity was safe and necessary for the heart's normal development (10). Unfortunately, prior to the middle of the twentieth century, there was little written about the role of physical activity and exercise in the prevention of chronic disease development or in the recovery process from disease conditions (4).

Late Twentieth-Century Influences

One of the most significant events in the development of clinical exercise physiology was a renowned study of coronary heart disease in London bus drivers and conductors performed by Jeremy Morris and his colleagues and published in 1953 (11). This study is considered the first to demonstrate the relationship between physical activity and the reduced risk of developing heart disease, and it helped initiate interest in disease risk reduction and public health epidemiology. The foundation for **cardiac rehabilitation** began to take shape in the 1950s, with the major focus on the restoration of functional capacity after a cardiovascular event. Around this time, Samuel Levine and Bernard Lown were the first to recommend armchair exercises for patients with heart disease (12), and Herman Hellerstein provided a step-by-step plan for the rehabilitation of cardiac patients (1). Although the use of physical activity and exercise in heart disease patients was not common prior to 1960 (13), the work of the research groups led by Hellerstein, Levine, Lown, and Morris created a foundation for using physical activity and exercise to promote the prevention of and recovery from various diseases. Throughout the 1960s and 1970s, experiments in both animals and humans led to using physical activity and exercise in the recovery process for patients with acute heart attacks (13). For example, several prospective studies supported the use of a structured physical activity and exercise program for reducing morbidity and mortality and improving the various clinical, medical, physiological, and psychological factors in patients with heart disease (14,15). The development of cardiac rehabilitation programs advanced significantly

Angina pectoris Severe chest pain caused by an insufficient supply of blood to the heart.
Cardiac rehabilitation A medically supervised program to help heart patients recover quickly and improve their overall physical and mental functioning.

during the 1970s when professional organizations such as the American Heart Association and the ACSM released textbooks addressing the proper procedures for the testing and training of healthy and diseased individuals (16–18). These books were very well received and remain in wide use.

The application of physical activity and exercise to promote recovery in patients with pulmonary disease is credited to Alvan Barach (6). Considered by many experts to be the "Father of modern day pulmonology," Barach used a variety of procedures and strategies to improve the conditions of individuals with congestive heart failure, pneumonia, and other pulmonary disorders (19). Beginning in 1958, William Miller and Thomas Petty published a series of papers promoting the use of physical activity and exercise for treating individuals with chronic pulmonary disorders, including airway obstruction and emphysema (20–22). The development of **pulmonary rehabilitation** as an appropriate part of the treatment for diseased individuals was further advanced when, in 1981, the American Thoracic Society released a statement supporting pulmonary rehabilitation as a necessary procedure for enhancing functional status in pulmonary patients (23).

In 1974, the *Journal of Cardiac Rehabilitation* began publication, providing practitioners access to scientific information vital to the rehabilitation of individuals with various forms of heart disease (6). In 1986, the content of the journal was expanded to include pulmonary rehabilitation, and the journal was renamed the *Journal of Cardiopulmonary Rehabilitation* (6). The development of a professional association for practicing cardiopulmonary rehabilitation specialists occurred in 1985, when the American Association of Cardiovascular and Pulmonary Rehabilitation (AACVPR) was established (6). The mission of the AACVPR is to reduce morbidity, mortality, and disability from cardiovascular and pulmonary diseases through education, prevention, rehabilitation, research, and disease management (6). The knowledge base in cardiopulmonary rehabilitation was expanded through research and the publication of guidelines and position statements by the AACVPR. Table 5.1 lists some important publications that helped further the development of cardiopulmonary rehabilitation. The AACVPR further enhanced the profession of cardiopulmonary rehabilitation with the establishment of certification for cardiac and pulmonary rehabilitation programs in 1996 (6).

The development of individual certification programs by a number of other professional organizations has led to further professional growth of cardiopulmonary rehabilitation. Major professional organizations such as the ACSM and the National Strength and Conditioning Association offer professional certifications to individuals desiring to work with healthy and diseased populations. Additional information about certification and licensure opportunities can be found in Chapter 12.

The endorsement of the health benefits obtained from physical activity and exercise by the United States Surgeon General in 1996 was a significant milestone in the promotion of physical activity and exercise for healthy and diseased individuals. The Surgeon General's report highlighted the positive health effects of physical activity and exercise on the

Pulmonary rehabilitation A medically supervised program to help patients with chronic respiratory disease stabilize or reverse systemic manifestations.

Table 5.1 Important Publications in Cardiopulmonary Rehabilitation

PUBLICATION	DATE
Outcome Measurement in Cardiac Rehabilitation (77)	2004
Clinical Competency Guidelines for Pulmonary Rehabilitation Professionals (78)	2007
Pulmonary Rehabilitation: Joint ACCP/AACVPR Evidence-Based Clinical Practice Guidelines (79)	2007
Core Competencies for Cardiac Rehabilitation/Secondary Prevention Professionals: 2010 Update (3)	2011
Guidelines for Pulmonary Rehabilitation Programs, 4th Edition (80)	2011
Guidelines for Cardiac Rehabilitation and Secondary Prevention Programs, 5th Edition (81)	2013
ACSM's Guidelines for Exercise Testing and Prescription, 11th Edition (29)	2021

musculoskeletal, cardiovascular, respiratory, and endocrine systems, including a reduced risk of premature mortality and reduced risks of coronary heart disease, hypertension, colon cancer, and diabetes mellitus. Recommendations for the appropriate amount of physical activity and exercise helped establish the standards for using exercise to assist in the treatment of diseased individuals (24,25).

Throughout the 1990s and early into the 21st century, government agencies such as the Department of Health and Human Services and the Centers for Disease Control and Prevention (CDC) established health promotion programs designed to reduce the risk of disease development in healthy individuals and improve the health of those with disease conditions. The National Institutes of Health and many professional organizations continue to promote research activities and public health programs designed to improve health and reduce disease risk. The Jump Rope for Heart program from the American Heart Association and the Exercise is Medicine® initiative from the ACSM are two examples of programs designed to promote exercise as a treatment for a number of chronic diseases.

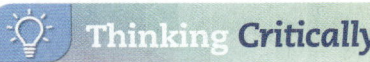

Thinking Critically

In what ways could the historic development of cardiopulmonary rehabilitation and the 1996 Surgeon General's report be used to promote participation in physical activity and exercise for the prevention and treatment of disease conditions?

Twenty-First-Century Influences

Into the 21st century, the AACVPR continues to be a leader in the field of clinical exercise physiology. The Clinical Exercise Physiology Association (CEPA), founded in 2008, is an affiliate society of ACSM with the mission of advancing the scientific and practical application of clinical exercise physiology for the betterment of the health, fitness, and quality of life for patients at high risk for or living with a chronic disease (26). CEPA first published the *Journal of Clinical Exercise Physiology* in 2012, which focuses on important topics of interest to clinical exercise physiologists.

In 2012, the Million Hearts initiative was launched by the CDC and the Centers for Medicare and Medicaid Services (CMS) with the goal of preventing 1 million heart attacks, strokes, and other acute cardiovascular events in a 5-year period (27). The initiative was renewed for a second 5-year term and is now called Million Hearts 2022. Although cardiac rehabilitation was not an initial focus of the Million Hearts initiative, it was identified as an important component of a comprehensive, evidence-based approach to secondary cardiovascular disease prevention. In conjunction, the Million Hearts Cardiac Rehabilitation Leadership Summit was held in 2015. The Summit brought together leaders in health care, health insurance, and cardiac rehabilitation with staff from CDC, CMS, and the National Institutes of Health to discuss strategies to increase cardiac rehabilitation referral and participation (27).

In 2014, individual certification was first made available in the form of the Certified Cardiac Rehabilitation Professional (CCRP) certificate. The CCRP is designed for cardiac rehabilitation professionals and is aligned with published cardiac rehabilitation competencies. The AACVPR is also a leading advocate for the practice of cardiovascular and pulmonary rehabilitation, promoting legislation and regulation that will have a positive impact on the care of clinical patients.

As we move further into the 21st century, additional education and health promotion programs will be instrumental in promoting the use of nutrition, stress management, weight management, physical activity, and exercise for promoting and ensuring good health and facilitating recovery from disease conditions. It is critical that the evaluation of the effectiveness of these programs is conducted on a regular basis so adjustments in programs can be made if warranted. For example, as part of the Exercise is Medicine® (EIM) initiative, the EIM Research Learning Collaborative was established to oversee the evaluation of the EIM initiative in collaboration with partnering health care systems, community organizations, the fitness industry, and technology businesses (28). Clinical exercise physiologists will be important contributors to this task. Table 5.2 provides a list of the significant events in the historic development of clinical exercise physiology.

Thinking Critically

In what ways do the professional organizations, AACVPR and CEPA, contribute to the development of clinical exercise physiology as both a discipline and a profession?

CLINICAL EXERCISE PHYSIOLOGISTS' DUTIES AND RESPONSIBILITIES

Clinical exercise physiologists take part in a number of significant duties and responsibilities to evaluate the health status of and prescribe physical activity and exercise to both healthy individuals and individuals with disease conditions. The primary duties of clinical exercise physiologists include conducting preexercise testing screening, performing exercise testing and evaluation, developing exercise prescriptions, instructing individuals in proper training techniques, and supervising safe and effective exercise programs in various health care, community, and employment settings. Clinical exercise physiologists must understand the normal physiologic responses of the body to acute and chronic physical activity and exercise to maximize the use of physical activity and exercise for the prevention, management, or rehabilitation of disease. It is also important to understand how different diseases and the medical management of the disease conditions affect the physiologic responses during rest and exercise (3,4).

Table 5.2 Significant Events in the History of Clinical Exercise Physiology

DATE	EVENT
1802	Heberden describes using physical exertion to treat angina pectoris
1854	Stokes recommends using physical activity and exercise during the recovery from heart disease
1948	Barach promotes the use of physical activity and exercise to promote recovery in individuals with pulmonary disease
1952	Levine and Lown recommend armchair exercises for patients with heart disease
1953	Morris demonstrates a relationship between physical activity and a reduced risk of developing heart disease
1957	Hellerstein and Ford outline a plan for rehabilitation of cardiac patients
1958	Miller and Petty publish a series of papers promoting the use of physical activity and exercise for treating individuals with chronic pulmonary disorders
1972 and 1975	American Heart Association and the American College of Sports Medicine release textbooks addressing the proper testing and training of healthy and diseased individuals
1974	Publication of the *Journal of Cardiac Rehabilitation*
1985	American Association of Cardiovascular and Pulmonary Rehabilitation is established
1996	Surgeon General's report highlights the positive health effects that physical activity has on the physiologic systems and the reduction of chronic disease risk
1996	American Association of Cardiovascular and Pulmonary Rehabilitation establishes certification for cardiac and pulmonary rehabilitation programs
2008	Exercise is Medicine® initiative launched by ACSM
2014	AACVPR introduces the Certified Cardiac Rehabilitation Professional certificate

Exercise Testing and Evaluation

Exercise testing is an important component of clinical exercise physiology as it is used to clear individuals for safe participation in physical activity and exercise and as a basis for developing an exercise prescription. Diagnostic testing and functional capacity testing are the two broad classifications of exercise testing and evaluation.

Diagnostic testing is commonly used to assess the presence of cardiovascular or pulmonary disease. Figure 5.1 illustrates the type of diagnostic exercise testing performed by clinical exercise physiologists. During diagnostic testing, measures of heart rate, electrical activity of the heart, and blood pressure are made. If an individual has symptoms of heart or lung disease, a history of a possible abnormal cardiac incident, abnormal electric activity of the heart,

Diagnostic testing Used to determine a specific disease condition or possible illness.

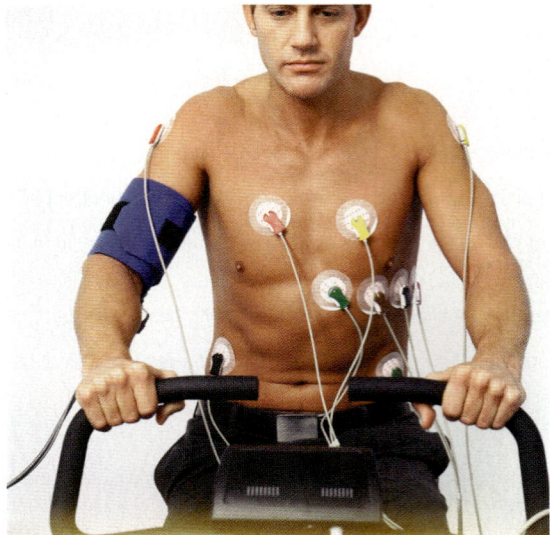

FIGURE 5.1 A diagnostic exercise test can be used to assess the presence of cardiovascular disease.

or a high probability of an underlying disease condition, then diagnostic testing is performed, typically under medical supervision. Exercise testing helps diagnose the presence of heart disease, primarily on the basis of abnormal changes of blood pressure, heart rate, or electrical activity during the test. During exercise, the increased demands placed on the heart allow abnormal responses and disease conditions to become more readily apparent. Although the clinical exercise physiologist plays an important role in administering the test, only a health care provider such as a medical doctor can provide a medical diagnosis of disease (4,29).

Functional capacity testing provides information about an individual's fitness level and capacity to participate in physical activity and exercise. The information obtained from testing can be used to prescribe an appropriate physical activity and exercise program to improve fitness and health. Functional capacity is usually determined by a submaximal or maximal exercise test that progressively increases in intensity. Functional capacity testing may also be used to determine whether an individual has normal cardiovascular and pulmonary responses to physical activity and exercise (29).

Both diagnostic and functional capacity testing are used to evaluate the cardiovascular and pulmonary responses to a standard exercise workload. Diagnostic and functional capacity tests, which use workloads that are incremental and increase in intensity, are referred to as graded exercise tests (GXTs). During a GXT, the intensity progresses in stages from light to maximal exertion or to a previously determined ending point. There are general guidelines to be followed when performing a GXT, which is typically conducted on a treadmill or a cycle ergometer. Several standard exercise protocols are available for use, and the protocol selected depends on the purpose of the test and the characteristics of the individual being tested (29).

Functional capacity testing Used to provide an objective measure of an individual's safe functional abilities.

Pretesting Procedures

Physical activity and exercise stress the systems of the body and increase the risk of musculoskeletal injury and abnormal cardiovascular and pulmonary events. Certain precautions must be taken prior to conducting a diagnostic or functional capacity GXT to reduce the risk of injury or an abnormal event. These precautions include pretesting screening, a physical examination, collection of health history information, and obtaining individual informed consent (29). Clinical exercise physiologists are instrumental in performing some of the pretesting procedures, including pretest screening for health risk, collection of health history information, and acquiring informed consent.

Pretest Screening for Health Risk

Determining whether physical activity, exercise, and exercise testing are appropriate for an individual is very important. For most healthy individuals, physical activity and exercise do not pose a safety risk if proper exercise techniques and principles are observed. However, physical activity and exercise may not be safe for everyone, especially if a preexisting medical condition such as cardiovascular, pulmonary, or metabolic disease exists. Some individuals may not be able to participate in exercise testing for specific medical reasons. A list of contraindications can be found in the *ACSM's Guidelines for Exercise Testing and Prescription* (29). The level of risk for an individual participating in physical activity and exercise must be determined before administering a diagnostic or functional capacity test and the start of an exercise program (4).

Physical Examination

Certain individuals who are physically inactive and have multiple risk factors for disease require a physician's referral before they can undergo exercise testing or begin an exercise program. In this situation, the physical examination is designed to assess the risk of an abnormal event while participating in exercise or exercise testing. The *ACSM's Guidelines for Exercise Testing and Prescription* can be used to assist clinical exercise physiologists and health care professionals in determining the safety of exercise for individuals (29).

Health Status

The assessment of an individual's personal health status and risk factors for cardiovascular disease is an important component of the pretesting screening phase of exercise testing and prescription. The assessment of individual health status is designed to

- identify individuals with **medical contraindications** to exercise
- identify individuals with clinically significant disease conditions who should be referred to a medically supervised exercise program

Medical contraindication A condition that makes a particular treatment or procedure inadvisable.

- identify individuals with symptoms and risk factors for a disease who should receive further medical evaluation before starting an exercise program
- identify individuals with special needs for safe exercise participation (*e.g.*, elderly, pregnant women) (29)

A number of instruments are available for recording and evaluating an individual's health history. The American Heart Association/ACSM Health/Fitness Facility Preparticipation Screening Questionnaire for Exercise Professionals and the Physical Activity Readiness Questionnaire (PAR-Q) are examples of instruments that can be used in the pretesting screening phase of exercise testing and prescription. Questionnaires should be designed to collect health history information about the individual and their family. Figure 5.2 shows an example of a questionnaire designed to determine health status (29).

Informed Consent

Informed consent is a process whereby the individual participating in the exercise test or receiving an exercise prescription is made aware of and understands the purposes, risks, and benefits associated with the test or exercise program. A signed informed consent document should be obtained from an individual prior to diagnostic or functional capacity testing. All of the procedures involved in the exercise test and the potential risks and benefits should be thoroughly explained before any testing is performed. During the collection of informed consent, participants should be encouraged to ask questions to clarify and resolve uncertainties about the testing procedures. Figure 5.3 shows an example of a comprehensive informed consent document for an exercise test (29).

Performing the Test

After the collection and evaluation of the health history information and the securing of informed consent, the GXT can begin if the individual being tested meets the accepted level of health and disease risk for participation in exercise testing. Figures 5.4 and 5.5 provide algorithm charts that can be used to identify those individuals who can participate in exercise and exercise testing and at what intensity of exercise they can safely participate (29). During the GXT, physiologic measures such as resting and exercise heart rate and blood pressure are collected. Additional measures collected often include the **rating of perceived exertion** (RPE), electric activity of the heart using an **electrocardiograph** (ECG), oxygen consumption ($\dot{V}O_2$) to determine **maximal oxygen consumption** ($\dot{V}O_{2max}$), and physical work capacity. A GXT can be either submaximal or maximal depending on the prescreening information and whether the test is for diagnostic or functional capacity assessment (29).

Rating of perceived exertion A subjective assessment of how hard an individual feels they are working.

Electrocardiograph An instrument that measures electric potentials on the body surface and generates a record of the electric currents associated with heart muscle activity.

Maximal oxygen consumption The maximal amount of oxygen used by the body during maximal effort exercise.

Exercise Preparticipation Health Screening Questionnaire for Exercise Professionals

Assess your client's health needs by marking all *true* statements.

Step 1

SIGNS AND SYMPTOMS
Does your client experience:
___ chest discomfort with exertion
___ unreasonable breathlessness
___ dizziness, fainting, blackouts
___ ankle swelling
___ unpleasant awareness of a forceful, rapid or irregular heart rate
___ burning or cramping sensations in your lower legs when walking short distance
___ known heart murmur

If you **did** mark any of these statements under the symptoms, **STOP**, your client should seek medical clearance before engaging in or resuming exercise. Your client may need to use a facility with a **medically qualified staff**.

If you **did not** mark any symptoms, continue to Steps 2 and 3.

Step 2

CURRENT ACTIVITY
Has your client performed planned, structured physical activity for at least 30 min at moderate intensity on at least 3 days per week for at least the last 3 months?

Yes ☐   No ☐

Continue to Step 3.

Step 3

MEDICAL CONDITIONS
Has your client had or do they currently have:
___ a heart attack
___ heart surgery, cardiac catheterization, or coronary angioplasty
___ pacemaker/implantable cardiac defibrillator/rhythm disturbance
___ heart valve disease
___ heart failure
___ heart transplantation
___ congenital heart disease
___ diabetes
___ renal disease

Evaluating Steps 2 and 3:

- If you **did not mark any of the statements in Step 3**, medical clearance is not necessary.
- If you marked Step 2 "**yes**" and **marked any of the statements in Step 3**, your client may continue to exercise at light to moderate intensity without medical clearance. However, medical clearance is recommended before engaging in vigorous exercise.
- If you marked Step 2 "**no**" and **marked any of the statements in Step 3**, medical clearance is recommended. Your client may need to use a facility with a **medically qualified staff**.

FIGURE 5.2 An Exercise Preparticipation Health Screening Questionnaire. (Reprinted from Magal M, Riebe D. New preparticipation health screening recommendations: what exercise professionals need to know. *ACSM Health Fit J*. 2016;20(3):22–7.)

Heart Rate

Resting heart rate is usually measured after the individual has been sitting quietly for 5 minutes or longer. An electronic heart rate monitor is often used to obtain heart rate in healthy individuals. In all clinical settings, however, the heart rate is determined from the

Informed Consent for an Exercise Test

1. **Purpose and Explanation of the Test**
   You will perform an exercise test on a cycle ergometer or a motor-driven treadmill. The exercise intensity will begin at a low level and will be advanced in stages depending on your fitness level. We may stop the test at any time because of signs of fatigue or changes in your heart rate, electrocardiogram, or blood pressure, or symptoms you may experience. It is important for you to realize that you may stop when you wish because of feelings of fatigue or any other discomfort.

2. **Attendant Risks and Discomforts**
   There exists the possibility of certain changes occurring during the test that increase risk. These include abnormal blood pressure; fainting; irregular, fast, or slow heart rhythm; and, in rare instances, heart attack, stroke, or death. Every effort will be made to minimize these risks by evaluation of preliminary information relating to your health and fitness and by careful observations during testing. Emergency equipment and trained personnel are available to deal with unusual situations that may arise.

3. **Responsibilities of the Participant**
   Information you possess about your health status or previous experiences of heart-related symptoms (*e.g.*, shortness of breath with low-level activity; pain; pressure; tightness; heaviness in the chest, neck, jaw, back and/or arms) with physical effort may affect the safety of your exercise test. Your prompt reporting of these and any other unusual feelings with effort during the exercise test itself is very important. You are responsible for fully disclosing your medical history as well as symptoms that may occur during the test. You are also expected to report all medications (including nonprescription) taken recently and, in particular, those taken today to the testing staff.

4. **Benefits to Be Expected**
   The results obtained from the exercise test may assist in the diagnosis of your illness, in evaluating the effect of your medications, or in evaluating what type of physical activities you might do with low risk.

5. **Inquiries**
   Any questions about the procedures used in the exercise test or the results of your test are encouraged. If you have any concerns or questions, please ask us for further explanations.

6. **Use of Medical Records**
   The information that is obtained during exercise testing will be treated as privileged and confidential as described in the Health Insurance Portability and Accountability Act of 1996. It is not to be released or revealed to any individual except your referring physician without your written consent. However, the information obtained may be used for statistical analysis or scientific purposes with your right to privacy retained.

7. **Freedom of Consent**
   I hereby consent to voluntarily engage in an exercise test to determine my exercise capacity and state of cardiovascular health. My permission to perform this exercise test is given voluntarily. I understand that I am free to stop the test at any point if I so desire.

I have read this form, and I understand the test procedures that I will perform and the attendant risks and discomforts. Knowing these risks and discomforts, and having had an opportunity to ask questions that have been answered to my satisfaction, I consent to participate in this test.

Date	Signature of Patient
Date	Signature of Witness
Date	Signature of Physician or Authorized Delegate

FIGURE 5.3 A comprehensive informed consent document for an exercise test (29).

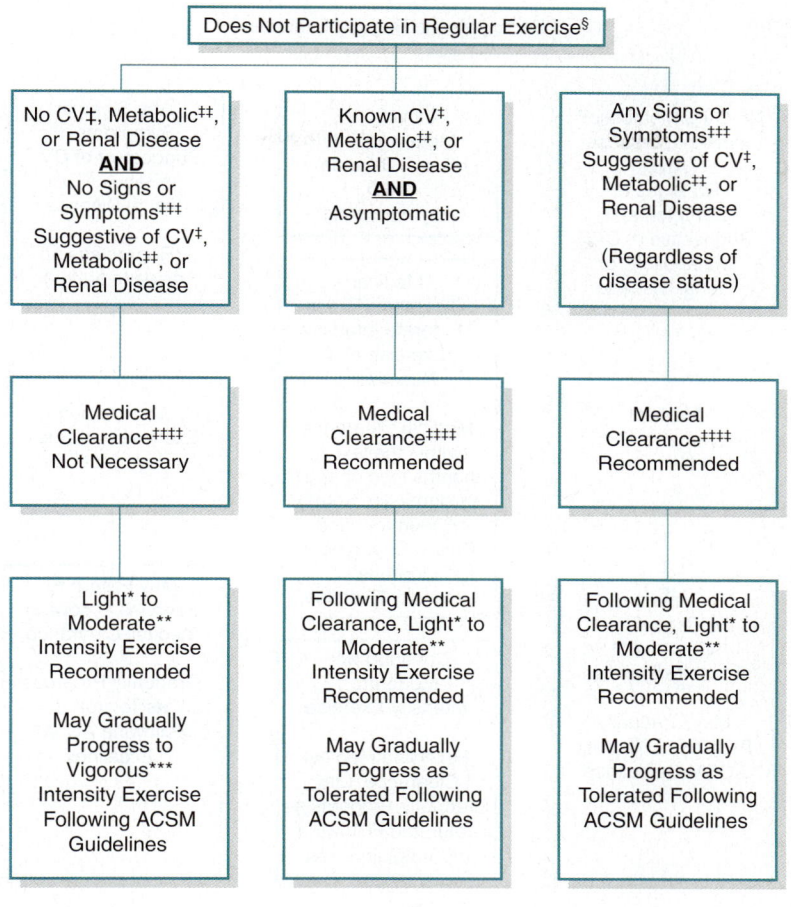

§Exercise Participation	Performing planned, structured physical activity at least 30 min at moderate intensity on at least 3 d · wk^{-1} for at least the last 3 mo	
*Light Intensity Exercise	30%–39% HRR or $\dot{V}O_2R$, 2–2.9 METs, RPE 9–11, an intensity that causes slight increases in HR and breathing	
**Moderate Intensity Exercise	40%–59% HRR or $\dot{V}O_2R$, 3–5.9 METs, RPE 12–13, an intensity that causes noticeable increases in HR and breathing	
***Vigorous Intensity Exercise	60% HRR or $\dot{V}O_2R$, 6 METs, RPE 14, an intensity that causes substantial increases in HR and breathing	
‡Cardiovascular (CV) Disease	Cardiac, peripheral vascular, or cerebrovascular disease	
‡‡Metabolic Disease	Type 1 and 2 diabetes mellitus	
‡‡‡Signs and Symptoms	At rest or during activity. Includes pain, discomfort in the chest, neck, jaw, arms, or other areas that may result from ischemia; shortness of breath at rest or with mild exertion; dizziness or syncope; orthopnea or paroxysmal nocturnal dyspnea; ankle edema; palpitations or tachycardia; intermittent claudication; known heart murmur; unusual fatigue or shortness of breath with usual activities.	
‡‡‡‡Medical Clearance	Approval from a health care professional to engage in exercise	
ACSM Guidelines	See the most current edition of *ACSM's Guidelines for Exercise Testing and Prescription*	

FIGURE 5.4 An algorithm for exercise testing and testing supervision for individuals who do not participate in regular exercise. (Used with permission from Riebe D, Franklin BA, Thompson PD, et al. Updating ACSM's recommendations for exercise preparticipation health screening. *Med Sci Sports Exerc.* 2015;47(11):2473–9.)

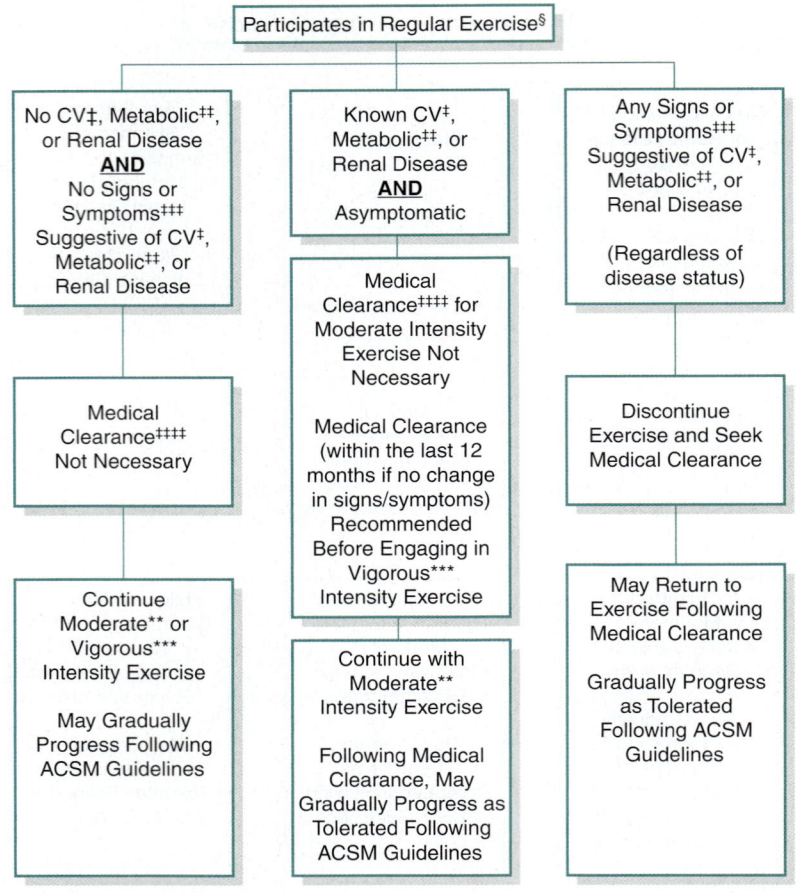

FIGURE 5.5 An algorithm for exercise testing and testing supervision for individuals who participate in regular exercise. (Used with permission from Riebe D, Franklin BA, Thompson PD, et al. Updating ACSM's recommendations for exercise preparticipation health screening. *Med Sci Sports Exerc.* 2015;47(11):2473–9.)

ECG recordings or directly off the digital display of the **oscilloscope**. Many factors can influence the resting and exercise heart rate, including smoking, caffeine ingestion, fever, high humidity, stress or anxiety, food digestion, certain medications, inflammation, and prior physical activity or exercise. That is why pretesting instructions to individuals often prohibit the use of caffeine, food consumption, or activity prior to exercise testing. During exercise, the heart rate is recorded periodically to ensure an appropriate cardiovascular response to exercise and for later use in developing an exercise prescription (29).

Blood Pressure

Blood pressure is measured after a period of quiet sitting and often at the same time as the resting heart rate. Blood pressure is the force exerting pressure against the walls of the blood vessels in the circulatory system. The highest pressure recorded during a **cardiac cycle** (one heart beat to the next heart beat) occurs during the contraction phase (systole) of the ventricles and is called the systolic blood pressure. A measurement of systolic blood pressure provides an estimation of the work of the heart, as well as the pressure exerted against the walls of the blood vessels. The period between heart contractions is called the relaxation phase of the heart (diastole), and the pressure recorded during this period is called the diastolic blood pressure. During diastole, the blood pressure is decreased, and this measurement gives an indirect indication of the ease with which blood flows through the circulatory system (29).

Blood pressure is an important indicator of overall health. When resting blood pressure is chronically elevated, a disease condition called hypertension exists. Individuals with hypertension have an increased risk of stroke and cardiovascular disease (30,31). In clinical exercise settings, blood pressure can be measured manually or by an automated analyzer. The manual method uses a sphygmomanometer (blood pressure cuff) and a stethoscope and is often performed by a clinical exercise physiologist. The automated analyzer eliminates much of the individual variability associated with the manual method assessment. Table 5.3 provides the categories for classification and management of blood pressure for adults (29).

Physical activity and exercise cause an increase in blood pressure. A GXT is used to evaluate the blood pressure response to increasing workloads and to help identify any abnormal responses that may occur. During testing, blood pressure is normally measured every 1 to 3 minutes. Systolic blood pressure increases linearly with workload and can reach approximately 200 mm Hg in healthy, fit men and women during maximal exercise. Diastolic pressure should remain the same or decrease slightly during a GXT. Abnormal systolic or diastolic blood pressure responses during exercise typically indicate a problem with the cardiovascular system (29).

Oscilloscope An electronic instrument that produces an instantaneous trace on the screen that corresponds to oscillations of voltage and current.
Blood pressure The force exerting pressure against the walls of the blood vessels.
Cardiac cycle A complete beat of the heart, including systole and diastole and the intervals between.

Table 5.3 Categories of Normal and Elevated Blood Pressure[a,b]

American College of Cardiology/American Heart Association (82)

BLOOD PRESSURE CLASSIFICATION	SYSTOLIC BLOOD PRESSURE (mm Hg)	DIASTOLIC BLOOD PRESSURE (mm Hg)
Normal	<120	and <80
Elevated	120–129	and <80
Hypertension, stage 1	130–139	or 80–89
Hypertension, stage 2	≥140	or ≥90

Joint National Committee on Prevention, Detection, Evaluation, and Treatment of High Blood Pressure (83)

BLOOD PRESSURE CLASSIFICATION	SYSTOLIC BLOOD PRESSURE (mm Hg)	DIASTOLIC BLOOD PRESSURE (mm Hg)
Normal	<120	<80
Prehypertension	120–139	80–89
Hypertension, stage 1	140–159	90–99
Hypertension, stage 2	≥160	≥100

[a]Individuals are classified in any given category by meeting either of SBP or DBP thresholds.
[b]Individuals with SBP and DBP in two classifications should be designated to the higher BP classification.

Rating of Perceived Exertion

During a GXT, an assessment of the psychological perception of the intensity of exercise is often made. The Borg Rating of Perceived Exertion (RPE) scale is commonly used to assess the subjective level of difficulty the individual is experiencing during exercise (32). This RPE scale provides a moderately accurate measure of how the individual feels in relation to the level of physical exertion. The RPE scale also allows the technician conducting the test to know when the individual being tested is nearing perceived maximal effort. Figure 5.6 illustrates the Borg RPE scales. The 6–20 point numerical RPE scale relates closely to the heart rates from rest to maximal exercise when multiplied by a factor of 10 (60–200 beats $\cdot$ min^{-1}) in nonmedicated individuals. This 15-point RPE scale is anchored with the terms very, very light and very, very hard. A revised RPE scale attempts to provide a category-ratio scale of the RPE values (ranging from 0 to 11) that are anchored with the terms "nothing at all" and "absolute maximum" (33). An additional perceived exertion scale called the Omni scale has been developed that uses pictures to help individuals identify the level of difficulty associated with exercise (34,35). An added benefit to collecting the RPE is that the information can be used in the development of an exercise prescription, because individuals can easily learn to exercise at a particular RPE that corresponds to a given exercise intensity (29).

Category Scale	Category-Ratio Scale†	
6	0 Nothing at all	"No I"
7 Very, very light	0.3	
8	0.5 Extremely weak	Just noticeable
9 Very light	0.7	
10	1 Very weak	
11 Fairly light	1.5	
12	2 Weak	Light
13 Somewhat hard	2.5	
14	3 Moderate	
15 Hard	4	
16	5 Strong	Heavy
17 Very Hard	6	
18	7 Very strong	
19 Very, very hard	8	
20	9	
	10 Extremely strong	"Strongest I"
	11	
	• Absolute maximum	Highest possible

*Copyright Gunnar Borg. Reproduced with permission. For correct usage of the Borg scales, it is necessary to follow the administration and instructions given in Borg G. *Borg's Perceived Exertion and Pain Scales.* Champaign (IL): Human Kinetics; 1998.

†Note: ON the Category-Ratio Scale, "I" represents intensity.

FIGURE 5.6 Borg scales for perceived exertion (29,33).

Electrocardiogram

The electrical activity of the heart is usually recorded using an electrocardiogram (ECG) during a diagnostic test. This information helps in the overall evaluation of health and disease status. The electrical activity of the heart can be recorded from electrodes placed on the surface of the chest. The number of electrodes used can range from 4 to 12, with the electrodes placed at specific locations on the upper torso of the individual during testing. The ECG is a valuable component of the exercise test because of its use in determining the presence of cardiovascular disease. The diagnosis of these conditions is made easier during exercise because many of these problems do not become apparent until the heart is required to beat faster and generate more force than when at rest (29). Some of the more common cardiovascular abnormalities include ST segment depression and atrial and ventricular arrhythmia. Figure 5.7 illustrates the assessments made during a GXT.

Echocardiography

The functional condition and disease status of the heart can also be assessed using **echocardiography**. An echocardiograph is an instrument that uses sound waves to create a moving picture of the heart (called an echocardiogram) that is much more detailed than an X-ray

Echocardiography The use of sound waves to create a moving picture of the heart.

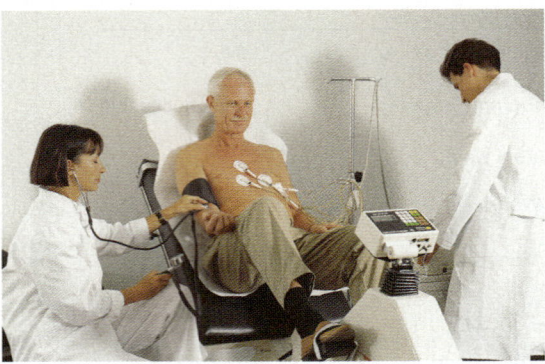

FIGURE 5.7 Assessments made during a graded exercise test include blood pressure, heart rate, and electrical activity of the heart.

image, and it involves no radiation exposure. The echocardiogram allows health care professionals to see the beating heart and to visualize many of its structures. Occasionally, a body structure such as the lungs or ribs may prevent the sound waves and echoes from providing a clear picture of heart function. In this instance, a small amount of contrast material (*i.e.*, dye) may be injected into a vein to provide a clearer picture of the vessels of the heart. An echocardiogram can also be combined with a GXT (often called a stress echocardiogram). This type of test allows health care professionals to learn how the heart functions when it is made to work harder. The echocardiogram combined with the GXT is especially useful in diagnosing coronary artery disease (36).

Oxygen Consumption and Functional Capacity

Functional capacity testing measures the degree to which a person can increase physical activity and exercise intensity and maintain this increased level. It is related to the ability of the body to deliver blood and oxygen to the working tissues of the body, which use the oxygen to change carbohydrates, fats, and proteins into energy for exercise. An individual's functional capacity is strongly influenced by the health and fitness level of the cardiovascular, respiratory, and muscular systems. The measurement of $\dot{V}O_{2max}$ is the best assessment of cardiovascular fitness and a good measure of overall physical fitness. The $\dot{V}O_{2max}$ is defined as the maximal rate at which oxygen can be taken up, distributed, and used by the body during exercise that involves large muscle mass (37). During a maximal GXT, the workload is gradually increased until the individual being tested can no longer exercise. The highest oxygen consumption measured is considered the $\dot{V}O_{2max}$. Specific test criteria have been established to ensure that a true maximal level of exercise and oxygen consumption has been achieved by individuals during the GXT (37).

Submaximal Graded Exercise Tests

A submaximal GXT can be used to evaluate the cardiovascular, respiratory, and muscular systems' responses to a standard submaximal exercise test. The submaximal GXT is conducted to an intensity that elicits between 70% and 85% of the age-predicted maximal heart rate. Information obtained during this test can be used to calculate an estimate of an individual's maximal fitness level. Cardiovascular fitness can be estimated from equations that predict $\dot{V}O_{2max}$ from the last workload achieved during the test, from the oxygen consumption requirement for horizontal and graded walking on a treadmill, or from the individual's heart rate response to a series of submaximal workloads (typically used during a cycle ergometer test). Heart rate and blood pressure are normally recorded at each stage of the test, with an ECG recording obtained, if necessary. Abnormal responses in heart rate, blood pressure, or the ECG indicate a cessation of testing and referral for further medical evaluation. A submaximal test is usually easier to administer, less costly, and safer than a maximal GXT because the individual is not required to exercise to maximal effort. In some exercise settings, an estimate of $\dot{V}O_{2max}$ is sufficient for approving exercise participation and in developing an individualized exercise prescription (29).

Maximal Graded Exercise Testing

A maximal GXT is often employed when a direct assessment of cardiovascular, respiratory, and muscular function is needed. A maximal GXT does not stop at a predetermined workload (*e.g.*, 70%–85% of age-predicted maximum) but is continued to the point of exhaustion or to the point at which abnormal physiological responses occur. If the test is diagnostic in nature, it may be conducted until abnormal responses in blood pressure and/or ECG activity as well as chest pain (angina pectoris), shortness of breath (**dyspnea**), or lightheadedness are observed. This type of test is often called a sign and symptom-limited stress test. It is often important to be able to take individuals to maximal effort exercise because many abnormal signs and symptoms of disease do not occur until the exercise workload is at a high intensity. The major concern associated with a maximal GXT is the level of physical stress placed on the participants, especially those who are not physically active, because this may increase the risk of an abnormal cardiovascular event (29).

Thinking Critically

Why is it important for a clinical exercise physiologist to collect information on heart rate, blood pressure, perceived exertion, and ECG activity during a graded exercise test in both healthy and diseased individuals?

Health-Related Physical Fitness Testing and Interpretation

The assessment of health-related physical fitness is a common and appropriate practice in preventive and rehabilitative physical activity and exercise programs. Health-related physical fitness testing can

- provide information to individuals about their current health-related fitness relative to standards and age- and gender-matched norms
- provide information that is helpful in developing exercise prescriptions
- allow the evaluation of progress by individuals in an exercise program
- enhance motivation by establishing fitness goals
- identify the level of risk for certain cardiovascular diseases (29)

Dyspnea A feeling of difficult or labored breathing.

Information obtained during health-related physical fitness testing can be combined with an individual's health and medical information to assess the level of risk for certain disease conditions and to develop appropriate exercise prescriptions. The tests selected should provide results that are indicative of the current state of physical fitness, sensitive enough to reflect changes from physical activity or exercise, and be directly comparable to **normative data** so that the level of fitness can be determined. Clinical exercise physiologists are often involved in health-related physical fitness assessment. Table 5.4 provides examples of common assessments used to determine health-related physical fitness (29).

Exercise Prescription

Increased levels of physical activity and regular exercise are useful for improving health and reducing the level of risk for a variety of disease conditions. After the completion of testing, clinical exercise physiologists play an important role in developing an exercise prescription specific for the individual. An exercise prescription is a plan for physical activity and exercise developed to achieve specific outcomes such as improvement in fitness, reduction in cardiovascular disease risk, or weight loss. For individuals with diagnosed disease, the exercise prescription should be individualized to optimize the probability of safe and effective exercise. Advances in basic and applied exercise physiology and clinical exercise physiology have led to a better understanding of the process of prescribing physical activity and exercise to healthy individuals and those with chronic diseases. Formulating a prescription that meets the interests, goals, health needs, and environment and clinical condition of an individual must be based on sound training principles and innovative programming (see Chapter 4). Exercise prescriptions should specify the type of activity and the intensity, duration, and

Table 5.4	Assessments Used to Determine Health-Related Physical Fitness (29)
HEALTH-RELATED PHYSICAL FITNESS COMPONENT	**EXAMPLES OF ASSESSMENT**
Body composition	Body mass index, waist circumference, waist/hip ratio, skinfold measurements, bioelectric impedance
Cardiovascular-respiratory fitness	Maximal and submaximal GXT using treadmill, and cycle ergometer, distance walking tests
Muscular strength	One repetition maximum in the bench press or leg press
Muscular endurance	Maximum number of repetitions performed with a set amount of weight in the bench press or leg press, submaximal weight held statically for time
Flexibility	Sit and reach test

Normative data Information generated that allows comparison of an individual to a group.

frequency of each training session (38). The results of the diagnostic or functional capacity exercise testing, health-related physical fitness testing, clinical evaluations, and individual goals should be used to develop the physical activity and exercise program (29).

SPECIFIC DISEASE CONDITIONS

Clinical exercise physiologists design physical activity and exercise programs to improve health, reduce disease risk, prevent complications associated with disease conditions, compensate for loss of anatomical or physiological function, and optimize functional capacity (4). Increased levels of physical activity and exercise are appropriate intervention strategies to achieve these goals because nearly 70% of disabling conditions limit mobility and function by interfering with the function of the cardiovascular, pulmonary, and musculoskeletal systems. Therefore, clinical exercise physiologists must have a comprehensive knowledge of the physiology of exercise and its application in the clinical environment. Clinical exercise physiologists must also be familiar with the drugs commonly used in the pharmacologic treatment of disease conditions because many of these drugs can affect the heart rate and blood pressure response to exercise (39). Treatment interventions for many health conditions include medical surgery, pharmacologic interventions, dietary alterations and therapy, lifestyle modifications (*e.g.*, stress reduction), weight loss, and exercise. Most disease conditions can be classified into the following groups: cardiovascular disease, respiratory disease, metabolic disease, neuromuscular disease, behavioral and mental health disease, and cancer (4,40). The following sections will provide a short overview of common disease conditions and how clinical exercise physiologists can use exercise to improve health and reduce disease risk.

Cardiovascular Disease

Cardiovascular disease is the leading cause of death in the United States and contributes to considerable morbidity. Figure 5.8 shows the prevalence of cardiovascular disease in the

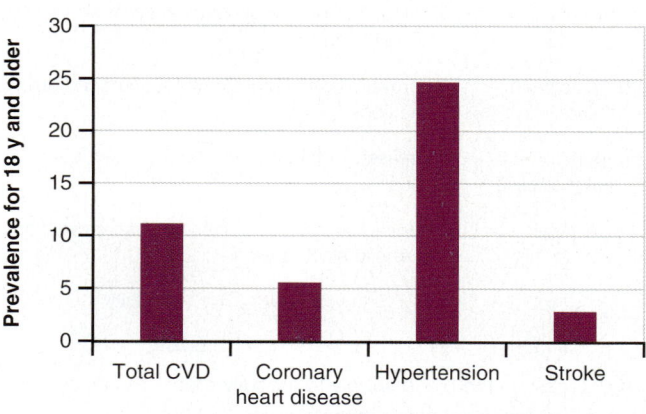

FIGURE 5.8 The prevalence of cardiovascular disease for individuals 18 years and older in the United States for 2017 (41).

United States (41,42). The prevalence of cardiovascular disease varies by age, gender, and race (43). For many of the cardiovascular disease conditions, increased levels of physical activity and exercise can play a beneficial role in restoring normal physical and physiological function. Table 5.5 illustrates the potential benefits of exercise for the primary cardiovascular disease conditions.

Myocardial Infarction

A myocardial infarction, also commonly referred to as a heart attack, occurs when an area of heart muscle is deprived of oxygen and tissue death occurs. This is usually caused by a blockage of a diseased coronary artery that results in a decreased blood flow (called ischemia) to an area of the heart supplied by those blood vessels (Figure 5.9). A myocardial infarction is typically accompanied by chest pain (called angina pectoris) radiating down one or both arms. There is usually damage to the cardiac muscle, the severity of

Table 5.5	Potential Benefits of Exercise for Cardiovascular Disease Conditions (44)
DISEASE CONDITION	PRIMARY BENEFITS OF EXERCISE
Myocardial infarction	Increased $\dot{V}O_{2max}$; reduction in heart rate, blood pressure, and myocardial oxygen requirements; improvements in blood lipid profile; improved well-being and self-efficacy; protection against triggers of a myocardial infarction
Coronary artery disease	Increased $\dot{V}O_{2max}$; reduction in heart rate, blood pressure, and myocardial oxygen requirements; improvements in blood lipid profile; improved well-being and self-efficacy; protection against triggers of a myocardial infarction
Angina pectoris	Reduction in myocardial oxygen demand
Cardiac arrhythmia	Reduction in heart rate at rest and submaximal workloads
Valvular heart disease	Improved working capacity of skeletal muscle improves ability to perform activities of daily living
Chronic heart failure	Improved skeletal muscle metabolism and distribution of blood to the tissues of the body
Peripheral vascular disease	Increased limb blood flow; better redistribution of blood flow; improvement of muscle function
Hypertension	Reduction in the rise of blood pressure over time; reduction in resting systolic and diastolic blood pressure

Ischemia A decrease in the blood supply to a bodily organ, tissue, or part caused by constriction or obstruction of the blood vessels.

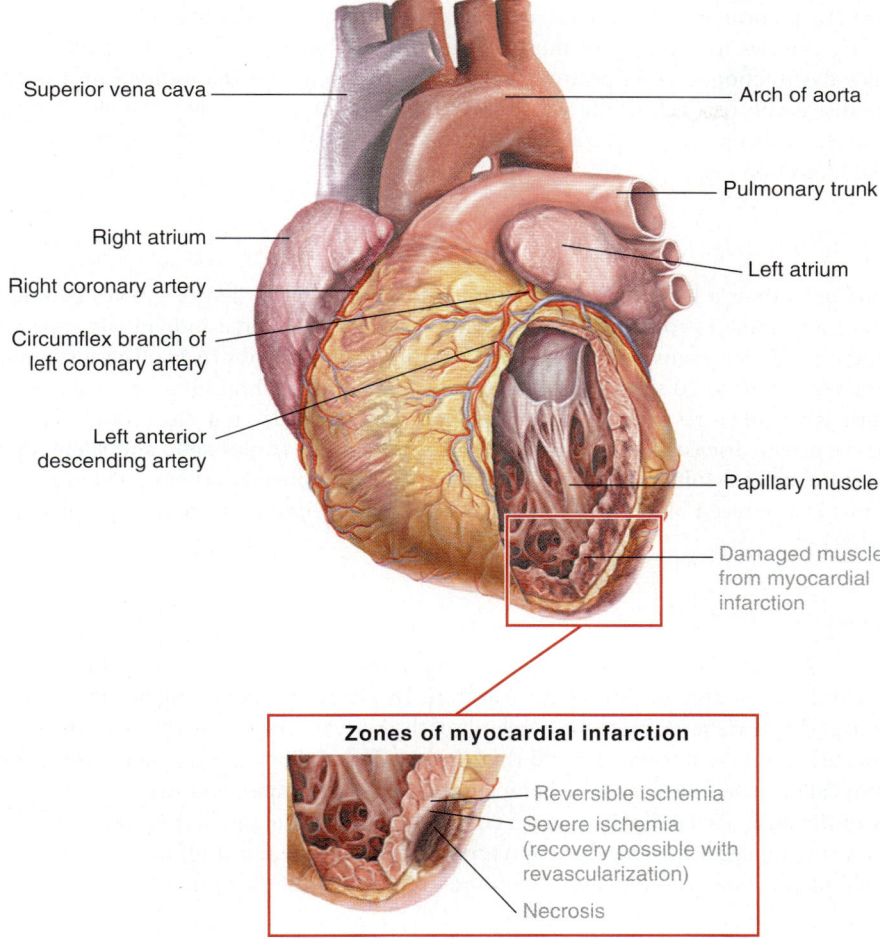

FIGURE 5.9 Myocardial ischemia and infarction. (Asset provided by Anatomical Chart Co.)

which varies with the extent and location of the damage, and this can contribute to a decreased functional ability of the heart (44). If too much damage to the heart occurs, death can result.

Coronary Artery Disease

Coronary artery disease involves a localized accumulation of fibrous tissue and, to a lesser extent, lipid matter within the coronary artery. The accumulation of lipoprotein droplets in the intima of the coronary vessels will attract leukocytes and macrophages, leading to the formation of foam cells. These foam cells form fatty streaks on the smooth muscle cells, which then cause a proliferation of lesions, leading to a buildup of materials that reduces blood flow through the vessel. The combination of fibrous tissue and lipid matter is called plaque. The resultant narrowing of the opening of the vessel (called the lumen) is often referred to as coronary **atherosclerosis**. This condition reduces the blood flow through the coronary arteries to the cardiac muscle. The reduced blood flow usually results in ventricular dysfunction, angina pectoris, and, frequently, a myocardial infarction. Coronary artery disease becomes clinically significant when approximately 75% of the blood vessel is obstructed. This blockage can lead to a reduced functional ability of the heart and a myocardial infarction (44).

Angina Pectoris

Angina pectoris is a feeling of pain or discomfort in the chest that originates behind the sternum and radiates to the shoulders, arms, neck, or jaw. Some individuals experience shortness of breath, nausea, or excessive sweating (called **diaphoresis**). These symptoms usually last for 10 to 20 seconds at a time but can occur for 30 minutes or longer. Angina pectoris is usually a response to reduced blood flow to cardiac muscle, brought about by coronary artery disease. There are two types of ischemia: symptomatic and silent. Symptomatic ischemia results in the development of the symptoms described previously. Silent ischemia is a reduced blood flow to the cardiac muscle without any accompanying symptoms (44).

Cardiac Arrhythmia

In the heart, nervous tissue in the right atrium is responsible for initiating contraction of the cardiac muscle and for setting the heart rate. In a normal functioning healthy heart, the rhythm and rate are set to meet the demands of the body for blood and oxygen. An arrhythmia occurs when the normal rate and rhythm are affected. For example, an abnormal heart rhythm called atrial fibrillation is characterized by chaotic, rapid, and irregular electric activity of the atria. Atrial fibrillation is a common **arrhythmia** and occurs more frequently with advancing age. The irregular contractions result in a reduced filling of the ventricles with blood and a decreased delivery of blood to the tissues of the body (44).

Valvular Heart Disease

The anatomical structure of the heart is designed such that one-way valves of fibrous tissue separate the ventricles (lower chambers) from the atria (upper chambers). These valves allow for coordinated and regulated blood flow through the heart and to the tissues of the body. Valvular heart disease can result from rheumatic fever, congenital abnormalities, infection, and aging. Valvular heart disease can result in a significant reduction of **cardiac output** leading to health complications, thereby affecting the ability to perform

Atherosclerosis A condition characterized by a reduced opening in the blood vessels.
Diaphoresis A condition of excessive sweating.
Arrhythmia Irregular electric activity of the heart.
Cardiac output The volume of blood ejected from the ventricles of the heart in one minute.

physical activity and exercise. The symptoms, limitations, and recommendations for physical activity and exercise in an individual with valvular heart disease depend on the heart valve that is affected, the condition of the valve, and the presence of any other disease condition (44).

Chronic Heart Failure

Chronic heart failure is characterized by the inability of the heart to deliver adequate amounts of blood to the tissues of the body. Chronic heart failure results in a depressed systolic function, depressed diastolic function, or a combination of both. Depressed systolic function occurs when the heart loses cardiac muscle (usually because of a myocardial infarction) or there is a reduction in the contractile force generated by the cardiac muscle. Depressed diastolic function is characterized by a reduced filling of the ventricles, typically owing to poor blood flow returning to the heart. Chronic heart failure is associated with changes to other tissues, including alterations in skeletal muscle metabolism, impaired **vasodilation** of blood vessels, and an inability of the kidneys to remove waste products from the body (called **renal insufficiency**) (44).

Peripheral Vascular Disease

Peripheral vascular disease can affect the arteries, the veins, or the lymph vessels of the body tissues outside the heart. The most common and important type of peripheral vascular disease is peripheral artery disease, which is a condition similar to coronary artery disease. In peripheral artery disease, fatty deposits build up in the inner linings of the artery walls, resulting in a blockage and restriction of blood flow, mainly in arteries leading to the kidneys, stomach, arms, legs, and feet (Figure 5.10). Peripheral vascular disease becomes more common with aging. Individuals with peripheral vascular disease have a four to five times higher risk of heart attack or stroke (44).

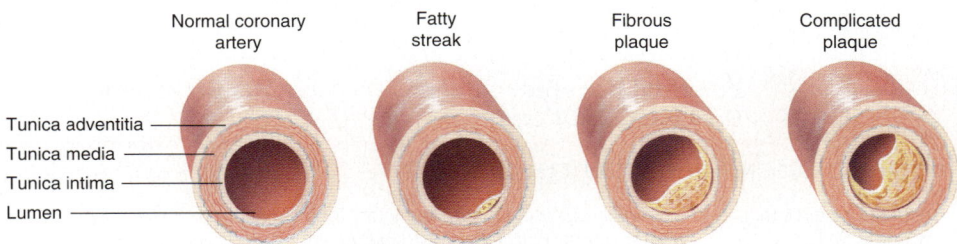

FIGURE 5.10 Progression of peripheral artery disease. (Asset provided by Anatomical Chart Co.)

Vasodilation The dilation of a blood vessel.
Renal insufficiency An inability of the kidneys to remove waste products from the body.

Hypertension

Hypertension is an abnormally high level of blood pressure. The high blood pressure results from an increased amount of blood pumped by the heart or from an increased resistance to the flow of blood through the arterial blood vessels of the body. Higher than normal blood pressure is generally defined as a blood pressure reading greater than 120 (systolic) over 80 (diastolic) mm Hg (see Table 5.3). Blood pressures considered elevated or prehypertensive should create greater awareness for allied health care professionals (33). In primary or essential hypertension, the cause of the condition is unknown. When the cause of the hypertension is known, (*e.g.*, a disorder of the adrenal glands, kidneys, or arteries), the condition is known as secondary hypertension. Several risk factors such as heredity, obesity, smoking, nutritional intake, and emotional stress are believed to contribute to the development of hypertension (44).

Respiratory Disease

Respiratory disease contributes to significant morbidity and mortality across gender, age, and race. For many respiratory disease conditions, increased levels of physical activity and regular exercise can play a beneficial role in restoring normal air flow into and out of the lungs. Table 5.6 illustrates the potential benefits of exercise for various respiratory disease conditions.

Obstructive Pulmonary Disease

Chronic obstructive pulmonary disease (COPD) affects more than 30 million Americans and is a leading cause of death in the United States (43). COPD results in an increased difficulty with breathing. Individuals with COPD typically have symptoms of both chronic bronchitis and emphysema (Figure 5.11). Chronic bronchitis is a condition characterized by a chronic inflammation of the lung and is accompanied by increased mucus production. Emphysema is a condition characterized by a loss of lung tissue. The majority of COPD is a result of tobacco abuse, although cystic fibrosis and other forms of lung disease may also

Table 5.6 Potential Benefits of Exercise for Various Respiratory Disease Conditions

DISEASE CONDITION	BENEFITS OF EXERCISE
Obstructive pulmonary disease	Improved cardiovascular and ventilatory function; increased muscular strength and flexibility; improved body composition
Restrictive pulmonary disease	Improved cardiovascular and ventilatory function; improved oxygen extraction in the lungs
Asthma	Improved overall fitness
Cystic fibrosis	Increased work capacity; improved ventilatory function; greater mucus clearance; delayed deterioration of pulmonary function

FIGURE 5.11 Chronic obstructive pulmonary disease as a result of emphysema.

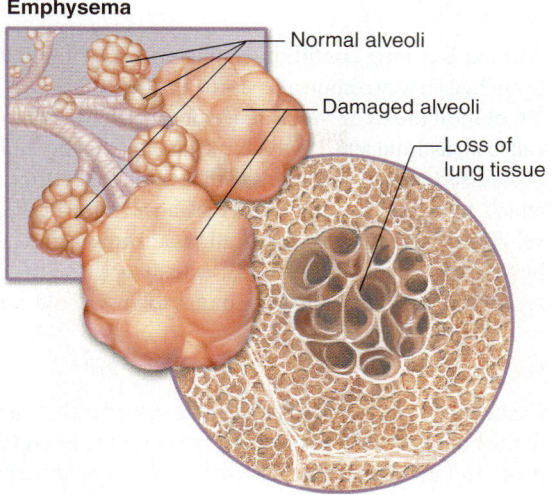

contribute to the development of COPD. Individuals with COPD are susceptible to many problems that can quickly lead to the development of other disease conditions. Ventilatory and gas exchange impairments in the lungs occur with COPD, but there can also be impairments of normal cardiovascular and muscular functions. Individuals with COPD also experience chronic anxiety and often depression that arises from the difficulty with breathing and performing physical activity and exercise (44).

Restrictive Pulmonary Disease

Restrictive lung diseases are characterized by reduced lung volume, caused by an alteration in the lung tissue or because of a disease associated with the lung tissue, chest wall, or neuromuscular breathing process. Restrictive lung diseases are distinguished by a decreased total lung capacity, vital capacity, or resting lung volume. If caused by **parenchymal lung disease**, restrictive lung disease is accompanied by reduced oxygen and carbon dioxide transfer between the lungs and the blood. The numerous disorders that cause a reduction or restriction of lung volumes may be divided into two groups based on anatomic structures. Intrinsic lung diseases (diseases of the lung tissue) cause inflammation or scarring of the lung tissue or result in filling of the air spaces with fluid and debris. Intrinsic lung diseases can be characterized according to causal factors and include lung diseases with no known cause, connective tissue diseases, drug-induced lung disease, and primary diseases of the lungs. Extrinsic lung diseases (diseases caused by factors outside of the lung) are the result of abnormal functioning of the chest wall, pleura membranes of the lung and chest cavity, and respiratory muscles. Diseases of these structures result in lung restriction, impaired ventilatory function, and respiratory failure (44).

Parenchymal lung disease A disease affecting the tissue of the lungs.

Asthma

Asthma is a lung condition characterized by reversible obstruction to airflow and increased bronchial airway responsiveness to a variety of stimuli, both **allergenic** and environmental. Over 7% of individuals 18 years of age or older have asthma (43). An acute worsening of asthma is called an asthma attack, which is characterized by shortness of breath, coughing, wheezing, and chest discomfort. Individuals with asthma experience a wide range of severity during an asthma attack, from very mild, provoked by an allergen or exercise, to very severe, which is largely irreversible despite optimal medication administration. Exercise-induced asthma is characterized by brief airway obstruction that usually occurs 5 to 15 minutes after the start of exercise. The symptoms may last up to 30 minutes following the completion of exercise (44).

Cystic Fibrosis

Cystic fibrosis is an inherited genetic disorder that causes the mucous secretions in many parts of the body to become thick and viscous. One in every 25 White people carry the cystic fibrosis gene, and a child will be afflicted if he or she inherits the gene from both parents. The thick mucus primarily affects the respiratory and digestive systems. Mucus accumulates in the respiratory airways, creating a reduced airflow and difficulty in getting air into the lungs. Lung tissue is extremely susceptible to infection, inflammation, and eventually **fibrosis** and irreversible loss of pulmonary function. The mucus also prevents the pancreatic enzymes from reaching the digestive tract, thereby causing difficulties with digestion and absorption of the macronutrients and micronutrients. Individuals with cystic fibrosis can be treated with daily therapy that can assist in the removal of mucus from the lungs and with oral supplements to replace the pancreatic enzymes. Despite aggressive treatment programs, individuals with cystic fibrosis usually experience early mortality (44).

Thinking Critically

What information should be used by a clinical exercise physiologist when developing a safe and effective exercise program for promoting health and reducing the risk of cardiopulmonary disease in a middle-aged adult?

Metabolic Disease

Various metabolic diseases contribute both directly and indirectly to increased levels of morbidity and mortality. Figure 5.12 shows the prevalence of the most common metabolic disease conditions: high blood cholesterol, diabetes mellitus, and overweight and obesity (43,45). These disease conditions can arise from a variety of sources, and treatment can include lifestyle and behavioral modification and pharmacologic interventions. Table 5.7 illustrates the potential benefits of exercise for each metabolic disease condition.

Diabetes Mellitus

Diabetes mellitus is a condition characterized by disordered metabolism and blood glucose levels consistently above normal. The percentage of individuals in the United States with

Allergenic A substance that causes an allergy.
Fibrosis The development of stiff cartilaginous tissue.

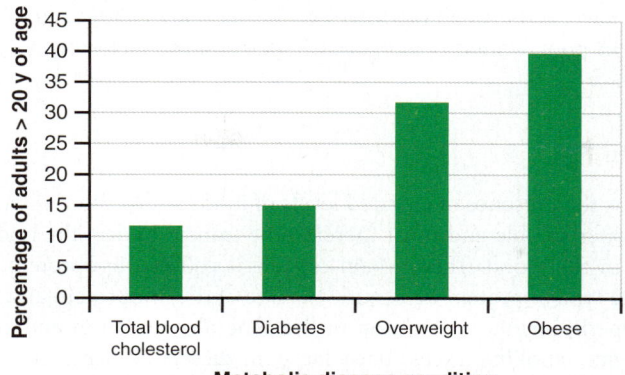

FIGURE 5.12 Prevalence of common metabolic disease conditions for individuals age 20+ years; total blood cholesterol (>240 mg/dL), diabetes mellitus (physician diagnosed and undiagnosed), overweight (BMI 25–30 kg/m²), and obese (BMI > 30 kg/m²) (43).

Table 5.7 Potential Benefits of Exercise for Various Metabolic Disease Conditions

DISEASE CONDITION	BENEFITS OF EXERCISE
Diabetes	Improved blood glucose control; improved fitness; reduction in body fat; reduction in stress
Hyperlipidemia	Improvement in blood lipid profile
Obesity	Reduction in body weight and percent body fat; improvement in fitness
Metabolic syndrome	Reduction in body weight and percent body fat; improvement in fitness; improvement in blood lipid profile; improved blood glucose control

diabetes (physician diagnosed or undiagnosed) remains around 15% (43). There are three types of diabetes: Type 1, Type 2, and gestational. Type 1 diabetes mellitus is characterized by an insufficient amount of insulin produced by the beta cells of the pancreas. Type 2 diabetes is characterized by insulin resistance in tissues, but there may also be some impairment of normal beta cell function. Gestational diabetes also involves insulin resistance, as some of the hormones secreted during pregnancy can impair normal glucose metabolism. Diabetes can cause many medical complications, including difficulties associated with acute **hypoglycemia** and **ketoacidosis**. Long-term complications include an increased risk of cardiovascular disease, chronic kidney failure, retinal damage of the eye, nerve tissue

Hypoglycemia Below normal levels of blood glucose.
Ketoacidosis An acidotic condition caused by the increased production of ketone bodies.

damage, and damage to the small blood vessels in the body. Inadequate blood flow to tissues may result in poor wound healing, particularly in the feet, which can often lead to amputation (44).

Hyperlipidemia

Hyperlipidemia is the presence of elevated levels of lipids and/or lipoproteins in the blood. Lipids are not water soluble and must therefore be transported in the body attached to a protein capsule in the blood. This protein capsule is called a lipoprotein. The density of the lipids and the type of protein determine the fate of the lipoprotein and its influence on metabolism. Hyperlipidemia is common in the general population and arises as a result of genetic influence, smoking, excess body fat, poor dietary choices, and physical inactivity. Hyperlipidemia is a strong risk factor for cardiovascular disease because it contributes to the development of atherosclerosis and coronary artery disease. The four classifications of lipoproteins include chylomicrons, very low-density lipoproteins (VLDL), low-density lipoproteins (LDL), and high-density lipoproteins (HDL). Hypertriglyceridemia is a condition of elevated triglyceride concentration, whereas hypercholesterolemia is a condition of elevated cholesterol concentration (44).

Obesity

Obesity is an excess amount of body fat that can result in a significant impairment of health and physical function. Excess body weight and body fat are the result of environmental influences, genetic factors, and metabolic disorders as well as increased calorie intake and low levels of physical activity and exercise. Obesity results in a variety of altered physiologic functions, including increased fasting insulin, increased insulin response to glucose, and decreased insulin sensitivity (46). Obesity is associated with high morbidity and mortality rates and increases the risk for diabetes mellitus, cancer, hypertension, and coronary heart disease. Typical treatment for obesity includes behavior modification, lifestyle management, nutritional modification, and increased physical activity and exercise. Pharmacologic and surgical interventions are available to those individuals who have been unsuccessful in repeated attempts to lose weight (47). Individuals who are obese and lose weight display a high rate of recidivism, a high percentage of them regaining the weight within 1 year (44).

Metabolic Syndrome

The metabolic syndrome is characterized by a clustering of metabolic risk factors, including abdominal obesity, **atherogenic dyslipidemia**, elevated blood pressure, insulin resistance or glucose intolerance, a **prothrombotic state** (*e.g.*, high fibrinogen or

Atherogenic dyslipidemia Abnormal levels of blood lipids that promote the development of atherosclerosis.
Prothrombotic state A condition of the body that favors the development of blood coagulation.

plasminogen activator inhibitor-1 in the blood), and a proinflammatory state (*e.g.*, elevated C-reactive protein in the blood) (48). Individuals with metabolic syndrome are at increased risk for coronary heart disease and other diseases related to atherosclerosis (*e.g.*, heart attack, stroke, and peripheral vascular disease) and Type 2 diabetes. Metabolic syndrome has become increasingly common in the United States, and ~23% of the adult population are estimated to have the condition (49). The dominant underlying risk factors for metabolic syndrome appear to be abdominal obesity and insulin resistance. Other conditions associated with the syndrome include physical inactivity, aging, hormonal imbalance, and genetic predisposition. There are no well-accepted criteria for diagnosing the metabolic syndrome, but many experts agree that the presence of three or more of the following indicates the presence of metabolic syndrome: elevated waist circumference, elevated triglycerides, reduced HDL cholesterol, elevated blood pressure, and elevated fasting glucose (50).

Orthopedic and Neuromuscular Disease

Orthopedic and neuromuscular disease can significantly affect participation in physical activity and exercise, contributing to considerable morbidity and mortality. Figure 5.13 shows the prevalence of the two most common orthopedic and neuromuscular diseases: arthritis and osteoporosis (43). The diseases provided in the following sections are some of the more common conditions encountered by clinical exercise physiologists in health care settings. Table 5.8 illustrates the potential benefits of exercise for each orthopedic and neuromuscular disease condition.

Arthritis

Arthritis is a painful condition affecting a single joint or numerous joints in the body. About 53 million individuals in the United States have arthritis, and it is a leading cause of disability in individuals older than age 55 years (51). The two most common arthritic conditions are

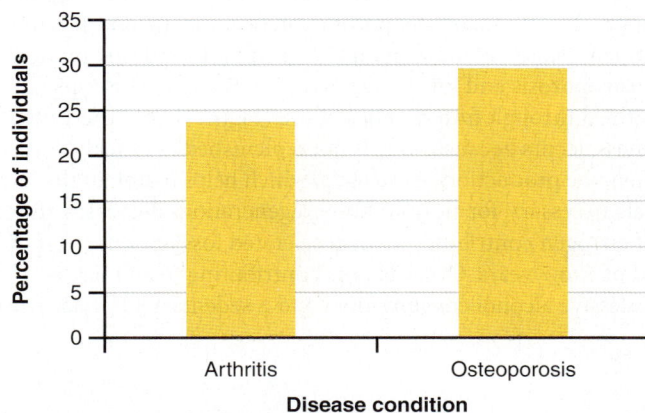

FIGURE 5.13 Prevalence of the two most common orthopedic and neuromuscular diseases; arthritis and osteoporosis (43).

Table 5.8	Potential Benefits of Exercise for Various Orthopedic and Neuromuscular Diseases
DISEASE CONDITION	BENEFITS OF EXERCISE
Arthritis	Improvement in fitness; decreased joint swelling and pain
Osteoporosis	Improvement in fitness; slowing of the age-related decline in bone mass
Muscular dystrophy	Improvement in strength and functional capacity
Multiple sclerosis	Improvement in short-term physical fitness and functional performance
Cerebral palsy	Improved fitness; increased sense of well-being

osteoarthritis, which is a degenerative joint disease, and rheumatoid arthritis, which is an inflammatory joint disease. Osteoarthritis is localized to the affected joint or joints and first appears as a deficit in the soft cartilage around the ends of the bones that constitute the joint. Also known as degenerative joint disease, osteoarthritis typically occurs following trauma to the joint, following an infection of the joint, or simply as a result of aging. Rheumatoid arthritis occurs as a result of an inflammatory response of the immune system against the joint tissue. In this case, the inflammatory response may affect numerous joints and other organ systems as well. Treatment of arthritic conditions varies depending on the type of arthritis but typically includes physical and occupational therapy, lifestyle modification to include increased physical activity and regular exercise, weight loss, and pharmacologic treatment. Joint replacement surgery may be required in severe cases involving joint deterioration (44).

Osteoporosis

Osteoporosis, a condition of decreased bone mass, occurs when the normal replenishment of bone tissue is severely disrupted, resulting in weakened bones and an increased risk of fracture. **Osteopenia** is a condition that occurs when bone-mass loss is significant but not as severe as in osteoporosis. Although osteoporosis can occur in anyone, it is most common in underweight postmenopausal White women (52). It is estimated that more than 54 million individuals have osteoporosis and low bone mass (52). Bone mass is typically at its greatest during a person's third through fifth decades of life (Figure 5.14). Thereafter, a gradual reduction in bone mass occurs because bone is not replenished as quickly as it is lost. In postmenopausal women, the production of estrogen, which helps maintain the levels of calcium and other minerals necessary for normal bone regeneration, decreases substantially. This decreased level of estrogen contributes to an accelerated loss of bone mass of up to 3% per year over a period of 5 to 7 years. Other factors contributing to an increase in bone loss include smoking, excessive alcohol consumption, and a sedentary lifestyle. The development

Osteopenia A condition of bone in which decreased calcification, decreased density, or reduced mass occurs.

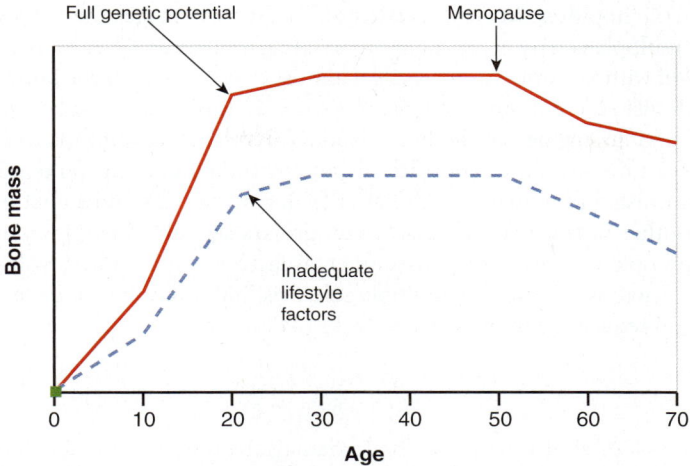

FIGURE 5.14 Acquisition and loss of bone mass during the lifespan. (Adapted from Heaney RP, Abrams S, Dawson-Hughes B, et al. Peak bone mass. *Osteoporos Int.* 2000;11:985–1009.)

of osteoporosis also has a genetic component. A receptor gene that affects calcium uptake and bone density has been identified, and the different forms of this gene appear to correlate with differences in levels of bone density among individuals with osteoporosis (44).

Muscular Dystrophy

Muscular dystrophy is an umbrella term used to describe several genetically inherited diseases characterized by the progressive wasting of skeletal muscles. There are five main forms of the disease, and they are classified according to the age at onset of symptoms, the pattern of inheritance, and the part of the body primarily affected. Muscular dystrophy is characterized by a progressive degeneration of muscle fibers that are replaced by fibrous tissue. Muscular dystrophy appears early in life and causes **symmetric** weakness and wasting of groups of muscles such as those of the lower limbs, shoulder girdle, and face. The most common form of the disease (*Duchene* type) is caused by a defect on the X chromosome and only affects boys. The muscle cell membrane lacks a specific protein called *dystrophin*, which normally prevents the muscle structure from being destroyed by its own contractions. There is no known treatment or cure for muscular dystrophy at this time. Supportive measures along with regular physical activity and exercise can improve the quality of life and preserve mobility for as long as possible (44).

Multiple Sclerosis

Multiple sclerosis is a chronic, slowly progressive **autoimmune** disease in which the immune system attacks the protective myelin sheaths that surround the nerve cells of the brain and spinal cord. This results in damaged areas of the nervous system that are unable

Symmetric Affecting corresponding parts simultaneously and similarly.

to transmit nerve impulses. Multiple sclerosis also slowly damages the nerves themselves. The onset of multiple sclerosis is usually between ages 20 and 40 years. The many symptoms associated with multiple sclerosis affect almost every system of the body and result in problems with vision, emotional disturbances, speech disorders, convulsions, paralysis or numbness of various regions of the body, bladder disturbances, and muscular weakness. The course of the disease varies considerably among individuals. In some individuals, the symptoms diminish and return, sometimes at frequent intervals and sometimes after several years. In other individuals, the disease progresses steadily. There is a genetic predisposition to multiple sclerosis, and environmental factors may also be involved in disease development. There is no cure for multiple sclerosis, but a number of drugs can slow its progression and reduce the frequency of attacks (44).

Cerebral Palsy

Cerebral palsy is a disability caused by brain damage before or during birth or in the first years of life, resulting in a loss of voluntary muscular control and coordination. Although the exact cause is unknown, several factors may predispose a child for developing cerebral palsy, including diseases such as rubella or genital herpes simplex, very low infant birth weight, injury or physical abuse, maternal smoking, alcohol consumption, and ingestion of certain drugs. Most cases of cerebral palsy are associated with prenatal problems. The severity of the affliction depends on the extent of the brain damage. Individuals with mild cases of cerebral palsy may have only a few muscles affected, whereas severe cases can result in total loss of coordination or even paralysis. There are many different forms of the disability, each caused by damage to a different area of the brain. Spastic cerebral palsy accounts for more than half of all cases and results from damage to the motor areas of the cerebral cortex. This condition causes the affected muscles to be contracted and overresponsive to stimuli. Other types of cerebral palsy include athetoid, choreic, and ataxic. The different types of cerebral palsy may occur singularly or in combination. Some individuals who are affected have a degree of mental disability, but in many individuals, the intellect is unimpaired. There is no cure for the disorder, and treatment usually includes physical, occupational, and speech therapy (44).

Behavioral and Mental Health Disease

Behavioral and mental health diseases are characterized by disturbance in thoughts, emotions, or behaviors significant enough to affect the individual's functioning. Numerous mental health conditions can significantly affect participation in physical activity and exercise, contributing to considerable morbidity and mortality. The two most common mental health conditions are anxiety disorders and depression (53). Participation in physical activity and exercise as adjunctive treatment can have positive outcomes on both anxiety and depression (40).

Anxiety Disorders

Anxiety disorders comprise a group of illnesses that include social phobias, panic disorder, obsessive-compulsive disorder, posttraumatic stress disorder, and generalized anxiety disorder (44). As many as 33% of adults are expected to experience an anxiety disorder during their lifetime (54). These disorders can have a significant influence on overall mental and physical health (53). Furthermore, anxiety disorders can increase sympathetic nervous system activity, resulting in increases in physiological functions such as blood pressure, heart rate, sweat rate, and blood flow to major muscle groups (44). This increased activity can be problematic for those individuals with comorbidities. Treatments for anxiety disorders include psychotherapy, medications, support groups, and stress management techniques. Medications used in the treatment of anxiety disorders can significantly influence physiological functions such as heart rate and blood pressure (53).

Depression

Depression (also known as major depressive disorder or clinical depression) is a common and serious mood disorder. As much as 7.1% of the U.S. adult population has experienced depression within the last year (53). Depression may cause severe symptoms that affect how individuals feel, think, and handle daily activities. To be diagnosed with depression, the symptoms must be present for at least 2 weeks (53). While depression does not directly affect physiological function, an individual with depression is likely to have a diminished interest in most activities, experience significant weight loss or weight gain, experience fatigue or loss of energy (44). These symptoms may have an indirect effect on whether a depressed individual participates in physical activity or exercise. The most common treatments for depression include psychotherapy and medications. Antidepressant medications may take several weeks before depressive symptoms begin to decrease, and often the dosage must be adjusted to maximize therapeutic effect (40).

Cancer

Cancer is an abnormal growth of tissue, characterized by a cellular growth that is more rapid than normal, shows a partial or complete lack of structural organization, lacks functional coordination with the normal tissue, and usually forms a distinct mass of tissue that may be either benign or malignant (55). Cancer is a leading cause of morbidity and mortality in the United States, and it ranks as the second leading cause of death (Figure 1.1). The amount of physical activity and exercise in which an individual is able to participate can be affected by the type and stage of cancer experienced by the individual. The seven most common types of cancer are breast, gynecologic, prostate, lung, colorectal, head and neck, and hematological (40). Cancer treatment varies according to the type and stage of the cancer, as well as other factors, including the individual's age and overall health. The most common treatments are surgery, chemotherapy, radiation therapy, and targeted biological, hormonal, and immunological therapies (40).

AREAS OF STUDY IN CLINICAL EXERCISE PHYSIOLOGY

Clinical exercise physiology encompasses numerous areas of research and scholarly inquiry with the goal of improving patient health and rehabilitation outcomes. Cardiac and pulmonary rehabilitation professionals are interested in the various responses of disease conditions to different exercise, nutritional, and behavioral interventions. The primary objectives are to identify the most effective and efficient rehabilitation programs that are individualized to the patient. The role of technology, exercise location, and exercise prescription are examples of some of the current primary interest areas in clinical exercise physiology. The areas selected are by no means meant to infer an inclusive list or indicate greater importance than a topic area not covered but are meant to provide a sampling of areas in which the knowledge of clinical exercise physiology is generated and used.

Use of Technology

The use of electronic technology in cardiac rehabilitation programs has great potential for improving individual patient compliance with exercise programs and thereby enhancing cardiovascular fitness and decreasing the risk of a future heart attack (56,57). Results have shown that patients and clinicians have found the use of smartphone devices to be feasible, safe, and an agreeable component of the cardiac rehabilitation caregiving process (56). An improved exercise capacity and a clinically significant improvement in functional capacity and participation rates have also been demonstrated (58,59). Telehealth, the remote delivery of health care via electronic communication methods is becoming a vital component for health care to individuals who cannot easily access health care face-to-face. Telehealth cardiac rehabilitation appears to be as equally effective as face-to-face cardiac rehabilitation for supporting improvements in factors that contribute to reducing cardiovascular disease risk, improving functional capacity, and overcoming common barriers that limit participation in a center-based program (60). The use of telehealth also has significant potential in pulmonary rehabilitation as it may help overcome access barriers for individuals living in rural areas and those unable to participate face-to-face owing to other commitments, such as childcare and employment (61).

Exercise Location and Modality

A prominent area of study in cardiac rehabilitation is the examination of the use of home-based versus center-based programs for enhancing exercise capacity in patients (62). There is sufficient evidence that cardiac rehabilitation improves physical fitness after a cardiac event. However, many eligible patients do not participate in cardiac rehabilitation (63), and, as a result, the physical and emotional benefits of cardiac rehabilitation are often not maintained over time. Several recent studies have found no significant differences between home-based training and center-based training on physical fitness, physical activity level or health-related quality of life in low-to-moderate cardiac risk patients entering cardiac rehabilitation. More importantly, home-based exercise was associated with a higher patient satisfaction and appears to be more cost-effective than center-based training (64,65). While home-based cardiac rehabilitation appears to hold promise in expanding the accessibility to qualified patients, additional research is still needed to clarify, strengthen, and extend the scientific evidence for crucial subgroups, including older adults, women, underrepresented minority groups, and other high-risk and understudied groups.

Similar to the studies conducted with heart disease patients, it is important to determine whether a home-based pulmonary rehabilitation program is effective for improving lung function in patients with COPD (66). In an effort to find alternative exercise programs for improving pulmonary function, researchers have examined whether traditional Chinese exercises such as "qigong" and "tai chi" provide therapeutic value in patients with COPD (67). Finally, researchers have recently determined that performing regular balance training can have a large effect on functional balance in patients with COPD (68).

Individualized Exercise

There are numerous factors that influence compliance with a cardiac rehabilitation program, including access to exercise facilities and exercise modality preference (29,69). Therefore, providing flexible models and a variety of exercise options are fundamental to supporting long-term exercise compliance in patients with cardiovascular disease. In an effort to provide patients with different options for exercise activities, researchers have examined the cardiovascular and respiratory responses during aquatic exercise because this exercise modality may appeal to many participants in the cardiac rehabilitation population (69–72). The results from several studies indicate that in heart disease patients, aquatic exercise can provide improvements in timed walking distance and speed, increased personal motivation for exercise, and safe and effective cardiovascular responses while exercising (69–72).

Identifying the appropriate individualized exercise program for use by cardiac rehabilitation patients is important for ensuring a safe and effective exercise program and a reduction in disease risk. For example, women enter cardiac rehabilitation programs with significantly lower baseline measures, and there is evidence that they do not increase their cardiorespiratory fitness as much as men (73,74). Factors that play a potential role in the lower improvement in women include lower exercise training intensity, a higher number of comorbidities, and the presence of diabetes mellitus (74,75). The study of how individual exercise prescription, especially for women, affects exercise capacity, improvement in cardiorespiratory fitness, and metabolic risk factors is needed and can provide valuable information for reducing disease risk (75,76).

Clinical exercise physiology also encompasses numerous areas of study that are mentioned in other chapters within the textbook. For example, exercise science professionals are also interested in the role of physical inactivity in disease risk, how genes regulate recovery from a cardiovascular disease, how exercise will slow the progression of muscular dystrophy, or how nutritional intake affects the development of osteoporosis. From a broader perspective, epidemiologists study the role of environmental, behavioral, and genetic factors on the risk and the prevalence of various disease conditions in large populations. The examples that have been identified are not meant to be an exhaustive list of research and scholarly activities, but merely a representation of areas of study related to clinical exercise physiology.

Thinking Critically

How might coursework in exercise science prepare an individual for a professional career as a clinical exercise physiologist and assist in the use of physical activity and exercise for the management of a chronic disease condition?

Interview

Mary Stauder, MS, RCEP, ACSM-CEP & EIM-3, NSCA-CPT, ACE-TES

Associate Director of Health & Healing at Canyon Ranch, Tucson, AZ

Brief Introduction – A native of Wisconsin, I received a BS in Exercise and Sport Science: Fitness emphasis and Nutrition minor and MS in Clinical Exercise Physiology from the University of Wisconsin-La Crosse. I completed my undergraduate internship at Duke University Diet and Fitness Center (DFC) in Durham, NC, and Master's in Clinical Internship at St. Luke's Medical Center in Milwaukee, WI, in cardiac rehabilitation. My passion is prescribing and educating individuals on how exercise is medicine to prevent, manage, and even treat chronic conditions. In addition to my position at Canyon Ranch, I am the founder and CEO of my independent practice, Rx Fit To Lead LLC, where I provide telehealth consultation for clients to "look past the workout and receive exercise as medicine" to enhance health and fitness.

Q: What are your most significant career experiences?

One of the most significant experiences was working as a personal training, group exercise instructor, and water safety instructor early on in my career, while completing my undergraduate degree. Being a full-time student and working multiple jobs to gain practical experience was invaluable when coupled with my academic education. Through these experiences, I realized that many of my clients had chronic conditions, which inspired me to pursue additional clinical exercise knowledge to provide safe and effective exercise prescriptions. Consequently, I sought after a clinically based integrative internship at the DFC and shortly thereafter worked toward a Master's degree in Clinical Exercise Physiology. I had my lightbulb moment during the internship, which changed my mindset of the exercise profession and patient journey forever. I realized that collaborative care teams are critical to impact the health and fitness of individuals. Consequently, I pursued opportunities that focused on integrative care teams after graduate school. Working in corporate wellness was a natural fit for me. I had the opportunity to be a fitness specialist and later come to pioneer the role of a Clinical Exercise Physiologist in a primary care–centered medical home model, where I additionally managed disease management programs and led fitness teams nationally. After traveling to Denmark for my husband's job and leaving my work in corporate wellness, I found myself seeking integrative wellness within the luxury hospitality industry at Canyon Ranch. There I have had the opportunity to transform lives as an exercise physiologist and now am the Associate Director of the Health & Healing division. Alongside integrative care teams, we offer lectures, curated programs, and second to none facilities and testing equipment. Managing a team of healers and caregivers to transform lives is incredibly rewarding and truly inspiring.

Q: Why did you choose to or how did you become an exercise professional?

Early on in my adolescence, I received the healing power of exercise as medicine that went far past just my physical fitness. After receiving this gift for myself, my personal mission was clear: share the medicinal benefits of exercise to as many individuals as I could, so they too could receive this healing power of exercise and look past the workout.

Q: What are the top two or three "best things" about your job?

One of the best things about my job is that I have the opportunity to educate individuals on how to exercise for lifelong fitness and as medicine. Oftentimes, individuals that work with me do not have their exercise "why" fully uncovered, which inhibits their ability to exercise for lifelong fitness. After education and coaching, it is incredible to see their transformation and adherence to sticking with an exercise program. Another thing that I love about my job is working with an integrative team. Working with nutritionists, behavioral health specialists, medical providers, athletic trainers, and spiritual wellness providers allows for a holistic approach to wellness for the client. The third thing that I love about my job is being able to design curated integrative programs focused on transformation of mind and spirit, nutrition and food, fitness and movement, and health and healing.

Q: What advice would you have for a student exploring a career in an exercise science/cardiac rehabilitation profession?

My advice to students is to shadow the roles in which they have interest in and establish a mentor. The exercise science field has so many different facets to it which a student might not be aware of. Oftentimes, students might think that a career in exercise science is just to become a personal trainer or cardiac rehabilitation exercise physiologist. There are many other fields where exercise professionals are present such as corporate wellness, disease management programs, wellness hospitality, and integrative primary care clinics. The key to my success has been following my passion and giving myself permission to change my mind at any stage of my career. The opportunities are endless! Gain experience, seek a mentor, and break the mold!

Interview

Dan Bayliss, RN, MS CES
Registered Nurse, Atlantic Coast Athletic Club, Richmond, VA

Brief Introduction – Originally from northeast Ohio, I now call Richmond, Virginia, my home. I received my Bachelor's degree in Medical Biology from Ohio Northern University, my Master's degree in Clinical Exercise Physiology/Adult Physical Fitness from Ball State University, and my nursing degree from Chesapeake College. Currently, I work as a registered nurse and exercise physiologist for a fitness and wellness business called Atlantic Coast Athletic Club (ACAC). As a former smoker and more-than-social drinker with a family history of diabetes, depression, cancer, and hypertension, I have changed my own life through physical fitness. In fact, finishing my first marathon was the most poignant point of my life. That was the day I started to *live*, not just exist, and it was only through exercise that I was able to change my life.

Q: What are your most significant career experiences?

After several years of working in the cardiopulmonary rehabilitation world, I had an opportunity to work for the University of Virginia Medical Center as the Clinical Director for the UVA SitFit Program, a clinical exercise program for dialysis patients during treatments in the hospital, one of only three dialysis exercise programs in the United States. It was a progressive, state-of-the-art program which we built into a nationally recognized program.

Q: Why did you choose to or how did you become a cardiac rehabilitation professional?

It's been an interesting life course for me because all I ever wanted to do is to become a pharmacist like both my parents. Unfortunately, organic chemistry and I didn't mix! So, all of a sudden, there I was without a direction in my life. I worked with an advisor at the student career center at my undergraduate college, who helped me find something that interested me. One day, I saw "exercise physiology," and that sparked my interest. It just so happened that my academic advisor personally knew the director of the internationally known Ball State Exercise Physiology Program. He made a call, and the next week we drove to Muncie, Indiana, and talked with him about the exercise physiology program. The next fall, I started in the program and subsequently graduated in 1999. In 2014, I had an opportunity to go to nursing school. However, from the start, I didn't want to become a hospital-based nurse. Rather, I wanted to combine my exercise background with the clinical world of nursing to carve out a special niche for my career. Since then, I have worked in a physician-referred exercise program (called P.R.E.P) at my fitness facility, guiding a 2-month clinical exercise program to people with chronic disease.

Q: What are the top two or three "best things" about your job?

I love that I'm working on the cutting edge in the medical fitness field. In my opinion, preventive medicine is going to be a critical part of our country's medical care in the near future, so I feel it's my responsibility to help educate the public about the benefits of being physically fit and how it correlates to health. I feel fortunate to help bridge the gap between the nursing and fitness worlds as well in that I get to combine clinical exercise programming with my nursing assessment skills to help provide a high level of care for people whom I work with every day.

Q: What advice would you have for a student exploring a career in an exercise science profession such as cardiac rehabilitation?

Well, if you're looking to make money, this is probably not the profession for you! But remember, there are more important things in the world than making money, like the opportunity to help change someone's life through physical fitness and to be able to see these lifestyle changes on a *daily* basis. Impacting someone's life in that way will bring you more satisfaction than any amount of money you can ever make. If you are in this profession with a high-level passion, hard work, and dedication, opportunities for growth will suddenly appear — this is not a career with a low ceiling. For me, exercise physiology was a great platform for other professions like nursing. But what it all comes down to is you. What Ted Kennedy once said was the "work of your own hands, matched to reason and principle, that will determine your destiny." Despite common belief, the possibilities in exercise physiology are endless, and I challenge you all to create your own destiny, push the boundaries to continue progressing this valuable field to new heights, and make a difference in our world.

SUMMARY

- Physical activity and exercise that can prevent or delay the onset of chronic disease in healthy individuals or provide therapeutic or functional benefits to individuals with disease conditions is the hallmark of clinical exercise physiology.
- Clinical exercise physiologists play an important role in the complete health care team by providing expertise in the prevention, treatment, and rehabilitation of numerous disease conditions and physical disabilities.
- Clinical exercise physiologists use physical activity and exercise as a means of evaluating functional capacity; assisting health care specialists in diagnostic testing; prescribing exercise based on individual needs, desires, and abilities; and instructing, supervising, and monitoring exercise programs in clinical settings.
- Cardiovascular, respiratory, metabolic, orthopedic, and neuromuscular disease conditions can be positively affected by the use of individualized physical activity and exercise program.

FOR REVIEW

1. What are the primary employment settings for a clinical exercise physiologist?
2. What significant events occurred during the 1950s that contributed to the understanding of the role of physical activity and exercise in the prevention of cardiovascular disease?
3. Define diagnostic and functional capacity testing.
4. What are primary functions of a graded exercise test?
5. What is the purpose of preexercise screening?
6. Why must an individual provide informed consent prior to participation in a graded exercise test?
7. Why is it important to collect the RPE, heart rate, blood pressure, and ECG during a graded exercise test?
8. What is the difference between a submaximal and maximal graded exercise test?
9. What are the components of health-related physical fitness testing?
10. Define the primary cardiovascular diseases:
    a. Myocardial infarction
    b. Coronary artery disease
    c. Angina pectoris
    d. Valvular heart disease
    e. Chronic heart failure
    f. Peripheral vascular disease
    g. Hypertension
11. Define the primary respiratory diseases:
    a. Obstructive pulmonary disease
    b. Restrictive pulmonary disease
    c. Asthma
    d. Cystic fibrosis
12. Define the primary metabolic diseases:
    a. Diabetes
    b. Hyperlipidemia
    c. Obesity
    d. Metabolic syndrome
13. Define the primary orthopedic and neuromuscular diseases:
    a. Arthritis
    b. Osteoporosis
    c. Muscular dystrophy
    d. Multiple sclerosis
    e. Cerebral palsy

Project-Based Learning

1. Identify a video of a graded exercise test (GXT) on YouTube. While watching the GXT, record the different measurements and how they are made on the exercising participant. Prepare a summary report highlighting the different disease conditions that could be identified from the GXT.

2. Using primary evidence–based literature sources, describe how regular physical activity and exercise could have a positive outcome on an individual who displays the following characteristics
   a. 40-year-old Black male with hypertension
   b. 65-year-old White female who is postmenopausal
   c. 18-year-old female with obesity and Type 2 diabetes
   d. 78-year-old male who has severe arthritis

REFERENCES

1. Hellerstein HK, Ford AB. Rehabilitation of the cardiac patient. *JAMA*. 1957;164(3):225–31.
2. Franklin BA, Fern A, Fowler A, Spring T, deJong A. Exercise physiologist's role in clinical practice. *Br J Sports Med*. 2009;43(2):93–8.
3. Hamm LF, Sanderson BK, Ades PA, Berra K, Kaminsky LA, Roitman JL, et al. Core competencies for cardiac rehabilitation/secondary prevention professionals: 2010 update: position statement of the American Association of Cardiovascular and Pulmonary Rehabilitation. *J Cardiopulm Rehabil Prev*. 2011;31(1):2–10.
4. Hornsby WG, Bryner RW. Clinical exercise physiology. In: Brown SP, editor. *Introduction to Exercise Science*. 1st ed. Philadelphia (PA): Lippincott, Williams & Wilkins; 2001. p. 212–34.
5. Berryman JW. Ancient and early influences. In: Tipton CM, editor. *Exercise Physiology: People and Ideas*. New York (NY): Oxford University Press; 2003. p. 1–38.
6. Wilson PK. AACVPR: The first 20 years. *J Cardiopulm Rehabil*. 2005;25:242–8.
7. Heberden W. Some accounts of a disorder of the chest. *Med Trans Coll Phys*. 1772;2:59–66.
8. Heberden W. *Commentaries on the History and Care of Disease*. London: T. Payne; 1802. 1802 p.
9. Stokes W. *Disease of the Heart and Aorta*. Philadelphia (PA): Lindsay; 1854.
10. Heggie V. A century of cardiomythology: exercise and the heart c.1880. *Soc Hist Med*. 2010;23(2):280–98.
11. Morris JN, Heady JA, Raffle PAB, Roberts CG, Parks JW. Coronary heart disease and physical activity of work. *Lancet*. 1953;262:1053–7.
12. Levine SA, Lown B. Armchair treatment of acute coronary thrombosis. *JAMA*. 1952;148:1365–7.
13. Froelicher VF. *Cardiac Rehabilitation. Exercise and the Heart: Clinical Concepts*. 2nd ed. Chicago (IL): Year Book Medical Publishers, Inc.; 1987. p. 423–86.
14. Cain HD, Frasher WG, Stiuelman R. Graded activity program for safe return to self care following myocardial infarction. *JAMA*. 1961;177:111–5.
15. Kallio V, Hamalainen H, Hakkila J, Luurila OJ. Reduction in sudden deaths by a multifactorial intervention programme after acute myocardial infarction. *Lancet*. 1979;2:1091–4.
16. American College of Sports Medicine. *Guidelines for Graded Exercise Testing and Prescription*. 1st ed. Philadelphia (PA): Lea and Febiger; 1975. 1975 p.
17. American Heart Association. *Exercise Testing and Training of Apparently Healthy Individuals: A Handbook for Physicians*. New York (NY): American Heart Association; 1972. 1972 p.
18. Erikssen G. Physical fitness and changes in mortality. *Sports Med*. 2001;31(8):571–6.
19. Barach AL. *Physiologic Therapy in Respiratory Disease*. Philadelphia (PA): J.B. Lippincott; 1948. 1948 p.
20. Miller WF. Physical therapeutic measures in the treatment of chronic bronchopulmonary disorders. *Am J Med*. 1958;24:929.
21. Miller WF. Rehabilitation of patients with chronic lung diseases. *Med Clin North Am*. 1967;51:349–56.
22. Petty TL, Nett LM, Finigan NM, Brink GA, Corsello PR. A comprehensive care program for chronic airway obstruction. *Ann Int Med*. 1969;70:1109–20.
23. Society AT. Pulmonary rehabilitation. *Am Rev Respir Dis*. 1981;24:663–6.
24. Pate RR, Pratt M, Blair SN, et al. A recommendation from the Centers for Disease Control and Prevention and the American College of Sports Medicine. *JAMA*. 1995;273(5):402–7.
25. Van Camp SP, Cantwell JD, Fletcher GF, Smith LK, Thompson PD. Exercise for patients with coronary artery disease. *Med Sci Sports Exerc*. 1994;26(3):i–v.
26. Clinical Exercise Physiology Association. 2021. Available from: https://cepa.clubexpress.com/.

27. Wall HK, Stolp H, Wright JS, et al. The Million Hearts initiative: catalyzing utilization of cardiac rehabilitation and accelerating implementation of new care models. *J Cardiopulm Rehabil Prev*. 2020;40(5):290–3.
28. Exercise is Medicine Web Site [Internet]. 2016. Available from: www.exerciseismedicine.org.
29. American College of Sports Medicine. *ACSM's Guidelines for Exercise Testing and Prescription*. 11th ed. Philadelphia (PA): Lippincott, Williams, and Wilkins; 2021.
30. James PA, Oparil S, Carter BL, et al. Evidence-based guidelines for the management of high blood pressure in adults. *JAMA*. 2014;311(5):507–20.
31. Joint National Committee on Prevention D, Evaluation, and Treatment of High Blood Pressure. The seventh report of the Joint National Committee on prevention, detection, evaluation, and treatment of high blood pressure (JNC VII). *Hypertension*. 2003;157:2413–46.
32. Borg GAV, Linderholm H. Perceived exertion and pulse rate during graded exercise in various age groups. *Acta Med Scand*. 1967;472:194–206.
33. Borg GAV. *Borg's Perceived Exertion and Pain Scales*. Champaign (IL): Human Kinetics; 1998. 1998 p.
34. Robertson RJ, Goss FL, Boer N, et al. OMNI scale perceived exertion at ventilatory breakpoint in children: response normalized. *Med Sci Sports Exerc*. 2001;33(11):1946–52.
35. Robertson RJ, Goss FL, Rutkowski J, et al. Concurrent validation of the OMNI perceived exertion scale for resistance exercise. *Med Sci Sports Exerc*. 2003;35(2):333–41.
36. Eisenmann JC, DuBose KD, Donnelly JE. Fatness, fitness, and insulin sensitivity among 7- to 9-year-old children. *Obesity*. 2007;15:2135–44.
37. Howley ET, Bassett DR, Welch HG. Criteria for maximal oxygen uptake: review and commentary. *Med Sci Sports Exerc*. 1995;27(9):1292–301.
38. Mezzani A, Hamm LF, Jones AM, et al. Aerobic exercise intensity assessment and prescription in cardiac rehabilitation: a joint position statement of the European Association for cardiovascular prevention and rehabilitation, the American Association of Cardiovascular and Pulmonary Rehabilitation, and the Canadian Association of Cardiac Rehabilitation. *J Cardiopulm Rehabil Prev*. 2012;32(6):327–50.
39. Kostoff D. Parmacotherapy. In: Ehrman JK, Gordon PM, Visich PS, Keteyian SJ, editors. *Clinical Exercise Physiology*. 2nd ed. Champaign (IL): Human Kinetics; 2009. p. 31–59.
40. Thompson WR. *ACSM's Clinical Exercise Physiology*. Philadelphia (PA): Wolters Kluwer Health; 2019. 792 p.
41. Centers for Disease Control and Prevention. *Leading Causes of Death: 2017*. 2020. Available from: http://www.cdc.gov/nchs/fastats/leading-causes-of-death.htm.
42. Mozaffarian D, Benjamin EJ, Go AS, et al. Heart disease and stroke statistics 2015 update: a report from the American Heart Association. *Circulation*. 2015;131(4):e29–e322.
43. Centers for Disease Control and Prevention. *National Center for Health Statistics 2021*. Available from: https://www.cdc.gov/nchs/index.htm.
44. American College of Sports Medicine. *ACSM's Exercise Management for Persons with Chronic Diseases and Disabilities*. 4th ed. Champaign (IL): Human Kinetics; 2009. 2009 p.
45. Centers for Disease Control and Prevention. *FastStats Homepage* 2020. Available from: www.cdc.gov/nchs/fastats/.
46. Allison DB, Downey M, Atkinson RL, et al. Obesity as a disease: a white paper on evidence and arguments commissioned by the Council of The Obesity Society. *Obesity*. 2008;16(6):1161–77.
47. Sugerman HJ, Kral JG. Evidence-based medicine reports on obesity surgery: a critique. *Int J Obes*. 2005;29:735–45.
48. Churilla JR, Fitzhugh EC, Thompson DL. The metabolic syndrome: how definition impacts the prevalence and risk in U.S. adults: 1999-2004 NHANES. *Metab Syndr Relat Disord*. 2007;5(4):331–42.
49. Beltran-Sanchez H, Harhay MO, Harhay MM, McElligott S. Prevalence and trends of metabolic syndrome in the adult U.S. population, 1999–2010. *J Am Coll Cardiol*. 2013;62(8):697–703.
50. Alberti KGMM, Eckel RH, Grundy SM, et al. Harmonizing the metabolic syndrome: a joint interim statement of the International Diabetes Federation Task Force on Epidemiology and Prevention; National Heart, Lung, and Blood Institute; American Heart Association; World Heart Federation; International Atherosclerosis Society; and International Association for the Study of Obesity. *Circulation*. 2009;120(16):1640–5.
51. Arthritis Foundation 2020 [cited 2020]. Available from: https://www.arthritis.org/.
52. National Osteoporosis Foundation. *Fast facts on osteoporosis*. Washington (DC): 2015. 2015 p. Available from: https://www.nof.org/.
53. National Institute of Mental Health. 2021 [cited 2021]. Available from: https://www.nimh.nih.gov.
54. Bandelow B, Michaelis S. Epidemiology of anxiety disorders in the 21st century. *Dialogues Clin Neurosci*. 2015;17(3):327–35.

55. Schairer JR, Keteyian SJ. Cancer. In: Ehrman JK, Gordon PM, Visich PS, Keteyian SJ, editors. *Clinical Exercise Physiology*. Champaign (IL): Human Kinetics; 2009.
56. Forman DE, LaFond K, Panch T, Allsup K, Manning K, Sattelmair J. Utility and efficacy of a smartphone application to enhance the learning and behavioral goals of traditional cardiac rehabilitation. *J Cardiopulm Rehabil Prev*. 2014;34(5):327–34.
57. Turk-Adawi K, Grace SL. Smartphone-based cardiac rehabilitation. *Heart*. 2014;100(22):1737–8.
58. Yudi M, Clark D, Tsang D, et al. SMARTphone-based, early cardiac REHABilitation in patients with acute coronary syndromes [SMART-REHAB Trial]: a randomised controlled trial. *Heart Lung Circ*. 2017;26:S349.
59. Meehan G, Koshy A, Kunniardy P, Murphy A, Farouque O, Yudi M. A systematic review and meta-analysis of randomized controlled trials assessing smartphone based cardiac rehabilitation in patients with coronary heart disease. *J Am Coll Cardiol*. 2020;75(11, Supplement 1):2004.
60. Rawstorn JC, Gant N, Direito A, Beckmann C, Maddison R. Telehealth exercise-based cardiac rehabilitation: a systematic review and meta-analysis. *Heart*. 2016;102(15):1183–92.
61. Inskip JA, Lauscher HN, Li LC, et al. Patient and health care professional perspectives on using telehealth to deliver pulmonary rehabilitation. *Chron Respir Dis*. 2017;15(1):71–80. doi:10.1177/1479972317709643.
62. Ramadi A, Haennel RG, Stone JA, et al. The sustainability of exercise capacity changes in home versus center-based cardiac rehabilitation. *J Cardiopulm Rehabil Prev*. 2015;35:21–8.
63. Schopfer DW, Nicosia FM, Ottoboni L, Whooley MA. Patient perspectives on declining to participate in home-based cardiac rehabilitation: a MIXED-METHODS STUDY. *J Cardiopulm Rehabil Prev*. 2020;40(5):335–40.
64. Kraal JJ, Van den Akker-Van Marle ME, Abu-Hanna A, Stut W, Peek N, Kemps HM. Clinical and cost-effectiveness of home-based cardiac rehabilitation compared to conventional, centre-based cardiac rehabilitation: results of the FIT@Home study. *Eur J Prev Cardiol*. 2017;24(12):1260–73.
65. Xu L, Cai Z, Xiong M, et al. Efficacy of an early home-based cardiac rehabilitation program for patients after acute myocardial infarction: a three-dimensional speckle tracking echocardiography randomized trial. *Medicine*. 2016;95(52):e5638.
66. de Sousa Pinto JM, Martin-Nogueras AM, Calvo-Arenillas JI, Ramos-Gonzalez J. Clinical benefits of home-based pulmonary rehabilitation in patients with chronic obstructive pulmonary disease. *J Cardiopulm Rehabil Prev*. 2014;34:355–9.
67. Ng BHP, Tsang HWH, Ng BFL, So CT. Traditional Chinese exercises for pulmonary rehabilitation. *J Cardiopulm Rehabil Prev*. 2014;34:367–77.
68. Marques A, Jacome C, Cruz J, Gabriel R, Figueiredo D. Effects of a pulmonary rehabilitation program with balance training on patients with COPD. *J Cardiopulm Rehabil Prev*. 2015;35:154–8.
69. Adsett JA, Morris NR, Kuys SS, Paratz JD, Mudge AM. Motivators and barriers for participation in aquatic and land-based exercise training programs for people with stable heart failure: a mixed methods approach. *Heart Lung*. 2019;48(4):287–93.
70. Adsett J, Morris N, Kuys S, et al. Aquatic exercise training is effective in maintaining exercise performance in trained heart failure patients: a randomised crossover pilot trial. *Heart Lung Circ*. 2017;26(6):572–9.
71. Dionne A, Leone M, Goulet S, Andrich DE, Pérusse L, Comtois A-S. Acute effects of water immersion on heart rate variability in participants with heart disease. *Clin Physiol Funct Imag*. 2018;38(2):233–9.
72. Choi JH, Kim BR, Joo SJ, et al. Comparison of cardiorespiratory responses during aquatic and land treadmill exercise in patients with coronary artery disease. *J Cardiopulm Rehabil Prev*. 2015;35:140–6.
73. Ades PA, Balady GJ, Berra K, et al. The *Journal of Cardiopulmonary Rehabilitation and Prevention* at 40 years and its role in the evolution of cardiac rehabilitation. *J Cardiopulm Rehabil Prev*. 2020;40(1):2–8.
74. Savage PD, Antkowiak M, Ades PA. Failure to improve cardiopulmonary fitness in cardiac rehabilitation. *J Cardiopulm Rehabil Prev*. 2009;29(5):284–91.
75. Rengo JL, Khadanga S, Savage PD, Ades PA. Response to exercise training during cardiac rehabilitation differs by sex. *J Cardiopulm Rehabil Prev*. 2020;40(5):319–24.
76. Hwang CL, Wu YT, Chou CH. Effect of aerobic interval training on exercise capacity and metabolic risk factors in people with cardiometabolic disorders: a meta-analysis. *J Cardiopulm Rehabil Prev*. 2011;31(6):378–85.
77. Sanderson BK, Southard D, Oldridge NB. Outcomes evaluation in cardiac rehabilitation/secondary prevention programs. *J Cardiopulm Rehabil*. 2004;24:68–79.
78. Nici L, Limberg T, Hilling L, et al. Clinical competency guidelines for pulmonary rehabilitation professionals. *J Cardiopulm Rehabil*. 2007;27:355–8.

79. Ries AL, Bauldoff GS, Carlin BW, et al. Pulmonary rehabilitation: joint ACCP/AACVPR evidence-based clinical practice guidelines. *Chest*. 2007;131(5):4S–42S.
80. American Association of Cardiovascular and Pulmonary Rehabilitation. *Guidelines for Pulmonary Rehabilitation Programs*. 4th ed. Champaign (IL): Human Kinetics; 2011.
81. American Association of Cardiovascular and Pulmonary Rehabilitation. *Guidelines for Cardiac Rehabilitation and Secondary Prevention Programs*. 5th ed. Champaign (IL): Human Kinetics; 2013.
82. Whelton PK, Carey RM, Aronow WS, et al. 2017 ACC/AHA/AAPA/ABC/ACPM/AGS/APhA/ASH/ASPC/NMA/PCNA guidelines for the prevention, detection, evaluation, and management of high blood pressure in adults: a report of the American College of Cardiology/American Heart Association Task Force on Clinical Practice Guidelines. *J Am Coll Cardiol*. 2018;71(19):e127–e248.
83. Chobanian AV, Hill M. National Heart, Lung, and Blood Institute workshop on sodium and blood pressure. *Hypertension*. 2000;35:858–63.

CHAPTER 6

Athletic Training and Sports Medicine

After completing this chapter, you will be able to:

1. Describe the importance of athletic training and sports medicine as they relate to enhancing the understanding of physical activity, exercise, sport, and athletic performance.

2. Describe the key highlights in the historic development of athletic training and sports medicine.

3. Identify the primary responsibilities of an athletic trainer and a sports medicine team physician.

4. Describe some of the important knowledge areas for athletic trainers and sports medicine physicians.

Athletic training is an area of exercise science that is involved in the prevention, treatment, and rehabilitation of injuries to physically active individuals and athletes (Figure 6.1). Many individuals often think of athletic training professionals as only working with athletes in a sport setting; however, athletic trainers also work closely with other allied health professionals in a clinical environment to provide care to anyone who may have an injury caused by participation in physical activity or exercise. Certified athletic trainers work under the direction of, or in collaboration with, a physician to perform professional practice in a variety of work environments, including secondary schools, colleges and universities, sports medicine clinics and hospitals, professional sports programs, performing arts and entertainment industry, the government, military and law enforcement settings, and other health care settings. Athletic trainers work closely with sports medicine physicians to form the primary athletic medicine team. Much of the work performed by athletic trainers is conducted in the athletic training clinic (1).

Sports medicine is an umbrella term used to describe all of the various issues interrelated among medicine, physical activity, exercise, sport performance enhancement, health promotion, and disease prevention (2). It is multidisciplinary and includes the medical, physiological, biomechanical, psychological, pathological, and rehabilitational aspects of physical activity, exercise, and sport and athletic performance. Sports medicine has four primary component areas:

- Medical supervision and care of recreational and competitive athletes
- Use of exercise and sports for people who are physically or mentally handicapped

FIGURE 6.1 Athletic trainers assist with injury prevention, diagnosis, and rehabilitation. (Shutterstock.)

Athletic training An area of exercise science that helps with the prevention, treatment, and rehabilitation of injuries to physically active individuals and athletes.
Sports medicine An umbrella term used to describe all things related to medicine, physical activity, exercise, sport performance enhancement, health promotion, and disease prevention.

- Helping people develop and maintain physical fitness and improve sport and athletic performance
- Use of exercise to treat and rehabilitate people who have been ill or injured.

The profession of sports medicine has created a balance between caring for competitive athletes and treating general patients by promoting exercise for health and disease prevention. The sports medicine physician may be either an athlete's primary care physician or a physician hired by the school, university or college, organization, or professional team to supervise all aspects of medical care to athletes. Sports medicine physicians, although at one time a group comprised primarily of orthopedic surgeons, now include those physicians who have successfully passed a subspecialty exam in sports medicine. The athletic trainer and sports medicine physician work together and often with other exercise science and allied health care professionals such as physician assistants and physical therapists to provide the best medical care and treatment possible to athletes and physically active individuals who have been injured (2).

Injuries are a common occurrence in individuals participating in physical activity and exercise programs as well as those actively involved in sport and athletic competition. Athletic trainers, sports medicine physicians, and other exercise science and allied health care professionals have a primary responsibility to help reduce the risk of injury among active individuals and also provide rehabilitation following the injury. Despite using many injury prevention precautions, individuals participating in sport and athletic competition are at risk for sport-related injuries. The following set of figures illustrates the incidence of injuries in certain high school and college sports and sport and recreational injuries treated at emergency rooms. Figure 6.2 provides information on the injury rates for the top five sports with the highest injury rate among high school athletes participating in sports during the 2018–19 school year (3). Figure 6.3 provides information on the injury rate during game competition and practice activities for the top five men and women National Collegiate Athletic Association–sponsored sports during the period 2014–15 through 2018–19 (4–13).

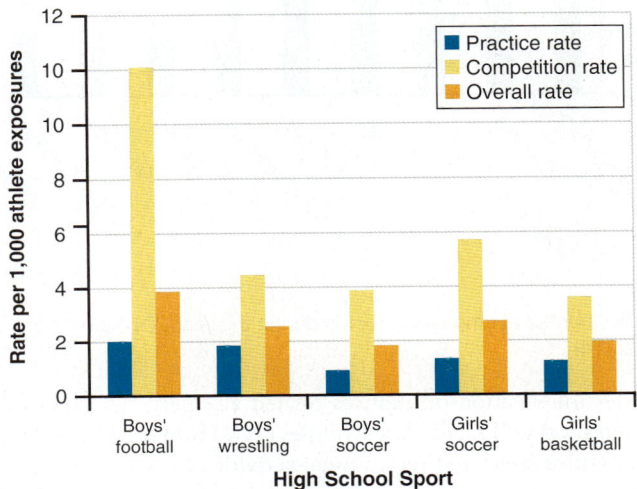

FIGURE 6.2 Overall injury rates for the top five sports with the highest injury rate among high school athletes participating in sports during the 2018–19 school year (3).

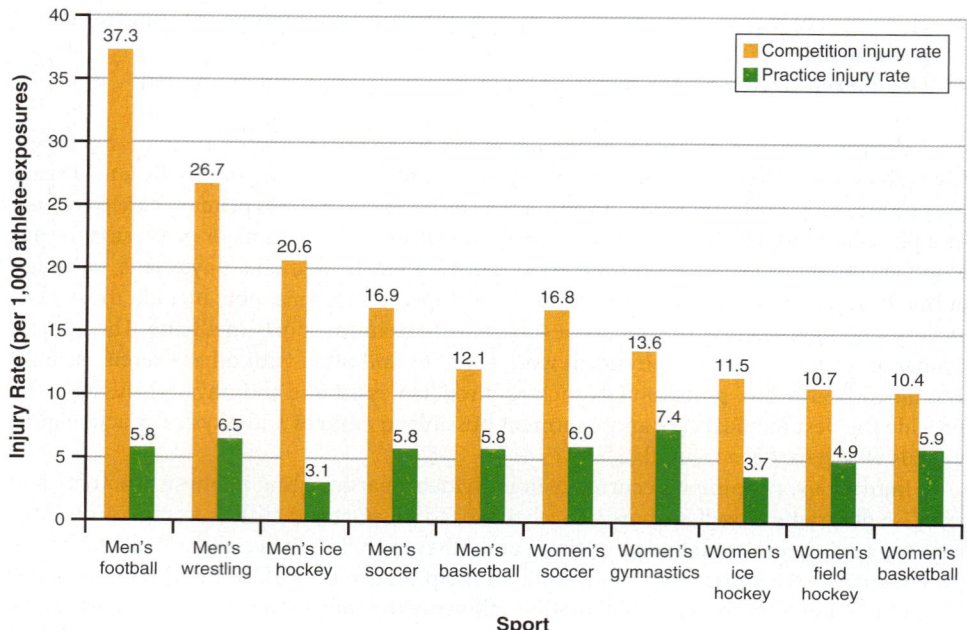

FIGURE 6.3 Overall injury rate during game competition and practice activities for top 10 National Collegiate Athletic Association–sponsored sports during the period from 2014–15 through 2018–19 (4–13).

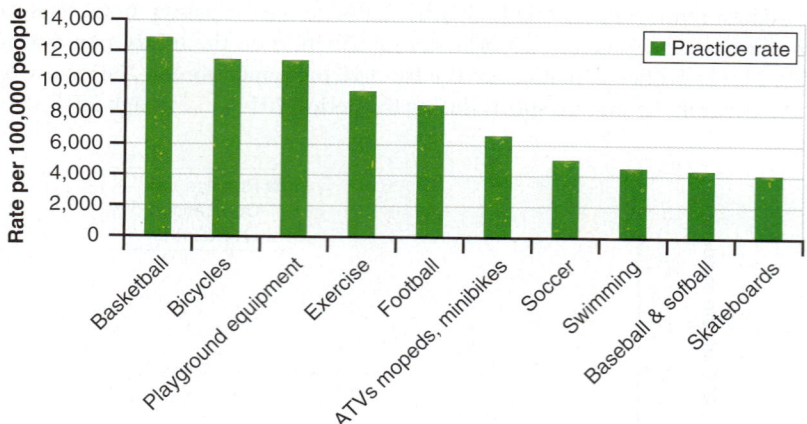

FIGURE 6.4 Injuries treated at emergency rooms during the year 2014 for the top 10 sport and recreation activities (14).

Figure 6.4 provides information on injuries treated at emergency rooms during the year 2019 for the top ten sport and recreation activities (14). From the figures, it is clear that participating in competitive sport and recreational activities carries an inherent risk of injury. The athletic trainer and sports medicine physician both play a coordinated and integral part in the treatment and rehabilitation of an individual from injury.

HISTORY OF ATHLETIC TRAINING AND SPORTS MEDICINE

The history of athletic training and that of sports medicine are intertwined because each of these areas grew primarily out of the professional service of providing care to injured athletes. This section begins with a brief history of athletic training and concludes with an overview of the historic development of sports medicine.

Early Influences of Athletic Training

The formal development of athletic training as a profession occurred in the early 1900s, and its growth coincided with the emergence of the sport of American tackle football (15). In reality, however, athletic training probably had its origin with primitive man. To survive, early man had to continuously develop physical skills and maintain fitness to meet the demands and dangers of the environment. These early hunters believed in a medicine man, called a "shaman," who was a specialist in healing. The shaman used herbs and heat to keep the hunter healthy so that he could continue his search for food and survival (15).

The Greek and Roman civilizations played an important role in the development of athletic training. The Greeks had trainers called Paidotribes, Aleittes, and Gymnastes. The meaning of Paidotribes and Aleittes were boy-rubber and anointer, respectively. The use of these terms in Greek writings suggests that massage was an important part of the duties of the trainer (16). Herodicus, of Megura, considered the greatest of all the Greek trainers, was also considered a physician. He was the advocate of modern gymnastics and the teacher of Hippocrates, the father of modern medicine. The medical gymnastae were not responsible for training individuals in a particular skill or sport activity but were expected to possess some idea of the effect of diet, rest, and exercise on the development of the body (15).

Claudius Galen (130–200 AD) is considered one of the first "athletic trainers" and team physicians. Galen wrote broadly about medicine and athletics, and his writings frequently criticized other athletic trainers and coaches for their practices and training of athletes. He recommended that patients exercise in the gymnasia to recover from sickness and fatigue. Unfortunately, from the time of Galen until early in the 20th century there is little written record of the work of athletic trainers (15).

Twentieth-Century Influences of Athletic Training

In the early 20th century, athletic trainers began to understand the need for the development and advancement of athletic training as a profession. A group of athletic trainers began to envision the development of a professional organization, which ultimately became the National Athletic Trainers' Association (NATA). The NATA has had a profound impact on the professional development of athletic training. The first national association for athletic trainers was organized in 1938; however, the challenges of travel, communication, financial limitations, and the influence of World War II led to the dissolution of the first NATA in 1944 (15).

The reemergence of the NATA was the result of the development of regional associations of athletic trainers within various collegiate athletic conferences. These associations eventually came together to form the distinct regional organizational structure of the NATA and included the following conference associations: Southern, Eastern, Pacific Coast, Southwest, and Southeastern. This original regional structure remained in place when the conference associations were designated as districts in the NATA (15).

In 1950, the NATA was officially formed by representatives from the regional collegiate conference associations. A total of nine districts comprised the new organizational structure of the NATA. The organizational leaders at the time wanted the association to build and strengthen the profession of athletic training through the exchange of ideas, knowledge, and methods of athletic training. Central to the development of the NATA was the Cramer Chemical Company, a producer of athletic training and sports medicine equipment that covered all the expenses of the association for the first 5 years of its existence (15). In 1956, in an effort to disseminate information about the profession of athletic training, the NATA began publication of a journal titled *The Journal of the National Athletic Trainers' Association*. The NATA later changed the name to *Athletic Training: The Journal of the National Athletic Trainers Association* before finally settling on the name *Journal of Athletic Training* (15).

From 1950 to 1975, the NATA steadily expanded its professional influence and standing in the athletic communities through two major accomplishments. In 1969, the NATA Committee on Professional Advancement developed and implemented rules and regulations for the certification of athletic trainers. A national certification examination was developed, and in 1970, the NATA began testing and certifying professional athletic trainers. Also during this period, a curricular program was approved by the NATA Professional Education Committee for a program of study in athletic training. This educational program was adopted as part of the college and university curricula for the professional preparation of athletic trainers. The Board of Certification, Inc. (BOC) provides a certification program for all entry-level athletic trainers. Only graduates of an athletic training program accredited by the Commission on Accreditation of Athletic Training Education (CAATE) may become a certified athletic trainer. In 1990, the American Medical Association recognized athletic training as an allied health profession (17). A major milestone was the eventual elimination of the internship route to certification in the late 1990s. This milestone made athletic training consistent with professional preparation in other allied health professions (17,18).

Twenty-First-Century Influences of Athletic Training

Through the early 21st century, athletic training programs were accredited by the Joint Review Committee on Educational Programs in Athletic Training. In 2007, the CAATE was officially recognized by the Council for Higher Education Accreditation (1). The influence of CAATE has had a significant impact on regulatory legislation, the practice of athletic training in nontraditional environments, and insurance considerations (1). Continued development of athletic training has led to increased regulation by individual state agencies (17). Each state has developed specific guidelines for licensure, certification, and registration of athletic training (see Chapter 12 for additional information). The NATA and it regulatory structures remain a key advocate for the development of athletic training as a profession.

The Youth Sport Safety Alliance (YSSA) was convened in 2009 by the NATA with the goal of making sports programs safer for young adults. The YSSA is comprised of members of parent advocate groups, research institutions, professional associations, health care organizations, and youth sports leagues. The YSSA has created a national action plan to give specific guidance to policymakers about the actions that will achieve that goal. The YSSA continues to hold regular summit meetings focusing on making youth sports safer (19).

In 2012, a "white paper," titled Professional Education in Athletic Training, was presented to the NATA that led to a change in the educational preparation for athletic training professionals, mandating that the entry-level for educational programs would move from the Bachelor's to the Master's level beginning in the Fall of 2022. After the Fall semester of 2022, baccalaureate programs may no longer admit students into the athletic training program. Students will be required to enroll in a Master's level program for professional preparation as an athletic trainer.

The NATA continues to be a leading advocate for the safety of individuals participating in physical activity, exercise, sport, and athletic performance. CAATE, BOC, NATA, and NATA Research and Education Foundation work together to promote and advance the athletic training profession. A summary of some of the key historic events in the development of athletic training is presented in Table 6.1.

Early Influences of Sports Medicine

Sports medicine, like many other disciplines related to health and medicine, had its origin in Greek and Roman times. For example, early writings refer to Claudius Galen serving as a physician to Greek gladiators (20). Much of the early written work, however, is centered more on the role of physicians in their dealings with health, physical activity, and exercise. During the period of enlightenment in the 18th century, there was a surge in the interest of orthopedic medicine. For example, in 1741, Nicolas Andry, a professor of medicine at the

Table 6.1 Key Historic Events in Athletic Training

DATE	DEVELOPMENT
1938	First attempt at formation of the NATA
1950	Formation of the NATA
1956	First publication of the *Journal of Athletic Training* as the *NATA Journal*
1969	Development of certification program for athletic trainers
1970	Development of accreditation guidelines for college and university athletic training education programs
1990	American Medical Association recognized athletic training as an allied health profession
2006	Commission on Accreditation of Athletic Training Education (CAATE) was established
2012	Recommendation for new educational preparation of athletic trainers was made to the NATA Board of Directors
2015	All athletic training education programs must be professional Master's programs by 2023

University of Paris, published the first edition of *Orthpaedia*. Andry developed the concept of the splinted tree for practicing orthopedic medicine in children and thereby helping to correct and prevent deformities. The concept for the splinted tree arises from the Greek meaning of the word orthopedic, *ortho* meaning straight and *pais* meaning child. From this origin, orthopedic sports medicine grew into the professional discipline that we know today (21).

Twentieth-Century Influences of Sports Medicine

In the first half of the 20th century, orthopedic medicine was a specialty area with a primary focus on children and only a limited role in the treatment of traumatic injuries. In 1928, a group of Olympic physicians at the Winter Olympic Games in St. Moritz, Switzerland, created the Association Internationale Médico-Sportive, which later became the International Federation of Sports Medicine (FIMS) (22). The primary purpose of FIMS was to promote the study and development of sports medicine throughout the world (1). Following World War II, orthopedic surgeons expanded their medical treatment of fractures and severe trauma through the use of advanced surgery. For example, the development of the total joint replacement procedure allowed reconstructive surgery to grow as a specialty area throughout the 1950s and 1960s. In the early 1970s, the birth of modern day sports medicine occurred. Initially, sports medicine was described as "locker room medicine" (23). Much of the science behind sports medicine was anecdotal, and the papers presented at conferences were mostly testimonials and observations of physicians and surgeons who were working with athletes. The practice of sports medicine initially focused on competitive athletes but gradually expanded its scope to include recreational athletes. The acceptance of primary care sports medicine for athletes signaled a major change in the field. A team approach to sports medicine developed in the 1980s and 1990s and incorporated orthopedists, primary care physicians, athletic trainers, physical therapists, exercise physiologists, cardiologists, nutritionists, and others (21).

Another key historic development in sports medicine occurred in 1989 with the recognition of sports medicine as a subspecialty by the American Board of Medical Specialties. The development of several professional organizations, such as the American Orthopaedic Society for Sports Medicine (AOSSM) and the American College of Sports Medicine (ACSM), have further facilitated the growth and maturation of sports medicine. The AOSSM was founded in 1972 and began publishing the *American Journal of Sports Medicine* in 1974. The ACSM serves sports medicine physicians as part of an overall organizational structure. The publication of the *Sports Medicine Bulletin* and the Team Physician Consensus Statements (24–31) has allowed the ACSM to distribute current information to physicians and help shape the care given to those individuals participating in competitive sports and athletics. The American Medical Society for Sports Medicine (AMSSM) was founded in 1991 and is a multidisciplinary organization that serves primary care physicians with fellowship training in sports medicine, nonsurgical sports medicine, and full-time team physician programs (32).

Several advancements in the treatment of injured athletes and other preventive measures have helped further shape sports medicine. One of the key technologic advancements was the development of the arthroscopic surgery procedure. First used in the early 1930s, arthroscopy grew so dramatically that by the late 1990s approximately 700,000 knee and shoulder arthroscopic procedures were being performed each year (21). A diagnostic and treatment intervention called soft-tissue endoscopy decreases the time it takes for athletes who have had tendon injuries to return to play. Once considered revolutionary, anterior cruciate ligament (ACL) surgery and ulnar collateral ligament reconstruction are now considered common in the sports medicine field. Future advances in sports medicine care will include treatments such as chondrocyte implantation, bone morphogenic protein, stem cell therapy, and gene therapy (33).

Several prominent orthopedic surgeons were instrumental in promoting the development of sports medicine. Frank Jobe performed the first ulnar collateral ligament reconstruction procedure on a baseball pitcher, Tommy John, in 1974. James Andrews is known for his pioneering work on the surgical correction of knee, elbow, and shoulder injuries. In 1984, Jack Hughston opened the Hughston Sports Medicine Hospital, the first facility of its kind in the country. Hughston is well known for his work in treating knee injuries and for being the first to establish postdoctoral fellowships in sports medicine. Hughston was also one of the founders of the AOSSM, and in 1972 he founded the *American Journal of Sports Medicine* (21).

Twenty-First-Century Influences of Sports Medicine

In 2012, the exposure of the depth and breadth of systematic performance enhancing in the Tour de France and, in particular, the case of Lance Armstrong created huge repercussions in sports medicine. The World Anti-Doping Agency (WADA), founded in 1999, became more instrumental in systematically testing and evaluating athletes from all sports in an effort to detect the use of performance enhancing drugs and eliminate them from athletic competition. The WADA regularly publishes two important documents, the World Anti-Doping Code, which brings consistency to antidoping rules, regulations, and policies worldwide and the Prohibited List, which identifies the substances and methods that are prohibited to athletes in- and out-of-competition (34).

One of the most striking developments in sports medicine in the early 21st century was the explosion of the use of ultrasound technology as both a diagnostic and a therapeutic modality. This has enabled clinical musculoskeletal diagnosis to increase procedural accuracy, lower cost, and improve patient satisfaction (35–37). Ultrasound technology can be used to diagnose and treat common musculoskeletal disorders, with emphasis on the shoulder, elbow, hip, knee, and foot and ankle (38) as well as to provide accurate guidance for completing injections into various joints in the body (39).

The Team Physician Consensus Statements, first published in 2000, continue to provide current information to sports medicine physicians so they may provide the proper care to athletes. Sports medicine professionals have improved care by further refining the preparticipation physical examination (PPE) (40) and providing guidelines for exercise during pregnancy (41). Sports medicine professionals have also advanced the understanding of the female athlete triad (24), the higher prevalence of ACL injuries in female athletes (42), and improved head concussion management and treatment (43). The generation of knowledge within the discipline has been enhanced by having sports medicine professionals embrace the scientific inquiry and evidence-based approach to the treatment and care of injured athletes (44). Leading organizations in sports medicine such ACSM and FIMS continue to offer national and international conferences designed to explore opportunities for collaborative research and

Table 6.2 Key Historic Events in Sports Medicine

DATE	DEVELOPMENT
1741	Andry published the first edition of *Orthpaedia*
1928	First FIMS International Congress was held
1934	Development of the arthroscopic surgery procedure
1972	Formation of the American Orthopaedic Society for Sports Medicine
1974	Jobe performed the first ulnar collateral ligament reconstruction procedure, also called Tommy John surgery
1989	Recognition of sports medicine as a subspecialty by the American Board of Medical Specialties
1991	Formation of the American Medical Society for Sports Medicine
2011	Publication of Concussion (Mild Traumatic Brain Injury) and the Team Physician: A Consensus Statement — 2011 Update
2017	Fifth International Conference on Concussion in Sport, Berlin, Germany

Thinking Critically

How has the historical development of athletic training and sports medicine contributed to providing better medical care for individuals injured during participation in exercise or sport activities?

the dissemination of educational information to sports medicine physicians and athletes (22). For example, in a collaborative effort, numerous sports medicine organizations developed a Consensus Statement on Concussion in Sport from discussions held at the Fifth International Conference on Concussion in Sport held in Berlin, 2017 (45). Table 6.2 provides some of the significant historic events in the development of sports medicine.

PRIMARY RESPONSIBILITY AREAS OF ATHLETIC TRAINING PROFESSIONALS

The core body of knowledge comprising athletic training has been divided into five domains by the BOC (1). Table 6.3 illustrates those domains. These domains are the guiding principles of athletic training education. Individuals studying in a CAATE-accredited program will develop knowledge, skills, and abilities in the following broad content areas.

1. Patient-centered care
2. Interprofessional practice and interprofessional education
3. Evidence-based practice
4. Health care informatics
5. Quality improvement
6. Professionalism

Table 6.3 Athletic Training Domain Description

Injury/Illness Prevention and Wellness Protection

Examination, Assessment, and Diagnosis

Immediate and Emergency Care

Therapeutic Intervention

Health Care Administration and Professional Responsibility

Data from Henderson J. *The 2015 athletic trainer practice analysis study.* Omaha (NE): Board of Certification; 2015.

Only those individuals graduating with a graduate degree from an accredited college or university athletic training program are eligible to take the national certification examination to become a certified athletic trainer. The following sections contain information about the primary responsibilities of an athletic trainer. You are encouraged to visit the following Web sites for updated information about the profession of athletic training: www.nata.org, www.bocatc.org and www.caate.net.

Injury/Illness Prevention and Wellness Protection

Participation in exercise and sport activities comes with an inherent risk of injury. This level of risk depends on a variety of factors, including the physical condition of the individual, the skill level of the individual, and the environment in which the individual is participating. One of the main responsibilities of an athletic trainer is to make sure that the level of risk of injury and illness is as low as possible. One of the most important preventive measures that physically active individuals can take to ensure a reduced risk of injury is to have a PPE performed prior to engaging in training for any sport or athletic competition (46). Athletic trainers, working with other sports medicine personnel, are an integral part of the complete PPE assessment team (1,40). The PPE typically assesses medical history, physical examination, cardiovascular screening, maturity assessment, orthopedic screening, and wellness screening (1).

Athletic trainers must also be aware of the physical and environmental conditions that may increase the risk of injury to an individual. Reducing the risk of injury may be as simple as making sure that equipment is in proper working order and meets health and safety guidelines for use. Performing a visual survey of the practice and playing areas for any equipment or conditions that can potentially increase the risk of injury is something that athletic trainers must be vigilant about. Athletic trainers should work with coaches to ensure that groups of athletes work in different areas of the practice facility to reduce the risk of unintended collisions. Athletic trainers can help reduce the risk of injury by making sure that individuals comply with safety equipment guidelines such as wearing mouthguards and using other equipment designed to enhance safety and reduce injury risk. If working in an industrial setting, athletic trainers may spend time with employees on job-specific training before a worker becomes part of the production community. Athletic trainers must also be aware of extreme environmental conditions such as hot temperatures, high relative humidity, cold temperatures with high wind velocity, and air pollution that can contribute to heat injury, hypothermia, and pulmonary disorders (1).

Examination, Assessment, and Diagnosis

Recognizing, evaluating, and assessing injuries to athletes that occur during training and competition are a primary responsibility of athletic trainers. This is probably the most visible responsibility because an athletic trainer is most likely the first person to intervene when an injury occurs to an athlete (Figure 6.5). Athletic trainers are responsible for taking a systematic approach to injury evaluation so that a comprehensive record can be generated for communication with physicians and other sports medicine and allied health personnel. The extensive preparation received in educational activities and field experiences provide the foundation for the successful functioning of athletic trainers in this role (1). A specific evaluation process is utilized when initially encountering an injured athlete. A primary survey is completed first, followed by a secondary survey. Both the primary and the secondary surveys have specific steps to follow in the process (1). Figure 6.6 provides an example of the steps used by athletic trainers when evaluating an injury situation.

Athletic trainers must be fully prepared to deal with a variety of injury situations. This includes having knowledge of emergency care and personal care procedures. Athletic trainers are required to be trained and current in adult and pediatric cardiopulmonary resuscitation (CPR), automated external defibrillators, 2nd Rescuer CPR, airway obstruction, and barrier devices. It is strongly encouraged that athletic trainers be certified in basic first aid (47). This is such a critical component of the skill set that they may also receive training and certification as emergency medical technicians. Athletic trainers should also be concerned about their own personal safety when performing duties, especially as it relates to exposure to blood-borne pathogens. Athletic trainers, as well as any medical professionals, must take precautionary measures to avoid contact with blood and other bodily fluids when performing an evaluation or first-aid procedures. Athletic trainers are required to use nonlatex gloves and may also use glasses and face shields. Proper disposal and cleanup procedures are required whenever blood or other fluids are present (48).

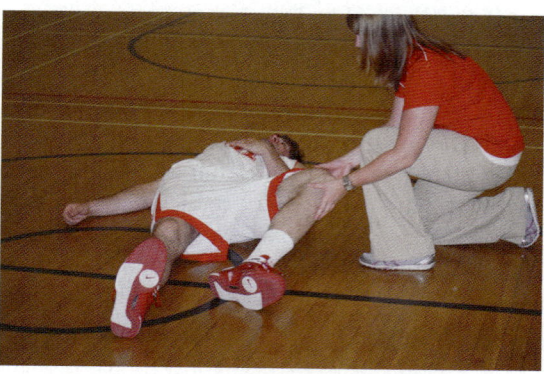

FIGURE 6.5 Evaluation of an acute injury by an athletic trainer.

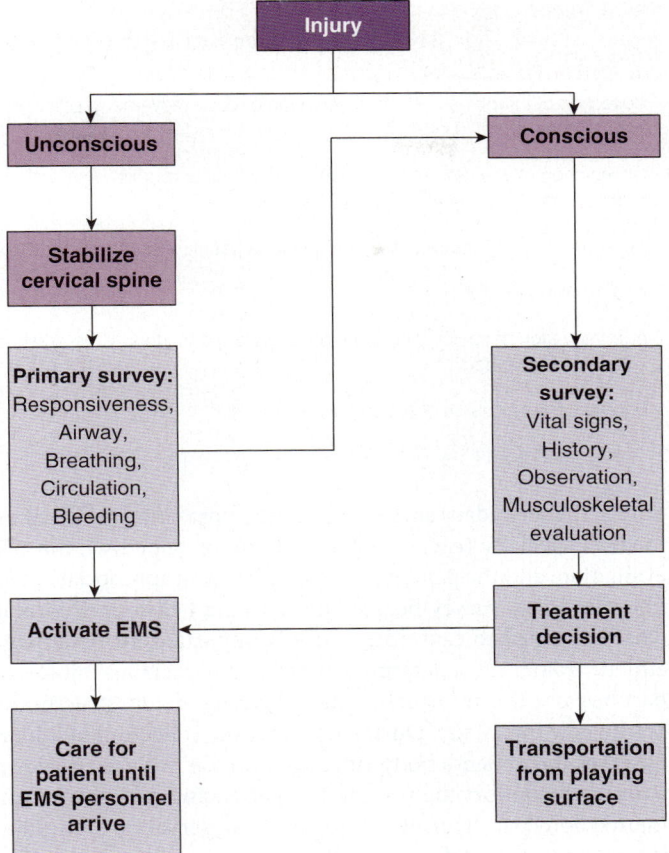

FIGURE 6.6 Steps involved in an injury evaluation process (1).

Primary Survey

Athletic trainers perform a primary survey when an athlete is injured to determine whether life-threatening injuries exist. This includes assessing the consciousness of the patient, the ABCs (airway, breathing, and circulation), determining whether the injured athlete is in shock, and activating the emergency medical service (EMS), if necessary. Unconscious patients are carefully monitored, the cervical spine is stabilized, and, if necessary, CPR or rescue breathing is initiated. Even if the injured patient is conscious, a primary survey is still a required part of the evaluation and assessment processes. After completion of the primary survey, the secondary survey is performed (1).

Secondary Survey

Table 6.4 provides the steps involved in the secondary survey. This process includes an extensive assessment of the patient and injured body area in an effort to provide a comprehensive evaluation and make a decision about the most appropriate course of action. If

Table 6.4 Secondary Survey of Injury Assessment

Collection of injury history and observation of body movement, pulse, respiration rate, blood pressure, and temperature

Observation of any deformities, change in skin color, pupils, and/or change in the shape and size of structures

Assessment of the injured area, including palpation of both bone and soft tissue

Assessment of ROM

Administration of specific structure tests designed to identify the presence of injury and assess functional performance

Decision on the course of action for the athlete

during the secondary survey any of the information gathered indicates a referral to another medical specialty (*e.g.*, emergency room, orthopedist), the athletic trainer terminates the evaluation, and the patient is transported to an appropriate medical facility for further care. The secondary survey begins with an injury history. Information about when, how, and what happened to cause the injury is important for obtaining an accurate diagnosis. The athletic trainer must determine whether any previous injuries occurred to the injured body part because this information has a bearing on any structural and functional testing done on the new injury site. During and after the collection of injury history, the athletic trainer observes the patient's body language because this can provide an indication of the level of pain or discomfort the patient is experiencing and how much the patient is favoring the injured body part. During a more formal observation of the injured area, the athletic trainer is looking for any deformities, change in skin color, and/or change in the shape and size of various structures. The next step is to **palpate** the injured area, including both bone and soft tissue. The injured patient is asked to identify the most painful area. The athletic trainer then palpates the body part distal to the area of pain, working toward the injury and finally arriving at the site of greatest discomfort (1).

If warranted, the athletic trainer can perform testing to establish the structural and functional integrities of an injured body area. The athletic trainer checks for circulation beyond the injury site, the injured athlete's response to touch, and the athlete's ability to activate the muscle of the injured area. If all of these are determined to be normal, the range of motion (ROM) of the body area is tested in three successive stages: active, passive, and resistive ROM. Active ROM requires the athlete to move the body part in response to the athletic trainer's instructions. Passive ROM involves a comparison of what ROM is achieved by the athlete compared to the ROM achieved by the athletic trainer when the muscles are relaxed. The third ROM assessment involves resistance applied by the athletic trainer as the athlete moves the injured body part through the normal ROM (1).

Palpate To examine by feeling and pressing with the palms of the hands and the fingers.

All of the information collected during the primary and secondary surveys is used to help determine which additional assessments will be made. Special tests may then be administered to determine whether specific musculoskeletal pathologies are present. If the assessments completed are normal, then functional testing involving activity-specific movements is used to determine whether the patient may safely return to full activity. During all testing and assessments, comparisons to the opposite side (bilateral) of the body are made to account for the variability found in individual people (1).

Course of Action

Upon completion of the injury evaluation, the athletic trainer must decide on the course of action. If the injury is mild, the patient may be released to return to activity. If the injury is moderate to severe, the patient is removed from the practice or competition and often referred to a health care provider or team physician for further evaluation and treatment. For some severe and all potentially catastrophic injuries, the EMS system is activated, and the patient is transported to the nearest appropriate facility for treatment (1).

Record Keeping

After completion of the injury evaluation, the athletic trainer completes a detailed record in either a paper or electronic format for documentation. The results of all assessments and tests are recorded for later use by medical personnel performing a further evaluation and treatment of the injury. The injury evaluation and treatment report can be used by other athletic trainers, sports medicine physicians, and allied health care personnel to supervise the rehabilitation of the injured patient. The SOAP notes method for recording the record is probably the most commonly used method for recording information. SOAP stands for Subjective, Objective, Assessment, and Plan (49). Table 6.5 provides a detailed description of the SOAP record keeping format.

Table 6.5		The SOAP Notes Method for Record Keeping (1)
S	Subjective information	What the athlete tells the athletic trainer
O	Objective information	Quantifiable information including signs, observations, palpation, and special tests performed by the athletic trainer
A	Assessment	Professional opinion of the athletic trainer or other health care professionals about the nature and extent of the injury
P	Plan	Includes all treatments rendered and disposition of the injury, whether referral or continued local intervention and rehabilitation

Immediate and Emergency Care

Athletic trainers are among the most common group of health care professionals to treat an acute injury. This means athletic trainers must be trained and ready to respond to any injury situation, whether commonplace or life threatening. The primary and secondary surveys provide information that dictates the course of action an athletic trainer will follow when responding to an injury. All athletic training settings should have a written emergency action plan (1). A well-designed and executed emergency action plan will greatly limit secondary injury caused by inappropriate movement, treatment, or time spent activating the EMS. If the EMS is activated by the athletic trainer, the EMS personnel take responsibility for the patient (1).

Acute musculoskeletal injuries are most commonly treated by using cold (*e.g.*, ice or vapocoolant spray), compression, elevation, and rest (1). Rest for the injured body part may mean removal from practice or competition for a specific period. If the injured body part requires immobilization, then crutches, a sling, or a cast can be used to limit and restrict movement. Ice packs and compression wraps are often applied to the injured body part. The ice and compression may help limit the swelling of the injured area. The application of a compression wrap and cold treatment is usually 20 minutes in duration, with removal of the cold treatment for a duration sufficient to allow the tissue to rewarm (50).

Therapeutic Intervention

Pain, swelling, decreased ROM, and loss of normal function are the immediate effects of an acute injury to a body part. Following an acute injury, there is usually a period of inactivity that, depending on the length of inactivity, can lead to significant changes in the muscle, including atrophy, and decreased strength, endurance, and neuromuscular coordination as well as reduced function of other body systems (1). A proper treatment, rehabilitation, and reconditioning program can reduce the amount of time required to return an athlete to practice and competition.

The various tissues (*e.g.*, tendons, ligaments, and bone) and systems (*e.g.*, muscular, cardiovascular, and respiratory) of the body all require regularly applied physical activity and exercise to maintain normal function and optimal performance (1,51). For example, the immobilization of a lower extremity can decrease various markers of cardiorespiratory function in a relatively short period (1,51). Muscle strength and performance will decrease as the number of days of inactivity increase (52). To reduce this loss of function, a systematic rehabilitation program must be followed. Active and passive exercises, as well as aerobic and resistance exercise should be used in a comprehensive rehabilitation program of an injured patient. Exercises should be selected that minimize stress to the injured body part, yet provide a complete body workout.

Restoring a normal ROM to an injured body part is critical for preparing an injured patient for returning to action. Completing physiological assessments and accessory movements is the key to increasing the ROM of a joint. Physiologic movements are those normally ascribed to a joint and include, for example, flexion and extension of the elbow joint. Accessory movements are small movements that reposition the bones for the maximum efficiency of physiologic movement and include additional movements. Normal function and ROM of a joint requires that both physiologic and accessory movements be properly coordinated. Large muscle activity through active, passive, and resistive ROM, performed throughout the entire ROM, is typically used to increase physiologic movements. Accessory movements of a joint are more commonly restored with joint mobilization techniques (1,53).

Exercise Activities

Athletic trainers can choose from a variety of muscle actions in the rehabilitation process, including isometric, isotonic, and isokinetic. **Isometric** muscle action involves contracting a muscle but not moving the joint involved with the muscle through a ROM. Force is generated by the muscle, but there is no movement of the muscle during the contraction process. Isometric muscle actions can increase muscle strength but only for about a 15-degree ROM (angle) around the joint position at which the muscle is contracted (54). To improve muscle strength through the entire ROM, muscle contractions must be performed at several joint angles. Isometric contractions are performed against a machine or individual (*e.g.*, athletic trainer) who provides resistance against the muscle contraction (53).

Isotonic muscle actions involve contracting a muscle against resistance and allowing the joint to move through the ROM. Force is generated by the muscle, and there is movement of the joint during the contraction process. Isotonic muscle actions are performed using either free weights or a machine to provide resistance against the muscle movement. During isotonic exercises, the resistance provided by the external force stays constant, but the force generated by the muscle changes as the joint angle changes. When the joint angle changes, there are changes in the lever action of the joint and the force that must be generated by the muscle (54). Isotonic muscle actions are performed at variable speeds against a fixed resistance.

Isokinetic muscle actions consist of contracting a muscle against a resistance and moving the joint through the ROM at a constant velocity. Specific machines called isokinetic dynamometers allow the muscle to contract at a controlled speed, generally somewhere between 0 and 360 degrees per second. At the same time, resistance is provided against the muscle movement, requiring the muscle to generate force (54). Isokinetic exercises are performed at a fixed speed with variable accommodating resistance.

Both isotonic and isokinetic exercises allow the muscle to perform concentric and eccentric muscle actions. **Concentric muscle actions** occur when a muscle shortens in length and develops tension. **Eccentric muscle actions** occur when a muscle increases in length and develops tension.

Athletic trainers also use **closed kinetic chain** and **open kinetic chain** activities to facilitate the rehabilitation of injured athletes. The functional anatomic relationships that exist in the upper and lower extremities provide the basis for kinetic chain exercises. When an individual is in a weight-bearing position, the lower extremity kinetic chain involves

Isometric The generation of force by a muscle without any movement of the joint.
Isotonic The generation of a constant force by a muscle and movement of the joint.
Isokinetic The generation of force by a muscle and movement of the joint at a constant speed or velocity.
Concentric muscle action When a muscle shortens in length and develops tension.
Eccentric muscle action When a muscle increases in length and develops tension.
Closed kinetic chain When forces along the body are transmitted to an adjacent structure, usually the floor or a piece of equipment.
Open kinetic chain When forces along the body are allowed to dissipate into the air.

the transmission of forces among the foot, ankle, lower leg, knee, thigh, and hip. The hand as an upper extremity weight-bearing surface transmits forces to the wrist, forearm, elbow, upper arm, and shoulder girdle (1,55). In a closed kinetic chain, the foot or hand is weight bearing, whereas in an open kinetic chain there is no hand or foot contact with a surface (1,55). In a closed kinetic chain, the forces begin at the contact point between the foot or hand and the surface and then work their way along each joint. In a closed kinetic chain, forces must be absorbed by various tissues and anatomic structures rather than simply dissipating as would occur in an open chain (56). Closed kinetic chain strengthening techniques have become an extremely popular rehabilitation treatment by many athletic trainers in part because they are more functional than are open kinetic chain activities (1). Closed kinetic chain exercises are also more sport- or activity-specific because these exercises involve movement that more closely approximates the desired activity, particularly for the lower extremities (1).

Therapeutic Modalities

Athletic trainers also use a variety of **therapeutic modalities** to assist in the rehabilitation process (Figure 6.7). Therapeutic modalities can provide effective support to the various techniques used in rehabilitation exercise (1). Appropriate rehabilitation protocols and proper progression of treatment must be based primarily on the physiological responses of the tissues to injury and on an understanding of how various tissues progress through the healing process. An athletic trainer must consider the signs, symptoms, and various phases of the healing process when making a decision on which therapeutic modalities may be best used to treat an injured athlete (1). The common therapeutic modalities used by an athletic trainer are contained in Table 6.6. Specific laws govern the use of therapeutic modalities, and athletic trainers must be aware of and follow those laws that dictate the use of the modalities. When used inappropriately, these modalities may cause significant harm and injury, and therefore athletic trainers must use great care when employing a therapeutic modality in the rehabilitation of an injured patient (1).

Health Care Administration and Professional Responsibility

Many athletic trainers often work in a position of autonomy. This requires the athletic trainer to be responsible for a variety of issues related to the organization and administration of an athletic training program and clinic, including personnel management, professional development, facility management and design, budgeting, PPEs, medical record keeping, insurance, and public relations (57). It is important for athletic trainers to be responsive to these issues so that the best possible care can be provided to the athletes and physically active individuals. The following sections provide an overview of some of the areas of responsibility.

Therapeutic modalities Machines, devices, or substances that are used to enhance recovery from an injury.

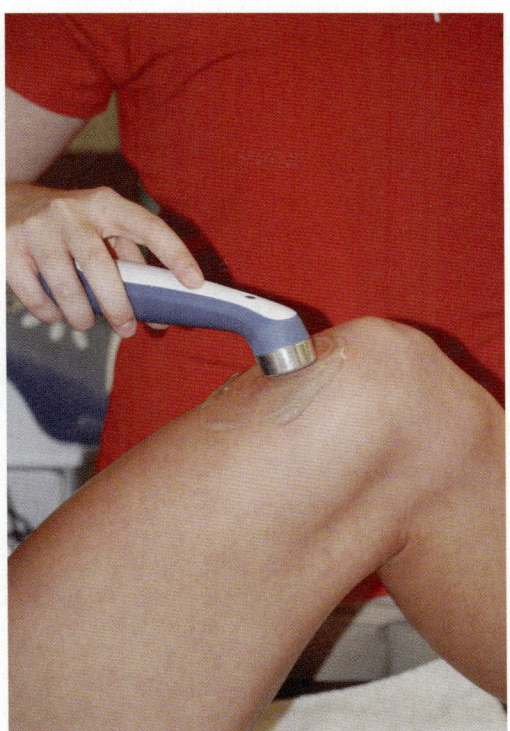

FIGURE 6.7 Ultrasound treatment of an injury by an athletic trainer.

Professional Development and Responsibility

Individuals who belong to a profession have a responsibility to maintain knowledge and skills at the current standard of care and practice. Athletic trainers must continue to advance their content knowledge and treatment skills through various continuing education activities and programs. For example, athletic trainers must maintain certification in emergency first aid and CPR to remain certified as athletic trainers. Furthermore, athletic trainers must continue to obtain continuing education units (CEUs) to satisfy the requirement of professional development established by the BOC. Athletic trainers also have a responsibility to promote the profession to the general public. Because of the close relationship established with patients, athletes, and family members, athletic trainers provide counsel to numerous individuals on prevention, rehabilitation, and treatment of injuries. These responsibilities are important in providing effective total care to the athlete (1).

Providing Coverage

Athletic trainers are responsible for providing medical coverage at the athletic training clinic and at practices and competitions. The athletic training clinic is generally a place where treatment and rehabilitation, preparation for competition, and injury management occurs. In college and university settings and often in commercial sports medicine facilities,

Table 6.6 Commonly Used Therapeutic Modalities (1)

MODALITY	PRINCIPLE	EXAMPLES
Cryotherapy	Cooling decreases physiologic function	Ice massage, cold or ice water immersion, ice packs, vapocoolant sprays
Cryokinetics	Cool the body part to analgesia, and then work to increase the ROM	Ice immersion, cold packs, ice massage
Thermotherapy	Heating increases physiologic function	Moist heat packs, whirlpool baths, paraffin bath, contrast bath, fluidotherapy
Ultrasound	Deep heats tissues to increase tissue temperatures	High frequency generator transmits continuous or pulsed ultrasound waves, phonophoresis
Electrotherapy	Increases the excitability of nerve tissue	Transcutaneous electric stimulators deliver biphasic, monophasic, or polyphasic currents
Massage	Manipulation of soft tissue causes mechanical, physiologic, and psychological responses	Effleurage, petrissage, friction, tapotement, vibration, deep friction massage, acupressure massage
Traction	Produces separation of vertebral bodies to reduce pressure on spinal column and associated tissues and joints	Manual, mechanical, positional, wall-mounted, and inverted traction
Intermittent compression	Increased pressure controls or reduces swelling and reduces edema	Pneumatic inflatable sleeve and cryo-cuff

the clinic may be open all day. In a high school setting, the clinic may be open only for a limited time. To provide the best care and protection to athletes, a certified athletic trainer should be in attendance at all practices and home and away competitions. Colleges and universities typically have sufficient personnel to provide concurrent coverage to several sports. In a high school setting, there are usually only one or two athletic trainers who must decide what sport practices or competitions will be covered during a specific time. These decisions are usually based on the number of athletes, the inherent risk of the sport, and whether the activity is a competition or a practice. In this situation, a schedule should be developed so that the athletic trainer is in attendance for critical components of the competition or practice (1).

Legal and Insurance Issues

Athletic trainers must keep current and accurate records as part of the comprehensive athletic training program or as part of a clinic. The essential components of the records to be kept include medical records, injury evaluations, injury reports, treatment logs, progress reports, and progress rates, supply and inventories, and annual program or clinic reports. A solid understanding of the laws governing confidentiality is critical to the effective functioning of the athletic trainer and athletic training program. The Health Insurance Portability and Accountability Act (HIPAA) and the Family Educational Rights and Privacy Act (FERPA) are responsible for protecting the medical and educational information of individuals, respectively. HIPAA regulates how anyone who has private health information about an individual (*e.g.*, athlete) can share that information with others (58). HIPAA ensures that all individuals have certain rights over the control and use of their medical records and provides a clear path of recourse if their medical privacy is compromised (1). FERPA is a law that protects the privacy of students' educational records. FERPA gives parents certain rights to their children's educational records until the child reaches 18 years of age or attends a postsecondary school. To release any information from a student's educational record, a school must have written permission from the parent or eligible student (1).

Given the significant changes in our health care system, it is becoming increasingly important for athletic trainers and other exercise science professionals to understand a variety of issues related to health care delivery. All athletes and physically active individuals should have health insurance that covers illness, hospitalization, and emergency care (1). It is important for school-sponsored activities to ensure that personal health insurance is arranged for or purchased by individuals not covered under a health insurance policy (59). Accident insurance is available to student athletes, and this covers accidents on school grounds while the student is in attendance at school. Catastrophic injury insurance is often available to athletes through sports-governing agencies to assist with the costs associated with a permanent disability. Although catastrophic injuries are extremely rare, the extensive medical and rehabilitation care associated with these types of injuries can create a financial burden on the athlete and their family (1).

Athletic trainers should also consider acquiring professional liability insurance. Professional liability insurance protects athletic trainers and other health care providers against damages that may arise from injuries occurring on school property and often covers claims of negligence on the part of individuals (60). This type of insurance helps protect professionals from assuming the full cost of defending against a negligence claim made by a client and potential damages that may be awarded in a legal action.

Athletic trainers may also acquire third-party payment for services rendered while working in a variety of settings, including hospitals, physician's offices, sports rehabilitation clinics, and colleges and university settings (61). In many states, certified athletic trainers are viewed as licensed health care professionals and therefore eligible for reimbursement for services by most third-party payers (*i.e.*, health insurance companies). To facilitate this process, the athletic trainer must file insurance claims immediately and correctly (62). It is important that the athletic trainer keep accurate and up-to-date records when billing for and receiving reimbursement for treating an injury sustained to an athlete.

Working with the Team Physician

In most high school, college, and professional sports settings, as well as clinic environments, the athletic trainer works primarily under the direction of, or in collaboration with, the team physician, usually a sports medicine physician (Figure 6.8). A working environment of respect and cooperation between the athletic trainer and the team physician ensures that patients receive the best care possible. Athletic trainers are often the

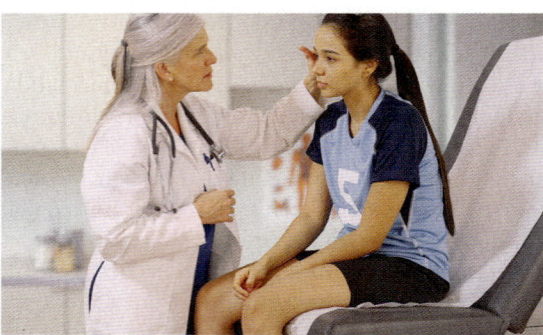

FIGURE 6.8 A team physician interviewing an athlete about an injury. (Shutterstock.)

first to initiate contact with an injured athlete. They must be clear and concise in relaying information about the nature of the injury to the athlete to the sports medicine or primary care physician. The team physician, who is ultimately responsible for directing the total health care of the athlete, will diagnose the extent of the injury and recommend follow-up care. Physicians in charge of sports teams should be aware of the progress of rehabilitation of the patient through the recovery period and make the final determination as to when an athlete can return to practice and competition (1). There are several roles and responsibilities that a team physician should assume with regard to injury prevention and the health care of the athlete that are presented in the next section of this chapter.

What personal qualities and professional characteristics do you believe are important for athletic trainers to possess in order to fulfill their primary responsibilities?

SPORTS MEDICINE

The professional practice of sports medicine has expanded dramatically since the early 1980s, mostly in response to the demand for high-quality health care for athletes and physically active individuals. The number of sport-related and physical activity–related injuries has increased, and many factors have contributed to this increase (63). Figures 6.2 and 6.3 provide the prevalence rates for injuries in high school and intercollegiate sports, respectively, whereas Figure 6.4 provides the injury rates for the top ten sport and recreation activities emergency room visits. Table 6.7 provides some suggested reasons for the increased number of sport-related injuries. Sports medicine physicians contribute to total athlete care by being an integral component of the primary athletic medicine team and working with other health care providers from orthopedics, physical medicine and rehabilitation, athletic training, biomechanics, cardiology, nutrition, optometry, pharmacology, physical therapy, exercise physiology, psychology, and podiatry (63).

Guidelines for Sports Medicine Physicians

Often, the sports medicine physician serves as the leader of the primary sports or athletic medicine team through their role as the team physician. The importance of this role has resulted in several organizations working together to provide several Team Physician Consensus

Table 6.7	Reasons for Increased Sports-Related Injuries (63)
Increased participation in sports	
Increased numbers of previously sedentary individuals becoming active	
Increased variety of sports available	
Increased opportunities for participation	
Increased sophistication of participants in sports	
Increased participation intensity, which often leads to increased risk of injury	
Athlete specialization at a young age leading to overuse injury	
Poor coaching and training methods leading to increased sport injuries	

Statements. These statements are used to guide the activities and responsibilities of the sports medicine team (24–31,43,64–71). A list of the current ACSM Team Physician Consensus Statements written in collaboration with numerous professional organizations is provided in Table 6.8.

The first Team Physician Consensus Statement published in 2000 and updated in 2013 provides physicians, school administrators, team owners, the general public, and individuals who are responsible for making decisions regarding the medical care of athletes and teams with guidelines for choosing a qualified team physician and an outline of the duties expected of a team physician (28). This consensus statement provides clear and concise information about the definition, qualifications, and duties of a team physician. Each of the subsequent consensus statements has identified important topics relevant to providing the best medical care for athletes at all levels of participation. Table 6.9 provides a summary of the responsibilities of the sports medicine team physician.

Advances in the Treatment of Sport-Related Injuries

The primary responsibility of sports medicine professionals is to provide the best care possible to physically active individuals and athletes, and several advances in the medical treatment of orthopedic injuries have accomplished that aim. In most cases, these advancements have resulted in the development of minimally invasive procedures and a shorter recovery time for the athlete. In some instances, the careers of athletes have been saved or extended as a result of these advancements. Examples of these procedures include arthroscopy, ACL reconstruction, ulnar collateral ligament reconstruction, and chondrocyte implantation. Each is discussed in the following sections. The examples provided are not meant to be an inclusive list of significant developments in athlete care but a sampling of advancements in sports medicine made over the years.

Thinking Critically

How does a coordinated effort among athletic trainers and sports medicine personnel provide the best medical care to an injured athlete?

Table 6.8 Team Physician Consensus Statements

TOPIC	YEAR PUBLISHED
Team Physician Consensus Statement (28)	2000
Sideline Preparedness for the Team Physician: A Consensus Statement (27)	2001
The Team Physician and Conditioning of Athletes for Sports: A Consensus Statement (26)	2001
The Team Physician and Return-To-Play Issues: A Consensus Statement (29)	2002
Female Athlete Issues for the Team Physician: A Consensus Statement (24)	2003
Mass Participation Event Management for the Team Physician: A Consensus Statement (25)	2004
Psychological Issues Related to Injury in Athletes and the Team Physician: A Consensus Statement (66)	2006
Selected Issues for the Adolescent Athlete and the Team Physician: A Consensus Statement (31)	2006
Selected Issues in Injury and Illness Prevention and the Team Physician: A Consensus Statement (135)	2007
Selected Issues for the Adolescent Athlete and the Team Physician: A Consensus Statement (31)	2008
Selected Issues for the Master Athlete and the Team Physician: A Consensus Statement (67)	2010
Concussion (Mild Traumatic Brain Injury) and the Team Physician: A Consensus Statement—2011 Update (30)	2011
The Team Physician and the Return-to-Play Decision: A Consensus Statement—2012 Update (69)	2012
Sideline Preparedness for the Team Physician: A Consensus Statement—2012 Update (68)	2012
Team Physician Consensus Statement: 2013 Update (70)	2013
Selected Issues for Nutrition and the Athlete: A Team Physician Consensus Statement (71)	2013
The Team Physician and Strength and Conditioning of Athletes for Sports: A Consensus Statement (136)	2015
Psychological Issues Related to Illness and Injury in Athletes and the Team Physician: A Consensus Statement—2016 Update (66)	2017
Female Athlete Issues for the Team Physician: A Consensus Statement V 2017 Update (137)	2018
Load, Overload, and Recovery in the Athlete: Select Issues for the Team Physician—A Consensus Statement (138)	2019
Select Issues in Pain Management for the Youth and Adolescent Athlete (139)	2020

Arthroscopy

Arthroscopy or arthroscopic surgery is a minimally invasive procedure used to examine and treat damage to the interior of a joint. The procedure is performed by a sports medicine physician, typically an orthopedic surgeon, using an arthroscope and is shown in Figure 6.9.

Arthroscopy A minimally invasive surgical procedure used to examine and treat damage to the interior of a joint.

Table 6.9 Responsibilities of the Team Physician (28,57,70)

RESPONSIBILITY	PHYSICIAN ACTIONS
Work with the athletic trainer	Supervise and advise athletic training staff; share philosophy regarding injury management and rehabilitation
Work with other sports medicine personnel	Provide the best possible care using the expertise of the primary sports medicine team
Compiling medical history	Oversee medical history; conduct physical examinations for each athlete
Diagnosing injury	Assume responsibility for diagnosing an injury; make recommendations for treatment to the athletic trainer and other sports medicine personnel
Deciding on disqualification and return to play	Determine when an athlete should be disqualified from competition and when that injured athlete may return to play
Attending practice and games	Attend as many practices, scrimmages, and competitions as possible; be available to other sports medicine personnel for consultation or advice
Commitment to sports and athletes	Demonstrate a strong affection and dedication to the athlete and sports
Academic Program Medical Director	Be responsible for the coordination and guidance of the medical aspects of an accredited athletic training education program

Arthroscopic surgery is used to evaluate and treat many orthopedic conditions such as torn floating cartilage, torn surface cartilage, and the ACL of the knee joint. Arthroscopic surgery does not require the treated joint to be surgically opened. Typically, three small incisions are made, one for the arthroscope, one for the surgical instruments, and one for fluid drainage. The physician views the joint area on a video monitor and can diagnose and repair torn joint tissue, including ligaments and cartilage. The postoperative recovery time is reduced, and the surgery success rate is usually increased because there is reduced trauma to the connective tissue of the joint. This procedure is especially useful for athletes who require a rapid recovery time following surgery. Arthroscopic surgery is commonly used for joints of the knee (72,73), shoulder (74,75), elbow (76,77), wrist (78,79), ankle (80), and hip (81–83).

Anterior Cruciate Ligament Reconstruction

Anterior cruciate ligament reconstruction (ACL reconstruction) is a surgical procedure that uses a graft replacement of a torn ACL in the knee (Figure 6.10). A torn ACL dramatically decreases the stability and functional ability of the knee joint and is usually treated medically. Torn ligaments do not heal, so surgery is often required to medically treat the injury. An ACL reconstruction requires a tissue graft from another part of the body, often

Anterior cruciate ligament reconstruction A surgical procedure that uses a graft replacement of a torn ACL in the knee.

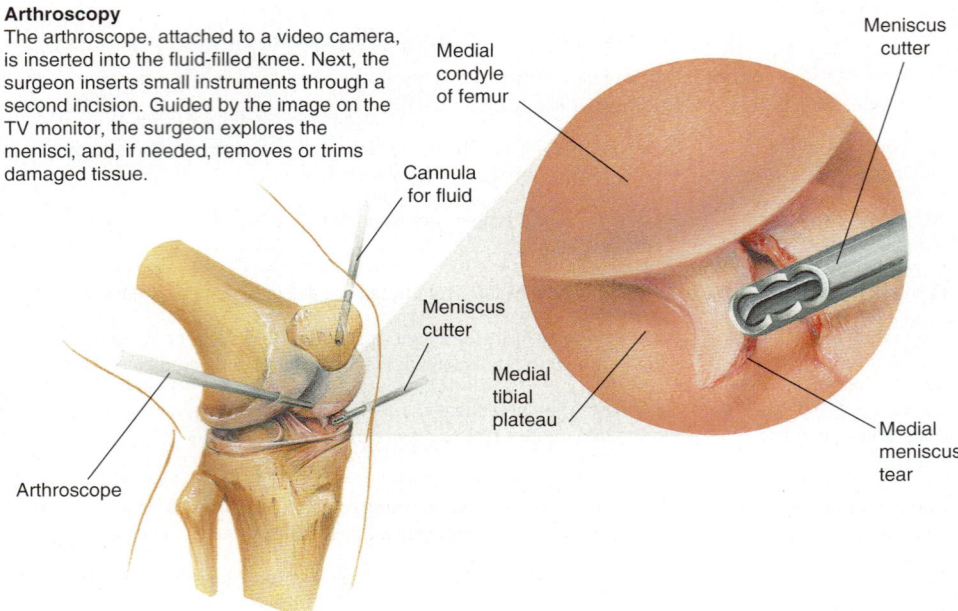

FIGURE 6.9 Arthroscopic surgery of the knee joint. (Asset provided by Anatomical Chart Co.)

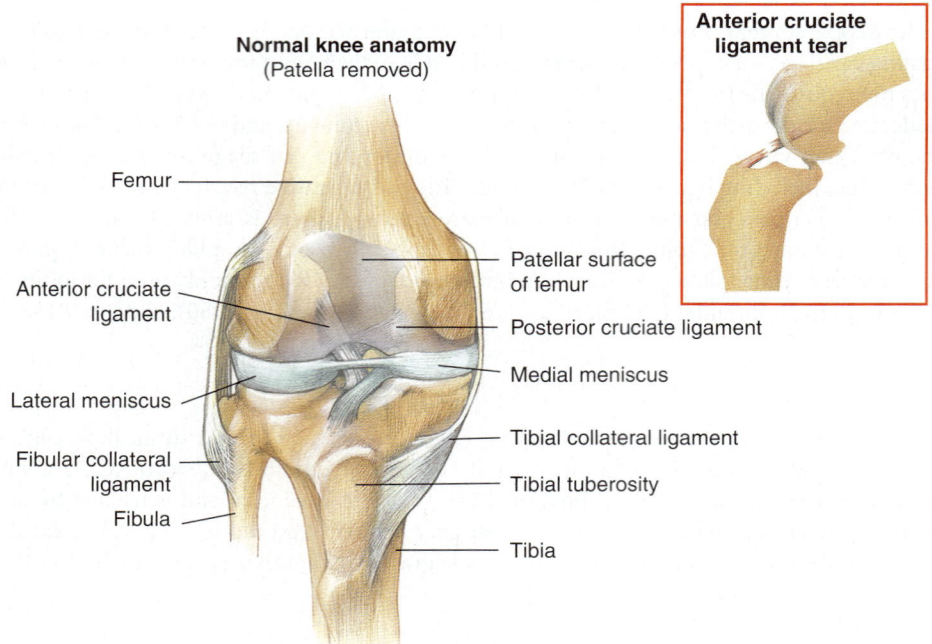

FIGURE 6.10 Anterior cruciate ligament tear. (Asset provided by Anatomical Chart Co.)

from the patella or hamstring tendon. The torn ligament is removed from the knee before the graft is inserted, and attachment is made to the tibia and the femur. The types of surgery differ mainly in the type of graft that is used. Part or all of the ACL reconstruction is performed using arthroscopic surgery (83). This procedure is especially useful for athletes who require extensive repair and a rapid recovery time following surgery (84,85).

Ulnar Collateral Ligament Reconstruction

Ulnar collateral ligament reconstruction is a surgical procedure in which a ligament in the medial elbow is replaced with a tendon from elsewhere in the body, usually from the forearm, hamstring, knee, or foot. Damage to the ulnar collateral ligament usually occurs in response to the stress of the throwing motion. In the procedure, the replacement tendon is woven in a figure-eight pattern through tunnels that have been drilled in the ulna and humerus bones that are part of the elbow joint. Ulnar collateral ligament reconstruction is also known as Tommy John surgery. John was a pitcher for the Los Angeles Dodgers professional baseball team and was the first professional athlete to undergo the surgery, which was performed by Dr. Frank Jobe. Following surgery, an extensive rehabilitation process is undertaken by the athlete. ROM and resistance training exercises are performed for about 6 months, after which the athlete can begin a throwing program (86). Reconstruction surgery is used on athletes of all ages in sport, with innovative techniques being continuously developed (87).

Autologous Chondrocyte Implantation

Articular cartilage covers the ends of the long bones of the body, is essentially frictionless, and provides a smooth surface for the contact and movement of bones. Articular cartilage is formed by cells called chondrocytes. **Autologous chondrocyte implantation** (ACI) is used to repair defects in the articular cartilage, usually in the knee. Patients eligible for treatment with ACI usually have joint pain, swelling, catching, or grinding. ACI is generally applied to patients between the ages of 15 and 55 years, with little or no additional damage to the knee joint. These are patients who do not have enough knee damage to need a total knee replacement but who are experiencing considerable pain that may be impairing their sport performance or quality of life. Clinically appropriate patients are identified through traditional diagnostic methods, such as magnetic resonance imaging, X-ray evaluation, and an arthroscopic examination (88).

Once a patient is determined to be eligible for ACI, a **biopsy** sample of between 200 and 300 mg of the patient's articular cartilage is collected. The tissue sample is then sent to a laboratory, where the chondrocytes are separated from their surrounding cartilage and cultured for 4 to 5 weeks, generating between 5 and 10 million cells. The implantation of the cells is a surgical procedure in which the patient's joint is exposed by the orthopedic

Ulnar collateral ligament reconstruction A surgical procedure in which a ligament in the medial elbow is replaced with a tendon from elsewhere in the body.
Autologous chondrocyte implantation A procedure that uses chondrocyte cells to repair defects in the articular cartilage of various joints.
Biopsy A procedure that collects a small sample of tissue using a specialized needle.

surgeon. The defect area is prepared by removing dead cartilage and smoothing the surrounding living cartilage. A piece of periosteum, the membrane that covers bone, is taken from the patient's tibia and attached over the prepared area. The cultured cells are injected by the surgeon under the periosteum, where they will grow and mature over time. In about 10 to 12 weeks, the patient can put full weight on the knee, but complete recovery may take up to 1 year (88). ACI has been shown to increase the speed at which athletes return to training and competition compared to other forms of cartilage repair (89,90).

AREAS OF STUDY IN ATHLETIC TRAINING AND SPORTS MEDICINE

With the rapid advances in the identification and treatment of sports-related injuries, athletic trainers, sports medicine physicians, and other allied health care professionals must maintain current knowledge to provide the best possible health care to the athletes they work with. To remain certified, an athletic trainer must accumulate 50 CEUs with at least 10 Evidence-Based Practice (EBP) CEUs. Sports medicine physicians must accumulate continuing medical education (CME) credits to remain licensed. The number of CMEs required varies by state licensing boards. The CEUs and CMEs can be obtained by participating in various educational activities. Many of these activities focus on issues currently being addressed by researchers, scholars, athletic trainers, and sports medicine professionals. Issues of particular importance to both athletic trainers and sports medicine physicians are wide ranging and include maintaining hydration status (91), preventing exertional heat illness (92), managing sudden cardiac arrest (93) and asthma (94) in athletes, and injury surveillance and prevention (95). The management of concussions sustained in sport competition and the increased incidence of ACL injuries in female athletes are particularly important and discussed in greater detail in the following section.

Concussion Management

One of the most potentially severe injuries that can be sustained by an athlete is a concussion to the brain. Athletic trainers and sports medicine physicians must use the most recent scientific and clinic-based evidence for managing sport-related concussion. From 1968 to 1990, there was a significant reduction in brain and cervical spine fatalities in high school and college football players (96,97). The decrease was attributed to a variety of factors, including rule changes, enhanced player education, improvements in equipment safety, and enhanced evaluation techniques by athletic trainers and sports medicine personnel (96). However, with increased attention focused on the relationship between concussion rates and mental illness (98) and cognitive function (99), there has been increased attention on the identification and removal from play of athletes who have sustained a concussion (43). This awareness of concussion incidence has spread to soccer (100), ice hockey (100,101), and other sports (102). Recent evidence would indicate that women are more likely to experience a higher risk of concussions in soccer and lacrosse owing to ball or equipment contact (103). In response, science- and clinic-based information has been used to develop guidelines for reducing the incidence and severity of sport-related concussion and improving return-to-participation decisions (34). Numerous professional organizations have position and consensus statements on the medical management of sport-related concussions (30,43,96,104,105).

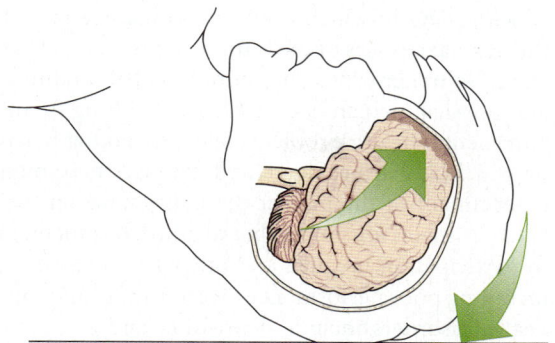

FIGURE 6.11 Example of how a concussion of the brain occurs.

The most common sport-related concussion is the diffuse brain injury. This type of injury occurs when a linear acceleration-deceleration motion (side-to-side or front-to-back) causes the brain to be shaken within the skull (Figure 6.11). A sudden momentum change by the brain can result in tissue damage. Cerebral concussion can best be classified as a mild, diffuse injury that results in one or more of the following conditions: headache, nausea, vomiting, dizziness, balance problems, feeling "slowed down," fatigue, trouble sleeping, drowsiness, sensitivity to light or noise, blurred vision, difficulty remembering, or difficulty concentrating (45). Even though the purpose of bone is to protect the brain mass, when the brain suddenly contacts the cranium concussion could result. Athletic trainers and sports medicine personnel are responsible for making on-site evaluations and often use a return-to-play protocol to assist with decision-making in regard to when participation is acceptable and appropriate (43,96). With the increased concern for athlete safety, new regulations employing an interprofessional/team-based care approach prevent a concussed athlete from returning to play during the same day (43).

Chronic traumatic encephalopathy (CTE) is a neurodegenerative disease associated with concussion, repetitive brain trauma, traumatic brain injury, and participation in contact and collision sports, including American football, boxing, soccer, rugby, and ice hockey (30,106–108). Numerous factors influence the development of CTE, among which length of time participating in contact and collision sports and the degree of exposure to the head impact play a significant role in the development of CTE (107,108). For example, participating in American football contributes to the risk of developing CTE, with the odds of developing CTE doubling for every 2.6 years played (106). Currently, CTE can be diagnosed only with a postmortem neuropathologic examination (108); however, a number of signs, symptoms, and biological markers may be present in individuals with CTE, making a clinical diagnosis more accurate (108). CTE contributes to uncontrollable aggressive and violent behavior, diminished attention, depression, executive dysfunction, and memory instabilities (109). These mental and psychological changes often lead to disruptions in normal behavior and premature death, making the diagnosis and treatment of CTE an important issue in contact and collision sport athletes.

> **Thinking Critically**
>
> Why is it important for athletic trainers and sports medicine personnel to adhere to the guidelines for identification and treatment of an individual who has sustained a concussion?

Mental Health Issues in Athletes

Mental health is critical to the overall well-being of individuals, and this is especially true of competitive athletes. The World Health Organization defines mental health as "a state of well-being in which every individual realizes his or her own potential, can cope with the normal stresses of life, can work productively and fruitfully, and is able to make a contribution to her or his community" (110). Training and competition place intense mental and physical demands on athletes, resulting in an increase in their susceptibility to certain mental health problems and risk-taking behaviors (111). The strategies and actions by which athletes evaluate and manage these mental health issues can be an influential determinant of the impact the issues have on their mental health and sporting success (112). Athletes are vulnerable to numerous mental health problems, which may be related to participation in sport and other factors associated with the demands of competition, including poor performance, overtraining, and injury (113,114). The predominant mental health disorders include depression and anxiety in males and females, and eating disorders/disordered eating in females (115). The effectiveness of treatment interventions varies considerably across athlete levels and types of competition (116). Institutional access to mental health counseling is critical to effective management and treatment (115). Recently, the AMSSM released a position statement detailing the specific mental health conditions in athletes, and the importance of recognizing and treating these conditions, such as eating disorders and disordered eating, depression and suicide, anxiety and stress, overtraining, sleep disorders and attention-deficit hyperactivity disorder, is reviewed with a focus on detection, management, and the effect on performance and prevention (117). Sports medicine physicians, athletic trainers, and other members involved in the care of the athlete are uniquely positioned to detect mental health issues early and make appropriate interventions (117).

Anterior Cruciate Ligament Injuries in Females

The passage of Title IX of the Educational Assistance Act of 1972 has resulted in the participation of more girls and women in sports. With the increased rate of participation, there has been an increased rate of injuries. Unfortunately, an increased number of ACL injuries have accompanied the increased participation in sports (42) and continues to rise (118). Additionally, there is a high probability that individuals who experienced an initial ACL injury will also have a second ACL injury within a few years of the first reconstructive surgery (119). Female athletes experience 4 to 10 times more ACL injuries than male athletes (120–122), and the number of ACL injuries in female athletes is higher than that in male athletes performing the same sports (42,123). The reasons for the different rates of injury in men and women are unclear but may be related to differences in structure and knee alignment (124), ligament laxity (42,123,125,126), and muscle strength (127).

In the knee joint, an intercondylar notch (compartment) lies between the two rounded ends of the thigh bone (femoral condyles). The ACL moves within this notch, connecting the femur (thigh bone) and tibia (shin bone) and providing stability to the knee. It prevents the tibia from moving too far forward and from rotating too far inward under the femur (Figure 6.12). Women have a narrower intercondylar notch than do men; therefore, the space for ACL movement is more limited in women than in men (124,128). Within this restricted space, the femoral condyles can more easily pinch the ACL as the knee bends and straightens out, especially during twisting and hyperextension movements. The pinching of the ACL in the joint can lead to its rupture or tear (128). Furthermore, an intercondylar notch that is stenotic is significantly correlated with ACL injury in females (129).

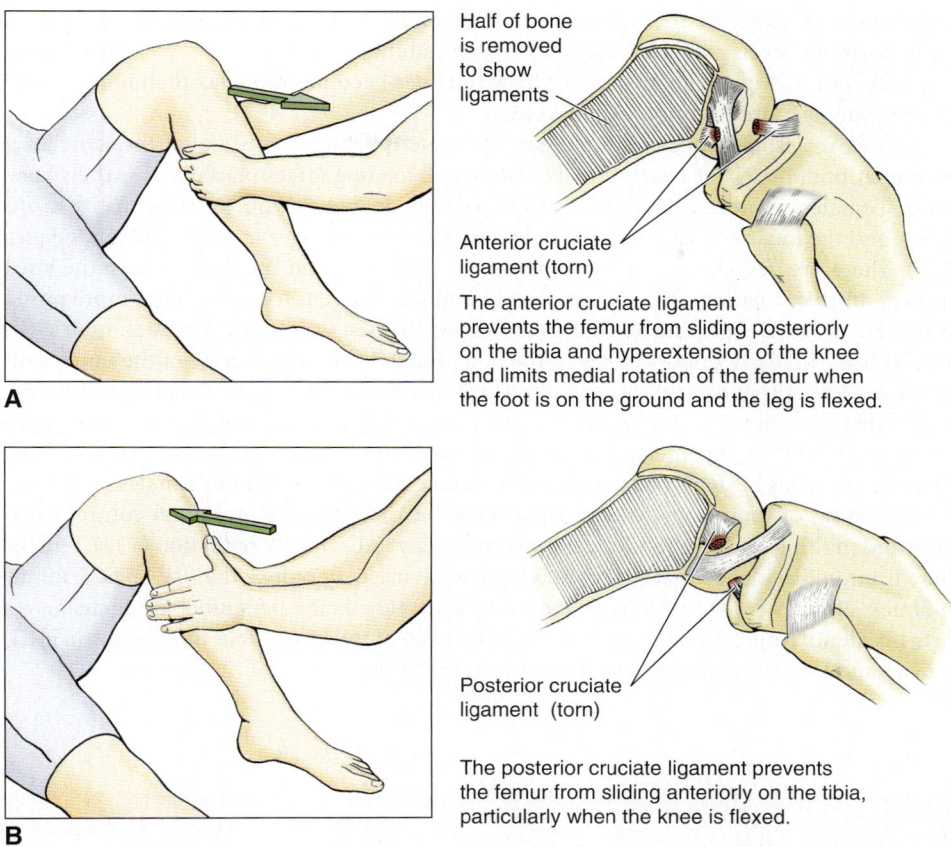

FIGURE 6.12 Mechanism for an anterior cruciate ligament (ACL) injury in the knee. Anterior drawer sign (ACL) (A) and posterior drawer sign (B). (Modified from Moore KL, Dalley AF, II. *Clinical Oriented Anatomy*. 4th ed. Baltimore (MD): Lippincott Williams & Wilkins; 1999.)

In the knee, the femur meets the tibia at an angle (called the quadriceps, or Q angle). The width of the pelvis determines the size of the Q angle. Females have a wider pelvis than males; therefore, the Q angle is greater in females than in males. At this greater angle, forces are concentrated on the ligament each time the knee twists, increasing the risk for an ACL tear. A twisting injury in the knee of a male may only stretch the ACL; however, because of the greater Q angle, the same type of twisting injury in the knee of a female may cause a complete ACL tear (124,128).

Female hormones such as estrogen and progesterone allow for greater flexibility and looseness of muscles, tendons, and ligaments. This looseness may help prevent injuries because it enables certain muscles and joints to absorb more impact before being damaged. This looseness, however, may contribute to ACL injuries in females. If the ligaments and muscles around the knee are too loose, they cannot absorb the stresses placed on them. In this situation, normal loads or forces may be transferred directly to the ACL, making it prone to tearing or rupture. In this situation, the ACL must maintain stability of the knee and also compensate for the instability in a generally loose knee. During the menstrual cycle, hormone levels vary and may affect knee stability. Recent studies have shown that, at specific points within the menstrual cycle, the knee becomes more lax than normal, and ACL rupture is more common (42,125,126).

When males and females compete in the same sporting events and at the same level of competition, they have nearly equal twisting and loading forces placed across their knee joints. Females, however, have less muscle strength in proportion to bone size than do males, and the muscles that help hold the knee stable are stronger in males than in females (127). Therefore, females rely less on the muscles and more on the ACL to hold the knee in place. In this situation, the ACL may have to absorb more forces, making it more prone to rupture (127). Developing an understanding of the best way to prevent ACL injuries in females is particularly important given the potentially debilitating nature of the injury and poor long-term prognosis following surgical reconstruction (130,131). Recent evidence indicates that preventive neuromuscular training can be effective in reducing ACL injuries in young females (132). Plyometric training combined with biomechanical analysis and technique training has been shown to effectively reduce ACL injury rates in females (130,133). Strengthening of the associated muscle groups, balance training, proximal control exercises, and multiple exercise modes demonstrate greater ACL injury reduction (132). Finally, there is limited evidence that oral contraceptive pill use may reduce the risk of ACL injury in female athletes (134). Athletic trainers and sports medicine personnel are focusing efforts on identifying those factors that make females more susceptible than males to ACL injuries and on developing interventions to aid in the prevention of ACL injuries.

Interview

Layci Harrison, PhD, LAT, ATC
Clinical Assistant Professor of Athletic Training, University of Houston

Brief Introduction – I graduated with a BS in Athletic Training from Lock Haven University. After my time at Lock Haven, I moved to Lebanon, TN, for an athletic training internship with Cumberland University Baseball. I spent 2 additional years at Cumberland University as a graduate assistant athletic trainer working with the wrestling team while completing an MS in Exercise Science. After leaving Cumberland, I spent 3 years teaching undergraduate exercise science courses while earning a PhD in Health and Human Performance from Middle Tennessee State University. My research focused on injury prevention and rehabilitation, including functional movement screenings and examining how the cross-over effect could be used to reduce asymmetries after injury. During my graduate education, I worked per diem providing medical care to cycling and recreational football leagues in the middle Tennessee area. I'm currently a Clinical Assistant Professor of Athletic Training at the University of Houston, where I coordinate interprofessional education and patient simulation.

Q: Why did you choose to become an athletic trainer?

In high school, anatomy was my favorite subject. I loved learning about the body and how performance could be impacted by injury or illness. Athletic training provided the perfect opportunity to study a topic that I enjoy while using that knowledge to help people get back to doing what they love. The variety of the job really interested me. Every day as an athletic trainer provides new opportunities and challenges. For example, maybe a patient has an ankle sprain and wants to get back to playing football or maybe that same patient wants to dance with their significant other at an upcoming event. Whatever the goal, athletic trainers have the unique challenge of getting the patient back to performing at levels beyond activities of daily living, and that is truly exciting.

Q: What are the top two or three "best things" about your job?

Throughout my career, I transitioned from a clinician to an educator, and both have been enjoyable. My favorite part of being a clinician was watching patients improve in all aspects of their lives. Being there at the onset of injury while the patient is dealing with physical and emotional distress is challenging, but working as part of a health care team and watching the patient return to activity and excel in other areas such as education, relationships, or professionally is amazing. Another great perk is that your "office" is constantly changing. Sometimes I was in a traditional office or clinic setting, but most days I was in a wrestling room, on a bus, at nationals, in a hotel lobby, or some other nontraditional setting. Providing health care from a sideline, stage, or industrial setting requires creativity. You won't get bored. Now that I'm teaching, my favorite part of my job is learning with my students. The medical field is constantly changing, and in order to teach my students to practice using evidence-based approaches to patient care, I have to stay up to date with new research. Every day is an opportunity to learn something new.

Q: What is important for an athletic trainer to have an understanding of exercise science?

All parts of the exercise science degree are important for athletic training. In fact, during class I often find myself saying "remember from your undergraduate exercise physiology class…" All health care providers have to understand how the body reacts to rest, exercise, disease, and injury. Athletic trainers will encounter patients with a variety of medical histories, and as a health care provider, you are responsible for keeping them safe during activity. You can't treat a patient if you don't understand how the body reacts and adapts.

Q: What advice would you have for an undergraduate student beginning to explore a career in exercise science or possibly as an athletic trainer?

Anyone considering a career in athletic training should observe athletic trainers in multiple settings and be willing to work as part of a health care team. Many students get into the profession not knowing that it is a medical field. Athletic trainers are health care providers and care for more than just athletes. Athletic trainers work in many nontraditional settings, including the military, public safety, industrial settings, National Aeronautics and Space Administration (NASA), performing arts, and pretty much any place where patients are trying to return to activity beyond those required for daily living. Athletic trainers also encounter patients with a variety of conditions. Be willing to learn from and about multiple health care professionals. The goal is to work as a team to provide the best care for your patients.

If you are interested in athletic training, start learning about the different programs. Athletic training requires a professional (Master's) degree, and program requirements may vary. Be aware of grade point average (GPA) minimums, specific grade requirements for anatomy and physiology, and observation hour requirements. Attend open houses, asks questions, and, most importantly, when accepted into a program, be open to new experiences and new settings.

Interview

Nailah Coleman, MD, FAAP, FACSM
Children's National Hospital

Brief Introduction – Currently, I practice Pediatrics and Primary Care Sports Medicine at Children's National Hospital. I completed my undergraduate studies at Emory University, receiving a BS in Biology with a double major in International Studies and a minor in Italian. I played varsity volleyball 1 year, was on the cheerleading squad for another, and even tried sports photography for another year. I remained at Emory University for medical school and earned my medical degree in 2000. I completed a pediatric residency program at Children's National Medical Center in Washington, DC in 2003. Following residency, I remained at Children's National as a Physician Analyst within the Information Technology Department. I also worked as a Hospitalist for incoming patients, a practicing physician in the Children's Health Clinic, and an on-call neonatal pediatrician at The George Washington University Medical Center. As a pediatrician working in multiple hospital environments, I had the opportunity to see children at different life stages and assess their growth and wellness. I observed a need to improve the physical health of the athletic and the nonathletic student during their various developmental stages. I believe that athletic performance, nutrition, and physical health plans should be developed and tracked for all students, starting in primary school and continuing through college. Sports medicine should include the athlete and the nonathlete. With these ideas in mind, I moved to Georgia and Phoebe Putney Memorial Hospital to complete a 1-year fellowship in Primary Care Sports Medicine, before returning to Children's in my current capacity, where I now work as a general pediatrician and a primary care sports medicine physician. I am board-certified in Pediatrics and Primary Care Sports Medicine. I am licensed in Washington, DC, Virginia, Maryland, and Georgia and a member of the American Medical Association, the American Academy of Pediatrics, the American College of Sports Medicine, and the American Medical Society of Sports Medicine. I also provide medical coverage and sports medicine information to local high schools and various athletic events in the DC area.

 Why did you choose to become a sports medicine physician?

I have always wanted to be a pediatrician; however, after residency, when I had become a pediatrician — finally — I was left wondering, "Now what?" I fortunately detoured into a position as an analyst in our IT Department, which gave me exposure to the entire hospital's systems, experience in the design and implementation of an EMR, and time to participate in hospital teaching sessions, including our weekly case conference. One day, while enjoying the lunch served at the conference — typically pizza — the expert physician was a pediatric sports physician. I thought instantly, "I want to do that!" I met with her later that week and applied for fellowships within the month. I have since realized that there were many signs that I had an interest in sports medicine. I played a variety of sports throughout my life and continue to be an active adult. I was always interested in what happened to the injured athlete on the field, even as a little child, watching the team physician run out to help a downed athlete — it was usually the only part of the game I watched. I enjoyed working in the ER and participating on the trauma surgery service in medical school. I have also been putting my parents on an eating and exercise plan since junior high, some more successful than others. With sports medicine, I can combine my love of medicine with my interests in sports, exercise, and well-being.

Q: What are the top two or three "best things" about your job?

I lean heavily toward extroversion. As such, my top two best things about my job are connecting and sharing. I love connecting with the patients (and their families). There is something beautiful about connecting with another human, and, as each human is unique, so is the connection we providers make with each of them; some are more comedic and joyful, while others can be more frustrating and sad; it is such an honor and privilege to help patients and families during one of their most vulnerable times. As one who loved "Show and Tell" in kindergarten, I also enjoy sharing my knowledge and information (with patients, families, students, staff, other providers), as well as strategies for improvement or resolution of a particular concern.

 Q: What advice would you have for an undergraduate student beginning to explore a career in exercise science or possibly as a sports medicine physician?

There are many avenues into the field of sport and exercise medicine. The primary driver should be a passion for the general area and an openness to experiencing all that is there. It is okay to decide…and then decide again (*i.e.*, change your mind).

SUMMARY

- Athletic trainers have an important role to play in keeping an athlete safe and injury free during training and competition.
- As part of the primary care sports medicine team, athletic trainers are often the first to diagnose and respond to an injury.
- Working in close conjunction with other sports medicine personnel, the athletic trainer develops an individual rehabilitation program to help an injured athlete return to practice and competition.
- Sports medicine physicians work closely with other allied health care professionals to ensure that athletes are provided with the best medical care possible.
- Numerous advancements in the identification, treatment, and rehabilitation of injuries have enhanced the overall care provided to individuals who are injured during physical activity, exercise, sport, and athletic competition.

FOR REVIEW

1. Describe the difference between athletic training and sports medicine.
2. Describe how the NATA was formed.
3. How has the development of the certification program by the BOC enhanced the profession of athletic training?
4. Name three prominent professional associations in the area of Sports Medicine.
5. What are the five domains of athletic training?
6. What is the difference between a primary and a secondary survey of an injured athlete?
7. Why is detailed record keeping an important aspect of an athletic training program?
8. What is the common initial treatment for an acute musculoskeletal injury?

9. Define the following muscle actions:
   a. Isometric
   b. Isotonic
   c. Isokinetic
   d. Concentric
   e. Eccentric
10. What is the difference between open and closed chain kinetic exercises?
11. What are therapeutic modalities used for?
12. What recent medical advances have allowed sports medicine to improve the opportunity for athletes to return to play quickly from an injury to the knee joint?
13. Describe the common symptoms exhibited by someone who may have experienced a concussion.
14. Discuss the key issues related to the higher incidence rate of ACL injuries to women.

Project-Based Learning

1. Create a decisional process from initial evaluation through to complete clearance for return to play for the following athlete injury situations:
   a. A possible sprained ankle
   b. A possible anterior cruciate ligament tear
   c. A possible concussion
2. Create a PowerPoint presentation that you can share with exercise science students who are interested in being recruited to the following:
   a. A graduate degree program in athletic training
   b. Medical school

REFERENCES

1. Prentice WE. *Principles of Athletic Training: A Competency-Based Approach*. 15th ed. New York, NY: McGraw-Hill Companies; 2014.
2. Matheson GO. Reflecting on 30 years of moving forward. *Phys Sports Med*. 2003;31(1):1–2.
3. Comstock RD, Currie DW, Pierpoint LA. *National High School Sport-Related Injury Surveillance Study*. Aurora, CO: 2015. Report No32.
4. Chandran A, Morris SN, Powell JR, Boltz AJ, Robison HJ, Collins CL. Epidemiology of injuries in National Collegiate Athletic Association men's football: 2014–2015 through 2018–2019. *J Athl Train*. 2021;56(7):643–50.
5. Powell JR, Boltz AJ, Robison HJ, Morris SN, Collins CL, Chandran A. Epidemiology of injuries in National Collegiate Athletic Association men's wrestling: 2014–2015 through 2018–2019. *J Athl Train*. 2021;56(7):727–33.
6. Boltz AJ, Nedimyer AK, Chandran A, Robison HJ, Collins CL, Morris SN. Epidemiology of injuries in National Collegiate Athletic Association men's ice hockey: 2014–2015 through 2018–2019. *J Athl Train*. 2021;56(7):703–10.
7. Chandran A, Morris SN, Boltz AJ, Robison HJ, Collins CL. Epidemiology of injuries in National Collegiate Athletic Association men's soccer: 2014–2015 through 2018–2019. *J Athl Train*. 2021;56(7):659–65.
8. Morris SN, Chandran A, Lempke LB, Boltz AJ, Robison HJ, Collins CL. Epidemiology of injuries in National Collegiate Athletic Association men's basketball: 2014–2015 through 2018–2019. *J Athl Train*. 2021;56(7):681–87.

9. Chandran A, Morris SN, Boltz AJ, Robison HJ, Collins CL. Epidemiology of injuries in National Collegiate Athletic Association women's soccer: 2014–2015 through 2018–2019. *J Athl Train*. 2021;56(7):651–58.
10. Chandran A, Roby PR, Boltz AJ, Robison HJ, Morris SN, Collins CL. Epidemiology of injuries in National Collegiate Athletic Association women's gymnastics: 2014–2015 through 2018–2019. *J Athl Train*. 2021;56(7):688–94.
11. Chandran A, Nedimyer AK, Boltz AJ, Robison HJ, Collins CL, Morris SN. Epidemiology of injuries in National Collegiate Athletic Association women's ice hockey: 2014–2015 through 2018–2019. *J Athl Train*. 2021;56(7):695–702.
12. Nedimyer AK, Boltz AJ, Robison HJ, Collins CL, Morris SN, Chandran A. Epidemiology of injuries in National Collegiate Athletic Association women's field hockey: 2014–2015 through 2018–2019. *J Athl Train*. 2021;56(7):636–42.
13. Lempke LB, Chandran A, Boltz AJ, Robison HJ, Collins CL, Morris SN. Epidemiology of injuries in National Collegiate Athletic Association women's basketball: 2014–2015 through 2018–2019. *J Athl Train*. 2021;56(7):674–80.
14. United States Consumer Product Safety Commission. 2015 [updated 2015]. *National Electronic Surveillance System*. Available from: http://www.cpsc.gov/en/Research--Statistics/NEISS-Injury-Data/.
15. O'Shea ME. *A History of the National Athletic Trainers Association*. Greenville (NC): National Athletic Trainers Association; 1980.
16. Harris HA. *Greek Athletes and Athletics*. London: Hutchinson of London; 1964. 1964.
17. Delforge GD, Behnke RS. The history and evolution of athletic training education in the United States. *J Athl Train*. 1999;34(1):53–61.
18. Weidner TG, Henning JM. Historical perspective of athletic training clinical education. *J Athl Train*. 2002;37(4 Suppl):S222–S8.
19. Youth Sports Safety Alliance. 2021. Available from: https://youthsportssafetyalliance.org/.
20. Berryman JW. Ancient and early influences. In: Tipton CM, editor. *Exercise Physiology: People and Ideas*. New York (NY): Oxford University Press; 2003. p. 1–38.
21. Twentieth century orthopaedics. *AAOS Bull*. 1999;47(6):35–41.
22. International Federation of Sports Medicine. 2015 [updated 2021]. Available from: www.fims.org.
23. Wappes JR. 30 years of sports medicine. *Phys Sports Med*. 2003;31(1):1–4.
24. Herring SA, Bergfeld JA, Boyajian-O'Neill L, et al. Female athlete issues for the team physician: a consensus statement. *Med Sci Sports Exerc*. 2003;35(10):1785–93.
25. Herring SA, Bergfeld JA, Boyajian-O'Neill L, et al. Mass participation event management for the team physician: a consensus statement. *Med Sci Sports Exerc*. 2004;36(11):2004–8.
26. Herring SA, Bergfeld JA, Boyd J, et al. The team physician and conditioning of athletes for sports: a consensus statement. *Med Sci Sports Exerc*. 2001;33(10):1789–93.
27. Herring SA, Bergfeld JA, Boyd J, et al. Sideline preparedness for the team physician: a consensus statement. *Med Sci Sports Exerc*. 2001;33(5):846–9.
28. Herring SA, Bergfeld JA, Boyd J, et al. Team physician consensus statement. *Med Sci Sports Exerc*. 2000;32(4):877–8.
29. Herring SA, Bergfeld JA, Boyd J, et al. The team physician and return-to-play issues: a consensus statement. *Med Sci Sports Exerc*. 2002;34(7):1212–4.
30. Herring SA, Cantu RC, Guskiewicz KM, Putukian M, Kibler WB. Concussion (mild traumatic brain injury) and the team physician: a consensus statement — 2011 update. *Med Sci Sports Exerc*. 2011;43(12):2412–22.
31. Herring SA, Bergfeld JA, Bernhardt DT, et al. Selected issues for the adolescent athlete and the team physician: a consensus statement. *Med Sci Sports Exerc*. 2008;40(11):1997–2012.
32. American Medical Society for Sports Medicine. 2020. Available from: www.amssm.org.
33. Schnirring L. Mending injured athletes: a track record of orthopedic advances. *Phys Sports Med*. 2003;31(9):1–3.
34. World Anti-Doping Agency. 2020 [cited 2020]. Available from: https://www.wada-ama.org/.
35. Tang L. Research on applying musculoskeletal ultrasound technology to real-time medical diagnosis of sports injuries. *J Med Imaging Health Inform*. 2020;10(4):837–41.
36. Wang WH, Wang XJ. Clinical application of high frequency ultrasound in diagnosis and treatment of lumbar muscle injury after strenuous exercise. *J Med Imaging Health Inform*. 2020;10(4):923–7.
37. Whittaker JL, Ellis R, Hodges PW, et al. Imaging with ultrasound in physical therapy: what is the PT's scope of practice? A competency-based educational model and training recommendations. *Br J Sports Med*. 2019;53(23):1447–53.

38. Lesniak BP, Loveland D, Jose J, Selley R, Jacobson JA, Bedi A. Use of ultrasonography as a diagnostic and therapeutic tool in sports medicine. *Arthroscopy*. 2014;30(2):260–70.
39. Finnoff JT, Hall MM, Adams E, et al. American Medical Society for Sports Medicine (AMSSM) position statement: interventional musculoskeletal ultrasound in sports medicine. *Br J Sports Med*. 2015;49(3):145–50.
40. Moeller JL, McKeag DB. Preparticipation screening. In: McKeag DB, Moeller JL, editors. *ACSM's Primary Care Sports Medicine*. Philadelphia (PA): Lippincott, Williams & Wilkins; 2007. p. 55–79.
41. Davies GAL, Wolfe LA, Mottola MF, MacKinnon C. Joint SOGC/CSEP clinical practice guideline: exercise in pregnancy and the postpartum period. *Can J Appl Physiol*. 2003;28(3):329–41.
42. Arendt E, Dick RW. Knee injury patterns among men and women in collegiate basketball and soccer: NCAA data and review of literature. *Am J Sports Med*. 1995;23(6):694–701.
43. Herring SA, Bergfeld JA, Boyland A, et al. Concussion (mild traumatic brain injury) and the team physician: a consensus statement. *Med Sci Sports Exerc*. 2006;38(2):395–9.
44. Roberts WO. Sports medicine's primary focus: health for all. *Phys Sports Med*. 2003;31(12):1–2.
45. McCrory P, Meeuwisse W, Dvorak J, et al. Consensus statement on concussion in sport-the 5th international conference on concussion in sport held in Berlin, October 2016. 2017;51:838.
46. Conley KM, Bolin DJ, Carek PJ, Konin JG, Neal TL, Violette D. National Athletic Trainer's Association position statement: preparticipation physical examinations and disqualifying conditions. *J Athl Train*. 2014;49(1):102–20.
47. Prentice WE. *On-the-Field Acute Care and Emergency Procedures. Arnheim's Principles of Athletic Training: A Competency Based Approach*. 14th ed. New York (NY): McGraw-Hill; 2011. p. 319–59.
48. Occupational Safety and Health Administration. The OSHA bloodborne pathogens standards. *Fed Regist*. 1991;55(235):64175.
49. Kettenbach G. *Writing SOAP Notes: With Patient/Client Management Format*. 3rd ed. Philadelphia (PA): F.A. Davis; 2003. 2003.
50. Knight KL. *Cryotherapy in Sports Injury Management*. Champaign (IL): Human Kinetics; 1995. 1995.
51. Brooks GA, Fahey TD, Baldwin KM. *Exercise Physiology: Human Bioenergetics and Its Applications*. 4th ed. Mountain View (CA): Mayfield; 2004.
52. Mujika I, Padilla S. Detraining: loss of training-induced physiological and performance adaptations. Part II. *Sports Med*. 2000;30(4):145–54.
53. Jackson MD. Rehabilitation. In: McKeag DB, Moeller JL, editors. *ACSM's Primary Care Sports Medicine*. 2nd ed. Philadelphia (PA): Lippincott, Williams & Wilkins; 2007. p. 563–93.
54. Fleck SJ, Kraemer WJ. *Designing Resistance Training Programs*. 3rd ed. Champaign (IL): Human Kinetics Books; 2004. 2004.
55. Prentice WE. Mobilization and traction techniques in rehabilitation. In: Prentice WE, editor. *Rehabilitation Techniques in Sports Medicine and Athletic Training*. St. Louis (MI): McGraw-Hill; 2004.
56. Rivera J. Open vs. closed kinetic chain rehabilitation of the lower extremity. *J Sport Rehabil*. 1994;3(2):154.
57. Prentice WE. *The Athletic Trainer as a Health Care Provider. Arnheim's Principles of Athletic Training*. 14th ed. New York (NY): McGraw-Hill; 2011. p. 1–43.
58. Hunt V. Meeting Clarifies HIPAA Restrictions. NATA News. 2003. Available from: www.nata.org/news-publications/publications/nata-news.
59. Rankin J, Ingersoll C. *Athletic Training Management: Concepts and Applications*. 3rd ed. New York (NY): McGraw-Hill; 2006. 2006.
60. Cotton DJ. What is covered by your liability insurance policy? A risk management essential. *Exerc Stand Malpract Rep*. 2001;15(4):54.
61. Hunt V. Reimbursement efforts continue steady progress. NATA News. 2002:10–2.
62. Ray R. *Management Strategies in Athletic Training*. 1st ed. Champaign (IL): Human Kinetics; 2000. 2000.
63. McKeag DB, Moeller JL. Primary care perspective. In: McKeag DB, Moeller JL, editors. *ACSM's Primary Care Sports Medicine*. 2nd ed. Philadelphia (PA): Lippincott, Williams & Wilkins; 2007. p. 3–9.
64. Selected issues in injury and illness prevention and the team physician: a consensus statement. *Med Sci Sports Exerc*. 2016;48(1):159–71.
65. Herring SA, Kibler WB, Putukian M. Psychological issues related to illness and injury in athletes and the team physician: a consensus statement—2016 update. *Med Sci Sports Exerc*. 2017;49(5):1043–54.
66. Herring SA, Boyajian-O'Neill L, Coppel DB, et al. Psychological issues related to injury in athletes and the team physician: a consensus statement. *Med Sci Sports Exerc*. 2006;38(11):2030–4.
67. Herring SA, Kibler WB, Putukian M. Selected issues for the master athlete and the team physician: a consensus statement. *Med Sci Sports Exerc*. 2010;42(4):820–33.

68. Herring SA, Kibler WB, Putukian M. Sideline preparedness for the team physician: a consensus statement: 2012 update. *Med Sci Sports Exerc*. 2012;44(12):2442–5.
69. Herring SA, Kibler WB, Putukian M. The team physician and the return-to-play decision: a consensus statement: 2012 update. *Med Sci Sports Exerc*. 2012;44(12):2446–8.
70. Herring SA, Kibler WB, Putukian M. Team physician consensus statement: 2013 Update. *Med Sci Sports Exerc*. 2013;45(8):1618–22.
71. Herring SA, Kibler WB, Putukian M, et al. Selected issues for nutrition and the athlete: a team physician consensus statement. *Med Sci Sports Exerc*. 2013;45(12):2378–86.
72. Mullins K, Hanlon M, Carton P. Arthroscopic correction of femoroacetabular impingement improves athletic performance in male athletes. *Knee Surg Sports Traumatol Arthrosc*. 2019;28(7):2285–94.
73. Pestka JM, Lang G, Maier D, Sudkamp NP, Ogon P, Izadpanah K. Arthroscopic patellar release allows timely return to performance in professional and amateur athletes with chronic patellar tendinopathy. *Knee Surg Sports Traumatol Arthrosc*. 2018;26(12):3553–9.
74. Tsikouris GD, Bolia IK, Vlaserou P, Odantzis N, Angelis K, Psychogios V. Shoulder arthroscopy with versus without suprascapular nerve release: clinical outcomes and return to sport rate in elite overhead athletes. *Arthroscopy*. 2018;34(9):2552–7.
75. Bradley JP, Arner JW, Jayakumar S, Vyas D. Revision arthroscopic posterior shoulder capsulolabral repair in contact athletes: risk factors and outcomes. *Arthroscopy*. 2020;36(3):660–5.
76. Cain EL, Moroski NM. Elbow surgery in athletes. *Sports Med Arthrosc Rev*. 2018;26(4):181–4.
77. Ciccotti MC, Stull JD, Buckley PS, Cohen SB. Correlation of MRI to arthroscopy in the elbow: thrower's elbow and ulnar collateral ligament injury. *Sports Med Arthrosc Rev*. 2017;25(4):191–8.
78. Gire JD, Yao J. Surgical techniques for the treatment of acute carpal ligament injuries in the athlete. *Clin Sports Med*. 2020;39(2):313–19.
79. Robertson G, Ang KK, Maffulli N, Simpson CK, Rust PA. Return to sport following surgical management of triangular fibrocartilage tears: a systematic review. *Br Med Bull*. 2019;130(1):89–103.
80. Guelfi M, Zamperetti M, Pantalone A, Usuelli FG, Salini V, Oliva XM. Open and arthroscopic lateral ligament repair for treatment of chronic ankle instability: A systematic review. *Foot Ankle Surg*. 2018;24(1):11–8.
81. Lovett-Carter D, Jawanda AS, Hannigan A. Meta-analysis of the surgical and rehabilitative outcomes of hip arthroscopy in athletes with femoroacetabular impingement. *Clin J Sport Med*. 2018;30(4):404–11.
82. McConkey MO, Chadayammuri V, Garabekyan T, Mayer SW, Kraeutler MJ, Mei-Dan O. Simultaneous bilateral hip arthroscopy in adolescent athletes with symptomatic femoroacetabular impingement. *J Pediatr Orthop*. 2019;39(4):193–7.
83. Brukner P, Khan K. *Clinical Sports Medicine*. 3rd ed. Sydney (Australia): McGraw-Hill; 2007. 2007.
84. King E, Richter C, Jackson M, et al. Factors influencing return to play and second anterior cruciate ligament injury rates in level 1 athletes after primary anterior cruciate ligament reconstruction: 2-year follow-up on 1432 reconstructions at a single center. *Am J Sports Med*. 2020;48(4):812–24.
85. Spindler KP, Huston LJ, Zajichek A, et al. Anterior cruciate ligament reconstruction in high school and college-aged athletes: does autograft choice influence anterior cruciate ligament revision rates? *Am J Sports Med*. 2020;48(2):298–309.
86. Vitale MA, Ahmad CS. The outcome of elbow ulnar collateral ligament reconstruction in overhead athletes: a systematic review. *Am J Sports Med*. 2008;36(6):1193–205.
87. Donohue BF, Lubitz MG, Kremchek TE. Elbow ulnar collateral ligament reconstruction using the novel docking plus technique in 324 athletes. *Sports Med Open*. 2019;5(1):1–9.
88. Brittberg M. Autologous chondrocyte implantation—technique and long-term follow-up. *Injury*. 2008;39(1):40–9.
89. Campbell AB, Pineda M, Harris JD, Flanigan DC. Return to sport after articular cartilage repair in athletes' knees: a systematic review. *Arthroscopy*. 2016;32(4):651–70.
90. Goldberg A, Mitchell K, Soans J, Kim L, Zaidi R. The use of mesenchymal stem cells for cartilage repair and regeneration: a systematic review. *J Orthop Surg Res*. 2017;12(1):1–30.
91. Casa DJ, Armstrong LE, Hillman S, et al. National Athletic Trainer's Association position statement: fluid replacement for athletes. *J Athl Train*. 2000;35(2):212–24.
92. Binkley HM, Beckett J, Casa DJ, Kleiner DM, Plummer PE. National Athletic Trainers Association position statement: exertional heat illness. *J Athl Train*. 2002;37(3):329.
93. Drezner JA, Courson RW, Roberts WO, Mosesso VN, Link MS, Maron BJ. Inter-association task force recommendations on emergency preparedness and management of sudden cardiac arrest in high school and college athletic programs: a consensus statement. *J Athl Train* . 2007;42(1):143–58.

94. Miller MG, Weiler JM, Baker R, Collins J, D'Alonzo G. National Athletic Trainers' Association position statement: management of asthma in athletes. *J Athl Train*. 2005;40(3):224–45.
95. Hootman JM, Dick RW, Agel J. Epidemiology of collegiate injuries for 15 sports: summary and recommendations for injury prevention initiatives. *J Athl Train*. 2007;42(2):311–9.
96. Guskiewicz KM, Bruce SL, Cantu RC, et al. National Athletic Trainers' Association position statement: management of sport-related concussion. *J Athl Train*. 2004;39(3):280–97.
97. Mueller FO, Cantu RC. *Nineteenth Annual Report of the National Center for Catastrophic Sports Injury Research: Fall 1982-Spring 2001*. Chapel Hill (NC): National Center for Catastrophic Sports Injury Research; 2002.
98. Guskiewicz KM, Marshall SW, Bailes J, et al. Recurrent concussion and risk of depression in retired professional football players. *Med Sci Sports Exerc*. 2007;39(6):903–9.
99. Baillargeon A, Lassonde M, Leclerc S, Ellemberg D. Neuropsychological and neurophysiological assessment of sport concussion in children, adolescents and adults. *Brain Inj*. 2012;26(3):211–20.
100. Delaney JS, Al-Kashmiri A, Correa JA. Mechanisms of injury for concussions in university football, ice hockey, and soccer. *Clin J Sport Med*. 2014;24(3).
101. Abbott K. Injuries in women's ice hockey: special considerations. *Curr Sports Med Rep*. 2014;13(6):377–82.
102. Zuckerman SL, Kerr ZY, Yengo-Kahn A, Wasserman E, Covassin T, Solomon GS. Epidemiology of sports-related concussion in NCAA Athletes from 2009–2010 to 2013–2014: incidence, recurrence, and mechanisms. *Am J Sports Med*. 2015.
103. Ling DI, Cheng J, Santiago K, et al. Women are at higher risk for concussions due to ball or equipment contact in soccer and lacrosse. *Clin Orthop Relat Res*. 2020;478(7):1469–79.
104. Harmon KG, Drezner J, Gammons M, et al. American Medical Society for Sports Medicine position statement: concussion in sport. *Clin J Sport Med*. 2013;23(1):1–18.
105. McCrory P, Meeuwisse W, Aubry M, et al. Consensus statement on concussion in sport: the 4th international conference on concussion in sport held in Zurich, November 2012. *Clin J Sport Med*. 2013;23(2):89–117.
106. Mez J, Daneshvar DH, Abdolmohammadi B, et al. Duration of American football play and chronic traumatic encephalopathy. *Ann Neurol*. 2020;87(1):116–31.
107. VanItallie TB. Traumatic brain injury (TBI) in collision sports: possible mechanisms of transformation into chronic traumatic encephalopathy (CTE). *Metabolism*. 2019;100:153943.
108. Asken BM, Sullan MJ, Snyder AR, et al. Factors influencing clinical correlates of chronic traumatic encephalopathy (CTE): a review. *Neuropsychol Rev*. 2016;26(4):340–63.
109. Huber BR, Alosco ML, Stein TD, McKee AC. Potential long-term consequences of concussive and subconcussive injury. *Phys Med Rehabil Clin N Am*. 2016;27(2):503–11.
110. World Health Organization. *World Health Report 2002*. Geneva (Switzerland): World Health Organization; 2002. Report No 6.
111. Hughes L, Leavey G. Setting the bar: athletes and vulnerability to mental illness. *Br J Psychiatry*. 2012;200(2):95–6.
112. Lazarus RS. How emotions influence performance in competitive sports. *Sport Psychol*. 2000;14(3):229–52.
113. Rice SM, Purcell R, De Silva S, Mawren D, McGorry PD, Parker AG. The mental health of elite athletes: a narrative systematic review. *Sports Med*. 2016;46(9):1333–53.
114. Reardon CL, Hainline B, Aron CM, et al. Mental health in elite athletes: International Olympic Committee consensus statement (2019). *Br J Sports Med*. 2019;53(11):667–99.
115. Kroshus E. Variability in institutional screening practices related to collegiate student-athlete mental health. *J Athl Train*. 2016;51(5):389–97.
116. Breslin G, Shannon S, Haughey T, Donnelly P, Leavey G. A systematic review of interventions to increase awareness of mental health and well-being in athletes, coaches and officials. *Syst Rev*. 2017;6(1):1–15.
117. Chang C, Putukian M, Aerni G, et al. Mental health issues and psychological factors in athletes: detection, management, effect on performance and prevention: American Medical Society for Sports Medicine Position Statement—executive summary. *Br J Sports Med*. 2020;54(4):216–20.
118. Zbrojkiewicz D, Vertullo C, Grayson JE. Increasing rates of anterior cruciate ligament reconstruction in young Australians, 2000–2015. *Med J Aust*. 2018;208(8):354–8.
119. Wiggins AJ, Grandhi RK, Schneider DK, Stanfield D, Webster KE, Myer GD. Risk of secondary injury in younger athletes after anterior cruciate ligament reconstruction: a systematic review and meta-analysis. *Am J Sports Med*. 2016;44(7):1861–76.
120. Chandy TA, Grana WA. Secondary school athletic injury in boys and girls: a 3-year comparison. *Phys Sports Med*. 1985;13:106–11.

121. Grindstaff TL, Hammill RR, Tuzson AE, Hertel J. Neuromuscular control training programs and noncontact anterior cruciate ligament injury rates in female athletes: a numbers-needed-to-treat analysis. *J Athl Train.* 2006;41(4):450–6.
122. Hutchinson MR, Ireland ML. Knee injuries in female athletes. *Sports Med.* 1995;19:288–302.
123. Stracciolini A, Stein CJ, Zurakowski D, Meehan WP, Myer GD, Micheli LJ. Anterior cruciate ligament injuries in pediatric athletes presenting to sports medicine clinic: a comparison of males and females through growth and development. *Sports Health.* 2014;7:1–7.
124. Uhorchak JM, Scoville CR, Williams GN, Arciero RA, St Pierre P, Taylor DC. Risk factors associated with noncontact injury of the anterior cruciate ligament: a prospective four-year evaluation. *Am J Sports Med.* 2003;31:831–42.
125. Boden BP, Griffin LY, Garrett WE Jr. Etiology and prevention of noncontact ACL injury. *Phys Sports Med.* 2000;28(4):53–60.
126. Wojtys EM, Huston LJ, Boynton MD, Spindler KP, Lindenfeld TN. The effect of the menstrual cycle on anterior cruciate ligament injuries in women as determined by hormone levels. *Am J Sports Med.* 2002;30:182–8.
127. McClay-Davis I, Ireland ML. ACL research retreat: the gender bias. *Clin Biomech.* 2001;16:937–59.
128. Swenson EJ Jr. Knee injuries. In: McKeag DB, Moeller JL, editors. *ACSM's Primary Care Sports Medicine.* Philadelphia (PA): Lippincott, Williams & Wilkins; 2007. p. 461–90.
129. Bouras T, Fennema P, Burke S, Bosman H. Stenotic intercondylar notch type is correlated with anterior cruciate ligament injury in female patients using magnetic resonance imaging. *Knee Surg Sports Traumatol Arthrosc.* 2018;26(4):1252–7.
130. Hewett TE, Myer GD. Reducing knee and anterior cruciate ligament injuries among female athletes. *J Knee Surg.* 2005;18(1):82–8.
131. McAlindon R. ACL injuries in women. Hughston Health Alert [Internet]. 1999. Available from: http://www.hughston.com/hha.
132. Sugimoto D, Myer GD, Barber Foss KD, Hewett TE. Specific exercise effects of preventive neuromuscular training intervention on anterior cruciate ligament injury risk reduction in young females: meta-analysis and subgroup analysis. *Br J Sports Med.* 2015;49(5):282–9.
133. Petushek EJ, Sugimoto D, Stoolmiller M, Smith G, Myer GD. Evidence-based best-practice guidelines for preventing anterior cruciate ligament injuries in young female athletes: a systematic review and meta-analysis. *Am J Sports Med.* 2019;47(7):1744–53.
134. Samuelson K, Balk EM, Sevetson EL, Fleming BC. Limited evidence suggests a protective association between oral contraceptive pill use and anterior cruciate ligament injuries in females: a systematic review. *Sports Health.* 2017;9(6):498–510.
135. Herring SA, Bernhardt DT, Boyajian-O'Neill L, et al. Selected issues in injury and illness prevention and the team physician. *Med Sci Sports Exerc.* 2007;39:2058–68.
136. Herring SA, Bergfeld JA, Boyajian-O'Neill L. The team physician and strength and conditioning of athletes for sports: a consensus statement. *Med Sci Sports Exerc.* 2015;47(2):440–5.
137. Herring SA, Kibler WB, Putukian M. Female athlete issues for the team physician: a consensus statement—2017 update. *Med Sci Sports Exerc.* 2018;50(5):1113–22.
138. Herring SA, Kibler WB, Putukian M. Load, overload, and recovery in the athlete: select issues for the team physician—a consensus statement. *Med Sci Sports Exerc.* 2019;51(4):821–8.
139. Herring SA, Kibler WB, Putukian M. Select issues in pain management for the youth and adolescent athlete. *Med Sci Sports Exerc.* 2020;52(9):2037–46.

CHAPTER 7

Exercise and Sport Nutrition

After completing this chapter, you will be able to:

1. Describe the importance of proper nutrition as it relates to enhancing health, physical activity, exercise, sport, and athletic performance.
2. Describe the key highlights in the historic development of nutrition and sport nutrition.
3. Identify the basic nutrients for healthy nutrition.
4. Explain the key issues in measuring nutritional intake.
5. Identify the key nutritional issues for an active individual.
6. Identify the key nutritional issues for a competitive athlete.

Proper nutrition is important for optimal health and successful performance in sport and athletic competition. As a society, we are increasingly aware of the role of good nutrition for decreasing the risk of various disease conditions and improving the overall health of individuals throughout the lifespan. Athletes, coaches, and sport nutritionists are also paying closer attention to the influence of proper nutrition for enhancing training and improving performance during various types of sport and athletic competitions.

Nutrition is defined as the science that interprets the connection between food intake and the function of the living organism (1). Nutrition, as a profession, consists of a number of subspecialty areas, including clinical nutrition, nutritional biochemistry, community nutrition, food science nutrition, nutritional management, nutritional counseling, nutrition for health, and sport nutrition (2). Nutrition for health promotion and sport nutrition are extremely important component of exercise science.

The terms diet and nutrition are often used interchangeably in today's society, although they have different meanings. In general, diet and nutrition are used to convey a description of the foods and beverages we consume (Figure 7.1). However, the word "diet" for many individuals means a restriction of food or energy intake, which often results in one of the most commonly used phrases in society, "I will start my diet tomorrow." Nutrition is a widely used term to describe all aspects related to food consumption (3). Nutrition is used in the medical and biological sciences and by politicians, economists, social and behavioral scientists, and consumers to describe all things related to food production and consumption (4).

Good health and a reduced risk of numerous diseases depend on proper nutrition (1). Nutritional intake (*i.e.*, food consumption) strongly influences the development and progression of chronic diseases and poor health conditions such as coronary heart disease, hypertension, osteoporosis, a variety of cancers, and obesity (5,6). Research evidence supports relationships between elevated serum cholesterol levels and coronary heart disease (7,8), reduced calcium intake and osteoporosis (9), consumption of dietary fats and certain cancers (4), and excess calorie intake and obesity (10). Figure 7.2 shows the relationship between nutritional intake and several common disease conditions.

Proper nutrition is also important for successful sport and athletic performance. Athletes and coaches are increasingly aware of how the macronutrients, vitamins, minerals, and fluid intake can improve sport and athletic performance during both training and

FIGURE 7.1 Optimal nutrition requires eating a variety of items from the food groups. (Shutterstock.)

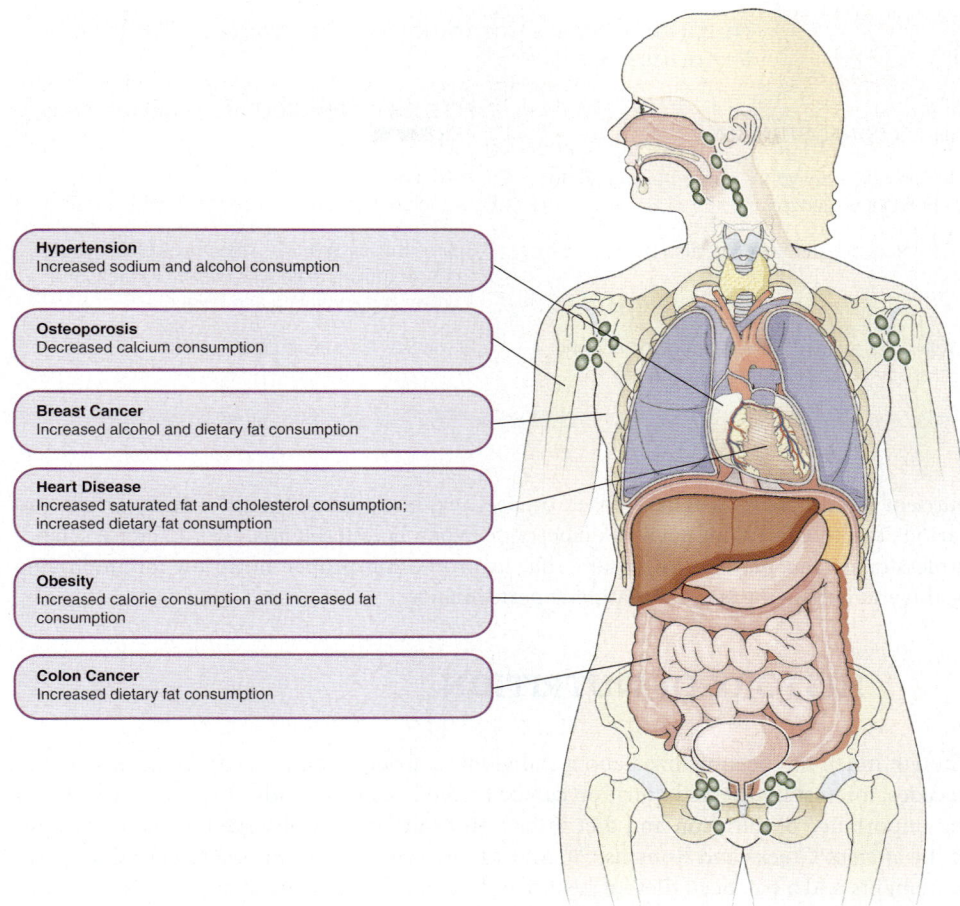

FIGURE 7.2 The relationship between nutritional intake and several disease conditions.

competition. Proper nutritional intake allows for the maintenance of appropriate training intensity, promotes recovery from training and competition, enhanced energy production, and the development of skeletal muscle tissue. For example, carbohydrate loading has been shown to increase muscle glycogen levels and improve certain types of endurance performance (11). Fluid and carbohydrate intake during prolonged exercise can enhance performance and prevent the adverse health effects of dehydration and carbohydrate depletion (12). Adequate protein intake has the potential to enhance skeletal muscle development and the performance of athletes who compete in certain types of events that rely on muscular strength and power production. Table 7.1 illustrates how certain nutritional strategies can enhance sport and athletic performance.

Proper nutritional intake consists of consuming the appropriate nutrients for tissue maintenance, repair, and growth and for providing the body with sufficient energy without an excess energy intake (13). There is no optimal nutrient intake for everyone as daily

Table 7.1	Nutritional Strategies for Enhancing Sport and Athletic Performance
NUTRITIONAL STRATEGY	EFFECTS ON PHYSIOLOGIC FUNCTION AND PERFORMANCE
Increased carbohydrate consumption prior to prolonged exercise	Maximizes muscle glycogen prior to exercise, which delays glycogen depletion and fatigue
Carbohydrate and fluid ingestion during exercise	Spares muscle glycogen, maintains blood glucose concentration, maintains plasma volume, and prevents dehydration and fatigue
Adequate protein intake when combined with a resistance exercise training program	Maximizes lean mass development

nutrient requirements will vary based on age, sex, body size, physical activity levels, and various health conditions such as diabetes or hypercholesterolemia (13). Exercise science professionals can play an important role in promoting proper nutrition for improving health and enhancing sport and athletic performance.

HISTORY OF NUTRITION

Though much has been realized about individual nutrient requirements in the last several decades, interest in diet and nutrition can be traced back thousands of years. Awareness of the importance of nutrition and diet in human health can be observed in the recordings of the ancient Greeks and Romans (3). Many Greek writings in this era refer to energy requirements and a balanced diet for health and the possibility to treat certain diseases with appropriate nutritional intake. For example, Hippocrates in the 4th century BC, formed a theory about the positive relationship between food consumption and health that was followed for centuries (14). These early writings in nutrition laid the foundation for the expansion of our understanding of how food intake affects health, sport, and athletic performance.

Early Influences of Nutrition for Health

The word "nutrition" in its various forms in the English language appears to have originated somewhere between the 15th and 16th centuries. Throughout the 17th and 18th centuries, physicians and scientists used nutrition interventions as part of experiments on diseased individuals. For example, it was observed during this period that increased iron intake could improve anemia and that citrus fruit consumption could cure scurvy. During the early 19th century, François Magendie noted that dogs fed only carbohydrate and fat lost body protein and died within a few weeks but that dogs fed a diet of carbohydrate, fat, and protein survived. This experiment demonstrated the importance of protein in the diet of animals (3).

Twentieth-Century Influences of Nutrition for Health

In the early 20th century, the terms "diet" and "dietetics" were used widely when referring to problems relating to food (3). Several important advancements were made. In 1903, W.O. Atwater and Francis Gano Benedict invented a respiration chamber (Figure 7.3) and performed very accurate **direct calorimetry** and **indirect calorimetry** measurements of food metabolism and energy expenditure (15). These experiments formed the foundation for future work in the areas of energy intake and energy expenditure. In 1936, Eugene Du Bois coined the term **basal metabolic rate** and examined the relationship between basal metabolic rate and age, sex, and weight (16). In 1937, Clive McCay demonstrated that restricting energy intake of rats by 33% led to increased longevity (by 25%), especially if simple carbohydrates were restricted (15). Ancel Keys studied the influence of diet on health, in particular, the effects of different kinds of dietary fat. Keys was closely associated with two famous diets: Keys rations (more commonly known as K-rations), formulated as balanced meals for combat soldiers in World War II, and the "Mediterranean diet," which he popularized in the 1950s and 1960s. Keys was also involved in the Minnesota Starvation Experiment, which provided considerable insight into the physiologic and psychological effects of severe and prolonged dietary restriction and the effectiveness of dietary rehabilitation strategies (17).

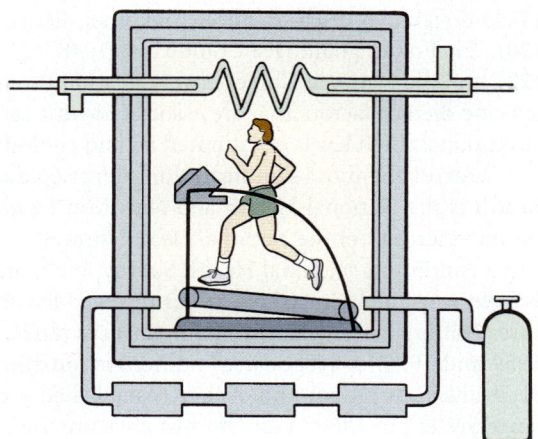

FIGURE 7.3 A human calorimeter is used to measure energy expenditure and substrate oxidation.

Direct calorimetry The measurement of heat produced by a chemical reaction or by the body.
Indirect calorimetry The measurement of energy production by the body using the amount of oxygen consumed and carbon dioxide produced.
Basal metabolic rate The level of metabolism, as measured by energy expenditure, required to maintain the normal physiologic functions of the body.

The recognition of nutrition as an academic discipline occurred in 1933, with the founding of the American Institute of Nutrition (AIN), which was instrumental in promoting nutrition as a science. The founding members of the AIN identified their disciplinary fields as nutrition, animal nutrition, chemistry, agricultural chemistry, biochemistry, physiological chemistry, physiology, and anatomy. This diversity of background remains evident today with individuals from a wide variety of disciplines working in the field of nutrition and nutrition science. Although AIN changed its name to the American Society for Nutrition in 2005, it remains a premier research society dedicated to improving the quality of life through the science of nutrition (3,18).

During the early 20th century, several laboratories were established to advance the understanding of nutrition. In 1904, The Nutrition Laboratory at the Carnegie Institute was created to study nutrition and energy metabolism. Established in 1927, the Harvard Fatigue Laboratory allowed for the further expansion of scientific research in the area of exercise and sport nutrition (19). Early scientific research in nutrition aimed to identify all the essential nutrients and the dietary requirements for each nutrient. Additional work was done to determine the distribution of each nutrient in various foods in an effort to define a nutritionally adequate diet or analyze a diet and determine whether it was nutritionally balanced for good health. This early research provided the foundation for the various databases that exist to provide both diet analysis and nutritional prescription (3).

The role of nutrition, particularly as it relates to chronic disease development, has received considerable attention over the last 60 years. **Epidemiological studies** have provided us with much understanding of how various nutrient patterns influence the development of cardiovascular disease, cancer, and other diseases affected by an individual's food intake (20). The Framingham Heart Study (est. 1948) (21), the Harvard Alumni Study (est. 1962) (22), and the National Cholesterol Education Program (est. 1985) (8) have helped identify specific dietary factors that are associated with cardiovascular disease, including the consumption of high levels of saturated fat and cholesterol (5,6).

One of the most significant long-term epidemiological studies about nutrition and health is the National Health and Nutrition Examination Survey (NHANES). NHANES began as a result of the National Health Survey Act of 1956, which was intended to establish a continuing National Health Survey to obtain information about the health status of U.S. citizens, including the services received for or because of health conditions. The first three National Health Examination Surveys (NHES I, II, and III) were conducted between 1959 and 1970. In response to numerous nutrition-related studies, the U.S. Department of Health, Education, and Welfare established a continuing National Nutrition Surveillance System in 1969 in an effort to measure the nutritional status of the U.S. population and monitor changes over time. The National Nutrition Surveillance System was merged with the National Health Examination Survey, creating NHANES (23). Table 7.2 provides the dates and specific target groups and foci of the various surveys. Data from NHANES have provided considerable insight into how nutritional patterns have changed during the past 60 years and how these changes have contributed to the development of disease conditions.

Epidemiological studies The study of factors affecting the health and disease of large groups of individuals.

Table 7.2 Overview of NHANES Surveys from 1959 to the Present

SURVEY AND YEARS	SPECIFIC FOCUS
NHES I (1959–62)	Selected chronic diseases of adults between 18 and 79 yr of age
NHES II (1963–65)	Growth and development of children between 6 and 11 yr of age
NHES III (1966–70)	Growth and development of adolescents between 12 and 17 yr of age
NHANES I (1971–75)	Extensive dietary intake and nutritional status were collected by interview, physical examination, and a battery of clinical tests and measurements
NHANES II (1976–80)	Expanded the age of the first NHANES sample by including individuals as young as 6 mo of age; children and adults living at or below the poverty level were sampled at higher rates than their proportions in the general population ("oversampled") because these individuals were thought to be at particular nutritional risk
NHANES III (1988–94)	Included infants as young as 2 mo of age, with no upper age limit on adults; African Americans, Mexican Americans, infants, children, and those over 60 yr old were oversampled; NHANES III also placed a greater emphasis on the effects of environment on health
1999, NHANES became a continuous survey	Surveys are conducted over a period of approximately 4 yr with a break of at least 1 yr between survey periods; surveys change focus to meet emerging needs in health and nutrition

Twenty-First-Century Influences of Nutrition for Health

The Academy of Nutrition and Dietetics (AND) was founded as the American Dietetic Association (ADA) in 1917 during a meeting of approximately 100 dieticians in Cleveland, Ohio. The first ADA president, Lulu Grace Graves, helped establish the initial areas of practice: (a) dietotherapy, (b) teaching, (c) social welfare, and (d) administration (14). These areas of practice remain at the heart of the mission of the AND, which strives to improve the nation's health and advance the profession of dietetics through research, education, and advocacy. The AND has been instrumental in promoting the dietetics profession, the enhanced understanding of nutrition, and the education of nutrition professionals. Federal, state, and many private foundations continue to support nutrition science research in an effort to help better understand the role of nutrition intake in disease development and disease risk reduction. The AND has been instrumental in developing recent position papers in certain areas of nutrition, namely, eating disorders (24), nutrition for children (25,26) and older adults (27), and food insecurity (28), and has partnered with the ACSM to release a position stand centered on Nutrition and Athletic Performance (29). A complete listing of position papers from the AND can be found at www.eatright.org. The AND continues to provide expert testimony at hearings, lobbies governmental agencies, comments on proposed federal and state regulations, and develops position statements on critical food and nutrition issues (30). The AND Foundation has created numerous programs to support current and future nutrition practitioners to optimize health around the world (30).

The ASN has also continued to be a world leader in advancing nutrition for health. The creation of two important newsletters, *Advances in Nutrition* (2010) and *Current Developments in Nutrition* (2017), by ASN have allowed the organization to disseminate timely nutritional information to its members.

Early History of Sport Nutrition

The role of nutrition for enhancing sport and athletic performance has a meaningful history. The areas that have received the greatest attention and contributed most significantly to the development of sport nutrition have been carbohydrate and protein consumption and vitamin and mineral supplementation. For example, historic recordings (ca. 500–400 BC) indicate that consumption of deer liver and lion heart would enhance bravery, speed, and strength in the athlete and warrior (31–33). Although early writings demonstrate people's awareness of the role of nutrition in promoting physical development, most of what we know about the role of nutrition for enhancing sport and athletic performance comes from more recent times.

Thinking Critically

In what ways has our knowledge about nutrition contributed to a broader understanding of how individuals can improve physical fitness and promote good health?

Twentieth-Century Influences of Sport Nutrition

Early in the 20th century, scholars described the importance of carbohydrate during prolonged exercise and the role of carbohydrates in maintaining adequate stores of muscle and liver glycogen (34,35). During the 1924 Boston Marathon, measures of blood glucose were made of the first 20 runners to cross the finish line. Many of the runners displayed hypoglycemia, symptoms of fatigue, stupor, and poor mental concentration (36). During the following year, runners were given large amounts of carbohydrate the day before the race and sugar candy during the race. The result of this nutritional strategy was a normalization of blood glucose levels and the alleviation of the symptoms of nervous system fatigue and poor mental concentration (37). The development and use of the **muscle biopsy** procedure by Swedish researchers in the 1960s allowed for the determination of how rapidly muscle glycogen was depleted during exercise (38). This information eventually led to the development of the carbohydrate loading procedure for improving endurance performance and the influence of carbohydrate consumption on muscle glycogen replenishment (39). Continued research in this area led to the use of sports drinks for delaying muscle fatigue and improving performance during prolonged exercise (40).

Experiments conducted in the 1940s demonstrated that increased protein consumption could enhance the development of skeletal muscle mass in individuals involved in resistance exercise training (31,41). Throughout the 1950s and 1960s, the increased consumption

Muscle biopsy A procedure whereby a small sample of muscle tissue is collected using a special needle.

of milk and beef products led to greater protein consumption (31). The development of isolated protein powders and amino acids in the 1970s and early 1980s resulted in athletes using these products for increasing dietary protein intake. Additional research in the 1990s led to many athletes closely matching amino acid intake and resistance exercise training in an effort to secure the greatest enhancement of lean muscle mass development (31).

Many water- and fat-soluble vitamins were discovered during the period of 1900 to 1930. The use of these compounds quickly spread through the sporting world so that by 1939 cyclists in the Tour de France reported performing better after taking vitamin supplements (32). Early scientific research did not support the use of vitamin supplements for enhancing athletic performance, but athletes remained committed to heavy vitamin supplementation. For example, during the 1972 Olympic Games athletes reported consuming large quantities of vitamins in an effort to enhance performance during competition (42). Athletes continue to use high doses of vitamin supplementation in an effort to improve performance or, at the very least, as an insurance mechanism to ensure adequate levels of vitamins in the body (13).

Within the AND, a dietary practice group called the Sports, Cardiovascular, and Wellness Nutritionists (SCAN) was established in 1981. The SCAN Dietetic Practice Group works to promote healthy, active lifestyles through excellence in dietetics practice in nutrition expertise in the areas of sports, physical activity, cardiovascular health, wellness, and the prevention and treatment of disordered eating and eating disorders. Registered dietician and nutritionists can also acquire board certification as a Specialist in Sports Dietetics, indicating they are specially trained to work with high school, collegiate, Olympic, and professional athletes to enhance sport performance (www.scandpg.org/).

In 1991, the *International Journal of Sport Nutrition* was first published and focused on sport nutrition from a variety of perspectives. The journal was renamed in 1999 as the *International Journal of Sport Nutrition and Exercise Metabolism* and is designed to provide for the publication of both scholarly and applied work relating to the nutritional, biochemical, and molecular aspects of exercise science.

Twenty-First-Century Influences of Sport Nutrition

The use of supplements by all types of athletes continues into the 21st century, with athletes mixing various vitamin, mineral, amino acids, and growth-promoting agents in an effort to maximize protein synthesis and skeletal muscle mass during training. This has led to an explosion of the supplement industry, with national and multinational companies promoting products genetically designed to enhance training and recovery for sport and athletic performance (43). Further advancements this century have come with blood biomarker profiles that allow for individualized nutritional interventions that target deficiencies in nutritional intake and promote a reduction in injury and disease risk and an enhancement in performance (43,44).

Several organizations continue to provide leadership in the 21st century through the dissemination of sport nutrition information to athletes and coaches at all levels of competition. The Gatorade Sports Science Institute, established in 1985, and currently under the direction of Global Senior Director Asker Jeukendrup continues to build on its rich history of research and education as well as increasing the services provided directly to athletes and coaches (45). The SCAN practice group has been instrumental in promoting sports performance and developing resources for athletes and coaches by working with other professional organizations such as the National Collegiate Athletic Association, the National Athletic Trainers Association, and the Professionals in Nutrition for Exercise and Sport.

Exercise and sport nutritionists working with other exercise science and allied health professionals continue to explore ways that nutrition can be used to promote good health and improve sport and athletic performance. Future research work and advancements in policy initiatives and educational program development will be a central focus as nutrition is used to address diet-related health problems and promoting improvement in sport and athletic performance. Table 7.3 provides a list of significant recent events in the historical development of nutrition as it relates to health, physical activity, exercise, sport, and athletic competition.

Thinking Critically

In what ways has our knowledge about nutrition contributed to a broader understanding of how to enhance sport and athletic performance?

Table 7.3 Significant Events in the Historical Development of Nutrition for Health and Sport

YEAR	EVENT
1903	W.O. Atwater and Francis Gano Benedict invented a respiration chamber
1904	The Nutrition Laboratory at the Carnegie Institute was created to study nutrition and energy metabolism
1917	American Dietetic Association was founded
1925	First experiment on carbohydrate supplementation during exercise was conducted at the Boston Marathon
1927	Harvard Fatigue Laboratory was established
1933	American Society for Nutrition was founded
1937	Clive McCay demonstrated that restricting energy intake of rats by 33% led to increased longevity
1939	First report of vitamin supplementation improving performance in cyclists in the Tour de France
1959	First edition of NHANES was administered
1981	The SCAN group of the ADA was formed
1999	NHANES became a continuous survey
2006	First exam given for the Certified Specialist in Sports Dietetics (CSSD) by the Commission on Dietetic Registration
2012	American Dietetic Association changed its name to the Academy of Nutrition and Dietetics
2013–2017	SCAN partners with NCAA, NATA, and PINES to offer enhanced services to athletes and coaches

NATA, National Athletic Trainers Association; NCAA, National Collegiate Athletic Association; PINES, Professionals in Nutrition for Exercise and Sport.

 BASIC NUTRIENTS

Each of us needs to consume adequate amounts of the **macronutrients** and **micronutrients** in our nutritional intake to ensure proper physiological and structural function and good health. Although individual circumstances and preferences will dictate food intake, many individuals will meet their dietary needs for carbohydrate, fat, protein, vitamins, and minerals by consuming a diet that is consistent with general nutrition guidelines. The **recommended dietary allowances (RDA)** and the **dietary reference intake (DRI)** are used by nutrition professionals to promote proper nutrient intake for diseased and healthy individuals. Proper nutrient intake promotes optimal growth and development, appropriate energy balance and body composition, health and longevity, and normal physiological function. Table 7.4 provides a list of the macro- and micronutrients and their primary functions related to physical activity, exercise, sport, and athletic competition. Exercise science professionals must have a sound knowledge of nutrient

Table 7.4 The Macronutrients and Micronutrients and Their Primary Functions Related to Physical Activity, Exercise, Sport, and Athletic Performance

NUTRIENT	PRIMARY FUNCTION RELATED TO PHYSICAL ACTIVITY, EXERCISE, SPORT, AND ATHLETIC PERFORMANCE
Carbohydrates	Provide energy during moderate to high-intensity physical activity or exercise
Fats	Provide energy during low to moderate intensity exercise
Protein	Important component of skeletal muscle
	Part of various compounds that regulate metabolism during rest and exercise
Vitamins	Important for controlling metabolic pathways that produce energy during rest and exercise
Minerals	Part of the structure of bone
	Part of various compounds that regulate metabolism during rest and exercise

Reprinted with permission from the Canadian Society for Exercise Physiology. *CSEP Physical Activity Training for Health® (CSEP-PATH®) Resource Manual.* 2nd ed. 2019.

Macronutrient A chemical substance (such as protein, carbohydrate, or fat) required in relatively large quantities in an individual's daily nutritional intake.

Micronutrients A chemical substance (such as a vitamin or mineral) required in small quantities in an individual's daily nutritional intake.

Recommended dietary allowance (RDA) The recommended intake level of a nutrient that is considered to meet the daily needs of nearly all healthy individuals.

Dietary reference intake (DRI) General term for a set of reference values used to plan and assess nutrient intakes of healthy people.

function in order to make appropriate recommendations for the promotion of good health and for the improvement of sport and athletic performance. The information contained in the following sections is designed to provide an overview of macro- and micronutrient function.

Carbohydrate

Carbohydrates are a macronutrient that provide energy to the body. Dietary carbohydrates exist in two forms: simple and complex. **Simple carbohydrates**, sometimes called simple sugars, are the carbohydrates naturally found in milk and fruit. A large percentage of the simple sugars consumed in the diet of individuals living in the United States is added to processed foods during the manufacturing process. These refined sugars, of which sucrose and high fructose corn syrup are most popular, are frequently added to soft drinks, fruit drinks, candies, and dessert items. **Complex carbohydrates**, also called starches, are found in whole grains (Figure 7.4) and vegetables, especially potatoes, beans, and peas. Complex carbohydrates are generally considered healthier and more beneficial when performing exercise and participating in sport and athletic activities. Foods with complex carbohydrates also contain large amounts of vitamins and minerals and result in a slower release of glucose into the body (1).

When carbohydrates are consumed in the diet, they are broken down in the gastrointestinal tract and absorbed in the small intestine as small six-carbon molecules, predominantly glucose, fructose, and galactose. Glucose is the most common and useful form of carbohydrate in the body. Glucose provides energy to the various tissues of the body as blood glucose and is stored after a meal as liver and muscle glycogen for future use of energy. Although almost all cells can use carbohydrate, fat, or protein for energy, the brain and nervous tissue depend almost exclusively on glucose to provide energy (1).

The storage form of glucose in the body is called glycogen. The liver and skeletal muscle are the primary tissues for glycogen synthesis and storage. Normal fasting blood glucose levels are between 70 and 100 $mg \cdot dL^{-1}$. When blood glucose decreases below 70 $mg \cdot dL^{-1}$

FIGURE 7.4 Examples of complex carbohydrates. (Shutterstock.)

Simple carbohydrate A carbohydrate, such as glucose or sucrose, that consists of a one or two monosaccharide units.

Complex carbohydrate A carbohydrate, such as starch, that consists of more than two monosaccharide units.

the condition of **hypoglycemia** occurs, and symptoms such as drowsiness, irritability, and fatigue may appear. The body responds to hypoglycemia primarily by breaking down liver glycogen into glucose and releasing it into the blood. At the same time, various signals from the brain stimulate hunger and the desire for the individual to eat. Combined, these two actions serve to elevate blood glucose levels (1).

When blood glucose levels increase after food consumption, the body releases insulin from the beta cells of the pancreas. Insulin works with protein receptors (called glucose transport proteins) in the tissues of the body to promote glucose uptake for immediate energy use or conversion to glycogen for storage. This process results in a return of blood glucose to normal levels. A disease condition called diabetes mellitus occurs when either the pancreas does not produce sufficient amounts of insulin (**Type 1 diabetes mellitus**) or the insulin does not facilitate glucose uptake into the tissues of the body (**Type 2 diabetes mellitus**). If the fasting blood glucose level is 100 to 125 mg·dL^{-1}, the individual may have impaired fasting glucose, commonly known as prediabetes. A fasting blood glucose level of 126 mg·dL^{-1} or higher is consistent with either Type 1 diabetes mellitus, especially when accompanied by classic signs and symptoms of diabetes, including increased thirst or hunger, frequent urination, weight loss, or blurred vision, or Type 2 diabetes mellitus (1).

Fat

Dietary fat and cholesterol are critical to the normal functioning of body tissues and overall good health of the body. Dietary fat is vital for the absorption of the fat-soluble vitamins (A, D, E, and K) and to provide key biochemical precursors that can be transformed into essential cellular products. Dietary fat also contributes to the flavor and texture of foods, and it is believed that fat maintains **satiety** and helps to keep us from being hungry. Fat also provides a concentrated form of energy for our body. Cholesterol, a subclass of fat, is important in the formation of cell membranes and some hormones in the body and is consumed in animal products in the diet and/or made by human cells. Despite the important functions of cholesterol, it is not necessary to consume dietary cholesterol because the body is capable of producing all that is needed (1).

Most consumed fats are broken down in the gastrointestinal tract into free fatty acids and **monoglycerides** for absorption in the small intestine. Dietary fat and cholesterol enter the blood stream and travel to the liver for further processing by the body. The fat absorbed from the small intestine is largely processed in the liver and stored in **adipose tissue** and is later used

Hypoglycemia An abnormally low level of sugar in the blood.
Type 1 diabetes mellitus Condition characterized by elevated blood glucose levels that are the result of a lack of insulin production by the pancreas.
Type 2 diabetes mellitus A metabolic disorder that is primarily characterized by insulin resistance, relative insulin deficiency, and hyperglycemia.
Satiety The feeling of fullness after eating.
Monoglycerides A chemical structure containing one glycerol molecule and one fatty acid molecule.
Adipose tissue The cells of the body that store fat derived from excess calorie intake.

in the body to provide energy at rest and during physical activity and exercise. Cholesterol travels through the circulatory system via lipoproteins for use by various tissues of the body (1).

Excessive intake of dietary fat can result in the development of assorted disease conditions, including atherosclerosis, hyperlipidemia, and obesity. The consumption of a high-fat diet, which often results in high total calorie intake, has been shown to be related to excessive weight gain and the development of obesity (46) and the prediction of weight gain (47). Individuals who are obese have a higher risk of developing cardiovascular disease, Type 2 diabetes, and certain types of cancer (48,49). The consumption of a high-fat diet can also lead to excessive levels of blood lipids. The conditions of **hyperlipidemia** and **hypercholesterolemia** can increase an individual's risk of cardiovascular disease and stroke (8,50).

Protein

Like carbohydrate and fat, protein is important for normal health and functioning of the body. Proteins are made from individual amino acids that are connected together to form chains. There are 22 common individual amino acids that can be combined to form several proteins for use by the body. Proteins are primarily used for tissue growth and repair. As all tissues of the body contain proteins, this is an extremely important function. Proteins also form hormones, enzymes, and protein receptors that work to control a range of physiologic functions of the body. Proteins can also be broken down into amino acids, which can be further metabolized to provide energy for the body (1).

Dietary proteins come from animal and certain vegetable sources (Figure 7.5). During the digestion process, these proteins break down into individual amino acids and small chain proteins called peptides (3–6 amino acids in length). Once absorbed into the gut, the peptides are broken down into individual amino acids and enter the circulatory system. Once in the circulatory system, these amino acids can be delivered to the tissues of the body. Our bodies are able to chemically transfer the amino group from one amino acid to another compound and create new amino acids, a process called **transamination**. Transamination

FIGURE 7.5 Different sources of dietary protein. (Shutterstock.)

Hyperlipidaemia The presence of high levels of fat in the blood.
Hypercholesterolemia The presence of high levels of cholesterol in the blood.
Transamination The transfer of an amino group from one chemical compound to another.

allows for the formation of **nonessential amino acids** in the body. **Essential amino acids** cannot be made in the body and must be consumed in the diet on a daily basis (1).

Normal physiologic function and growth and development require the consumption of about 0.8 g of protein per kg body mass per day. Insufficient dietary consumption of protein can result in protein **catabolism**. During catabolism, the body breaks down tissue proteins, typically skeletal muscle, to ensure the formation of the essential proteins of the body (*e.g.*, hormones and neurotransmitters). Higher levels of protein consumption (>0.8 g $\cdot$ kg^{-1} body weight per day) can support a state of increased **anabolism**, especially if the individual participates in a regular resistance exercise training program to enhance muscle growth. In an anabolic state, the body uses the amino acids to form proteins, including enzymes, hormones, and skeletal muscle (1,51).

Thinking Critically

Where can an athlete or coach receive knowledge about proper nutrition for increasing lean body mass and enhancing muscular strength?

Vitamins

Vitamins are organic substances required by the body in very small amounts to perform vital physiological functions. Vitamins have no common chemical structure, do not supply energy, and do not contribute to the total mass of the body. The body cannot produce a sufficient quantity of vitamins, and therefore vitamin needs must be supplied in the foods we consume or through dietary supplementation. There are 13 different vitamins, and they are classified as either fat soluble or water soluble (1).

Fat-soluble vitamins are contained in dietary fat and are dissolved and stored in the fat tissues of the body. This storage process means that it may take years for a deficiency to develop, although this could be accelerated by consuming an extremely low-fat or fat-free diet. Conversely, excessive intake of fat-soluble vitamins can be dangerous and lead to toxic effects. The consumption of fat-soluble vitamins — above the recommended level is believed to be of no additional health or athletic performance benefit (13).

Water-soluble vitamins are typically grouped together as the B-complex vitamins and vitamin C. Many water-soluble vitamins work with large protein compounds to form active enzymes that help regulate the chemical reactions in the body (1). Water-soluble vitamins are not stored in the body and must therefore be consumed in the diet on a regular basis. Excessive intake of vitamins, as is often the case with supplementation, leads to the excretion of the excess vitamins in the urine (13).

Nonessential amino acids An amino acid that can be synthesized by the body from other amino acids.

Essential amino acids An amino acid that cannot be synthesized by the body and must be consumed in the diet.

Catabolism Metabolic breakdown of complex molecules into simpler ones, often resulting in a release of energy.

Anabolism Phase of metabolism in which complex molecules, such as the proteins and fats that make up body tissues, are formed from simpler ones.

Fat-soluble vitamins Compounds that dissolve and are stored in fat.

Water-soluble vitamins Compounds that are easily dissolved in water.

Minerals

Dietary minerals are inorganic elements required for normal physiologic function and are used as components or cofactors of enzymes, hormones, and vitamins (13). Of the more than 20 minerals known or suspected to be essential for humans, 15 have established RDAs or adequate intake (AI). Minerals appear either in combination with organic compounds or as free minerals in body fluids. Minerals present in large quantities with known biologic function are classified as major minerals; those present in very small quantities (<0.05% of body mass) are referred to as trace minerals. Excess minerals in the body provide no known biologic function and in some instances may even be harmful (1). Most minerals are found in water, topsoil of the ground, the root systems of plants and trees, and the tissues of animals that consume plants and water containing the minerals (1). Minerals play key roles in the function of the body because they provide for structure (*e.g.*, bones and teeth), function (*e.g.*, cardiac rhythm, muscle contraction), and regulation (*e.g.*, energy metabolism) within the body. Minerals are also important for the synthesis of biologic nutrients, including glycogen, triglycerides, proteins, and hormones (1).

Water

Water plays an important part in maintaining health and proper physiological function. Water does not contribute to the nutrient value of food, but the energy content of a specific food is generally inversely related to its water content. An individual's body weight is between 40% and 60% water (52). Water content in tissue ranges from approximately 25% in adipose tissue to 75% in lean tissue such as skeletal muscle (52). Total body water content is, therefore, a function of the body composition (*i.e.*, amount of adipose tissue vs. muscle tissue). The intracellular and extracellular compartments of the body contain water. Intracellular water comprises the fluid matrix that is the interior of the cell. The extracellular compartment comprises all the fluid that is external or outside the cell membrane and includes, for example, the blood plasma, lymph fluid, saliva, fluids of the eyes, and fluids secreted by glands and the intestines. Appropriate water intake in the form of liquids and foods is critical for health and normal body functioning (1).

MEASURING NUTRITIONAL INTAKE

An accurate assessment of energy and nutrient intake is critically important for assuring the optimal health and fitness of individuals and the best performance of athletes during training and competition. This is particularly important for ensuring energy balance and maintaining appropriate nutrient intake. Measuring habitual food consumption in humans, however, is one of the most difficult aspects of assessing nutritional intake (53). Two primary challenges for determining an accurate assessment that researchers, clinicians, and exercise science professionals face are the following: (a) obtaining a precise determination of an individual's normal food intake and (b) converting the information to nutrient and energy intake (53). When assessing dietary intake, it is important to verify that the technique used does not interfere with the individual's normal nutritional habits and thus influence the factor being measured. The two most common methods of measuring food intake at the individual level are the dietary recall and the dietary record (54).

Dietary Recall and Dietary Record

The **dietary recall** method requires an individual to report intake over the previous 24-hour period (called a 24-h recall) or report the customary intake over the previous time period up to the past year (called a food frequency questionnaire). Use of food models, volume and size models, and pictures of food items can increase the accuracy of dietary recall. The **dietary record** method requires individuals to record the types and amounts of all foods being consumed over a period of time (*e.g.*, 3 or 7 d). Figure 7.6 provides an example of the information collected on a dietary record form. Computer software programs can also be

FOOD INTAKE RECORD Date:_____

Name:_____

Day/Time	Activity While Eating	Place of Eating	Food — Quality — Brand

FIGURE 7.6 A dietary record form used to record food intake.

Dietary recall A process that requires individuals to recall from memory food items consumed during a prior period of time.

Dietary record A process that requires individuals to record food items consumed during a designated period of time.

used to accurately record food intake. It is important that the foods and fluids consumed by the individual are weighed or recorded in available household measures like spoons, cups, bowls, plates, serving size, or product size. During data analysis, the recorded information is converted to a weight or volume by the actual measuring device used or adopting standard values from reference tables (53). Less used alternatives to the dietary recall, food frequency questionnaire, and dietary record methods include the direct observation with weighted food records, the double-portion technique, and the food supply technique (53). Table 7.5 illustrates important advantages and disadvantages of the dietary recall and dietary record food measurement techniques (55).

It is unclear how long food intake should be measured to determine customary nutritional consumption. If individuals live by a regular activity pattern (*e.g.*, 5 d of work or school and 2 d of nonwork or nonschool), it seems reasonable to suppose that their social and dietary habits are determined in part by this pattern of activity. The measurement of food intake should include samples of both workday/school day and nonwork/nonschool day intake. If possible, food intake is measured for the whole week. The length of the measurement period is determined by the level of day-to-day variability and the level of accuracy desired in the assessment. Generally, longer periods of measurement lead to a more accurate assessment of habitual food intake (53).

Once dietary intake data have been collected, conversion of this information to nutrient and energy intake occurs, typically using electronic databases that contain the nutritional information of food samples. A primary challenge for obtaining accurate nutritional information lies in the interpretation of the food item from the dietary record or recall and the selection of the matching food item from the database. Verifying specific information of the food item (size, volume, weight, brand, etc.) on the food intake record prior to entering the food into the database can improve accuracy. Additional accuracy can be obtained by using a database product that has several individual food items and also provides nutritional information on foods available

Table 7.5 Advantages and Disadvantages of the Dietary Recall and Dietary Record Food Measurement Techniques (55)

TYPE OF MEASUREMENT	ADVANTAGES	DISADVANTAGES
Dietary recall — 24 h Trained interviewer elicits food, portion size, place, and timing of meals eaten within past 24 h	Easy to complete Interview in person or by phone	Relies on short-term memory May not reflect typical intake Data entry may be time-intensive
Dietary record Trained interviewer instructs person to make detailed list of foods consumed, including preparation methods and brand names	Does not rely on short-term memory Accurate estimate of portion sizes with use of food models Can include culture-specific foods	Requires motivation for prolonged period of record keeping Person may alter typical diet Cost of carefully training participants is high

in restaurants or prepackaged items from grocery stores. The information obtained from the nutritional assessment is then used to make recommendations for either adjusting or enhancing the nutritional intake of the individual (53).

NUTRITION FOR HEALTH

Habitual nutrient intake plays an important role in health promotion and disease prevention. Data from NHANES and other epidemiologic studies have demonstrated relationships between diet and increased risk for cardiovascular disease, hypertension, obesity, diabetes mellitus, osteoporosis, and certain forms of cancer (56–62). A primary treatment intervention for these lifestyle diseases and conditions is changing an individual's nutritional intake. Table 7.6 illustrates the potential beneficial effects of changing dietary intake on common disease conditions (1,63).

The importance of proper nutrition for optimal health cannot be underestimated. The federal government has used scientific research to develop numerous promotional campaigns and programs aimed at improving nutritional intake in both the general population and in specific groups (*e.g.*, those with high blood pressure, diabetes mellitus, or limited income). Examples of some of these programs include the following:

- The DASH diet, which stands for **D**ietary **A**pproaches to **S**top **H**ypertension, is designed to help individuals modify their diets in an effort to reduce blood pressure. The National Heart, Lung, and Blood Institute developed the DASH diet in the early 1990s to prevent and treat high blood pressure.
- Fruits & Veggies — More Matters is a public health initiative designed to increase the consumption of fruits and vegetables. Headed by the Centers for Disease Control and Prevention in 2007, it is designed to increase the consumption of fruits and vegetables.
- We Can!, which stands for **W**ays to **E**nhance **C**hildren's **A**ctivity & **N**utrition, is a national program designed for families and communities to help children maintain a

Table 7.6 Dietary Changes and Risk for Common Disease Conditions

DIETARY CHANGE	DISEASE CONDITION CHANGE
Decreased sodium intake	Decreased blood pressure in hypertensive individuals
Taking a daily multivitamin that includes folic acid and limiting their intake of alcohol	Decreases excess risk of colon cancer associated with a family history of the disease
Decreased saturated fat and cholesterol intake	Decreased cardiovascular disease
Decreased calcium intake	Increased risk of osteoporosis
Decreased simple carbohydrate intake	Decreased risk of Type 2 diabetes
Increased fruit and vegetable intake	Decreased risk of colon cancer

healthy weight. Spearheaded by the National Heart, Lung, and Blood Institute in 2005, it is designed to help youth ages 8 to 13 years old stay at a healthy body weight.

- The **S**upplemental **N**utrition **A**ssistance **P**rogram **Ed**ucation (SNAP-Ed) teaches people to shop for and cook healthy meals and learn how to make their SNAP dollars stretch.
- The **E**xpanded **F**ood and **N**utrition **E**ducation **P**rogram (EFNEP) brings together federal, state, and local resources to improve the health and well-being of limited-resource families and youth. The EFNEP was created in 1969 and is administered by the National Institute of Food and Agriculture.
- The **N**ational **S**chool **L**unch **P**rogram (NSLP) is a federally assisted meal program operating in public and nonprofit private schools and residential child care institutions. Established in 1946 under the National School Lunch Act, the NSLP provides nutritionally balanced and low-cost or free lunches to children each school day.
- Private organizations and foundations have also provided support to the important role that nutrition plays in optimizing health. Examples include the Bill and Melinda Gates Foundation and the Robert Wood Johnson Foundation. Additional information can be obtained by consulting the abundance of credible dietary self-help books, Internet Web sites, and commercial products for improving nutritional intake and improving health.

Dietary Guidelines for Health

The Dietary Guidelines for Americans, first published in 1980, provide science-based evidence to promote health and to reduce risk for chronic diseases through changing nutritional intake and physical activity patterns. The U.S. Department of Health and Human Services (HHS) and the U.S. Department of Agriculture (USDA) are responsible for developing and establishing the Dietary Guidelines of Americans. The process has evolved to include three stages for guideline development that are illustrated below (64):

- Stage 1 — An expert panel conducts an analysis of new scientific information on nutritional intake and health and issues a detailed report.
- Stage 2 — Key dietary recommendations are formed based on the report and from public and agency comments.
- Stage 3 — Communication of the recommendations (*i.e.*, Dietary Guidelines) is made to the general public.

Every 5 years, the Dietary Guidelines Advisory Committee (DGAC) analyzes new scientific information during the revision of the Dietary Guidelines. This analysis is published in the DGAC Report (available at https://www.dietaryguidelines.gov/). The DGAC Report is used to form the recommendations that are used by the USDA and HHS for programs and public policy development (64).

The Dietary Guidelines summarize and synthesize knowledge regarding nutrient and food components into recommendations for a pattern of eating that can be adapted by the general public. The Dietary Guidelines provide key recommendations based on the scientific evidence for lowering the risk of chronic disease and promoting health. A basic premise of the Dietary Guidelines is that nutrient needs should be met primarily through food consumption. Foods provide various nutrients and other compounds that may have beneficial effects on health and reduce disease risk. Fortified foods and dietary supplements may be useful sources for one or more nutrients that otherwise might be consumed in less than recommended amounts. Though recommended in some instances, dietary supplements cannot replace a healthful diet (64).

The USDA Food Guide (64) and the DASH Eating Plan (65) are intended to combine the dietary recommendations from the Dietary Guidelines into a healthy way of eating for most individuals. The USDA Food Guide and DASH Eating Plan are based on age and sex and have a wide range of energy levels to meet the needs of different groups of people. The USDA Food Guide uses population food intake to create the nutritional content for the different food groups (64). The DASH Eating Plan is based on selected foods chosen for a sample 7-day menu (65).

Recommended energy intake for each plan will differ for individuals based on age, sex, and activity levels. Individuals who eat nutrient-dense foods may be able to meet their DRI of nutrients without consuming their full calorie allowance. The Dietary Guidelines are organized by specific themes and provide key recommendations for specific population groups that are used together to plan an overall healthy diet. The Dietary Guidelines for 2020 are organized to address the following topics (64):

1. Current Dietary Intakes through the Life Course
2. Diet and Health Relationships: Pregnancy and Lactation
3. Diet and Health Relationships: Birth to Age 24 Months
4. Diet and Health Relationships: Individuals Ages 2 years and Older

The Dietary Guidelines are used to develop programs for improving the health of the general public. The most recent guidelines can be found at http://health.gov/dietaryguidelines/. The creation of ChooseMyPlate in 2011 allows individuals to create diets and alter nutritional intake to promote health. Figure 7.7 illustrates the specific components of ChooseMyPlate. Specific options on ChooseMyPlate include information about the

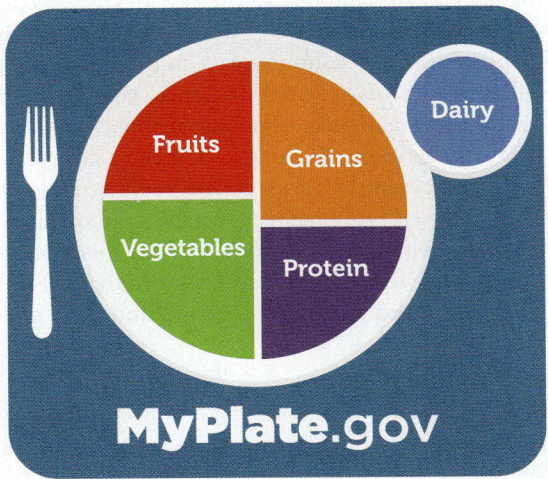

FIGURE 7.7 ChooseMyPlate: The current recommendations for healthy eating. (From ChooseMyPlate.gov)

Thinking Critically

What nutritional information might a registered dietician provide to an individual just beginning an exercise program to improve health and fitness?

following: fruits, vegetables, grains, protein foods, dairy, and oils. There is also information on weight management, calories, and physical activity. Children often have different nutritional requirements than adults and require information to be presented in a format that is easy to understand. As a result, the ChooseMyPlate contains ways to promote healthy eating and active living among children aged 2 to 11 years.

AREAS OF STUDY IN NUTRITION FOR HEALTH

Exercise science professionals and nutritionists play important roles in determining how nutritional intake influences health and disease risk. The role of nutrition as it affects health can be studied from multiple disciplines and numerous perspectives, including the most effective diet for weight loss, effective strategies for promoting healthy weight loss, and food deserts and food swamps on disparities in diet and diet-related health outcomes. In this section, we will examine several primary areas of study in nutrition and health and disease risk. The selection of these topics is not meant to minimize the importance of other areas of nutrition and health, but to provide you with some examples of popular areas of study.

Effective Diets for Weight Loss

The effectiveness of diets with different macronutrient composition (*e.g.*, low carbohydrate or high protein) in achieving and maintaining a healthy body weight and reducing the risk of disease has received considerable investigation. The search for the optimal weight loss diet has resulted in seemingly thousands of different diets promoting effective combinations of altered and/or reduced energy, carbohydrate, fat, and protein intake. Identifying which weight loss diet or program is most appropriate and efficient is critically important for individuals who struggle to lose excess body weight and maintain the weight loss in order to maintain health and reduce the risk for disease conditions associated with overweight and obesity (66). Comparison of weight loss diets against one another, as well as against control diet conditions, is commonplace in the scientific literature. The three primary factors that affect weight loss — energy balance, genetics, and behavior — are extremely difficult to control and manipulate, making comparisons across the scientific literature challenging (67). Furthermore, the length of dietary compliance and duration of follow-up after the dietary intervention varies considerably across diets. Fortunately, the statistical procedure meta-analysis (see Chapter 2) can be used to evaluate which diets may be the most effective. An analysis of 48 registered clinical trials indicates that low-carbohydrate and low-fat diets appear to be the most effective, with individuals losing approximately 8 kg after 6 months when compared with a control condition of no dietary changes (68). Typical weight regain

Meta-analysis The process of statistically analyzing data from previously published research studies.

after 12 months is approximately 1 to 2 kg (68). While there were differences between diets, those differences were small and likely to be unimportant for weight loss (68). Often, there are competing health benefits when employing certain diets. For example, while individuals consuming low-carbohydrate diets experience weight loss, there are also increases in LDL cholesterol, making low-carbohydrate diets problematic for any individual attempting to manage blood cholesterol levels and reduce the risk of cardiovascular disease (69). The popular Mediterranean diet has been investigated extensively owing to its well-established health benefits (70) and is effective for weight loss in overweight and obese individuals (70). What remains to be determined is how individuals attempting to lose weight can effectively maintain the weight loss after terminating any particular diet (66).

Strategies for Promoting Healthy Weight Loss

While it is important to understand the appropriate energy, carbohydrate, fat, and protein intake, it is also critical to identify the most suitable and effective educational strategies for promoting healthy eating by individuals, in general, and society, as a whole. Dietary habits are shaped at a young age and continue over time (71,72), making changing dietary behavior a challenging task. Dietary habits vary by social group, and different social groups often adopt unhealthy eating patterns (73). For example, individuals within a low socioeconomic status in high-income countries tend to have a higher intake of processed sugar, fat, and sodium (74), whereas those individuals with a high socioeconomic status in low-income countries also tend to have a high intake of processed sugar, fat, and sodium (75). Numerous factors such as maternal and paternal influence, food preferences, and exposure to media and educational opportunities can have a significant impact and interactive effect on nutritional behavior (72). Strategies to improve young children's eating habits include avoiding the use of food as a part of a reward system, promoting self-regulation, and having family meals with no media interaction present during eating (72).

The use of behavior change therapy for modifying nutritional intake has received considerable investigation (76,77). The results from a recent meta-analysis identify critical themes important in the maintenance of dietary change. Those behavior change therapies that facilitate goal setting, self-regulated behavior, and an understanding of the underlying nature of the motivation responsible for maintaining the new behavior over time seem necessary when working with overweight and obese adults attempting to lose weight (76).

Food Deserts and Food Swamps

Food deserts, areas characterized by poor access to healthy and affordable food, and **food swamps**, areas characterized by access to high-calorie fast food and junk food, may contribute to disparities in diet and diet-related health outcomes. Food swamps are a separate phenomenon

Food desert Geographical areas characterized by poor access to healthy and affordable food.
Food swamp Geographical areas characterized by access to high-calorie fast food and junk food.

from food deserts and may actually be a better predictor of overweight and obesity (78) and subsequent morbidity and mortality. It is important to understand how food deserts and food swamps influence morbidity and mortality in those individuals residing in these areas (79). It is clear that evidence exists for disparities in food access in the United States by income and race (79) and that these disparities should be examined by factors such as socioeconomic status, race, gender, and age. For example, cardiovascular health in young adults can be strongly influenced by living in a food desert and appears most prevalent in low socioeconomic residents (80), although these same findings may not extend to the elderly (81,82). What also remains unclear is how the introduction of healthier food markets into food desert areas influences nutritional and health outcomes. Several studies that have assessed the influence of food deserts and food swamps on dietary intake and health have found that a full-service food market can improve economic well-being, health, and improved healthy food selection in residents living in a low socioeconomic neighborhood (83–85). The combination of affordability, easy access to and availability of healthy foods, combined with interactive and engaging nutrition information, can assist customers with the purchase and increased consumption of healthy foods (84,85). However, other studies have suggested limited effectiveness of food retail interventions on improving health behaviors in residents of a low-income neighborhood (86,87). This is clearly an area of research that has significant implications for individual and population health, public health policy, and environmental influence. Exercise science professionals can play an important role in promoting healthy nutrition because of the close interaction with people, young and old, who are interested in making positive changes in their nutritional intake and eating behaviors.

Thinking Critically

Why should individuals be careful about where nutritional information is obtained from when trying to consume a healthier diet?

NUTRITION FOR SPORT

In order for athletes to perform at optimal levels, proper nutrient intake during both training and competition is very important. Athletes across the sports spectrum, from ultra-endurance performers to those who must rely on high levels of strength and power production to be successful, work closely with sport nutritionists and exercise science professionals to ensure proper macronutrient and micronutrient intake. For example, during training, athletes must have adequate energy supplies to ensure that the proper intensity and duration of individual workouts can be maintained so that improvements in performance can occur. In various sports, it is critical that prior to and during competition fluid and food intake strategies be appropriate for maintaining peak performance.

Carbohydrate Intake

Carbohydrates in the form of blood glucose and **muscle glycogen** provide energy for muscle contraction. Adequate daily intake of dietary carbohydrate is also important for athletes participating in sports that are endurance sports or those sports that rely on strength and

Muscle glycogen The storage form of glucose in skeletal muscle.

power production by the muscles of the body. This is especially true during moderate- to very high-intensity exercise and high-volume training (88). Sufficient daily carbohydrate intake is necessary to replenish muscle glycogen levels following training and help create an anabolic environment that will promote skeletal muscle repair and protein synthesis (89–91). It is generally recommended that 60% to 70% of an athlete's total energy intake should consist of carbohydrates (13,92). Other recommendations suggest that the replenishment of muscle glycogen is enhanced within 24 hours if between 5 and 12 g of carbohydrate per kg body mass is consumed (93). This amount of dietary carbohydrate intake provides sufficient carbohydrate for energy production and saves amino acids for protein synthesis in the body (93).

Complex carbohydrates, including grains, pasta, whole grain breads, potatoes, brown rice, and whole grain bagels, are some of the best food sources for replenishing muscle glycogen and providing other nutrients needed by the body. Complex carbohydrates provide a sustained release of glucose into the blood, resulting in a lower insulin release from the pancreas, and provide more nutrients than do simple carbohydrates. An insufficient supply of carbohydrates will potentially leave an athlete feeling fatigued, lethargic, and unable to train at the desired intensity and duration (92).

During exercise, carbohydrate ingestion is important for maintaining blood glucose concentrations within normal ranges, especially when the exercise duration is long (>90 min). During prolonged exercise, the body uses its available stores of muscle glycogen and increasingly relies on blood glucose supplied by the liver. As liver glycogen stores are depleted, the athlete may experience low blood glucose levels (94). This condition, called hypoglycemia, can result in feelings of anxiety, nervousness, and tremor, and can affect the functioning of the central nervous system (95). If blood glucose and muscle glycogen levels become depleted, then exercise intensity typically is decreased, resulting in a reduction of exercise performance. Glucose ingestion during long-duration exercise (over 60 min) has been shown to reduce the rate of muscle and liver glycogen depletion, maintain normal blood glucose concentration, delay the onset of fatigue, and improve endurance performance (94). Table 7.7 provides some of the beneficial effects of consuming carbohydrates prior to and during prolonged exercise.

Table 7.7 Benefits of Carbohydrate Consumption Prior to and During Prolonged Exercise

CONSUMPTION OF CARBOHYDRATE	BENEFICIAL EFFECTS
Prior to exercise	Increases muscle and liver glycogen concentrations
	Delays muscle and liver glycogen depletion
	Improves exercise performance
During exercise	Delays muscle and liver glycogen depletion
	Maintains blood glucose concentration
	Improves exercise performance
Following exercise	Enhances muscle and liver glycogen synthesis
	Improves recovery from exercise

Protein Intake

The regular consumption of adequate amounts and appropriate types of protein is important for ensuring the optimal performance of individuals during sport and athletic competition. The RDA for protein is 0.8 g protein per kg body mass per day (1). This level, however, may not be sufficient for athletes who need more protein than sedentary individuals. This greater requirement arises because athletes have increased needs for energy during training and competition, must repair tissue damaged during exercise, and must build new skeletal muscle to meet the demands of exercise training (96,97). Endurance athletes may require between 1.2 to 1.4 g of protein per kg body mass per day, whereas strength and power athletes may require between 1.6 and to 1.7 g of protein per kg body mass per day (98). Table 7.8 illustrates the differences in protein requirements for a nonathletic person, an endurance athlete, and an offensive lineman in football using the protein requirements discussed above. The amount of daily protein intake required varies by individual and is dependent on the factors shown in Figure 7.8 (51).

Foods high in protein content include many animal and dairy products. It is probably best to consume protein in several meals throughout the day so that the individual amino acids are readily available to the body tissues for continued protein synthesis (51,96,99). This practice is common place for athletes who are trying to either maintain or increase skeletal muscle mass. Dietary protein intake should closely match protein needs. Too much protein in the diet may lead to increased urinary calcium excretion (100,101), but there is no clear consensus on the significance of this issue on health and performance (102).

Processed protein supplements are frequently viewed by athletes and coaches as an economical and convenient source of dietary protein (Figure 7.9). This is particularly true of athletes who find that "mixing a shake" easier than preparing or purchasing a meal. Commercial companies often claim that processed protein supplements provide improved and faster absorption of the amino acids over food protein. These processed protein supplements are available in the form of whey protein powders, free hydrolyzed amino acids, and free form amino acids. Though protein supplements may be a convenient source of protein, good food sources provide the necessary dietary protein, are usually less expensive, and contain other nutrients that can help athletes maximize their sport and athletic

Table 7.8 Daily Protein Requirements for a Nonathletic Person, an Endurance Athlete, and an Offensive Lineman in Football

INDIVIDUAL	BODY WEIGHT	PROTEIN REQUIREMENTS
Nonathlete	150 lb (68 kg)	Approx. 55 g protein/day (0.8 g · kg^{-1})
Endurance athlete	140 lb (63.5 kg)	76–90 g protein/day (1.2–1.4 g · kg^{-1})
Offensive lineman in American football	300 lb (136 kg)	218–231 g protein/day (1.6–1.7 g · kg^{-1})

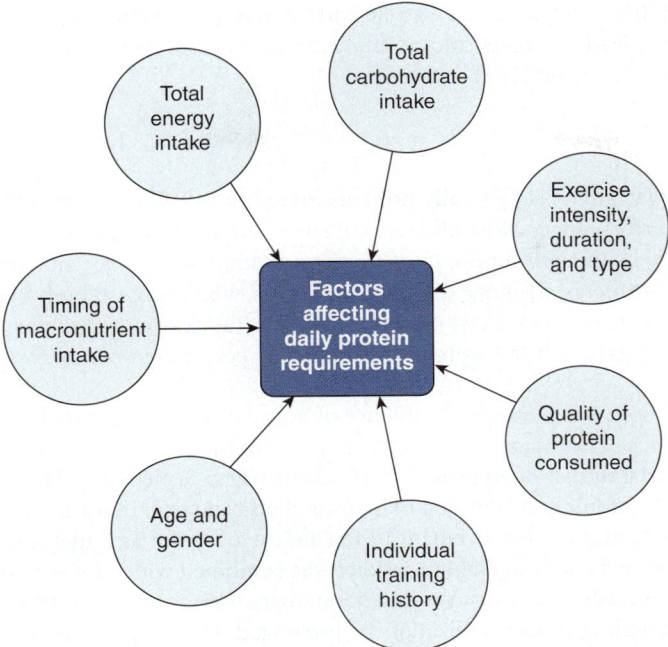

FIGURE 7.8 Factors affecting daily protein requirements. (Adapted from (51).)

FIGURE 7.9 Processed protein supplements.

performance (103). Athletes must also be careful about consuming too much protein as this practice can lead to excess calorie intake, reduced carbohydrate intake, and possibly abnormal kidney function (104).

Fat Intake

Excess dietary fat intake is generally not considered to be of benefit for enhancing sport and athletic performance. Most athletes consume sufficient amounts of dietary fat and therefore receive sufficient amounts of the fat-soluble vitamins for normal physiologic function (13). Athletes who are involved in sports where excess body weight may be detrimental to performance must closely monitor total calorie and dietary fat intake in an effort to maintain a body weight and body fat percentage conducive to successful performance.

There has been some interest in the use of high-fat diets for improving long-duration aerobic endurance performance (105–107). Fatigue during aerobic endurance events, such as a marathon or triathlon, is often related to muscle glycogen depletion and low blood glucose levels (108). The chronic consumption of high-fat diets can result in the increase of fat stores in skeletal muscle and an improvement in the ability to use fat as a fuel source (107,109). For example, when a 2 to 4 week high-fat diet was combined with a high carbohydrate diet for 1 to 3 days immediately prior to competition, there were increases in fat oxidation and a reduction in muscle glycogen utilization in prolonged aerobic endurance events such as a marathon or ultra-endurance events lasting longer than 4 hours (106,107,110). There is a need to better understand the overall dietary composition (*e.g.*, level of carbohydrate restriction) necessary to elicit changes in fat and carbohydrate metabolism and performance, as well as the factors contributing to individual variability in responses to diets of this nature (107). Continued study and research in this subject by exercise science professionals should provide additional insight into this area of nutritional intake.

Vitamin and Mineral Intake

Vitamins and minerals are not a direct source of energy for the body. They are important, however, in the regulation and control of the metabolic reactions in the body. Vitamins and minerals are contained in the various foods we consume. Most athletes consuming an adequate amount of total energy receive sufficient amounts of vitamins and minerals in their diets (13). Athletes who are consuming a low-calorie diet or a specialty diet (*e.g.*, vegetarian) must be careful to eat foods from a variety of different food groups to ensure that they are receiving the vitamins and minerals needed for normal physiologic function (13).

Vitamins play an important role in energy production and tissue metabolism. The B-complex vitamins, niacin, pantothenic acid, folate, biotin, and vitamin C play essential roles in using carbohydrate and fat for energy. Vitamins B_6 and B_{12}, vitamin C, folate, and biotin play an essential role in protein metabolism. In general, vitamin supplementation in excess of the daily recommended intake has not been shown to improve exercise performance in individuals consuming a nutritionally well-balanced diet (13). Athletes on a low-calorie diet or a vegetarian diet, however, may become deficient in certain vitamins if appropriate food choices are not made (13).

Vitamins E, C, and beta-carotene are referred to as **antioxidant vitamins** because of their ability to protect the body from damage caused by **oxygen free radicals**. Free radicals are produced during cellular metabolism when an oxygen molecule is combined with an unpaired electron making them highly reactive to other compounds. Free radicals attack the cellular membrane and have been linked to damage associated with aging, cancer, coronary artery disease, and other chronic diseases (111). Free radicals are also important as signaling molecules and help regulate numerous cell functions. Antioxidants, such as vitamin E, vitamin C, and glutathione and enzymes such as glutathione peroxidase, catalase, and superoxide dismutase, protect the tissues of the body (112). These vitamins and enzymes react directly with free radicals to reduce their reactivity and thus help to protect the cells of the body from damage (111). Table 7.9 provides some examples of vitamins that may be important for improving athletic performance.

Minerals are an important component of numerous metabolic reactions in the body including energy production and muscle contraction (1,13). Minerals serve three broad roles in the body: structural, functional, and regulatory. Structural roles for minerals include the formation of bones and teeth. Functional roles include maintaining normal heart rhythm, initiating muscle contraction, creating nervous system conductivity, and promoting normal acid-base balance. Regulatory roles of minerals include becoming components

Table 7.9 Vitamins and Potential Benefits to Athletic Performance

VITAMIN	ROLE IN ATHLETIC PERFORMANCE
E	Functions as an antioxidant to prevent cell damage
B_1 thiamine	Involved in carbohydrate metabolism
B_2 riboflavin	Involved in carbohydrate metabolism
B_3 niacin	Involved in energy metabolism
B_6 pyridoxine	Involved in amino acid and glycogen metabolism
Pantothenic acid	Involved in energy metabolism
Folate	Important in amino acid metabolism
B_{12} cobalamin	Important in amino acid metabolism
Biotin	Involved in amino acid and glycogen metabolism
Vitamin C	Functions as an antioxidant to prevent cell damage

Antioxidant vitamins Compounds that work to limit the formation of oxygen free radicals.
Oxygen free radicals Compounds produced during cellular metabolism when an oxygen molecule is combined with an unpaired electron making them highly reactive and potentially damaging to the cell.

of the enzymes and hormones that moderate cell activity (13). In general, athletes do not require more minerals than healthy physically active people (13). However, athletes who do not receive sufficient minerals from foods owing to low energy intake or specialty diets may be at risk for certain disease conditions. Osteoporosis and anemia are two common health conditions experienced by athletes not consuming sufficient calcium and iron.

Osteoporosis can result from insufficient calcium intake and is of particular concern among athletes involved in endurance and weight control sports such as long distance running, dance, and gymnastics (113). The **female athlete triad** is a condition linked with eating disorders and menstrual irregularities and with the development of osteoporosis at a young age (Figure 4.7). A key issue in the prevention of osteoporosis is the regular sufficient intake of calcium. The ACSM has published position stands on this potentially very serious health condition (113–115). A more detailed discussion of osteoporosis and the female athlete triad is found in Chapter 4.

The mineral iron is very important for ensuring proper oxygen transport to tissues of the body (1). Iron is an important component of **hemoglobin**, which is a fundamental component of red blood cells. An insufficient intake of iron can lead to reduced hemoglobin concentration, red blood cell number, and anemia (1). Female endurance athletes frequently exhibit low levels of hemoglobin, which can result in **sports anemia** (116,117). Iron supplementation to iron deficient athletes has been shown to improve blood measures of iron status, maximal oxygen consumption, and endurance performance times (118). Iron supplementation that improves the iron status of athletes has the potential to improve aerobic endurance performance (119).

Hydration Status and Fluid Replacement

Normal physiologic function depends on a proper water and electrolyte balance. Water is the medium in which all cells exist and functions occur. Even small decreases in total body water can impair sport and athletic performance (120,121). For example, a loss of just 2% of total body water can significantly impact athletic performance (122). Electrolytes important for normal body function include sodium, potassium, and chloride (122). Electrolytes are lost predominately in sweat and a significant reduction in electrolytes can impair sport and athletic performance (121). If an individual loses too much body water or electrolytes and cannot regulate body temperature adequately, there is the potential for serious injury arising from heat exhaustion and heat stroke (121). Water loss during breathing also contributes to fluid loss and can vary depending on temperature, humidity and the volume of pulmonary ventilation. The amount of water loss during physical exercise is linearly related to heart rate and is approximately four times higher at 140 bpm than at rest.

Female athlete triad A combination of disordered eating, amenorrhea, and osteoporosis that is prevalent in female athletes who participated in sports where low body weight is an important factor in success.

Hemoglobin An iron-containing protein present in red blood cells that carries oxygen.

Sports anemia A condition of low blood hemoglobin that is the result of increases in blood volume.

FIGURE 7.10 Fluid replacement is important for maintaining normal hydration status. (Photo from Comstock/Getty Images.)

Athletes sweat considerably during training and competition and therefore must pay close attention to replacing fluids (Figure 7.10). Athletes train and compete during hot and humid weather conditions and athletes who participate in prolonged endurance events such as marathons, ultramarathons, triathlons, and long distance cycling require both water and electrolytes to maintain normal hydration levels (120,121). Consuming only water during prolonged training and competition may help maintain **euhydration**, but it will not replace the electrolytes lost in sweat (121). A condition called **hyponatremia** (low sodium concentration) can arise if levels of sodium in the blood get too low. Hyponatremia can result in seizures, respiratory arrest, very low blood pressure, coma, and even death (121). It is important that athletes consume carbohydrate and electrolyte beverages when competing in long-duration activities, especially in hot and/or humid conditions.

Athletes engaged in training and competition that result in excessive sweating can monitor hydration status by performing two simple tasks. First, athletes should monitor their weight before and after practice and throughout the day. For each pound of weight lost during practice, approximately 16 oz of water or fluid should be consumed. A second strategy for maintaining hydration status is to monitor the frequency of urination and the color of the urine. Repeated urination that is of a light color typically indicates that the athlete is euhydrated. Athletes who urinate infrequently and who have dark-colored urine are generally dehydrated (121).

AREAS OF STUDY IN SPORT NUTRITION

Exercise science professionals, as well as sport nutritionists, play important roles in determining how nutritional intake influences sports and athletic performance. The role of nutrition as it affects sport and athletic performance can be studied from several perspectives including training versus competition; macronutrient intake versus micronutrient intake; endurance athlete versus strength athlete. In this section, we will examine several primary areas of sport nutrition: the influence of dietary carbohydrate loading and supplementation,

Euhydration A state of normal levels of body water.
Hyponatremia An abnormally low concentration of sodium ions in blood.

fluid intake and replacement, and ergogenic aids as they relate to improving sport and athletic performance. The selection of these topics is not meant to minimize the importance of other areas of sport nutrition, but to provide you with some examples of popular areas of study in sport nutrition.

Carbohydrate Loading and Supplementation

Through research in the early 20th century, it became apparent that the availability of carbohydrate to working muscles was a limiting factor in prolonged endurance performance (36,37,123). The research work of many prominent exercise science professionals demonstrated that glycogen depletion in contracting skeletal muscle resulted in a decrease in exercise intensity or even a cessation of exercise (39,124). Through daily dietary manipulation, the storage of muscle glycogen can be returned to normal levels or elevated above normal levels (called muscle glycogen loading). This elevation in muscle glycogen can result in an improvement of exercise performance and a subsequent delay in fatigue (108). Important nutrition and exercise components of muscle glycogen loading for competition include the following (93):

- A tapered reduction in the intensity and duration of training the week prior to the competition
- Consumption of dietary carbohydrate at 5 to 7 g CHO per kg body mass per day for the week prior to competition
- Consumption of dietary carbohydrate at 7 to 12 g CHO per kg body mass the day prior to competition
- Rest the day prior to competition
- High carbohydrate meal before the competition

Athletes who participate in non-endurance sports may also benefit from strategic carbohydrate replacement and supplementation (125). Training and competition in sports such as soccer, basketball, and American football require athletes to exercise at a high intensity and cover a high volume of distance during participation for upwards of 60 to 90 minutes at a time. The predominate energy sources during these types of sports activities are intramuscular ATP, creatine phosphate, and glycogen, as well as blood glucose. The total energy expenditure is significant and training and competition performance can likely benefit from having adequate storage levels of muscle and liver glycogen prior to exercise, as well as consuming carbohydrate during exercise (125). Several investigations have demonstrated improved team sport performance following ingestion of carbohydrate enriched diets and supplements for ice hockey (126), soccer (127–129), and training and conditioning (130,131). Continued research in this area of study will most certainly lead to a greater understanding of how to best provide carbohydrate to athletes participating in a wide range of sports and athletic competitions.

Fluid Intake and Replacement

Maintaining appropriate hydration status is critical to the health, safety, and performance success of athletes. Dehydration increases the physiological strain on the cardiovascular system causing an elevation in heart rate and core and skin body temperature (121). Additionally, the greater the total body water loss, the greater the increase in physiological strain which results in a further decrease in sweat rate, evaporative heat loss and an increase in heat storage in the body (132–134). Heat stress combined with dehydration further exacerbates these cardiovascular responses because it creates competition between the central and peripheral circulation for limited blood volume (52) causing additional strain on the cardiovascular system (132,133).

An athlete's health and performance can be adversely affected by dehydration. Dehydration increases the risk for heat exhaustion (121) and is a risk factor for heat stroke (121) both which can lead to significant health issues and possible death. Dehydration can also exacerbate the level of **rhabdomyolysis** (121) and has been associated with altered intracranial volume (135) and a decrease in cerebral blood flow velocity in response to changes in body position (134). Skeletal muscle cramps associated with dehydration, electrolyte deficits and muscle fatigue are common in athletes participating in American football, tennis, prolonged endurance events such as marathons and triathlons, soccer, volleyball, cross-country ski-racers, and ice hockey goalies (121). Factors affecting the risk level of dehydration include sex, age, and dietary intake (121).

Drinking to thirst and planning to drink are the primary strategies for maintaining appropriate hydration during training and competition (136). While the sensation of thirst works well for maintaining hydration at rest (137), it is less sensitive during exercise as mechanisms that stimulate thirst sensation are affected by numerous influences (138). Even when drinking *ad libitum*, individuals feel more thirsty when compared to planned drinking trials (139). Thirst is alleviated before complete rehydration is achieved (140) and even when exercise commences in euhydrated state, accumulated fluid loss and the subsequent development of thirst sensations can take time and depend on factors such as environmental

Table 7.10	Fluid Replacement Recommendations to Maintain Normal Hydration Status (121)
TIME PERIOD	RECOMMENDATION
Before exercise	Prehydrate at least several hours before exercise to allow fluid absorption and urine output to return to normal
	Consuming beverages with sodium and/or small snacks or meals may stimulate thirst and fluid intake
During exercise	Develop an individualized fluid replacement strategy that will prevent excessive dehydration or hyponatremia
	Consuming beverages with electrolytes and carbohydrates can help maintain fluid and electrolyte balance
After exercise	Consume 1.5 L of fluid for each kilogram of body weight lost
	Consuming beverages with sodium and/or small snacks or meals may stimulate thirst and fluid retention

Rhabdomyolysis A breakdown of skeletal muscle fibers with leakage of muscle fiber contents into the circulation.

Ad libitum A condition of drinking or eating as much or as often and necessary of desired.

conditions, exercise intensity and duration, and sweat rate (121). A drinking to thirst fluid replacement strategy that maintains hydration state within ±2% of body weight appears to be effective in the preservation of physiological function and exercise performance especially at lower exercise intensities and durations and in cooler weather temperatures (136). There are also conditions where fluid intake requirements are challenged and a customized planned drinking strategy will need to be employed to avoid potential health risks from thermoregulatory and cardiovascular impairments and a decrease in exercise performance. These conditions include exercise training or competition that is longer in duration (>90 min), higher in exercise intensity, and occur in warm or hot environmental conditions (136). The impact of dehydration on team sport performance is less clear and more research is needed to understand how performance is impacted (141).

The ACSM position stand on exercise and fluid replacement is an excellent resource for understanding the implications of dehydration and the correct procedures for maintaining euhydration, especially during exercise (121). Athletes should develop individualized fluid and electrolyte replacement strategies to maintain euhydration and reduce the risk for dehydration (121). Table 7.10 provides fluid replacement strategies to maintain normal hydration status.

Ergogenic Aids

Ergogenic aids are substances or devices that work to improve performance during training or competition. Ergogenic aids may be classified as biomechanical, nutritional, pharmacologic, physiologic, and psychological (142). The use of many ergogenic aids is considered illegal by various sports governing associations. Consequently, many sports governing associations have lists of banned substances, equipment, or aids. Many athletes, however, continue to use ergogenic aids as a means to improve performance during training and competition.

Nutritional ergogenic aids work to enhance sport and athletic performance by improving energy production, enhancing anabolic activities in the body, influencing exercise metabolism, and aiding in recovery from exercise (103,143). Carbohydrate, protein, vitamin, and mineral supplementation, as well as other ingestible compounds are often considered nutritional ergogenic aids. Caffeine, creatine phosphate, and polyphenols are examples of commonly used nutritional ergogenic aids that may enhance athletic performance if consumed under the right circumstances.

Caffeine ingestion is a commonly used ergogenic aid that athletes consume in an effort to increase energy, promote alertness, counteract sleep deprivation, and promote cognitive and mood enhancement (144). Studies using moderate to high doses of caffeine (5–9 mg · kg^{-1} body mass) have reported ergogenic effects in endurance activities and distinct physiological responses including increased heart rate, higher epinephrine and norepinephrine concentrations, higher blood lactic acid, free fatty acid, and glycerol levels (145,146). Moderate to high doses of caffeine also produced unwanted side effects, including gastrointestinal distress, increased nervousness and mental confusion, an inability to concentrate, and disturbed sleeping patterns (147). The administration of low doses of caffeine (3 mg · kg^{-1} body mass) can also produce an ergogenic effect with few if any side effects as long as there are marked increases in plasma caffeine concentrations (145,147). Low doses of caffeine can provide an ergogenic effect when given prior to exercise in endurance events (148–150), resistance exercise (151,152), team sports (147,153), and individual sports (154,155). Additional research is needed to determine the relationship between genetic predisposition and caffeine intake, whether caffeine can still be an effective ergogenic aid with different delivery methods, and what dosage of caffeine can provide an ergogenic effect in fine motor skilled sports and athletic competitions.

Creatine phosphate is a compound found in skeletal muscle that is critically important in the production of energy during high-intensity muscle contraction. It has been demonstrated that increased creatine monohydrate consumption can lead to higher concentrations of creatine and creatine phosphate in skeletal muscle (156). Increases in muscle creatine can be accomplished with a high-dose, short-duration ingestion period (~20 g · d^{-1} for 5 d) or a lower-dose, longer-duration ingestion period (3–5 g · d^{-1} for ~30 d) (157). Creatine supplementation that results in elevated muscle creatine typically improves sport and athletic performance lasting less than 30 seconds, especially if there are repeated bouts of high-intensity exercise (158–160). Creatine supplementation has also been shown to increase lean body mass, which in turn likely contributes to the improved performance observed with resistance exercise training and competition (161–163). Additional research is needed to determine the relationship between genetic predisposition and creatine supplementation, whether creatine supplementation can still be an effective ergogenic aid in fine motor skilled sports and athletic competitions, and what potential health risks are associated with long-term creatine supplementation.

Polyphenols are micronutrients derived from plants and increases in the diet can be attained by consumption of fruits and vegetables or from supplements (164). Polyphenols consumed in the diet eventually appear as various metabolites in the body. The primary functions of polyphenols are to serve as antioxidants responsible for scavenging free radicals and as anti-inflammatory compounds (164). Fruit-derived polyphenols, including those consumed from cherries, blueberries, blackcurrant, pomegranate, and cocoa, lead to lower plasma markers of oxidative damage and inflammation (164). Acute polyphenol supplementation within 1 hour prior to exercise may improve exercise performance in recreationally active subjects (165–167) but not trained athletes (165,168–170). Chronic supplementation of different polyphenols for at least 7 days has been shown to improve cycling time trial performance (171,172) and increase high-intensity intermittent running distance to exhaustion (173). Consumption of polyphenol for at least 7 days appears to generate performance improvements for recreationally active subjects and, to a lesser amount, for trained athletes. Additional research is needed to determine if polyphenol supplementation is beneficial for recovery from training and competition, how highly trained athletes are affected by supplementation and how athletes consuming sufficient dietary polyphenols may benefit from enhanced supplementation.

Athletes and coaches training to enhance sport and athletic performance work with sport nutrition and other exercise science professionals to examine the influence of various nutritional ergogenic aids on physiologic, biochemical, and performance factors. Conversely, other exercise science and sports medicine professionals continue to work to detect the use of illegal and banned ergogenic aids by athletes. Numerous ergogenic aids are classified as nutritional, and some of the more commonly used ones are presented in Table 7.11 (103,143).

Thinking Critically

How might coursework in nutrition prepare an individual for a career as a registered dietician and nutritionist, exercise physiologist, strength and conditioning coach, or a fitness instructor?

Table 7.11 Nutritional Ergogenic Aids (103,143)

ERGOGENIC AID	POTENTIAL BASIS FOR ENHANCING PERFORMANCE
Creatine	Improves energy production during high-intensity exercise
	Increases body weight and lean body mass
Caffeine	Increases alertness and wards off drowsiness
	Increases fat oxidation and reduces carbohydrate utilization
Sodium bicarbonate and sodium citrate	Increase the body's ability to buffer lactic acid production
L-carnitine	Improves fat oxidation
	Decreases lactic acid formation
Aspartic acid	Decreases ammonia formation in muscle
Ginseng	Increases fat oxidation and reduces muscle glycogen utilization
Omega-3 fatty acids	Improve oxygen delivery to muscles thereby enhancing aerobic metabolism
Antioxidants (superoxide dismutase and catalase) and polyphenols	Protect tissues from the damage caused by oxygen free radicals
Coenzyme Q_{10}	Increases aerobic energy production
Glycerol	Improves hydration and may decrease dehydration
Chromium picolinate	Enhances muscle protein development

Interview

Lisa Heaton MS, RD, CSSD, LDN
Senior Scientist, Gatorade Sports Science Institute

Brief Introduction – I earned a Bachelor's degree in Foods, Nutrition and Dietetics from Bradley University. I completed my dietetic internship and earned my Master's degree in Exercise Science from the University of Houston. I earned and have maintained the credential of Board Certified Specialist in Sports

Dietetics (CSSD) since 2009 and have focused my career on improving athlete health, safety, and performance through both research and education in sports nutrition, recovery, hydration, and performance.

My sports nutrition career started at the National Institute for Athletic Health and Performance, serving as a research assistant and sports dietitian. I supported research initiatives focused on hydration and thermoregulation throughout my time there. My work also involved individual nutrition counseling, body composition testing, cardiovascular fitness testing, working with several NCAA D-II and D-III athletic programs in the area, and serving as the sports dietitian for the Sioux Falls Sky Force basketball team.

As a senior scientist and sports dietitian at the Gatorade Sports Science Institute (GSSI), one of my primary focuses has been to translate the body of research on various sports nutrition topics into nutrition-related education materials for practitioners as well as academics and researchers. I support GSSI's elite athlete testing program and manage athlete testing that occurs at the GSSI in Barrington, Illinois. I also served as the consulting Sports Dietitian for the Chicago Bulls for the 2015–16 and 2016–17 seasons.

Q: Why did you choose to become involved in your work as an exercise and sport nutritionist?

I knew in high school that I wanted to become a registered dietitian. At that point I thought I would pursue a career helping those with eating disorders. It wasn't until my junior year of college, when I took a sports nutrition course, that I realized what I truly wanted to do. Sports dietetics combines evaluating the athlete's diet from an overall health standpoint, as well as how to tailor recommendations to help them perform to the best of their ability in their sport. This audience, whether recreational or professional, tends to be motivated to make training and dietary changes to feel and perform at their best. Many competitive athletes, from high school to the professional level, look for competitive advantages they can gain over the competition. Having a solid nutrition foundation is an important part of this desire. I enjoy meeting athletes where they are with their nutrition and making changes to improve overall nutrient intake and tailor pre-, during, and post-nutrition intake to maximize performance as well as recovery and training adaptations. New sports nutrition-related research is continually being published, so the field is constantly evolving, which makes it a very exciting area in which to work.

Q: Why it is important for exercise science students to have an understanding of "nutrition"?

Physiology, fitness, and movement are just one part of overall health in an individual. Nutrition is a related discipline and one can have a direct impact upon the other. Understanding the basic principles of how nutrients function within the body gives a more holistic view of the individual and their needs. The nutrition needs and recommendations during and around exercise are different than those compared to rest. It's important for an exercise scientist to understand the impact of exercise on nutrient metabolism, and how nutrients can support the energy needs of exercise. Having this understanding can improve not just the care of the individual but also how you interact with a multidisciplinary team and potentially identify when an individual athlete would benefit from additional support from other members of your team. Training plans and strategies can be coordinated to give the individual the best possible strategy to achieve their goals.

Q: What advice would you have for an undergraduate student beginning to explore a career in exercise science?

The beauty of an exercise science degree is that it allows you to pursue advanced degrees or certificates in multiple disciplines such as athletic training, sports dietetics, and physical therapy. I would encourage any undergraduate student to spend time talking with or shadowing someone within any profession that they may be interested in pursuing. Understanding the real life demands and duties of those professions can really help an undergraduate student decide if that is a career they want to pursue. If you are interested in pursuing a research-based Master's or doctoral degree, look into the research topic area in which your potential advisor may focus. This can be beneficial when narrowing down to which college or university you apply. If the degree requires a research thesis or dissertation, focusing your work on a topic that you are truly interested in can be helpful in your overall career development.

Interview

D. Enette Larson-Meyer, PhD, RD, CSSD, FACSM
Virginia Tech University

Brief Introduction – I am in the Department of Human Nutrition, Foods, and Exercise where I serve as Director of the Master of Science program in Nutrition and Dietetics and also conduct human subjects research. For 16 years, I was at the University of Wyoming where I taught and conducted research on how nutrition influences the health and performance of active individuals at all stages of the lifecycle and all levels of performance. I have over 90 scientific publications on nutrition, exercise, and health including a book "*Plant-Based Sports Nutrition. Expert Fueling Strategies for Training, Recovery, and Performance.*" I have served on several International Olympic Committee Panels, including the 2011 Sports Nutrition Consensus Panel and the Expert Panel for Dietary Supplements and the High-Performance Athlete. I am also a former sports nutritionist for the University of Alabama at Birmingham. I have been active in ACSM as well as PINES (Professionals in Nutrition for Exercise and Sport) and SCAN (the Sports, Cardiovascular and Wellness Nutrition practice group of the Academy of Nutrition and Dietetics). I completed my Bachelor of Science degree from the University of Wyoming and my dietetic internship and Master's degree at Massachusetts General Hospital in Boston. I worked as the research dietitian for the Clinical Diabetes and Nutrition Section of the National Institutes of Health, National Institute of Diabetes and Digestive and Kidney Diseases in Phoenix, AZ, before completing my doctoral and postdoctoral studies at the University of Alabama at Birmingham.

 Why did you choose to become an "exercise and sport nutritionist"?

My father had a copy of *The Complete Book for Running* by the late Jim Fixx. The cover had a photo of Jim's legs and I distinctly remember wanting to have legs like him. I started weight training in high school and running in college and Jim's premature death in 1984 from a heart attack had a major impact on me. Jim had a strong family history of cardiovascular disease but his lifestyle intervention focused on exercise and apparently ignored nutrition. With this in the back of my mind I became interested in biomedical research and the importance of *both* regular exercise and good nutrition.

 Why it is important for exercise science students to understand "nutrition"?

Good eating habits and regular exercise/physical activity work together—along with other health habits including adequate sleep and stress reduction—to promote and maintain health, reduce chronic disease, and optimize physical performance. Exercise science students should understand the foundations of a good healthy diet so they can help reinforce its importance and understand when a diet fad or trend may not be appropriate (or even harmful) for the athlete, avid exerciser or patient. Diet fads are very common in the exercise and sports world but there are no short cuts for eating healthy.

 What advice would you have for a student exploring a career in any exercise science profession?

Students should gain a strong foundation in the basic sciences including biology, chemistry, biochemistry, physics, immunology, anatomy, and physiology. These sciences establish the foundation for our profession and it is hard to understand the depth of exercise science without this foundation. Students should also gain an understanding of the research process while in school and be a lifelong learner afterwards. Join and be active in ACSM and your regional chapter and keep up with the changing knowledge in the field. Don't be afraid to critically evaluate what we know and always ask "why or how do we know that?"

SUMMARY

- The consumption of appropriate amounts of the macronutrients and micronutrients is important for promoting health and successful performance in sport and athletic competition.
- Proper nutrient intake can reduce the risk of certain diseases and allow individuals to derive health benefits from participating in physical activity and exercise.
- Dietary guidelines have been established to help individuals make food choices that will enhance health and reduce disease risk.
- Aerobic endurance and strength-power athletes have an increased need for some nutrients including carbohydrate, protein, and certain vitamins and minerals.
- It is important for athletes to monitor nutrition and water intake so that successful performance can be derived during training and competition.

FOR REVIEW

1. What are some nutritional intake habits that could lead to an increased risk for hypertension, heart disease, breast cancer, colon cancer, osteoporosis, and obesity?
2. How has NHANES contributed to the understanding of nutritional patterns of people living in the United States?
3. What is the main function of the ADA and the SCAN?
4. What are the differences between simple and complex carbohydrates?
5. What is the difference between Type 1 diabetes mellitus and Type 2 diabetes mellitus?
6. What role do vitamins and minerals play in enhancing health?
7. Describe the differences between a dietary recall and a dietary record.
8. Why must endurance and strength and power athletes be concerned about daily carbohydrate intake?
9. Why do endurance and strength and power athletes need more protein than the RDI?
10. Why are vitamins E, C, and beta-carotene considered antioxidants?
11. What benefits are derived from consuming carbohydrate during prolonged exercise?
12. Which ergogenic aids would be beneficial for an endurance athlete?
13. Which ergogenic aids would be beneficial for a strength athlete?

Project-Based Learning

1. Identify an individual who does not eat healthy and does not participate in regular physical activity or exercise. Prepare a presentation that includes at least five key nutritional recommendations that a registered dietitian would give to that individual in an effort to improve their nutritional intake. What are those key points and how does the health nutrition literature support the recommendations?
2. Identify an athlete who does not have a proper nutritional intake for the sport in which they participate. Prepare a presentation that includes at least five key points that a

registered dietitian would give to this athlete who is struggling with their nutritional intake. What are those key points and how does the sport nutrition literature support the recommendations?

REFERENCES

1. Gropper SS, Smith JL, Carr TP. *Advanced Nutrition and Human Metabolism*. 7th ed. Boston (MA): Cengage; 2018.
2. Volpe SL. Sports nutrition. In: Brown SP, editor. *Introduction to Exercise Science*. 1st ed. Philadelphia (PA): Lippincott, Williams & Wilkins; 2001. p. 162–91.
3. Todhunter EN. Reflections on nutrition history. *J Nutr*. 1983;113:1681–5.
4. Sanjoaquin MA, Appleby PN, Thorogood M, Mann JI, Key TJ. Nutrition, lifestyle and colorectal cancer incidence: a prospective investigation of 10998 vegetarians and non-vegetarians in the United Kingdom. *Br J Cancer*. 2004;90:118–21.
5. McCullough ML, Feskanich D, Rimm EB, et al. Adherence to the Dietary Guidelines for Americans and risk of major chronic disease in men. *Am J Clin Nutr*. 2000;72:1223–31.
6. McCullough ML, Feskanich D, Stampfer MJ, et al. Adherence to the Dietary Guidelines for Americans and risk of major chronic disease in women. *Am J Clin Nutr*. 2000;72:1214–22.
7. Glueck CJ, Gartside P, Laskarzewski PM, Khoury P, Tyroler HA. High-density lipoprotein cholesterol in blacks and whites: potential ramifications for coronary heart disease. *Am Heart J*. 1984;108(3 Pt 2):815–26.
8. Yu-Poth S, Zhao G, Etherton T, Naglak M, Jonnalagadda S, Kris-Therton PM. Effects of the National Cholesterol Education Program's step I and step II dietary intervention programs on cardiovascular disease risk factors: a meta-analysis. *Am J Clin Nutr*. 1999;69:632–46.
9. Heaney RP. Calcium, dairy products, and osteoporosis. *J Am Coll Nutr*. 2000;19:88S–99S.
10. Binkley JK, Eales J, Jekanowski M. The relation between dietary change and rising US obesity. *Int J Obes*. 2000;24:1032–9.
11. Ivy JL, Lee MC, Brozinick JT, Reed MJ. Muscle glycogen storage after different amounts of carbohydrate ingestion. *J Appl Physiol*. 1988;65(5):2018–23.
12. Coyle EF, Montain SJ. Benefits of fluid replacement with carbohydrate during exercise. *Med Sci Sports Exerc*. 1992;24(9):S324–S30.
13. McArdle WD, Katch FI, Katch VL. *Sports and Exercise Nutrition*. 4th ed. Baltimore (MD): Lippincott Williams & Wilkins; 2012.
14. Cassell JA. *Carry the Flame: A History of the American Dietetic Association*. 1st ed. Sudbury (MA): Jones and Bartlett; 1990.
15. Nichols BL. Atwater and USDA nutrition research and service: a prologue of the past century. *J Nutr*. 1994;124:1718S–27S.
16. Sawin CT. Eugene F. DuBois (1882–1959), basal metabolism, and the thyroid. *Endocrinologist*. 2003;13(5):369–71.
17. Keys A, Brozek J, Henschel A, Mickelsen O, Taylor HL. *The Biology of Human Starvation*. 1st ed. Minneapolis (MN): The University of Minnesota Press; 1950.
18. American Society for Nutrition. Web site [Internet]. 2021. Available from: https://nutrition.org/
19. Tipton CM. Exercise Physiology, part II: a contemporary historical perspective. In: Massengale JD, Swanson RA, editors. *The History of Exercise and Sport Science*. Champaign (IL): Human Kinetics; 1997. p. 396–438.
20. Hegsted DM. A look back at lessons learned and not learned. *J Nutr*. 1994;124:1867S–70S.
21. Kannel WB, Larson M. Long-term epidemiologic prediction of coronary disease. *Cardiology*. 1993;82:137–52.
22. Paffenbarger RS, Wing AL, Hyde RT. Physical activity as an index of heart attack risk in college alumni. *J Epidemiol*. 1978;108(3):161–75.
23. Centers for Disease Control and Prevention Web site [Internet]. National Health and Nutrition Examination Survey. 2015. Available from: http://www.cdc.gov/nchs/nhanes.htm
24. American Dietetic Association. Position of the American Dietetic Association: nutrition intervention in the treatment of eating disorders. *J Am Diet Assoc*. 2011;111(1236):1241.
25. American Dietetic Association. Position of the American Dietetic Association: local support of nutrition integrity in schools. *J Am Diet Assoc*. 2010;110:1244–54.

26. American Dietetic Association. Position of the American Dietetic Association: benchmarks for nutrition in child care. *J Am Diet Assoc.* 2011;111:607–15.
27. American Dietetic Association. Position of the American Dietetic Association: individualized nutrition approaches for older adults in health care communities. *J Am Diet Assoc.* 2010;110:1549–53.
28. American Dietetic Association. Position of the American Dietetic Association: food insecurity in the United States. *J Am Diet Assoc.* 2010;110:1368–77.
29. American College of Sports Medicine, American Dietetic Association, Dietitians of Canada. Nutrition and athletic performance. *Med Sci Sports Exerc.* 2009;41(3):709–31.
30. Academy of Nutrition and Dietetics. Academy of Nutrition and Dietetics Web site [Internet]. 2016. Available from: www.eatright.org
31. Applegate EA, Grivetti LE. Search for the competitive edge: a history of dietary fads and supplements. *J Nutr.* 1997;127:869S–73S.
32. Mayer J, Bullen B. Nutrition and athletic performance. *Physiol Rev.* 1960;40:369–97.
33. Van Itallie TB, Sinisterra L, Stare FJ. Nutrition and athletic performance. *JAMA.* 1956;162:1120–6.
34. Courtice FC, Douglas CG. The effect of prolonged muscular exercise on the metabolism. *Proc R Soc London.* 1935;119:381–3.
35. Krogh A, Lindhard J. The relative value of fats and carbohydrates as sources of muscular energy. *Biochem J.* 1919;14:290–4.
36. Levine SA, Gordon B, Derick CL. Some changes in the chemical constituents of the blood following a marathon race. *JAMA.* 1924;82:1778–9.
37. Gordon B, Kohn LA, Levine SA, Matton M, Scriver WDM, Whiting WB. Sugar content of the blood in runners following a marathon race. *JAMA.* 1925;85:508–9.
38. Ahlborg B, Bergstrom J, Ekelund LG, Hultman E. Muscle glycogen and muscle electrolytes during prolonged physical exercise. *Acta Physiol Scand.* 1967;70:129–42.
39. Bergstrom J, Hermansen L, Hultman E, Saltin B. Diet, muscle glycogen and physical performance. *Acta Physiol Scand.* 1967;71:140–50.
40. Wright DA, Sherman WM, Dernbach AR. Carbohydrate feedings before, during or in combination improve cycling endurance performance. *J Appl Physiol.* 1991;71(3):1082–8.
41. Kraut H, Muller EA, Muller-Wecker H. Die abhangigkeit des muskeitrainings und eiweissbestand des korpers. *Biochem Z.* 1954;324:280–94.
42. Darden E. Olympic athletes view vitamins and victory. *J Home Econ.* 1973;65:8–11.
43. Pedlar C, Newell J, Lewis NA. Blood biomarker analysis for the high-performance athlete. *Gatorade Sports Sci Exch.* 2020;29(204):1–5.
44. Pedlar CR, Newell J, Lewis NA. Blood biomarker profiling and monitoring for high-performance physiology and nutrition: current perspectives, limitations and recommendations. *Sports Med.* 2019;49(S2):185–98.
45. Gatorade Sport Science Institute Web site [Internet]. 2021. Available from: https://www.gssiweb.org/en
46. Astrup A, Buemann B, Western P, Toubro S, Raben A, Christensen NJ. Obesity as an adaptation to a high-fat diet: evidence from a cross-sectional study. *Am J Clin Nutr.* 1994;59(2):350–5.
47. Stinson EJ, Piaggi P, Ibrahim M, Venti C, Krakoff J, Votruba SB. High fat and sugar consumption during ad libitum intake predicts weight gain. *Obesity.* 2018;26(4):689–95.
48. Calle EE, Rodriguez C, Walker-Thurmond K, Thun MJ. Overweight, obesity, and mortality from cancer in a prospectively studied cohort of U.S. adults. *N Engl J Med.* 2003;348(17):1625–38.
49. Flegal KM, Williamson DF, Pamuk ER, Rosenberg HM. Estimating deaths attributable to obesity in the United States. *Am J Public Health.* 2004;94(9):1486–9.
50. Després JP, Moorjani S, Lupien PJ, Tremblay A, Nadeau A, Bouchard C. Regional distribution of body fat, plasma lipoproteins, and cardiovascular disease. *Arteriosclerosis.* 1990;10:497–511.
51. Lemon PWR. Beyond the zone: protein needs of active individuals. *J Am Coll Nutr.* 2000;19(5):513S–21S.
52. Guyton AC, Hall JE. *Textbook of Medical Physiology.* 13th ed. Oxford (UK): Elsevier; 2016.
53. Westerterp KR. The assessment of energy and nutrient intake in humans. In: Bouchard C, editor. *Physical Activity and Obesity.* Champaign (IL): Human Kinetics Publishers, Inc; 2000. p. 133–49.
54. Kirkpatrick SI, Baranowski T, Subar AF, Tooze JA, Frongillo EA. Best practices for conducting and interpreting studies to validate self-report dietary assessment methods. *J Acad Nutr Diet.* 2019;119(11):1801–16.
55. Fowles ER, Sterling BS, Walker LO. Measuring dietary intake in nursing research. *Can J Nurs Res.* 2007;39(2):146–65.
56. Abbasi F, McLaughlin T, Lamendola C, et al. High carbohydrate diets, triglyceride rich lipoproteins, and coronary heart disease risk. *Am J Cardiol.* 2000;85(1):45–8.

57. Fuchs CS, Willett WC, Colditz GA, et al. The influence of folate and multivitamin use on the familial risk of colon cancer in women. *Cancer Epidemiol Biomarkers Prev*. 2002;11(3):227–34.
58. Hu FB, Rimm EB, Stampfer MJ, Ascherio A, Spiegelman D, Willett WC. Prospective study of major dietary patterns and risk of coronary heart disease in men. *Am J Clin Nutr*. 2000;72:912–21.
59. Koushik A, Hunter DJ, Spiegelman D, et al. Fruits, vegetables, and colon cancer risk in a pooled analysis of 14 cohort studies. *J Natl Cancer Inst*. 2007;99(19):1471–83.
60. Lindquist CH, Gower BA, Goran MI. Role of dietary factors in ethnic differences in early risk of cardiovascular disease and type 2 diabetes. *Am J Clin Nutr*. 2000;71:725–32.
61. Manore MM. Health consequences of chronic dieting in active women. *Health Fitness*. 1998;2(2):24–31.
62. Slattery ML, Schumacher MC, Smith KR, West DW, Abd-Elghany N. Physical activity, diet, and risk of colon cancer in Utah. *Am J Epidemiol*. 1988;128(5):989–99.
63. Nestle M, Dixon LB. *Taking Sides: Clashing Views on Controversial Issues in Food and Nutrition*. Guilford (CT): McGraw-Hill/Dushkin; 2004.
64. United States Department of Health and Human Services, United States Department of Agriculture. *Dietary Guidelines for Americans, 2015*. Washington (D.C): U.S. Government Printing Office; 2005 [updated 2005]. Available from: http://health.gov/DietaryGuidelines/
65. Karanja NM, Obarzanek E, Lin PH, et al. Descriptive characteristics of the dietary patterns used in the Dietary Approaches to Stop Hypertension Trial. *J Am Diet Assoc*. 1999;99:S19–S27.
66. Donnelly JE, Blair SN, Jakicic JM, Manore MM, Rankin JW, Smith BK. Appropriate physical activity intervention strategies for weight loss and prevention of weight regain for adults. *Med Sci Sports Exerc*. 2009;41(2):459–71.
67. Thom G, Lean M. is there an optimal diet for weight management and metabolic health? *Gastroenterology*. 2017;152(7):1739–51.
68. Johnston BC, Kanters S, Bandayrel K, et al. Comparison of weight loss among named diet programs in overweight and obese adults: a meta-analysis. *JAMA*. 2014;312(9):923–33.
69. Mansoor N, Vinknes KJ, Veierød MB, Retterstøl K. Effects of low-carbohydrate diets v. low-fat diets on body weight and cardiovascular risk factors: a meta-analysis of randomised controlled trials. *Br J Nutr*. 2016;115(3):466–79.
70. Mancini JG, Filion KB, Atallah R, Eisenberg MJ. Systematic review of the Mediterranean diet for long-term weight loss. *Am J Med*. 2016;129(4):407–15.e4.
71. Montaño Z, Smith JD, Dishion TJ, Shaw DS, Wilson MN. Longitudinal relations between observed parenting behaviors and dietary quality of meals from ages 2 to 5. *Appetite*. 2015;87:324–9.
72. Scaglioni S, De Cosmi V, Ciappolino V, Parazzini F, Brambilla P, Agostoni C. Factors influencing children's eating behaviours. *Nutrients*. 2018;10(6):706.
73. Mayén A-L, de Mestral C, Zamora G, et al. Interventions promoting healthy eating as a tool for reducing social inequalities in diet in low-and middle-income countries: a systematic review. *Int J Equity Health*. 2016;15(1):205.
74. Darmon N, Drewnowski A. Does social class predict diet quality? *Am J Clin Nutr*. 2008;87(5):1107–17.
75. Popkin BM, Adair LS, Ng SW. The global nutrition transition: the pandemic of obesity in developing countries. *Nutr Rev*. 2012;70(1):3–21.
76. Samdal GB, Eide GE, Barth T, Williams G, Meland E. Effective behaviour change techniques for physical activity and healthy eating in overweight and obese adults; systematic review and meta-regression analyses. *Int J Behav Nutr Phys Act*. 2017;14(1):42.
77. McDermott MS, Oliver M, Iverson D, Sharma R. Effective techniques for changing physical activity and healthy eating intentions and behaviour: a systematic review and meta-analysis. *Br J Health Psychol*. 2016;21(4):827–41.
78. Cooksey-Stowers K, Schwartz M, Brownell K. Food swamps predict obesity rates better than food deserts in the United States. *Int J Environ Res Public Health*. 2017;14(11):1366.
79. Beaulac J, Kristjansson E, Cummins S. A systematic review of food deserts, 1966–2007. *Prev Chronic Dis*. 2009;6(3):1–9.
80. Testa A, Jackson DB, Semenza DC, Vaughn MG. Food deserts and cardiovascular health among young adults. *Public Health Nutr*. 2021;24:117–24.
81. Fitzpatrick K, Greenhalgh-Stanley N, Ver Ploeg M. Food deserts and diet-related health outcomes of the elderly. *Food Policy*. 2019;87:101747.
82. Fitzpatrick K, Greenhalgh-Stanley N, Ver Ploeg M. The impact of food deserts on food insufficiency and SNAP participation among the elderly. *Am J Agric Econ*. 2016;98(1):19–40.

83. Richardson AS, Ghosh-Dastidar M, Beckman R, et al. Can the introduction of a full-service supermarket in a food desert improve residents' economic status and health? *Ann Epidemiol.* 2017;27(12):771–6.
84. Adam A, Jensen JD. What is the effectiveness of obesity related interventions at retail grocery stores and supermarkets?—a systematic review. *BMC Public Health.* 2016;16(1):1247.
85. Glanz K, Yaroch AL. Strategies for increasing fruit and vegetable intake in grocery stores and communities: policy, pricing, and environmental change. *Prev Med.* 2004;39:75–80.
86. Ortega AN, Albert SL, Chan-Golston AM, et al. Substantial improvements not seen in health behaviors following corner store conversions in two Latino food swamps. *BMC Public Health.* 2016;16(1):389.
87. Dubowitz T, Ghosh-Dastidar M, Cohen DA, et al. Diet and perceptions change with supermarket introduction in a food desert, but not because of supermarket use. *Health Aff.* 2015;34(11):1858–68.
88. Gollnick PD. Metabolism of substrates: energy substrate metabolism during exercise and as modified by training. *Fed Proc.* 1983;44(2):353–7.
89. Chandler RM, Byrne HK, Patterson JG, Ivy JL. Dietary supplements affect the anabolic hormones after weight-training exercise. *J Appl Physiol.* 1994;76(2):839–45.
90. Haff GG, Koch AJ, Potteiger JA, et al. Carbohydrate supplementation attenuates muscle glycogen loss during acute bouts of resistance exercise. *Int J Sport Nutr Exerc Metab.* 2000;10:326–39.
91. Volek JS, Kraemer WJ, Bush JA, Incledon T, Boetes M. Testosterone and cortisol in relationship to dietary nutrients and resistance exercise. *J Appl Physiol.* 1997;82(1):49–54.
92. Costill DL. Carbohydrates for exercise: dietary demands for optimal performance. *Int J Sports Med.* 1988;9:1–18.
93. Burke LM, Kiens B, Ivy JL. Carbohydrates and fat for training and recovery. *J Sports Sci.* 2004;22:15–30.
94. Maughan RJ. Physiology and nutrition for middle distance and long distance running. In: Lamb DL, Knuttgen HG, Murray R, editors. *Perspectives in Exercise Science and Sports Medicine: Physiology and Nutrition for Competitive Sport.* 7th ed. Carmel (IN): Cooper Publishing Group; 1994. p. 329–71.
95. Levin BE, Dunn-Meynell AA, Routh VH. Brain glucose sensing and body energy homeostasis: role in obesity and diabetes. *Am J Physiol.* 1999;276:R1223–31.
96. Butterfield G. Amino acids and high protein diets. In: Lamb DL, Williams MH, editors. *Ergogenics: Enhancement of Performance in Exercise and Sport. Perspectives in Exercise Science and Sports Medicine.* Ann Arbor (MI): Wm C. Brown; 1991. p. 87–122.
97. Mettler SU, Mitchell NM, Tipton KD. Increased protein intake reduces lean body mass loss during weight loss in athletes. *Med Sci Sports Exerc.* 2010;42(2):326–37.
98. Manore MM, Barr SI, Butterfield GE, American College of Sports Medicine, American Dietetic Association, Dietitians of Canada. Nutrition and athletic performance. *Med Sci Sports Exerc.* 2000;32(12):2130–45.
99. Rennie MJ, Tipton KD. Protein and amino acid metabolism during and after exercise and the effects of nutrition. *Ann Rev Nutr.* 2000;20:457–83.
100. Itoh R, Nishiyama N, Suyama Y. Dietary protein intake and urinary excretion of calcium: a cross-sectional study in a healthy Japanese population. *Am J Clin Nutr.* 1998;67(3):438–44.
101. Kerstetter JE, O'Brien KO, Insogna KL. Dietary protein, calcium metabolism, and skeletal homeostasis revisited. *Am J Clin Nutr.* 2003;78(3):584S–92S.
102. Dawson-Hughes B, Harris SS, Rasmussen H, Song L, Dallal GE. Effect of dietary protein supplements on calcium excretion in healthy older men and women. *J Clin Endocrinol Metab.* 2004;89(3):1169–73.
103. Applegate E. Effective nutritional ergogenic aids. *Int J Sport Nutr.* 1999;9:229–39.
104. Lemon PWR. Do athletes need more dietary protein and amino acids? *Int J Sport Nutr.* 1995;5:S39–61.
105. Burke LM, Hawley JA, Angus DJ, et al. Adaptations to short-term high-fat diet persist during exercise despite high carbohydrate availability. *Med Sci Sports Exerc.* 2002;34(1):83–91.
106. Rowlands DS, Hopkins WG. Effects of high-fat and high-carbohydrate diets on metabolism and performance in cycling. *Metabolism.* 2002;51(6):678–90.
107. Volek JS, Noakes T, Phinney SD. Rethinking fat as a fuel for endurance exercise. *Eur J Sport Sci.* 2014:1–8.
108. Sherman WM. Carbohydrate feedings before and after exercise. In: Lamb DL, Williams MR, editors. *Perspectives in Exercise Science and Sports Medicine. Volume 4: Ergogenics—Enhancement of Performance in Exercise and Sport.* New York (NY): McGraw-Hill Companies; 1991.
109. Goedecke JH, Christie C, Wilson G, et al. Metabolic adaptations to a high-fat diet in endurance cyclists. *Metabolism.* 1999;48(12):1509–17.
110. Lambert EV, Goedecke JH, van Zyl C, et al. High-fat diet versus habitual diet prior to carbohydrate loading: effects on exercise metabolism and cycling performance. *Int J Sport Nutr Exerc Metab.* 2001;11:209–25.

111. Close DC, Hagerman AE. Chapter 1: Chemistry of reactive oxygen species and antioxidants. In: Alessio HM, Hagerman AE, editors. *Oxidative Stress, Exercise and Aging*. 1st ed. London: Imperial College Press; 2006. p. 1–8.
112. Quindry J, Powers SK. Aging, exercise, antioxidants, and cardioprotection. In: Alessio HM, Hagerman AE, editors. *Oxidative Stress, Exercise and Aging*. 1st ed. London: Imperial College Press; 2006. p. 125–44.
113. Kohrt WM, Bloomfield SA, Little KD, Nelson ME, Yingling VR. Physical activity and bone health. *Med Sci Sports Exerc*. 2004;36(11):1985–96.
114. American College of Sports Medicine. Osteoporosis and exercise. *Med Sci Sports Exerc*. 1995;27(4):i–vii.
115. Nattiv A, Loucks AB, Manore MM, Sanborn CF, Sundgot-Borgen J, Warren MP. The female athlete triad. *Med Sci Sports Exerc*. 2007;29(5):1867–82.
116. Balaban EP, Cox JV, Snell P, Vaughan RH, Frenkel EP. The frequency of anemia and iron deficiency in the runner. *Med Sci Sports Exerc*. 1989;21(6):643–8.
117. Eichner ER. Fatigue of anemia. *Nutr Rev*. 2001;59:S17–9.
118. Hinton PA, Giordano C, Brownlie T, Haas JD. Iron supplementation improves endurance after training in iron-depleted, nonanemic women. *J Appl Physiol*. 2000;88:1103–11.
119. Nielsen P, Nachtigall D. Iron supplementation in athletes. *Sports Med*. 1998;25:207–16.
120. Casa DJ, Armstrong LE, Hillman S, et al. National Athletic Trainer's Association position statement: fluid replacement for athletes. *J Athl Train*. 2000;35(2):212–24.
121. Sawka MN, Burke LM, Eichner ER, Maughan RJ, Montain SJ, Stachenfeld NS. Exercise and fluid replacement. *Med Sci Sports Exerc*. 2007;39(2):377–90.
122. Barr SI, Costill DL, Fink WJ. Fluid replacement during prolonged exercise: effects of water, saline, or no fluid. *Med Sci Sports Exerc*. 1991;23(7):811–7.
123. Christensen EH, Hansen O. Respiratorischen quotient und O2-aufnahme. *Skand Arch Physiol*. 1939;81:180–9.
124. Bergstrom J, Hultman E. Nutrition for maximizing sports performance. *JAMA*. 1972;221:999–1006.
125. Williams C, Rollo I. Carbohydrate nutrition and team sports performance. *Gatorade Sports Sci Exch*. 2015;28(140):1–7.
126. Akermark C, Jacobs I, Rasmusson M, Karlsson J. Diet and muscle glycogen concentration in relation to physical performance in Swedish elite ice hockey players. *Int J Sport Nutr*. 1996;6(3):272–84.
127. Balsom P, Wood K, Olsson P, Ekblom B. Carbohydrate intake and multiple sprint sports: with special reference to Football (Soccer). *Int J Sports Med*. 2007;20(1):48–52.
128. Currell K, Conway S, Jeukendrup AE. Carbohydrate ingestion improves performance of a new reliable test of soccer performance. *Int J Sport Nutr Exerc Metab*. 2009;19(1):34–46.
129. Russell M, Kingsley M. The efficacy of acute nutritional interventions on soccer skill performance. *Sports Med*. 2014;44(7):957–70.
130. Phillips SM, Phillips SM, Turner AP, et al. Carbohydrate gel ingestion significantly improves the intermittent endurance capacity, but not sprint performance, of adolescent team games players during a simulated team games protocol. *Eur J Appl Physiol*. 2012;112(3):1133–41.
131. Winnick JJ, Davis JM, Welsh RS, Carmichael MD, Murphy EA, Blackmon JA. Carbohydrate feedings during team sport exercise preserve physical and CNS function. *Med Sci Sports Exerc*. 2005;37(2):306–15.
132. Montain SJ, Coyle EF. Influence of graded dehydration on hyperthermia and cardiovascular drift during exercise. *J Appl Physiol*. 1992;73(4):1340–50.
133. Montain SJ, Latzka WA, Sawka MN. Control of thermoregulatory sweating is altered by hydration level and exercise intensity. *J Appl Physiol*. 1995;79(5):1434–9.
134. Carter R, Cheuvront SN, Vernieuw CR, Sawka MN. Hypohydration and prior heat stress exacerbates decreases in cerebral blood flow velocity during standing. *J Appl Physiol*. 2006;101(6):1744–50.
135. Dickson JM, Weavers HM, Mitchell N, et al. The effects of dehydration on brain volume—preliminary results. *Int J Sports Med*. 2005;26(6):481–5.
136. Kenefick RW. Drinking strategies: planned drinking versus drinking to thirst. *Sports Med*. 2018;48(S1):31–7.
137. Greenleaf JE, Sargent F. Voluntary dehydration in man. *J Appl Physiol*. 1965;20(4):719–24.
138. Greenleaf JE, Castle BL. Exercise temperature regulation in man during hypohydration and hyperhydration. *J Appl Physiol*. 1971;30:847–53.
139. Dion T, Savoie FA, Asselin A, Gariepy C, Goulet EDB. Half-marathon running performance is not improved by a rate of fluid intake above that dictated by thirst sensation in trained distance runners. *Eur J Appl Physiol*. 2013;113(12):3011–20.
140. Greenleaf JE. Problem–thirst, drinking behavior, and involuntary dehydration. *Med Sci Sports Exerc*. 1992;24(6):645–56.

141. Cheuvront SN, Kenefick RW. Dehydration: physiology, assessment, and performance effects. *Compr Physiol.* 2014;4(1):257–85.
142. Williams MH. *Beyond Training: How Athletes Enhance Performance Legally and Illegally.* 1st ed. Champaign (IL): Leisure Press; 1989.
143. Butterfield G. Ergogenic aids: evaluating sport nutrition products. *Int J Sport Nutr.* 1996;6:191–7.
144. Rosenbloom C. Energy drinks, caffeine, and athletics. *Nutr Today.* 2014;49(2):49–54.
145. Graham TE, Spriet LL. Metabolic, catecholamine, and exercise performance responses to various doses of caffeine. *J Appl Physiol.* 1995;78(3):867–74.
146. Pasman WJ, van Baak MA, Jeukendrup AE, de Haan A. The effects of different dosages of caffeine on endurance performance time. *Int J Sports Med.* 1995;16:225–30.
147. Spriet LL. Caffeine and exercise performance: an update. *Gatorade Sports Sci Exch.* 2020;29(203):1–5.
148. Lane SC, Hawley JA, Desbrow B, et al. Single and combined effects of beetroot juice and caffeine supplementation on cycling time trial performance. *Appl Physiol Nutr Metab.* 2014;39(9):1050–7.
149. Pitchford NW, Fell JW, Leveritt MD, Desbrow B, Shing CM. Effect of caffeine on cycling time-trial performance in the heat. *J Sci Med Sport.* 2013;17(4):445–9.
150. Skinner TL, Desbrow B, Arapova J, et al. Women experience the same ergogenic response to caffeine as men. *Med Sci Sports Exerc.* 2019;51(6):1195–202.
151. Grgic J, Grgic I, Pickering C, Schoenfeld BJ, Bishop DJ, Pedisic Z. Wake up and smell the coffee: caffeine supplementation and exercise performance—an umbrella review of 21 published meta-analyses. *Br J Sports Med.* 2020;54(11):681–8. doi:10.1136/bjsports-2018-100278
152. Bellar DM, Kamimori G, Judge L, et al. Effects of low-dose caffeine supplementation on early morning performance in the standing shot put throw. *Eur J Sport Sci.* 2012;12(1):57–61.
153. Chia JS, Barrett LA, Chow JY, Burns SF. Effects of caffeine supplementation on performance in ball games. *Sports Med.* 2017;47(12):2453–71.
154. Lara B, Ruiz-Vicente D, Areces F, et al. Acute consumption of a caffeinated energy drink enhances aspects of performance in sprint swimmers. *Br J Nutr.* 2015;114(6):908–14.
155. Gallo-Salazar C, Areces F, Abián-Vicén J, et al. Enhancing physical performance in elite junior tennis players with a caffeinated energy drink. *Int J Sports Physiol Perform.* 2015;10(3):305–10.
156. Harris RC, Soderlund K, Hultman E. Elevation of creatine in resting and exercised muscle of normal subjects by creatine supplementation. *Clin Sci.* 1992;83:367–74.
157. Hultman E, Soderlund K, Timmons J, Cederblad G, Greenhaff PL. Muscle creatine loading in men. *J Appl Physiol.* 1996;81(1):232–7.
158. Branch JD. Effect of creatine supplementation on body composition and performance: a meta-analysis. *Int J Sport Nutr Exerc Metab.* 2003;13(2):198–226.
159. Cox G, Mujika I, Tumilty D, Burke L. Acute creatine supplementation and performance during a field test simulating match play in elite female soccer players. *Int J Sport Nutr Exerc Metab.* 2002;12(1):33–46.
160. Rawson ES. The safety and efficacy of creatine monohydrate supplementation what we have learned from the past 25 years of research. *Sports Sci Exch.* 2018;29(186):1–6.
161. Rawson ES, Volek JS. Effects of creatine supplementation and resistance training on muscle strength and weightlifting performance. *J Strength Cond Res.* 2003;17(4):822–31.
162. Lanhers C, Lanhers C, Pereira B, et al. Creatine supplementation and upper limb strength performance: a systematic review and meta-analysis. *Sports Med.* 2017;47(1):163–73.
163. Lanhers C, Pereira B, Naughton G, Trousselard M, Lesage F-X, Dutheil F. Creatine supplementation and lower limb strength performance: a systematic review and meta-analyses. *Sports Med.* 2015;45(9):1285–94.
164. Bowtell JL, Wangdi JT, Kelly VG. Fruit-derived polyphenol supplementation for performance and recovery. *Gatorade Sports Sci Exch.* 2019;29(195):1–5.
165. Roelofs EJ, Smith-Ryan AE, Trexler ET, Hirsch Katie R, Mock MG. Effects of pomegranate extract on blood flow and vessel diameter after high-intensity exercise in young, healthy adults. *Eur J Sport Sci.* 2016;17(3):317–25.
166. Cases J, Romain C, Marin-Pagan C, et al. Supplementation with a polyphenol-rich extract, perfload (R), improves physical performance during high-intensity exercise: a randomized, double blind, crossover trial. *Nutrients.* 2017;9(4):421.
167. Deley G, Guillemet D, François-Andre A, Babault N. An acute dose of specific grape and apple polyphenols improves endurance performance: a randomized, crossover, double-blind versus placebo controlled study. *Nutrients.* 2017;9(8):917.

168. Decroix L, Tonoli C, Soares DD, et al. Acute cocoa Flavanols intake has minimal effects on exercise-induced oxidative stress and nitric oxide production in healthy cyclists: a randomized controlled trial. *J Int Soc Sports Nutr*. 2017;14(1):28–11.
169. Crum EM, Muhamed AMC, Barnes M, Stannard SR. The effect of acute pomegranate extract supplementation on oxygen uptake in highly-trained cyclists during high-intensity exercise in a high altitude environment. *J Int Soc Sports Nutr*. 2017;14(1):14–9.
170. Trexler ET, Smith-Ryan AE, Melvin MN, Roelofs EJ, Wingfield HL. Effects of pomegranate extract on blood flow and running time to exhaustion. *Appl Physiol Nutr Metab*. 2014;39(9):1038–42.
171. Cook MD, Myers SD, Blacker SD, Willems MET. New Zealand blackcurrant extract improves cycling performance and fat oxidation in cyclists. *Eur J Appl Physiol*. 2015;115(11):2357–65.
172. Morgan PT, Barton MJ, Bowtell JL. Montmorency cherry supplementation improves 15-km cycling time-trial performance. *Eur J Appl Physiol*. 2019;119(3):675–84.
173. Perkins IC, Vine SA, Blacker SD, Willems MET. New Zealand blackcurrant extract improves high-intensity intermittent running. *Int J Sport Nutr Exerc Metab*. 2015;25(5):487–93.

CHAPTER 8

Exercise and Sport Psychology

After completing this chapter, you will be able to:

1. Define exercise and sport psychology and provide examples of how each area contributes to the understanding of physical activity, exercise, sport, and athletic performance.

2. Identify the important historic events in the development of exercise and sport psychology.

3. Discuss the different areas of study that are related to exercise and sport psychology.

4. Discuss the different psychological factors that influence participation in regular physical activity and exercise.

Exercise and sport psychology are areas of study concerned with the behavior, thoughts, and feelings of healthy, disabled, and diseased individuals engaging in physical activity, exercise, sport, and athletic competition. Many of the theories and methodologies from the parent discipline of psychology are used in the basic and applied studies of exercise and sport psychology. The American Psychological Association (APA) defines exercise and sport psychology as "the scientific study of the psychological factors that are associated with participation and performance in sport, exercise, and other types of physical activity" (1). A commonly used conceptual framework for exercise and sport psychology is shown in Figure 8.1 (2,3). The relationship between exercise and sport psychology and the parent discipline of psychology can be delineated using the following assumptions:

1. Content knowledge in exercise and sport psychology is fundamentally linked to the discipline of psychology.

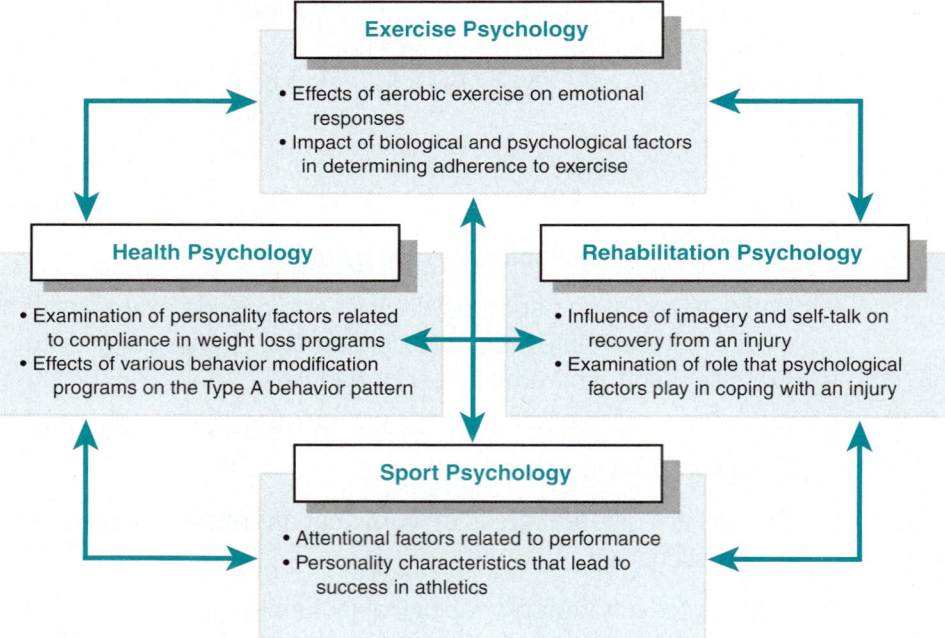

FIGURE 8.1 Conceptual framework of exercise and sport psychology (2,3). (From Brown SP. *Introduction to Exercise Science*. Baltimore (MD): Lippincott Williams & Wilkins; 2000. 312 p. Originally modified from Rejeski WJ, Brawley LR. Defining the boundaries of sport psychology. *Sport Psychol*. 1988;2:231–42.)

Exercise psychology Concerned with the cognitions, emotions, and behaviors that are related to perceptual and objective changes in cardiovascular fitness, muscular strength and endurance, flexibility, and body composition.

Sport psychology Concerned with the application of psychological principles to performance in the areas of sport and athletic competition.

2. Examination and study of relevant issues in exercise and sport psychology involves the use of a wide range of models and techniques from various aspects of psychology.
3. Exercise and sport psychology has several distinct areas of study, but those areas also have important relationships with each other (2,4).

Exercise and sport psychology use the educational, scientific, and professional contributions of psychology to help understand the mental aspects of physical activity, exercise, sport, and athletic competition. One area of exercise and sport psychology is the study of the **cognitions**, emotions, and behaviors that are related to the **perceptual** and **objective** changes in cardiovascular fitness, muscular strength and endurance, flexibility, and body composition that result from the participation in physical activity and exercise (2). Another area is specifically concerned with the application of psychological principles to performance in the areas of sport and athletic competition (2,3). A more recently studied area focuses on the psychological aspects of physical activity and health and the role of behavioral strategies in physical activity and health promotion (5).

The principles of exercise and sport psychology can be used on both the individual and the group levels in a variety of professional environments. Exercise science professionals working to generate new knowledge in the areas of exercise and sport psychology may be concerned with developing methods for enhancing exercise adherence, self-esteem, leadership skills, and group or team cohesion. Exercise science and allied health care professionals working with patients in a clinical setting may use various principles from exercise and sport psychology to enhance exercise program adherence or reduce anxiety and depression associated with participation in a cardiopulmonary rehabilitation program. Personal trainers and coaches working with healthy individuals or athletes in an applied environment may employ the principles of exercise and sport psychology to promote improvement in psychological well-being or performance during training and competition (4,6). Collectively, these examples of professional practice could be referred to as performance psychology (7). Hays (8) has described performance psychology as helping individuals learn how to better accomplish and more consistently complete endeavors where excellence counts. Performance psychologists are distinctively trained and specialized to participate in a broad range of activities to help individuals function in the upper range of their abilities (7).

Psychological phenomena are examined using a variety of methods not often employed in other areas of exercise science. For example, some psychological methods give considerable emphasis to the individual's subjective experience. Self-report measures, using standardized questionnaires or psychological inventories, are commonly used in exercise and sport psychology to measure thoughts, feelings, or behaviors (9). Individuals are frequently asked to complete questionnaires prior to and following participation in an exercise session, athletic practice, or competition, or as part of an acute or chronic intervention strategy (4).

Cognition Mental faculty of knowing, which includes perceiving, recognizing, conceiving, judging, reasoning, and imagining.
Perceptual Recognition and interpretation of sensory stimuli based chiefly on memory.
Objective Interpretation of facts based on observable phenomena.

Observing individuals and recording what they do during exercise, sport, or athletic activities has also been used to enhance the understanding of behavior and psychological processes. Interviews and other qualitative methods can be used in exercise and sport psychology when a greater understanding of beliefs, experiences, or values of the individual is required. Observations and interviews, like other scientific methods of collecting information, must be structured to be consistent and systematic, and individuals must be trained in the proper use of interview techniques to ensure accuracy and effectiveness (10). Exercise and sport psychology has developed into a well-defined discipline and commonly used professional field of study, and the basis for that development has a rich history.

HISTORY OF EXERCISE AND SPORT PSYCHOLOGY

Early writings by the ancient Greeks extolled the virtues of a strong relationship between the mind and the body (11,12). It was not until the 20th century, however, that a fundamental knowledge base in exercise and sport psychology was established. Most of the significant developments of exercise and sport psychology have occurred since the mid-1960s leading to exercise and sport psychology emerging as an area of study in exercise science and as an important component for improving participation and performance in physical activity, exercise, sport, and athletic competition (11).

Early Influences

The foundation for the development of exercise and sport psychology as a field of study and practice originated during the late nineteenth and early 20th centuries. The discipline of psychology arose from the discipline of philosophy during the mid-19th century and ultimately became the parent discipline to exercise and sport psychology (12). Early work establishing relationships between the mind and the body was largely a result of individuals trained in psychology examining factors related to physical activity and exercise. For example, in 1884, Conrad Rieger published what is considered to be the first article related to psychology and exercise, in which he reported that the mental state of **hypnotic catalepsy** enhanced muscular endurance (13). Shortly thereafter, Norman Triplett published the first true experimental study that was directly linked to exercise and sport psychology (13). Triplett, a bicycling enthusiast, was interested in how direct competition between two individuals affected performance. Triplett observed that when two individuals competed against each other, a social factor that seemed to motivate cyclists to perform better existed that was not observed when individuals performed alone (11).

Hypnotic catalepsy A state of physical rigidity induced by hypnosis.

Twentieth-Century Influences

In the early 20th century, several scholars studied various relationships between physical activity and sport and the responses of the brain and the nervous system. In 1908, the president of the APA, G. Stanley Hall, highlighted the advancements in these areas in a report that described the psychological benefits from participating in physical education (14).

One of the early true pioneers of exercise and sport psychology was Coleman Griffith (Figure 8.2), who in 1925 became the director of the Research in Athletics Laboratory at the University of Illinois (12). Griffith, who is often recognized as the "Father of North American Sport Psychology" studied various psychological factors related to participation in American football and basketball. Griffith published two classic textbooks about his work with athletes and coaches; *Psychology of Coaching,* in 1926, and *Psychology and Athletics,* in 1928. Likely the first practicing sport psychologist, Griffith was hired in 1938 by Philip Wrigley, then owner of the Chicago Cubs, to help improve the performance of the professional baseball team (15). It should also be noted that Griffith maintained an active research lab and conducted numerous studies related to psychology of exercise and motor performance during his entire career.

The expansion of the foundation of exercise and sport psychology occurred throughout the mid-20th century. In 1938, Franklin Henry established a graduate program in the psychology of physical activity at the University of California-Berkley. Henry, a scholar in motor behavior, advocated for the scientific development of sport and exercise psychology (6). During the early years of World War II, Dorothy Hazeltine Yates engaged in mental training interventions with boxers and aviators, focusing primarily on a "relaxation set-method" and mental preparation for performance (12). Yates published two books and developed a psychology course for athletes and aviators at San Jose State University (12). Throughout the 1940s and 1950s, research in the areas of physical activity, exercise, sport, and athletic competition examined personality traits in athletes, as well as emotions and stress related to youth sport competition, college athletic performance and competition, motor performance, and exercise (12).

The period from 1965 to 1980 saw tremendous advances in exercise and sport psychology, including the development of a research knowledge base that was distinct and separate

FIGURE 8.2 Coleman R. Griffith (ca. 1920). (Photo used with permission from University Archives, University of Illinois, Urbana.)

from the closely related disciplines of motor development and motor learning (12). Contributing to this growth were several important historic events. The International Society of Sport Psychology (founded in 1965) and the meeting of the First World Congress of Sport Psychology in Rome, Italy, in the same year, were instrumental in bringing together scholars interested in the use of psychological techniques for improving sport and athletic performance. Professional organizations also emerged in America to help promote exercise and sport psychology. The first meeting of the North American Society for the Psychology of Sport and Physical Activity (NASPSPA) occurred during the 1967 American Alliance for Health, Physical Education, and Recreation (AAHPER) National Conference in Las Vegas, Nevada. Early on, NASPSPA continued to meet during the annual AAHPER conferences until 1973, when it began to hold independent meetings (12).

Many national and international scholars played key roles in the development of exercise and sport psychology from the 1960s through the 1980s. Noteworthy first generation American scholars included Rainer Martens, Dan Landers, William Morgan, Bruce Ogilvie, and Dorothy Harris. Prominent international scholars included Paul Kunath, Peter Roudik, Miroslav Vanek, Morgan Olsen, and John Kane. Of particular importance is Ferruccio Antonelli, who was instrumental in the establishment of the International Society of Sport Psychologists and spearheaded the first sport psychology research journal, the *International Journal of Sport Psychology*. Additional prominent journals helped to disseminate exercise and sport psychology research and facilitate its development as a field of study, particularly the *Journal of Sport Psychology* (1979), which was renamed the *Journal of Sport and Exercise Psychology* in 1988; *The Sport Psychologist* (1987); and *Journal of Applied Sport Psychology* (1989) (12).

During the mid-1980s, there was a shift in the research and application interests of many exercise and sport psychologists from laboratory activities to more field-based experimentation and practice (11). This shift resulted in a significant increase in the number of individuals working to enhance psychological skills training with competitive athletes. As a result of this movement, the Association for Applied Sport Psychology was established in 1985. Another key event in the development of exercise and sport psychology occurred in 1986 when the APA recognized Division 47 — Exercise and Sport Psychology as a formal division within the APA. At about the same time, there was an increase in the publication of research on exercise and psychology (12). Systematic research investigations provided evidence that exercise decreased stress, anxiety, and depression (16), improved mood and positive emotion (17), and enhanced self-efficacy, self-concept, and self-esteem (18). Additional research in the area of exercise adherence (19) and interventions to change

Self-efficacy An impression that an individual is capable of performing in a certain manner or attaining certain goals.

Self-concept The sum total of an individual's knowledge and understanding of his or her self.

Self-esteem The way individuals think and feel about themselves and how well they do things that are important to them.

physical activity behavior (20) was also being performed. A key step in the professionalization of exercise and sport psychology occurred in 1991, when the Association of Applied Sport Psychology began to confer the title "Certified Consultant" on individuals who met specific training criteria (11). This certification designation limits the certified consultant's role to an educational one, emphasizing psychological skills training (11). Additional information on the Association of Applied Sport Psychology can be found at www.appliedsportpsych.org/

Twenty-First-Century Influences

At the end of the 20th century and into the 21 century, exercise and sport psychology has continued to further define the knowledge base, expand into new areas of study, and clarify professional practice-related issues (12). Early in the century, specific disciplinary study and scholarly activity was conducted to better understand interpretive approaches to research and knowledge construction (21), feminist methodology (22), and **ecological models** and **metatheoretical approaches** (23). In 2012, the journal *Sport, Exercise, and Performance Psychology* was first published as the official journal of the Exercise and Sport Psychology Division of the APA. Discussion about the appropriate competencies required for training in the area of performance psychology continues among the leaders of the profession (24,25). The subareas of exercise and sport psychology continue to expand and advance our understanding of the mental and psychological factors that affect performance and behavior in physical activity, exercise, sport, and athletic competition.

In 2015, the Exercise and Sport Psychology Division of the APA changed its name to the Society for Sport, Exercise, and Performance Psychology. This change better reflects the holistic approach employed in the discipline and the need for providing knowledge and services to the exercising, athletic, and general populations (1). Professionals in the field of exercise psychology have expanded their practice to include promoting exercise and physical activity, exploring the relationship of physical activity and exercise to quality of life among clinical populations and older adults, and the role of reducing sedentary behavior in academic and cognitive performance (1). Table 8.1 provides some of the significant events in the historical development of exercise and sport psychology.

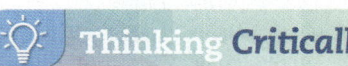
Thinking Critically

In what ways has exercise and sport psychology contributed to a broader understanding of physical fitness and health?

In what ways has exercise and sport psychology contributed to a broader understanding of sport and athletic performance?

Ecological models Belief in an understanding that the individual is part of a larger group and that this influences the actions and behaviors of the individual.

Metatheoretical approach The discussion of the fundamentals, structure, or relationship of a specific theory of knowledge.

Table 8.1	Significant Events in the Development of Exercise and Sport Psychology
DATE	SIGNIFICANT EVENT
1884	Conrad Rieger published the first article related to psychology and exercise
1898	Norman Triplett published the first true experimental study related to exercise and sport psychology
1908	APA President G. Stanley Hall issued a report highlighting the psychological benefits from participating in physical education
1925	Coleman R. Griffith established the Research in Athletics Laboratory at the University of Illinois and studied exercise and sport psychology
1938	Franklin Henry established a graduate program in the psychology of physical activity at the University of California-Berkley
1965	Formation of the International Society of Sport Psychology and the First World Congress of Sport Psychology in Rome, Italy
1967	NASPSPA was founded
1970	First publication of the *International Journal of Sport Psychology*
1979	First publication of the *Journal of Sport Psychology*
1985	Formation of the Association for the Advancement of Applied Sport Psychology
1986	APA formally recognized Division 47 — Exercise and Sport Psychology
1988	*Journal of Sport Psychology* began publication as the *Journal of Sport and Exercise Psychology*
1989	First publication of the *Journal of Applied Sport Psychology*
1991	Association for the Advancement of Applied Sport Psychology began to confer the title Certified Consultant, AAASP
2006	AAASP was renamed the Association for Applied Sport Psychology
2012	Publication of the journal *Sport, Exercise, and Performance Psychology* by the APA
2015	Division 47 of the APA changed its name to the Society for Sport, Exercise, and Performance Psychology

STUDY OF THE MIND AND BODY

Participation in physical activity, exercise, sport, and athletic competition involves coordinated efforts of the body and the mind. The study of the mental aspects is the foundation of exercise and sport psychology. Exercise and sport psychology includes the study and application of the affect (*i.e.*, emotion), behavior, and cognition (*i.e.*, thought) of individuals as they prepare to engage in planned movement. Exercise and sport psychology continues to be closely connected to the parent discipline of psychology specifically as it relates to cognitive and **behavioral psychology**. Exercise and sport psychology has three primary objectives linked to physical activity, exercise, sport, and athletic competition:

1. Understanding the social-psychological factors that influence individual behavior
2. Understanding the psychological effects of participation
3. Enhancing the experiences of individuals prior to, during, and following participation (10)

To accomplish these objectives, exercise and sport psychology researchers and practicing professionals focus primarily on the following areas: personality, motivation, emotion and performance, attention, and cognition. A solid understanding of these areas will assist many exercise science professionals in their attempts to fulfill their professional job responsibilities.

Personality

Personality is described as the entire set of qualities and traits, including character and dispositions that are specific to someone. Individual personality plays an important role in the behaviors that individuals exhibit during participation in physical activity, exercise, sport, and athletic competition. Personality research in exercise and sport psychology has been a popular area of study (12) and underlies much of what exercise and sport psychology professionals research and practice (26). Personality is conceptualized in different ways; an example of a model is provided in Figure 8.3 (27).

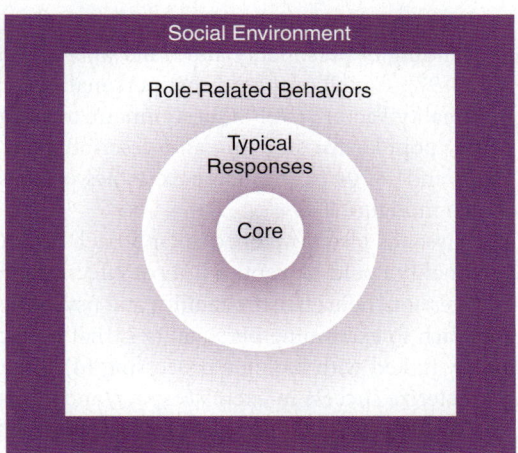

FIGURE 8.3 Conceptualization of personality. (From Brown SP. *Introduction to Exercise Science*. Baltimore (MD): Lippincott Williams & Wilkins; 2000. 314 p.)

Behavioral psychology A branch of psychology based on the proposition that all things that organisms do can and should be regarded as behaviors.

Personality The complete qualities and traits, including character and behavior, that are specific to a person.

At the center of an individual's personality is the psychological core, thought to be the most stable and least modifiable aspect of personality. The psychological core is developed from early interactions with the environment (parents and objects) and includes such aspects of personality as perceptions of the external world, perceptions of self and basic attitudes, values, interests, and motives (4).

Our core personality is the basis for our thoughts, feelings, and behaviors. Arising from the core are actions and typical responses that are consistent with the core and usually fairly consistent over time. Our behaviors can change as a result of influences from the social environment, such as being a member of an amateur or recreational sport team, and these are often referred to as role-related behaviors. These role-related behaviors represent the most dynamic aspect of an individual's personality because behaviors can vary based on the situation or surroundings the individual is involved in at a particular moment in time. These role-related behaviors remain consistent with an individual's psychological core. In general, personality is relatively stable over time but can be changed gradually over time and modified according to the environment and situations. Common approaches to studying personality include the dispositional approach, which is focused on the person, and the learning approach, which is focused on the environment (4).

Dispositional Approach

One approach to studying personality is the trait approach. **Traits** are enduring and consistent internal attributes that an individual possesses and exhibits. Considerable information has been generated on individual traits, and individuals can be described as having traits such as temperamental, nervous, sensitive, restless, confident, dynamic, gregarious, lighthearted, composed, and poised. Two individuals who helped develop our current understanding of personality and traits were Raymond Cattell (1905–98) and Hans Eysenck (1916–97). Cattell proposed that personality consisted of 16 factors and developed the 16 Personality Factor (16PF) Questionnaire to assess them (28). Use of this questionnaire was widely popular for studying and describing personality in the sports domain during the 1960s and 1970s (4,26). Some examples of questions used in the 16PF questionnaire are shown in Figure 8.4.

Eysenck also developed an approach to examining personality (29). He believed that personality could be captured most effectively with only three dimensions: extroversion-introversion, neuroticism-stability, and psychoticism-superego. An important aspect in this approach to examining personality is that each dimension has a biologic basis and is intimately linked with biologic processing (4). Eysenck's model has been used in attempts to characterize success in exercise, sport, and athletic performance. For example, individuals who are extroverts tend to seek out sensory stimulation and are well able to tolerate pain. Consequently, it was hypothesized that extroverts would be more likely to participate in sport and athletic competition and would be more successful in sports than individuals who are introverts (30). The current research indicates, however, that no distinguishable

Traits Relatively enduring, highly consistent internal attributes that an individual possesses and exhibits.

1. On social occasions I:
   - ☐ Readily come forward
   - ☐ In between
   - ☐ Prefer to stay quietly in the background
2. I sometimes cannot get to sleep because an idea keeps running through my head.
   - ☐ True
   - ☐ Uncertain
   - ☐ False
3. In my personal life, I reach the goals I set almost all of the time.
   - ☐ True
   - ☐ Uncertain
   - ☐ False
4. I would prefer to have an office of my own, not sharing it with another person.
   - ☐ Yes
   - ☐ Uncertain
   - ☐ No
5. When I am in a small group, I am content to sit back and let others do most of the talking.
   - ☐ True
   - ☐ Uncertain
   - ☐ False

FIGURE 8.4 Examples of the 16PF Questionnaire (28).

personality exists for athletes. Furthermore, there appear to be no consistent personality differences between athletic subgroups (*e.g.*, team athletes vs. individual sport athletes, contact sport vs. noncontact sport). Several differences have been identified in personality characteristics, particularly role-related characteristics and traits, between successful and unsuccessful athletes (4). Successful athletes, through training and practice, can develop and become:

- More self-confident
- Better able to retain optimal competition focus in response to obstacles and distractions
- Efficiently self-regulate activation
- Have more positive thoughts, images, and feelings about sport
- More highly determined and committed to excellence in their sport (10)

Exercise and sport psychology is also concerned with whether participation in sports and athletic competition can influence personality development and change. For example, are independent, extroverted individuals developed through sport and athletic participation, or are independent, extroverted people attracted toward sport and athletics? Research suggests that independent, extroverted people are more interested in participating in sport and athletics. Furthermore, it appears that engaging in structured sports programs can also lead to positive changes in personality and behavior (4,6).

With respect to physical activity and exercise, no set of personality characteristics have been identified that predict adherence to a physical activity or exercise program. Personality type cannot be used to predict regular exercisers from sedentary individuals; however, two personality characteristics are strong predictors of exercise behavior. Individuals more confident in their physical abilities are likely to exercise more than those who are less physically confident. Additionally, individuals who express self-motivation are more likely to begin and continue with regular exercise, and less motivated individuals are more likely to never start regular exercise or discontinue a program that has been started (10).

Participation in regular physical activity or exercise programs is associated with positive changes in mental health. Negative factors (*e.g.*, neuroticism) are reduced and positive factors (*e.g.*, extroversion) are enhanced following participation in regular exercise programs (4). Participation in long-term physical activity and exercise programs can result in reductions in both anxiety and neuroticism (31). Children, adolescents, and adults (32) show improvement in self-esteem with improvement in physical fitness parameters (33). Regular physical activity is also associated with decreases in depression and a reduction of depressive symptoms in individuals who are clinically depressed at the beginning of the exercise treatment (4,10).

Thinking Critically

In what ways might exercise enhance the mental state of an individual?

Motivation

Motivation is an important component for participation in physical activity, exercise, sport, and athletic competition. Motivation is a complex set of internal and external forces that influence individuals to behave in certain ways (10). Extrinsic motivation is the predominant influence that occurs when individuals engage in a certain behavior to gain some external reward from that participation. Intrinsic motivation is the predominant factor when an individual engages in behavior because the individual enjoys the process and gains pleasure and satisfaction from that participation.

Numerous social-psychological theories of motivation have been developed. Many of these theories use a cognitive approach to motivation, such that behavior is assumed to be determined by cognitive mechanisms. Self-confidence or an individual's perception of their own ability to perform a skill or activity is one of the most important cognitive concepts. Common social-psychological theories in exercise and sport psychology are Bandura's social cognitive theory (34), Weiner's attribution theory (35), and self-determination theory by Deci and Ryan (36).

Social Cognitive Theory

Social cognitive theory suggests that the individual, environment, and behavior are all related to each other and that components of an individual's knowledge attainment can

Motivation Psychological feature that arouses an organism to action toward a desired goal.
Extrinsic motivation The reasons for motivation come from factors outside an individual.
Intrinsic motivation The reasons for motivation come from factors within an individual.
Social cognitive theory Belief that portions of an individual's knowledge acquisition can be directly related to observing others within the context of social interactions, experiences, and outside media influences.

be directly related to observing others within the context of social interactions, experiences, and outside media influences (37). Social cognitive theory developed from the earlier self-efficacy theory, and self-efficacy is a major component of the social cognitive theory (37). Self-efficacy is the cognitive mechanism mediating motivation and thus an individual's behavior. The convictions or beliefs an individual has that they are capable of performing in a certain manner or attaining certain goals provide consistent support for that individual to be successful. Self-efficacy is not a measure of the skills that individuals actually possess but instead a measure of an individual's own judgment of what they can do with those skills and the self-confidence specific to a particular situation. Self-efficacy has been shown to be an important factor in an individual's choice of activity, effort exerted in those activities, and persistence in the activity when faced with challenges (4). A model of self-efficacy is presented in Figure 8.5 (4).

Self-efficacy is derived from four factors: past performance, observing others, social persuasion, and physiologic arousal. The most prominent of these factors is past performance in an activity. Past performance is the most dependable and influential factor affecting self-efficacy. Past success leads to an increased feeling of self-efficacy, whereas past failures, especially repeated failures, lead to decreased feelings of self-efficacy. Observing others engaging in the same physical activity, exercise, sport, or athletic competition can also affect efficacy judgments. This is particularly important if an individual has limited experience with the movement skill or activity. Social persuasion, although a relatively weak source, is the influence of a social situation on self-efficacy. The final influence on efficacy is physiological arousal, which is the appraisal by individuals of their own physiological states. An individual may have an increased heart rate, nervousness, anxiety, muscle fatigue, or pain prior to physical performance, and each of these may lead to a decrease in self-efficacy (4).

Self-efficacy can determine individual behavior as it relates to choice, effort, persistence, thoughts, and emotional reactions. A key aspect of social cognitive theory is the

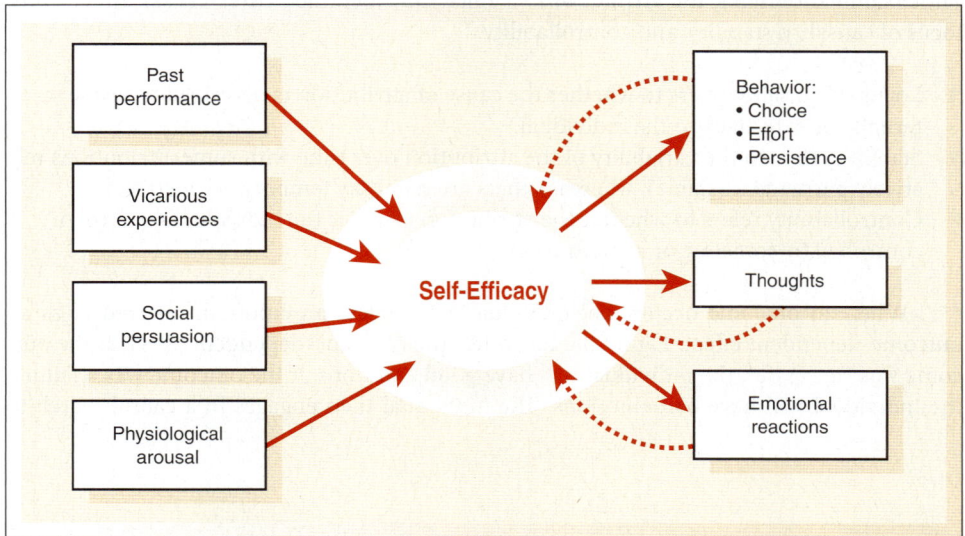

FIGURE 8.5 Model of self-efficacy. (From Brown SP. *Introduction to Exercise Science*. Baltimore (MD): Lippincott Williams & Wilkins; 2000. 318 p.)

three-pronged relationship of the individual (specifically self-efficacy), the environment, and behavior. Behaviors, thoughts, and feelings have a reciprocal influence on self-efficacy, which can, in turn, influence the sources of efficacy information. Self-efficacy is dynamic and can be constantly changed as new information is presented to the individual (4). Self-efficacy is an important determinant of behavior in physical activity, exercise, sport, and athletic performance. It accounts for approximately 25% of all the available possibilities for explaining individual performance (37). Self-efficacy has been shown to predict the adoption and the maintenance of moderate and vigorous physical activity and exercise programs in a variety of populations (4,37).

Exercise and sport psychology professionals, as well other exercise science professionals, can use the knowledge about self-efficacy to enhance performance in physical activity, exercise, sport, and athletic competition. Using appropriate interventions to enhance self-efficacy increases the probability that an individual will adopt physical activity and exercise behaviors that are consistent with good health. Additionally, the enhancement of self-efficacy in an athlete will likely improve that individual's performance during a sport or athletic competition.

Attribution Theory

Another prominent cognitive motivation theory is Weiner's **attribution theory**. This theory explains how an individual interprets achievement outcomes and how that interpretation influences future behavior (38). The basic principle of attribution theory claims that after engaging in a behavior that results in some outcome, an individual begins to search for reasons why the outcome happened as it did (4). These reasons for the outcome are referred to as causal attributions. Individuals predominately use four common attributions: ability, effort, task difficulty, and luck (4). The attribution theory focuses on identifying common dimensions underlying the attributions. Weiner (35) identified three causal dimensions: locus of causality, stability, and controllability:

- Locus of causality refers to whether the cause of attribution is perceived to reside internally or externally to the individual.
- Stability refers to the variability of the attribution over time with some attributions relatively permanent (stable), whereas others are relatively temporary (unstable).
- Controllability refers to whether the attribution is under the individual's control or controlled by someone or something else.

When an outcome occurs, an individual experiences an emotion, referred to as an outcome-dependent effect. Figure 8.6 illustrates the outcome-dependent effect. If the outcome was successful, the individual will have good emotions; if the outcome was a failure, the individual will have bad emotions. The individual then engages in a causal search to

Attribution theory Belief that explains how an individual interprets achievement outcomes and how that interpretation influences future behavior.

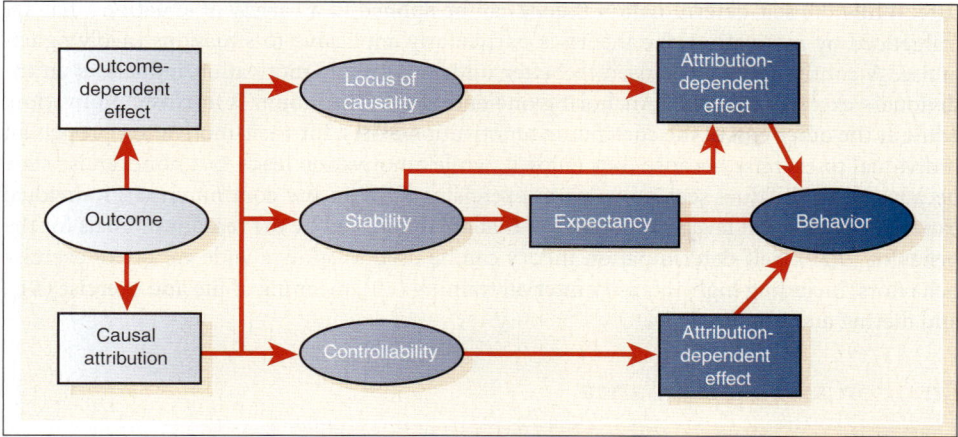

FIGURE 8.6 The outcome-dependent effect. (From Brown SP. *Introduction to Exercise Science*. Baltimore (MD): Lippincott Williams & Wilkins; 2000. 319 p.)

determine the factors that were responsible for the outcome. The factors or causes are then processed in terms of placement along the three dimensions (causality, stability, and controllability) (4).

Individuals who are characterized as chronic exercisers have higher perceptions of individual control over their own health, an internal locus of causality, and more control over exercise behavior. Each of these three causal dimensions can predict negative emotional reactions to discontinuing an exercise program (4). The attributional processes may be significant components of complex behaviors like exercise. In addition to the social cognitive and attribution theories, other popular topics of motivational study have included intrinsic motivation, social-psychological theories of intentions, and achievement goal orientations (4,6). The promotion of positive outcomes during physical activity, exercise, sport, and athletic competitions will result in a positive emotional response of the individual, providing further positive support for continued participation. Exercise science professionals engage in all manner of supportive behaviors in attempts to try and enhance the experience of individuals participating in physical activity, exercise, sport, and athletic competition.

Self-Determination Theory

This theory proposes that the type and quality of motivation the individual experiences is very important in predicting behavior (9). The self-determination theory is broad based and can address issues ranging from personality development to the impact of the social environment on motivation, affect, behavior, and well-being (9,39). There are four subcomponents that constitute this broad theory: basic needs, cognitive evaluation, organismic integration, and causality orientation (9). A key aspect of self-determination theory is the assumption that autonomy, competence, and relatedness are fundamental psychological needs and the degree to which these needs are satisfied or dissatisfied is the basis for individual differences in motivation (9).

While the self-determination theory can be applied to a variety of situations that are influenced by motivation, the theory is particularly applicable to situations involving exercise. A continuum can be used to better understand how motivation influences an individual's exercise behavior. Anchoring one end of the continuum is intrinsic motivation, while at the other end of the continuum amotivation exists. Intrinsic motivation propels an individual to exercise because they enjoy it, while amotivation leads to a nonexercise state because the individual sees no value in exercising. Within the continuum, an individual experiences different levels of motivation that are influenced by the reasons or goals for the behavior (9,39). Self-determination theory can be used to study a wide variety of exercise behaviors, including high-intensity interval training (40), meaning of life and exercise (41), and dieting and exercise (42).

Emotion and Performance

Affect is the term for feeling states that include emotion and moods. For example, **arousal** is a state of heightened physiological and psychological activity. Participation in physical activity, exercise, sport, and athletic competition requires a level of arousal to energize an individual for movement (Figure 8.7). An individual's state of arousal varies along a

FIGURE 8.7 Arousal in an athletic competition. (Photo by Digital Vision/Ryan McVay/Getty Images.)

Arousal A state of elevated responsiveness to stimuli.

continuum from deep sleep at one end to extreme excitement or agitation at the other end. Arousal can also be viewed as a construct with multiple dimensions — having at least the dimensions of valence (degree of attraction or aversion that an individual feels toward a specific object or event) and intensity (the strength of the state of arousal) (4,38).

An individual's level of affect and arousal is constantly changing depending on the situation and environment. The central nervous system, predominately the brain, contains the physical structures involved in the control of arousal. Interactions among the reticular-activating system, cerebral cortex, and hypothalamus in the brain interact with the peripheral nervous system, the somatic nervous system, and the autonomic nervous system to regulate arousal. The brain exerts a strong influence on the adrenal glands, which are responsible for releasing epinephrine and norepinephrine into the circulatory system. These hormones are responsible for helping to increase the level of activity (*i.e.*, arousal) in various tissues of the body (4).

A variety of factors can influence the level of arousal. For example, a perceived stressor might initiate the fight or flight response, which increases the level of arousal. Some level of stress is beneficial for enhancing physiological and psychological functions. Many individuals, however, equate all stressors with negative reactions. This has resulted in arousal being equated with anxiety. Not all stressors should be viewed as creating a negative response in the body. Instead, an individual's interpretation of a stressor determines whether it is viewed as a threat or a challenge (24). If an individual believes that a stressor is a threat, arousal leads to anxiety (4,10).

Physiological and biochemical measures such as brain activity, heart rate, and the stress hormone cortisol have been used to measure arousal. Additionally, questionnaires have been used to specifically measure individual perceptions of arousal. The study of arousal in sport and athletic competition is important because of the role arousal plays in affecting performance. Major models used to explain the effects of arousal on performance include drive theory and inverted U hypothesis (4) as well as the circumplex model (43).

Drive Theory

Drive theory is used to describe the relationship between an individual's level of arousal and performance. In general, as arousal increases, performance increases in a linear fashion. Figure 8.8 shows the relationship between arousal and performance. Drive theory predicts that individual motor movements or skill performance is a function of the interaction between habit and arousal. Habit is used to describe the dominance of the most well-learned response of an individual to a given situation, even if it is an incorrect response (44). Increasing arousal will increase the probability that an individual will select the dominant response for the situation. If it is a correct response, then performance is successful. When a new skill is being learned, there is a higher probability that the dominant response will likely be an incorrect one (4). As an individual becomes better at performing the skill, the dominant response becomes the correct response. In this instance, the increased arousal actually facilitates successful performance (4,6).

For gross motor skills that require muscular strength, speed, and/or endurance, a positive linear relationship between arousal and performance exists (4). For example, during the assessment of muscular strength, it is important for the individual being tested to be aroused so that maximal force production is generated during the muscular movement.

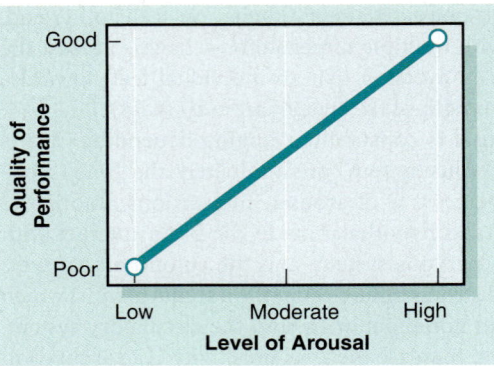

FIGURE 8.8 Relationship between arousal and performance using the drive theory. (Based on Brown SP. *Introduction to Exercise Science*. Baltimore (MD): Lippincott Williams & Wilkins; 2000. 322 p.)

Professional scholars disagree over whether a positive linear relationship exists for those activities requiring the accuracy of fine motor skills. If an individual becomes too aroused, this might result in a decrease in performance. For example, an amateur golfer playing for the local club championship does not want to become too aroused when attempting a short putt that could result in winning the tournament. There appears to be a point of diminishing returns regarding arousal, so at the high-intensity end of the arousal continuum, performance is adversely affected by too much arousal (4,45).

Inverted U Hypothesis

The inverted U hypothesis is used to explain changes in performance when there are changes in the level of arousal. As arousal increases from low to moderate levels, there is an increase in skill performance. As arousal continues to increase, there is a point when performance begins to decline. Figure 8.9 shows the relationship between arousal and performance using the inverted U hypothesis (4,45).

To achieve the best possible performance during sport or athletic competitions, optimal arousal must be achieved. At least two factors contribute to obtaining optimal levels of arousal and performance: task characteristics and individual differences. The main features of task characteristics are task complexity and type of task. A simple movement skill requiring few decisions will be less affected by higher levels of arousal than a complex movement skill requiring many decisions. Conversely, a fine movement skill requiring great accuracy or precision would require much less arousal for optimal performance than a gross movement skill requiring strength, speed, or endurance (4).

Experience and personality factors can also affect individual performance. A highly experienced athlete can endure a higher level of arousal without an adverse affect on performance. In this instance, the optimal level of arousal is shifted to the right on the arousal continuum. Personality factors associated with arousal have the most influence on the relationship between arousal and performance. Individuals who are more highly aroused in normal situations would not be able to tolerate much additional arousal without an adverse effect on performance. Much of the performance-enhancement work in exercise and sport psychology deals with helping athletes determine their optimal zones for effective performance and then teaching them skills to assist in achieving and maintaining that optimal level (4,45).

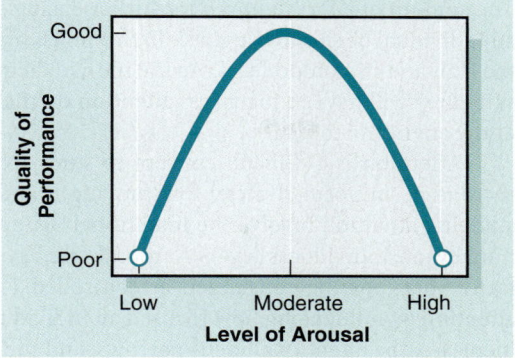

FIGURE 8.9 Relationship between arousal and performance using the inverted U hypothesis (4,45). (Based on Brown SP. *Introduction to Exercise Science*. Baltimore (MD): Lippincott Williams & Wilkins; 2000. 322 p.)

Circumplex Model

The circumplex model has been proposed as a useful conceptual and measurement framework for investigating the effects of exercise on affect (43). The circumplex model uses a circle and arranges the affective states around the perimeter (see Ekkekakis and Petruzzello 2002 for an example of the model) (46). Valance and activation in both positive and negative forms divide the circle into four quadrants. The four quadrants can be characterized by the various blends of valence and activation. The circumplex model provides a more general depiction of affect, allows for both positive and negative affective states, and is a more conservative approach to studying affect in the exercise field (43,47).

Attention and Cognition

Attention, as used in exercise and sport psychology, is defined as the ability to focus on a specific skill or activity. Attention may also be more broadly defined as concentration (48) with four components:

- Focusing on the relevant clues in the environment (selective attention)
- Maintaining the attentional focus over time
- Having awareness of the situation
- Shifting attentional focus when necessary (6)

Being able to focus one's attention greatly contributes to success in exercise, sport, and athletic competition. Individuals and especially athletes who can concentrate on certain relevant environmental stimuli, while ignoring irrelevant stimuli, will have a greater chance at successful performance than individuals who are distracted and unable to concentrate.

Attention Process whereby a person concentrates on some features of the environment to the (relative) exclusion of others.

The amount of effort required for such focusing of attention can be important. Typically, as an individual begins to learn a skill, they must make a conscious effort to focus attention on the skill and its components. As the individual becomes more proficient at performing the skill, they will have to focus less attention on the skill and can focus more on other aspects of the environment (4,6).

Attention is a difficult concept to study in sport and athletic environments because there is no uniform strategy for studying it and because many of the methodologies for studying attention involve the intentional disruption of performance at some level. Additionally, an individual's level of arousal can have an impact on attention, adding another factor that must be considered or controlled. Arousal has its most significant effects on attention by influencing how focused an individual can be. In essence, as individual arousal increases, the attentional focus narrows, and the individual's attention becomes more concentrated. This is required for effective performance. As arousal increases from low to moderate intensity, the individual's attentional field becomes more focused. Performance is generally improved because the narrowing of attention allows the individual to eliminate needless stimuli from the field of attention. However, if arousal is too high it can have an adverse effect on performance. In this instance, attention can become so narrow that the individual misses important information that is central to the performance of the skill or task at hand (4,6).

Exercise and sport psychology professionals use a variety of techniques and strategies to enhance attentional skills and performance within individuals. Attention can be indirectly assessed using several measurement procedures, including behavioral measures of attention, self-report measures of attention, and psychophysiological measures of attention (4). Table 8.2 provides three of the most common ways to assess attention and limitations associated with each (4,6).

Thinking Critically

What specific information about an individual would you need to help that individual improve performance during a sport or athletic competition?

EXERCISE PSYCHOLOGY

Mental health problems account for significant hospitalization and medical costs in the United States (6,49,50). About 14% of all Americans use mental health services, with an estimated 18.1% suffering from **anxiety** and 9.5% suffering from a mood disorder (49,51). Anxiety and **depression** are significant mental health problems that can also lead to other adverse health conditions (49). Acute and chronic exercise has the potential to influence psychological moods and emotions. Regular exercise also has the ability to enhance mental health by reducing anxiety and depression (4,52,53) and enhancing psychological well-being (54).

Anxiety State of uneasiness and apprehension related to future uncertainties.
Depression A reduction in physiological and psychological activity.

Table 8.2 Common Ways to Assess Attention and Limitations Associated with Each (4,6)

MEASURE OF ATTENTION	DESCRIPTION	LIMITATION
1. Dual-task paradigm	• Individual is asked to perform two tasks at the same time with the belief that the primary task will require the majority of attention, allowing performance on the second task to be assessed	• The manipulation required fundamentally disrupts performance • It is unclear whether there is a limit to attentional capacity, the assumption of which underlies the technique
2. Self-report	• Individual is asked to provide information about what they were focusing on during performance of a skill or task	• It is unclear whether athletes can actually access the cognitive processes that occur during attentionally demanding activities and then put those operations into words • Athletes may not complete assessment questionnaires right before a competition
3. Psychophysiologic measurements	• The psychological construct of attention can be assessed based on physiologic responses of the body immediately prior to performance	• Measurement of a physiologic variable does necessarily provide the assessment of a psychological variable

Anxiety

Anxiety is a state of uneasiness and apprehension related to future uncertainties. Acute bouts of physical activity and exercise and participation in regular exercise programs reduce state and trait levels of anxiety. Positive benefits occur in those individuals suffering from normal to moderately high levels of anxiety. Both aerobic exercise and resistance exercise training can reduce anxiety levels, with aerobic exercise producing anxiety reductions similar to other commonly used anxiety treatments (4,6,31). Exercise reduces anxiety regardless of an individual's age, gender, fitness level, and health status or anxiety level. Although anxiety levels can be reduced through exercise, there are some general exercise guidelines that should be followed to maximize the influence:

- Exercise intensity should be at least 30% to 40% of maximal heart rate.
- Exercise durations of up to 30 minutes provide the greatest effects.
- Participation in longer training programs has more effect than shorter programs (6,31).

The total volume of exercise and the type of exercise required for decreasing anxiety is not entirely clear. Individuals can vary the intensity, duration, and frequency of exercise in an effort to affect psychological status. Both aerobic exercise and resistance exercise training have been shown to have positive effects on anxiety (9,47) and, depending on the level of anxiety, as little as 5 minutes of exercise may have a positive impact on anxiety (31).

Depression

Mental depression is a state of general emotional dejection and withdrawal. Although the standard treatment for mental depression is psychotherapy or prescription medications, many individuals use exercise as an effective alternative treatment choice (6). Physical inactivity is also related to higher levels of depression (55). Both aerobic exercise and resistance exercise training can reduce depression, and the reduction occurs for all types of individuals across age, gender, and health status. Exercise produces larger antidepressant effects when the training program is at least 9 weeks in duration and does not depend on changes in fitness level (4,31,55,56). Exercise may work to alleviate depression through psychological or neurobiological mechanisms (57). The psychological mechanisms might include enhanced self-efficacy, self-esteem, positive well-being, and improved social support (58). The neurobiological mechanisms might include changes in neurotransmitter substances such as norepinephrine, serotonin, and tryptophan, as well as changes in the secretion of hormones from the hypothalamus–pituitary–adrenal axis (56). It appears that a dose–response effect can be found for the effect of exercise on depression, with higher intensity exercise more effective than lower intensity exercise (57), but the exact volume of exercise required to influence depression has not been completely determined (9,47). Individuals who maintain regular exercise programs are likely to have significantly lower relapse rates for depression (59). Future work in this area should lead to additional insight into the role of physical activity and exercise as part of the complete treatment regimen for individuals with depression.

Emotional Well-Being

Although much of the past study of psychological changes associated with physical activity and exercise has focused on reductions in negative emotions, there is a significant amount of support for exercise being effective in enhancing and improving many positive psychological states (Figure 8.10) (4). Exercise can have positive effects on mood states such as vigor, clear thinking, energy, alertness, and well-being (6,58). In some instances, exercise performed for as little as 10 minutes can produce positive psychological benefits (60). One of the most consistently reported effects from acute and chronic exercise is increased feelings of energy (38,61). Physical activity and exercise have also been shown to increase self-confidence, self-esteem, and cognitive function (6,62,63). Although it is clear that exercise is related to positive changes in mood (64), it is unclear whether exercise actually causes the enhanced mood (6). The following characteristics and guidelines for physical activity and exercise appear to have the greatest impact on changing an individual's mood:

- Performing rhythmic abdominal breathing
- Relative absence of interpersonal communication
- Performing closed and predictable activities that allow for preplanned movement
- Performing rhythmic and repetitive movements that allow the mind to focus on important issues

FIGURE 8.10 Exercise and relaxation. (Photo by Shutterstock.)

- Performing 20 minutes of a moderate intensity physical activity and exercise
- Performing moderate intensity physical activity or exercise at least two to three times per week
- Performing activities that are enjoyable (5,55,58,65)

Explanations of Exercise and Psychological Well-Being

Numerous explanations and theories, both physiological and psychological, have been proposed on how physical activity and exercise enhance mental health. Table 8.3 provides a list of the physiological and psychological factors that could play a role in facilitating improvements in mental health from exercise (6). Although no single theory has support as the sole or primary mechanism producing these positive changes, four explanations are frequently mentioned (61). These include the distraction hypothesis, endorphin hypothesis, thermogenic hypothesis, and monoamine hypothesis (4).

The *distraction hypothesis* is one of the most common psychological explanations among those hypotheses commonly proposed. This hypothesis suggests that the reason for an improved emotional profile after exercise is that exercise provides a distraction from the normal everyday occurrences that often lead to stress and negative emotions. This allows

Table 8.3	Physiological and Psychological Factors That Could Play a Role in the Mental Health Effects of Exercise (4,6)
PHYSIOLOGIC FACTORS	**PSYCHOLOGICAL FACTORS**
Increases in cerebral blood flow	Enhanced feeling of control
Changes in brain neurotransmitters	Feeling of competency and self-efficacy
Increases in maximal oxygen consumption and delivery of oxygen to brain	Positive social interactions
Reduction in muscle tension	Improved self-concept and self-esteem
Structural changes in the brain	Opportunities for fun and enjoyment

the individual to focus on things other than those factors leading to stress and negative emotions (66). The *endorphin hypothesis* is named after a class of stress hormones called endorphins. During exercise, endorphin concentrations are increased and remain elevated for some time after the exercise is finished. Elevated endorphin levels have been positively correlated with individuals feeling better, and it has become popular to claim that the endorphins are responsible for this improved mood (67). It has been difficult to prove, however, that the elevation of endorphin levels is responsible for making individuals feel better following exercise. The *thermogenic hypothesis* suggests that exercise of sufficient intensity and/or duration will result in an elevation of body temperature. It is thought that an elevated body temperature will result in a variety of positive changes such as a reduction in muscle tension after exercise and other psychological changes (68). The *monoamine hypothesis* suggests that changes in brain neurotransmitters can result in exercise-induced emotional changes. The neurotransmitters norepinephrine, dopamine, and serotonin are often localized to brain structures known to have an important role in emotion and are altered with exercise (69). Although each of these hypotheses provides indirect support for a role of physical activity and exercise reducing stress and negative emotions, much more work needs to be done to better understand the mechanisms for how physical activity and exercise enhances mental health.

Exercise and Cognitive Function

Physical activity and exercise impacts every system in the body, including the brain where it enhances and protects brain function. Aging causes dysfunction and degeneration in neurons and personality changes in some instances. Changes in brain **neurotransmitters**

Neurotransmitter A chemical substance that is produced and secreted by a nerve ending and then diffuses across a synapse to cause the excitation or inhibition of another nerve.

determine how factors like brain blood flow, brain electrical activity, and by-products of brain neurotransmitters are linked with emotions and thoughts (70). Blood flow to the brain changes with exercise. As exercise intensity increases from low to moderate levels, the temporal, parietal, and frontal regions of the brain have an increase in blood flow (71). Recorded electrical activity of the brain has been shown to predict how an individual will feel after that exercise (72). Higher-level functions of the brain, especially those associated with frontal lobe and hippocampal regions of the brain, may be selectively maintained or enhanced in humans with higher levels of fitness (4,73).

Regular exercise training can also influence brain function, in both children and preadolescence (9,47), as well as in older adults (9,62,63,74). Lifetime exercise enhances several aspects of cognition, brain structures, and brain functions (74). Much of this research has employed aerobic exercise such as walking, running, bicycling, and swimming. Cross-sectional studies have shown that aerobic exercise improves both the peripheral and central nervous system components of reaction, resulting in a reduced time to recognize and respond to a stimulus. Individuals who have engaged in exercise for a lengthy period of their lives respond more quickly to the presentation of auditory or visual stimuli, discriminate between multiple stimuli, and make faster movements. Additionally, exercisers can outperform nonexercisers on tasks such as reasoning, working memory, and fluid intelligence tests (9,47,75). Differences in performance on seemingly similar tasks between lifetime exercisers and nonexercisers, however, have not always been found (4,73). Much work needs to be done in order to identify the appropriate exercise dose and prescription for improving cognitive function for individuals across the lifespan (47).

Thinking Critically

How do the various content areas of exercise and sport psychology provide for improvements in health care for healthy and diseased individuals?

EXERCISE BEHAVIOR

Physical inactivity and poor fitness are strong predictors of disease risk. For example, individuals who are physically inactive have a higher risk for certain types of cardiovascular disease and cancer (76). Being physically active promotes good health and decreases the risk of morbidity and mortality (54,77–79). Despite national efforts at promoting physical activity, morbidity and mortality rates from lifestyle-related diseases remain high (76,80–83). There are numerous physiological and psychological benefits to be derived from being physically active. Regular physical activity has been shown to:

- Reduce the risk of cardiovascular disease, including stroke, high blood pressure, and coronary artery disease
- Reduce the risk of certain cancers, including colorectal and breast
- Contribute to weight loss and maintaining a healthy body weight
- Improve psychological factors such as a reduction in stress and depression and an increase in self-esteem (6)

Despite the many benefits derived from participating in regular programs of physical activity and exercise, approximately 50% of American men and women do not perform sufficient regular physical activities to meet the *Healthy People 2030* objectives for substantial health benefits (84). Many barriers to exercise participation are within the control of the individual and hence amenable to change (6). The major barriers to participating in regular physical activity and exercise programs are usually lack of time, lack of energy, and lack of motivation (6,85,86). If an individual does overcome the barriers and begins an exercise program, the next obstacle they will encounter is to continue regular participation with the program. **Exercise adherence** is challenging for most individuals, and approximately 50% of those individuals who start a program discontinue or "drop out" within the first 6 months (19). Individuals cite a variety of reasons for discontinuing regular physical activity and exercise (19,87). Investigations have identified several effective strategies for increasing exercise adherence, including financial incentives (88,89), counseling (90), and self-regulatory strategies (91). In an effort to better understand why people either do not begin a regular program or discontinue a program, exercise psychology professionals have promoted several theories and models of exercise behavior. It is important for exercise science professionals to have a solid understanding of how these theories influence exercise behavior.

Theories and Models of Exercise Behavior

It is believed that improvements in exercise adoption and adherence can occur if a better understanding of how and why individuals participate in physical activity and exercise is achieved. Four prominent models are used to help explain the process of exercise adoption and adherence (6,92,93). Table 8.4 illustrates those models of exercise behavior and the theory associated with each model. Each of the discussed models (**Health Behavior**, **Theory of Planned Behavior/Reasoned Action**, Social Cognitive Theory, and **Transtheoretical Model of Behavior**) has been used in different settings to help understand why individuals may or may not initiate and continue participation in a regular physical activity and exercise program (94). Other models such as the Social Ecological Model, the Relapse Prevention Model, and the Habit Theory are also considered important in the understanding of how and why individuals participate in regular physical activity and exercise (9,47).

The *Health Behavior Model* is a psychological model that attempts to explain and predict individual health behaviors (93). The original model provides four constructs representing the perceived threat and net benefits of an action or behavior: perceived susceptibility, perceived severity, perceived benefits, and perceived barriers. These concepts

Exercise adherence The behavior of continuing participation in a regular program of exercise.

Health Behavior A theoretical model used to explain and predict individual health behaviors using six constructs of behavior.

Theory of Planned Behavior/Reason Action A theory used to describe and predict deliberate and planned individual behavior.

Transtheoretical Model of Behavior A model used to understand the stages that individuals progress through and the cognitive and behavioral processes they use while changing health behaviors.

Table 8.4 — Models of Exercise Behavior and the Theory Associated with Each Model (6)

MODEL	THEORY
Health Belief Model	• Probability of an individual engaging in exercise depends on the person's perception of the severity of the potential illness as well as the appraisal of the costs and benefits of exercise (131)
Theory of Planned Behavior/Reasoned Action	• Probability of an individual engaging in exercise depends on the individual's attitudes toward a particular behavior and the individual's perceptions of their ability to perform the behavior (132)
Social Cognitive Theory	• Probability of an individual engaging in exercise depends on the personal, behavioral, and environmental factors that interact in a reciprocal manner (34)
Transtheoretical Model	• Probability of an individual engaging in exercise depends on the stage of change the individual is currently in for establishing and maintaining a lifestyle change (130)

were proposed as an explanation for an individual's readiness to take action and change behavior. Two additional concepts were subsequently added to the model: cues to action and self-efficacy. Cues to action would activate the readiness of an individual and stimulate overt behavior. The most recent addition to the model is self-efficacy, or an individual's confidence in the ability to successfully perform an action. This concept was added to help the Health Behavior Model better fit the challenges of changing habitual unhealthy behaviors, such as being sedentary, smoking, or overeating (95).

The *Theory of Planned Behavior/Reasoned Action* is used to describe and predict deliberate behavior because behavior can be deliberative and planned (93). The Theory of Planned Behavior/Reasoned Action is based on the belief that an individual's behavior is determined by their intention to carry out the behavior and that this intention is, in turn, a function of their attitude about the behavior and their subjective norm. The best predictor of an individual's behavior is the intention to perform the behavior. Intention is the cognitive representation of an individual's readiness to perform a given behavior, and it is considered to be the immediate antecedent of behavior. This intention is determined by three factors: attitude toward the specific behavior, subjective norms, and perceived behavioral control. The Theory of Planned Behavior/Reasoned Action maintains that only specific attitudes toward the behavior in question can be expected to predict that behavior (96).

Social Cognitive Theory explains how individuals acquire and maintain certain behavioral patterns and can also be used to provide the basis for intervention strategies (97). Evaluating behavioral change depends on the environment, the individual, and the specific behavior. This theory provides a framework for designing, implementing, and evaluating exercise programs. The environment refers to the factors, both social and physical, that can affect an individual's behavior. The social environment refers to individual and group interactions that could influence behavior and includes family members, friends, and colleagues. The physical environment refers to the factors in the built environment the individual encounters and includes, for example, the size and shape of a room, the weather conditions, or the availability of specific foods. The environment combined with the individual situation provides the framework for understanding an individual's behavior. The situation refers to the mental representations of the environment that may affect an individual's behavior. The situation includes an individual's perception of the place, time, physical features, and activity (98). The social and physical environments constantly influence each other. Individual behavior is not simply the result of the environment and the individual (98). The environment provides models for behavior, and individuals do certain things in response to the environmental and social factors (97).

The *Transtheoretical Model* has been used to understand the stages that individuals progress through, and the cognitive and behavioral processes they use while changing health behaviors (81,93). The Transtheoretical model is effective at predicting and explaining the exercise behaviors of individuals (99). In this model, there are five stages of change through which individuals progress as a behavior is adopted (99). The five stages are shown in Figure 8.11. The model postulates that individuals engaging in a new behavior move through the stages of precontemplation, contemplation, preparation, action, and maintenance. Movement through these stages does not always occur in a linear manner, but may also be cyclic because many individuals must make several attempts at behavioral change before the behavior is adopted and their goals are realized. The amount of progress people make as a result of an intervention tends to be a function of the stage they are in at the start of treatment. Instruments have been developed to measure the stages and processes of exercise adoption and maintenance and the related constructs of exercise-specific self-efficacy and decision making (98). Movement through the stages can be affected by many factors outside a person's control. Individuals use numerous strategies and techniques to change behaviors, and the strategies are their processes of change. Effective use of behavioral strategies that work on the environment (*e.g.*, reinforcement, social support) and cognitive strategies that focus on thoughts and perceptions (*e.g.*, education, media campaigns) can help people move through the stages and maintain regular physical activity (6).

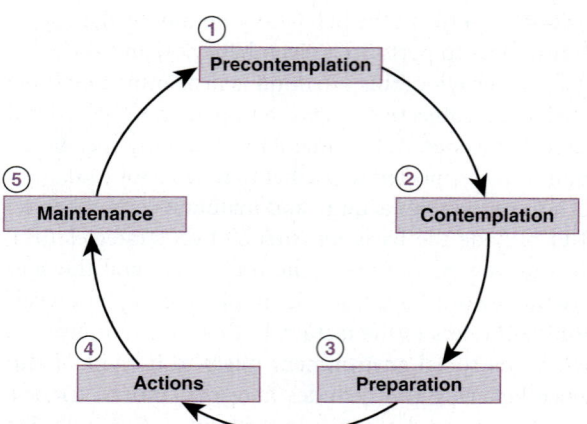

FIGURE 8.11 Cyclic pattern of stages of change (6,130).

AREAS OF STUDY IN EXERCISE AND SPORT PSYCHOLOGY

Exercise and sport psychology encompass numerous areas of research and scholarly inquiry with the goal of improving performance in health, physical activity, exercise, sport, and athletic performance. Exercise and sport psychology professionals are interested in the various mental and psychological responses that result in an improvement in the outcomes associated with exercise and sport. The role of the determinants of exercise adherence, injury and mental health, and imagery and performance enhancement are examples of some of the primary interest areas in exercise and sport psychology. The areas selected are by no means meant to infer an inclusive list or indicate greater importance than a topic area not covered but rather to provide a sampling of areas in which the knowledge of exercise and sport psychology is generated and used.

Determinants of Exercise Adherence

Exercise and sport psychology is used to help better understand the thought process of individuals adopting and maintaining a regular exercise and training program. This is critical for developing strategies that will enhance individual exercise adherence. The natural progression for an individual wishing to participate in a physical activity and exercise program is illustrated in Figure 8.12. Movement into and out of each major phase can be affected by numerous factors (100). These determinants can be categorized into personal and environmental factors that can have positive, negative, or no influence on exercise adherence (6,93,101,102). The personal determinants can be further divided into demographic, cognitive and personality variables, and behaviors, whereas the environmental determinants can be divided into social environment, physical environment, and physical activity characteristics (6). Examples of personal determinants include age, education level, exercise knowledge, attitude, and level of self-confidence. Examples of environmental determinants include access to facilities, cost, family influences, and perceived available time (100).

Using the information about an individual's personal and environmental determinants in conjunction with the theories of behavior change, exercise and sport psychology professionals as well as other exercise science professionals can develop effective strategies

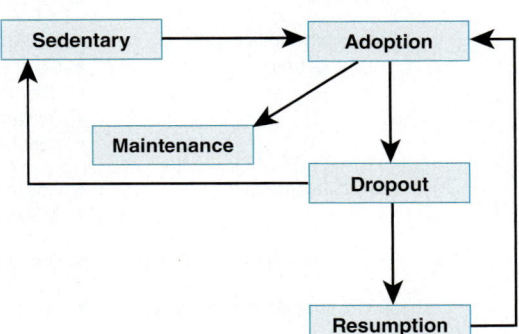

FIGURE 8.12 Natural progression phases of exercise (100).

for enhancing exercise adherence. For example, research has indicated that both personal habits and physical activity intensions should be considered when promoting exercise adherence strategies in adults (103). Table 8.5 illustrates some of the major approaches and strategies for enhancing exercise adherence (6). It is important to effectively match the theories and models of exercise behavior with the appropriate strategy for enhancing adherence to increase the probability of an individual adopting regular exercise as a healthy behavior. Effectively promoting adherence increases the likelihood that an individual will continue a regular program of physical activity and exercise and thereby derive the associated health benefits (4,6).

Exercise adherence has been studied in a variety of environmental settings with individuals of all ages and characteristics. For example, the importance of exercise adherence is critical to individuals with a variety of health concerns and disease conditions. The ability of individuals with either heart disease or cancer to maintain participation in an exercise program is critical to the overall morbidity and mortality rate for those disease conditions. One of the most significant factors affecting exercise adherence in heart failure patients is poor social support. Patients with higher perceived social support have been shown to exercise more and have better outcomes than those with lower perceived social support (104). Participation in a physical activity and exercise program following treatment for various types of cancer is critical for maintaining an active and independent lifestyle (105). Higher levels of self-efficacy are significantly associated with high participation rates and compliance with endurance exercises, whereas lower psychological distress is significantly associated with high participation rates with resistance exercise (106).

Injury and Mental Health

Participation in sport and athletic competition is associated with a level of injury risk that is a function of numerous factors. Both intrinsic (*e.g.*, age, neuromuscular control, previous injury, strength) and extrinsic (*e.g.*, equipment, environment) factors contribute to making an athlete susceptible to a physical injury during training and/or competition (107). The

Table 8.5 Major Approaches and Strategies for Enhancing Exercise Adherence (6)

STRATEGY	EXAMPLES
Behavior modification approaches	• Prompts • Contracting • Charting attendance and participation • Rewarding attendance and participation • Feedback
Cognitive approaches	• Goal setting • Association and dissociation
Decision-making approaches	• Decision balance sheet
Social support approaches	• Social and family actions

severity of the injury will impact the length of time the athlete is prohibited from engaging in practice or training as well as participating in competition. While the physical aspect of the injury is often visible or describable to the athlete, coach, and sports medicine professional, there is often a psychological aspect of the injury that is less visible or describable. Any psychological component associated with a physical injury requires attention and intervention. An untreated psychological component associated with an injury can contribute to a delay in the complete recovery from the injury, poor practice and training quality, poor performance during competition, and an inability to participate in the sport or athletic competition (108).

To an athlete, an injury can be a significant life event, which may affect the cognition, emotions, and behaviors of the athlete. Cognitive assessment and evaluation of the injury may determine the emotional response of the athlete, which may in turn affect behavioral aspects of the athlete such as goal setting, motivation for training and competition, compliance with coach expectations, and adherence to the rehabilitation treatment (108). Pain perception, optimism/self-efficacy, and depression/stress are prominent psychological factors that have been shown to be important in illness and injury treatment (108). Suggested psychological strategies such as counseling, goal setting, positive self-talk, cognitive restructuring, and imagery/visualization are associated with faster recovery from a musculoskeletal sports injury (109–112). These psychological interventions enable positive mood changes, effective pain management, increased exercise compliance, and improved rehabilitation adherence (109,112). Psychological interventions can facilitate recovery for injured athletes by promoting a positive emotional state and rehabilitation adherence. Additional research is needed to determine the most effective interventions for specific psychological factors, the appropriate duration of interventions, the best method of implementation following a sports injury, and the impact of these interventions on an athlete's ability to return to practice and competition (109,112).

Common psychological health issues in both injury-free and injured athletes include stress/anxiety (113), depression (114), disordered eating/eating disorders (115), and substance use disorder (108,116). All of these psychological health issues have specific signs and symptoms with treatment management that is specific to the mental health issue, especially as it relates to injury recovery. Treatment approaches consist of monitoring mental health status (117), discussions around demystifying mental health issues (117,118), providing positive support interventions (119), and specific behavioral intervention programs such as mindfulness and resilience training (117,120).

Imagery and Performance Enhancement

Imagery is a behavior executed in the brain using some or all of the body's senses (47). Imagery can enhance performance by directly priming physical movement patterns or

Imagery A behavior executed in the brain using some or all of the body's senses to enhance performance.

indirectly through altering constructs and dispositions associated with successful performance (*e.g.*, enriching confidence, controlling anxiety) (121,122). Imagery is a fundamental psychological skill used to enhance performance in physical activity, exercise, sport, and athletic performance (121–123). Imagery is an effective technique in enabling the acquiring of movement skills, learning sport strategies, and improving sport performance. Imagery can also have positive effects on competitive anxiety, motivation, self-efficacy, and confidence (124). Athletes use imagery to learn, refine, and/or review sport skills and strategies; to increase motivation by envisioning successful sport outcomes; to increase confidence, focus, and/or mental toughness; and to calm oneself down or to energize oneself (125).

The influence of various forms and types of imagery on sport and athletic performance has received considerable investigation. The speed of imagery can have a significant influence on sport and athletic performance outcomes. Athletes can employ slow-motion, real-time, and fast-motion imagery in an effort to improve performance, with each speed affecting various psychological aspects of the learning and performance strategy (124). Athletes incorporating slow-motion and/or fast-motion imagery speeds when learning complete sport movements may benefit performance when the slow and fast imaging movements are paired with real-time speed images (124). Positive imagery can neutralize the effect of negative imagery and directly influence the level of stress and anxiety (126). When used as combined interventions, both visual and kinesthetic imagery are advantageous to athletes, whereas separate use of those two modalities of imagery may seem less efficient (127). Imagery combined with self-talk and relaxation improves mental toughness and specialized skill performance (128,129).

How might coursework in exercise and sport psychology prepare an individual for a career as a health care specialist, personal trainer, or athletic coach?

Interview

Jennifer Etnier, PhD, FACSM

Distinguished Professor in the Department of Kinesiology at the University of North Carolina at Greensboro

Brief Introduction – Born into a family of physical activity enthusiasts, I grew up playing team and individual sports and enjoying outdoor recreational pursuits. I initially majored in Mathematics and Computer Science at the University of Tennessee, but after spending an entire summer doing computer programming, I had a change of heart and quickly added a minor in Psychology to my BS. I then more fully acknowledged my passion for physical activity and for working with people and obtained two graduate degrees in Sport Psychology (MA at the University of North Carolina at Chapel Hill; PhD from Arizona State University). In my current position, I conduct research focused on the cognitive benefits of physical activity across the lifespan and am currently the Principal Investigator on a National Institutes of Health grant entitled Physical Activity and Alzheimer's Disease. In addition to my research and teaching, I am also committed to improving the youth sport experience and have written two books with that purpose in mind (*Bring Your "A" Game*; *Coaching for the Love of the Game*).

 What are two or three of your most significant career experiences?

One significant career experience was that I had the opportunity to interact with Dr. Gene Glass at Arizona State University. Dr. Glass is the person credited with coining the term "meta-analysis," and he brought this technique to the behavioral sciences. As a doctoral student, I was invited to be the lead author on a meta-analytic review of the literature on exercise and cognition and benefited enormously from having access to Dr. Glass for conversations about this statistical technique. This meta-analysis was foundational for me in pursuing a decades-long interest in exercise and cognition. And this experience was also important because it helped me to realize that many scholars with international reputations are also down-to-earth and willing to be of help to less experienced academics. A second significant experience was having the opportunity to serve as the President of the NASPSPA. This was important because it gave me the chance to interact with and learn from scholars from universities across North America through our work on the executive committee and our commitment to the organization. Being in a position to provide service to the profession is a win-win, and I encourage students and young professionals to take advantage of service opportunities because they will bring a sense of accomplishment, will help with networking, and will contribute to professional growth.

 Why did you choose to become an "exercise and sport psychologist"?

In pursuing my Master's degree, my plan was to become a sport psychology consultant who would help athletes achieve their sport-related goals. But as I watched the cohort in front of me graduate and have difficulties gaining employment in the field, I decided to pursue a PhD and to consider a career in academia. In my faculty position, I have the privilege of being able to pursue my interests in exercise psychology (conducting research focused on the mental health benefits of exercise and mentoring the next generation of scholars) while also maintaining my interests in helping athletes have better sport experiences (through publishing books, providing educational seminars, and consulting). In the end, I think it goes back to my early exposure to and love for physical activity. As a sport and exercise psychologist, I have the opportunity to generate new knowledge relative to the benefits of physical activity and to improve the sport experience with an ultimate overarching goal of encouraging more people to be physically active from childhood through adulthood.

 Why is it important for exercise science students to have an understanding of exercise and sport psychology?

Exercise science students will one day be exercise science professionals working with clients, patients, athletes, coaches, and students. Exercise and sport psychology provides exercise science professionals with the evidence and tools needed to promote physical activity and to encourage the initiation and maintenance of this behavior. By understanding the myriad mental health benefits that accrue in response to physical activity, information can be used to help people recognize the short- and long-term benefits that physical activity provides, thus motivating participation. Through theories of behavioral change taught in exercise and sport psychology courses, exercise science professionals are given the tools to help people initiate and maintain physical activity behaviors. Finally, exercise and sport psychology provides the skill set needed to help athletes and exercisers achieve their goals in part through performance-enhancement techniques, but also through the development of coping and mental skills that can be applied to sport, academics, and life.

 What advice would you have for a student exploring a career in any exercise science profession?

For students interested in exercise science, my advice is to reach out to practicing professionals who are doing what you think you want to do. The goals in talking with these professionals are to learn more about their educational and experiential paths, to learn more about the benefits and shortcomings of their position, and to ask for advice as to how to most effectively pursue a similar professional position. In thinking about your future, it is important to ensure that the career you have identified will be a good fit for you and then to learn what steps you need to take to attain that position. I would also encourage students to take advantage of opportunities at their institution to get hands-on experience or to have out-of-class interactions with faculty members. These experiences will help you to learn more about the various career options available in the field of exercise science.

Interview

Panteleimon Ekkekakis, PhD, FACSM, FNAK

Professor, Department of Kinesiology at Iowa State University

Brief Introduction – I am a native of Greece. I migrated to the United States after completing my undergraduate degree in Exercise and Sport Science at the University of Athens and an extra (fifth) year as an exchange student at the University of Liverpool in the United Kingdom. My late father was a physical education teacher and track-and-field coach, so I was raised in stadiums, always surrounded by athletes. Getting into the "family business" was almost inescapable. Yet, around my senior year as an undergraduate, I realized that I was more passionate about exercise for health than about athletic performance. So I got into the then-new field of exercise psychology and felt that I had found my calling. After completing my MS degree at Kansas State University and my PhD at the University of Illinois, I was hired as an Assistant Professor at Iowa State University and have been there for over 20 years.

 What are your most significant career experiences?

You can only identify the period during which you experienced your largest intellectual growth in retrospect, almost never while it's happening. For me, this time was the four-and-a-half years I spent as a doctoral student at the University of Illinois in Urbana-Champaign. As a student, I was aware that UIUC was traditionally the "epicenter" of American sport psychology and later exercise psychology. But I was less aware of the reputation of my teachers outside of kinesiology, such as the people teaching me psychophysiology, emotions, or psychometrics. Years after I graduated and delved more deeply into these literatures, I started to realize that my teachers were some of the most influential minds in their respective fields.

 Why did you choose to become an "exercise and sport psychologist"?

I got an unusual start in this field, in the sense that I announced to my family that I was going to become a "sport psychologist" when I was still an adolescent, well before I had a clear idea of what exactly that meant. I only knew I had a fascination with sports and with psychology, so I thought it would be great to combine these two areas of interest. On the other hand, choosing between sport and exercise psychology, and picking exercise psychology, was a more informed and deliberate decision. Not only was exercise psychology a newer field, with more opportunities for growth, but I also realized that I could potentially help many more people, namely all those who are not necessarily young, healthy, or athletic.

Q: Why is it important for exercise science students to have an understanding of exercise and sport psychology?

Thanks to the astonishing discoveries made by exercise scientists and others since the 1950s, it has now become "common knowledge" that exercise is "good for you," benefiting nearly every aspect of human health across the lifespan. Where we're still falling short as a scientific and professional field is in finding ways to convert this knowledge to an increased number of people who are physically active. I show my students survey results suggesting that 97% of American adults recognize that the absence of regular physical activity is an important health risk factor but 97% do less than the recommended minimum amount of physical activity. I then ask my students to "put their skeptical hats on" and to "think like a psychologist" as we analyze common professional practices, such as the messages we choose in promoting physical activity, how we prescribe exercise, how we structure the physical and social environments of gyms, how we do physical education. While a few become defensive, most realize that we can do better.

Q: What advice would you have for a student exploring a career in any exercise science profession?

Although this is probably going to sound like a cliché, it's a cliché for a reason. You have to follow your heart, and you have to do what makes you feel most passionate. For some, like me, this happens to be finding ways to make the exercise and physical activity experience more appealing, more enjoyable, and something that more people want to come back and repeat, ideally, every day of their lives. This makes me happy and gives me a sense of fulfillment, such that I do not foresee ever running out of enthusiasm and the desire to share what I am learning with others. I hope that all my students, and all exercise science students, will find a similar passion that will drive them throughout their careers.

SUMMARY

- Exercise and sport psychology is an area of study concerned with the behavior, thoughts, and feelings of healthy, disabled, and diseased individuals engaging in physical activity, exercise, sport, and athletic competition.
- Individual factors of personality, motivation, emotion, and attention each have a strong influence on the successful performance of exercise and sport.
- Regular physical activity and exercise can influence mental health by reducing anxiety and depression, enhancing psychological well-being, and improving various aspects of brain function.
- Exercise adherence, which is important for achieving significant physical and mental health benefits, is affected by several personal and environmental factors.

For Review

1. What are the specific areas of study in exercise and sport psychology?
2. Describe the first experimental research study that directly impacted exercise and sport psychology.
3. What are the primary professional organizations in exercise and sport psychology?
4. Name the primary research journals in exercise and sport psychology.

5. Describe how personality plays a role in exercise, sport, and athletic competition.
6. What two personality characteristics are strong predictors of exercise behavior?
7. Describe the difference between extrinsic and intrinsic motivation as they relate to participation in physical activity and exercise.
8. How does self-efficacy influence performance in sport and athletic competition?
9. What individual characteristics do chronic exercisers display?
10. Describe why a level of arousal that is too high can adversely affect performance.
11. Why is attention difficult to study during a sport or athletic competition?
12. Describe how acute and chronic exercise might influence an individual's level of anxiety and depression.
13. List five factors or characteristics that physical activity and exercise should possess to impact an individual's mood.
14. Describe how exercise enhances psychological well-being according to each of the following theories:
    a. Distraction
    b. Endorphin
    c. Thermogenic
    d. Monoamine
15. How does the transtheoretical model predict and explain the exercise behavior of individuals?
16. As an athlete, what factors influence psychological recovery from an injury?

Project-Based Learning

1. Identify an individual who does not participate in regular physical activity or exercise. Prepare a presentation that includes at least five key recommendations that you would give to that individual in an effort to improve their mental approach to physical activity and exercise. What are those key points, and how does the exercise psychology literature support your recommendations?
2. Identify an athlete that you believe demonstrates some of the characteristics discussed in this chapter. Prepare a presentation that includes at least five points that you would give to another athlete who is struggling with their mental approach to training and competition. What are those key points, and how does the sport psychology literature support your recommendations?

REFERENCES

1. American Psychology Association. APA Division 47 2020. Available from: https://www.apadivisions.org/division-47.
2. Rejeski WJ, Brawley LR. Defining the boundaries of sport psychology. *Sport Psychol.* 1988;2:231–42.
3. Rejeski WJ, Thompson A. Historical and conceptual roots of exercise psychology. In: Seraganian P, editor. *Exercise Psychology: The Influence of Physical Exercise on Psychological Processes.* 1st ed. New York (NY): John Wiley & Sons; 1993. p. 3–35.

4. Petruzzello SJ. Exercise and sports psychology. In: Brown SP, editor. *Introduction to Exercise Science*. 1st ed. Philadelphia (PA): Lippincott, Williams & Wilkins; 2001. p. 310–33.
5. Hazelton G, Williams JW, Wakefield J, Perlman A, Kraus WE, Wolever RQ. Psychosocial benefits of cardiac rehabilitation among women compared with men. *J Cardiopulm Rehabil Prev*. 2014;34(1):21–8.
6. Weinberg RS, Gould D. *Foundations of Sport and Exercise Psychology*. 6th ed. Champaign (IL): Human Kinetics; 2015.
7. Portenga ST, Aoyagi MW, Balague G, Cohen A, Harmison B. *Defining the Practice of Sport and Performance Psychology*. American Psychological Association; 2015.
8. Hays KF. Being fit: the ethics of practice diversification in performance psychology. *Prof Psychol Res Pract*. 2006;37(3):223–32.
9. Buckworth J, Dishman RK, O'Connor PJ, Tomporowski PD. *Exercise Psychology*. Champaign (IL): Human Kinetics; 2013.
10. Vealey RS. Sport and exercise psychology. In: Hoffman S, editor. *Introduction to Kinesiology*. 3nd ed. Champaign (IL): Human Kinetics; 2005. p. 269–300.
11. Gill DL. Sport and exercise psychology. In: Massengale JD, Swanson RA, editors. *The History of Exercise and Sport*. 1st ed. Champaign (IL): Human Kinetics, Inc; 2003. p. 293–320.
12. Vealey RS. Smocks and jocks outside the box: the paradigmatic evolution of sport and exercise psychology. *Quest*. 2006:128–59.
13. Morgan WP. Hypnosis and muscular performance. In: Morgan WP, editor. *Ergogenic Aids and Muscular Performance*. 1st ed. New York (NY): Academic Press; 1972. p. 193–233.
14. Hall GS. *Physical Education in Colleges: Report of the National Education Association*. Chicago; 1908.
15. Green CD. Psychology strikes out: Coleman R. Griffith and the Chicago Cubs. *Hist Psychol*. 2003;6:267–83.
16. Crews DJ, Landers DM. A meta-analytic review of aerobic fitness and reactivity to psychological stressors. *Med Sci Sports Exerc*. 1987;19:114–20.
17. Berger BG, Owen DR. Stress reduction and mood enhancement in four exercise modes: swimming, body conditioning, hatha yoga, and fencing. *Res Q Exerc Sport*. 1988;59:148–59.
18. Sonstroem RJ, Morgan WP. Exercise and self-esteem: rationale and model. *Med Sci Sports Exer*. 1989;21:329–37.
19. Dishman RK. *Exercise Adherence: Its Impact on Public Health*. 2nd ed. Champaign (IL): Human Kinetics; 1994.
20. Dzewaltowski DA. Toward a model of exercise motivation. *J Sport Exerc Psychol*. 1989;11:251–69.
21. Brustad R. A critical analysis of knowledge construction in sport psychology. In: Horn TS, editor. *Advances in Sport Psychology*. 2nd ed. Champaign (IL): Human Kinetics; 2002. p. 21–37.
22. Gill DL. Feminist sport psychology: a guide for our journey. *Sport Psychol*. 2001;15:363–72.
23. Dzewaltowski DA. The ecology of physical activity and sport: merging science and practice. *J Appl Sport Psychol*. 1997;9:254–76.
24. Aoyagi MW, Portenga ST, Poczwardowski A, Cohen AB, Statler T. Reflections and directions: the profession on sport psychology past, present, and future. *Prof Psychol Res Pract*. 2011;43(1):32–8.
25. Fletcher D, Maher J. Toward a competency-based understanding of the training and development of applied sport psychologists. *Sport Exerc Perform Psychol*. 2013;2(4):265–80.
26. Vealey RS. Personality and sport: A comprehensive view. In: Horn TS, editor. *Advances in Sport Psychology*. 2nd ed. Champaign (IL): Human Kinetics; 2002.
27. Hollander EP. *Principles and Methods of Social Psychology*. New York (NY): Oxford University Press; 1967.
28. Cattell RB. *The Scientific Analysis of Personality*. Baltimore (MD): Penguin; 1965.
29. Eysenck HJ, Eysenck SBG. *Eysenck Personality Inventory Manual*. London (UK): University of London Press; 1968.
30. Eysenck HJ, Nias DK, Cox DN. Sport and personality. *Adv Behav Res Ther*. 1982;4:1–56.
31. Petruzzello SJ, Landers DM, Hatfield BD, Kubitz KA, Salazar W. A meta-analysis on the anxiety reducing effects of acute and chronic exercise: outcomes and mechanisms. *Sports Med*. 1991;11:143–82.
32. Moore JB, Mitchell NG, Beets MW, Bartholomew JB. Physical self-esteem in older adults: a test of the indirect effect of physical activity. *Sport Exerc Perform Psychol*. 2012;1(4):231–41.
33. Fox KR. Self-esteem, self-perceptions and exercise. *Int J Sport Psychol*. 2000;31:228–40.
34. Bandura A. Social foundations of thought and actions: a social cognitive theory. Englewood Cliffs (NJ): Prentice-Hall; 1986.
35. Weiner B. *An Attributional Theory of Motivation and Emotion*. New York (NY): Springer; 1986.

36. Deci EL, Ryan RM. *Intrinsic Motivation and Self-Determination in Human Behavior*. New York (NY): Springer; 1985.
37. Feltz DL. Self-confidence and sports performance. In: Pandolf KB, editor. *Exercise and Sport Science Reviews*. New York (NY): MacMillan; 1988. p. 423–57.
38. Thayer RE. *The Biopsychology of Mood and Arousal*. New York (NY): Oxford University Press; 1989.
39. Deci EL, Ryan RM. Self-determination theory: a macrotheory of human motivation, development, and health. *Can Psychol*. 2008;49:182–5.
40. Burn N, Niven A. Why do they do (h)it? Using self-determination theory to understand why people start and continue to do high-intensity interval training group exercise classes. *Int J Sport Exerc Psychol*. 2019; 17(5):537–51.
41. Hooker SA, Masters KS, Ranby KW. Integrating meaning in life and self-determination theory to predict physical activity adoption in previously inactive exercise initiates enrolled in a randomized trial. *Psychol Sport Exerc*. 2020;49:1–10.
42. Benau EM, Plumhoff J, Timko CA. Women's dieting goals (weight loss, weight maintenance, or not dieting) predict exercise motivation, goals, and engagement in undergraduate women: a self-determination theory framework. *Int J Sport Exerc Psychol*. 2019;17(6):553–67.
43. Ekkekakis P, Petruzzello SJ. Acute aerobic exercise and affect. *Sports Med*. 1999;5:337–74.
44. Spence JT, Spence KW. The motivational components of manifest anxiety: drive and drive stimuli. In: Spielberger CD, editor. *Anxiety and Behavior*. New York (NY): Academic; 1966.
45. Landers DM, Boutcher SH. Arousal performance relationships. In: Williams JM, editor. *Applied Sport Psychology: Personal Growth to Peak Performance*. Mountain View (CA): Mayfield; 1993. p. 197–218.
46. Ekkekakis P, Petruzzello SJ. Analysis of the affect measurement conundrum in exercise psychology: IV. A conceptual case for the affect circumplex. *Psychol Sport Exer*. 2002;3(1):35–63.
47. Lox CL, Burns SP, Treasure DC, Wasley DA. Physical and psychological predictors of exercise dosage in healthy adults. *Med Sci Sports Exerc*. 1999;31(7):1060–4.
48. Solso RL. *Cognitive Psychology*. 4th ed. Boston (MA): Allyn & Bacon; 1995.
49. Kessler RC, Chiu WT, Demler O, Walters EE. Prevalence, severity, and comorbidity of twelve-month DSM-IV disorders in the National Comorbidity Survey Replication. *Arch Gen Psychiatry*. 2005;62(6):617–27.
50. Thorpe KE, Florence CS, Joski P. Which medical conditions account for the rise in health care spending? *Health Aff*. 2004:437–45.
51. Substance Abuse and Mental Health Services Administration. Results from the 2008 National Survey on Drug Use and Health: National Findings. Available from: https://www.dpft.org/resources/NSDUHresults2008.pdf.
52. Landers DM, Petruzzello SJ. Physical activity, fitness, and anxiety. In: Bouchard C, Shephard RJ, Stephens T, editors. *Physical Activity, Fitness, and Health*. 1st ed. Champaign (IL): Human Kinetics; 1994. p. 868–82.
53. Morgan WP. Physical activity, fitness, and depression. In: Bouchard C, Shephard RJ, Stephens T, editors. *Physical Activity, Fitness, and Health*. 1st ed. Champaign (IL): Human Kinetics; 1994. p. 851–67.
54. Pate RR, Pratt M, Blair SN, et al. A recommendation from the Centers for Disease Control and Prevention and the American College of Sports Medicine. *JAMA*. 1995;273(5):402–7.
55. Dunn AL, Trivedi MH, O'Neal HA. Physical activity dose–response effects on outcomes of depression and anxiety. *Med Sci Sports Exerc*. 2001;33(6):S587–97.
56. Knubben K, Reischies FM, Adli M, Schlattmann P, Bauer M, Dimeo F. A randomised, controlled study on the effects of a short-term endurance training programme in patients with major depression. *Br J Sports Med*. 2007;41(1):29–33.
57. Stathopoulou G, Powers MB, Berry AC, Smits JAJ, Otto MW. Exercise interventions for mental health: a quantitative and qualitative review. *Clin Psychol Sci Pract*. 2006;13(2):179–93.
58. Bartholomew JB, Morrison D, Ciccolo JT. Effects of acute exercise on mood and well-being in patients with major depressive disorder. *Med Sci Sports Exerc*. 2005;37(12):2032–7.
59. Babyak M, Blumenthal JA, Herman S, et al. Exercise treatment for major depression: maintenance of therapeutic benefit at 10 months. *Psychosom Med*. 2000;62(5):633–8.
60. Hansen CJ, Stevens LC, Coast JR. Exercise duration and mood state: how much is enough to feel better? *Health Psychol*. 2001;20:267–75.
61. Morgan WP, O'Connor PJ. Exercise and mental health. In: Dishman RK, editor. *Exercise Adherence: Its Impact on Public Health*. 1st ed. Champaign (IL): Human Kinetics; 1988. p. 91–121.
62. Chang YK, Pan CY, Chen FT, Tsai CL, Huang CC. Effect of resistance-exercise training on cognitive function in healthy older adults: a review. *J Aging Phys Act*. 2012;20:497–517.

63. Larson EB, Wang L, Bowen JD, et al. Exercise is associated with reduced risk for incident dementia among persons 65 years of age and older. *Ann Intern Med.* 2006;144(2):73–81.
64. Gauvin L, Rejeski WJ, Reboussin BA. Contributions of acute bouts of vigorous physical activity to explaining diurnal variations in feeling states in active middle-aged women. *Health Psychol.* 2000;19:265–75.
65. Berger BG, Motl RW. Physical activity and quality of life. In: Singer R, Hausenblas HA, Janelle CM, editors. *Handbook of Sport Psychology.* 2nd ed. New York (NY): Wiley; 2001. p. 636–70.
66. Morgan WP. Affective beneficence of vigorous physical activity. *Med Sci Sports Exerc.* 1985;6:422–5.
67. Steinberg H, Sykes EA. Introduction to symposium on endorphins and behavioral processes: review of literature on endorphins and exercise. *Pharmacol Biochem Behav.* 1985;23:857–62.
68. Bulbulian R, Darabos BL. Motor neuron excitability: the Hoffman reflex following exercise of high and low intensity. *Med Sci Sports Exerc.* 1986;18:697–702.
69. Ransford CP. A role for amines in the antidepressive effect of exercise: a review. *Med Sci Sports Exerc.* 1982;14:1–10.
70. Cotman CW, Engesser-Cesar C. Exercise enhances and protects brain function. *Exerc Sport Sci Rev.* 2002;30(2):75–9.
71. Delp MD, Armstrong RB, Godfrey DA, Laughlin MH, Ross CD, Wilkerson MK. Exercise increases blood flow to locomotor, vestibular, cardiorespiratory and visual regions of the brain in miniature swine. *J Physiol.* 2001;533:849–59.
72. Petruzzello SJ, Tate AK. Brain activation, affect, and aerobic exercise: an examination of both state-independent and state-dependent relationships. *Psychophysiology.* 1997;34:527–33.
73. Churchill JD, Galvez R, Colcombe S, Swain RA, Kramer AF, Greenough WT. Exercise, experience and the aging brain. *Neurobiol Aging.* 2002;23:941–55.
74. Bherer L, Erickson KI, Liu-Ambrose T. A review of the effects of physical activity and exercise on cognitive and brain functions in older adults. *J Aging Res.* 2013;2013:1–8.
75. Langlois F, Vu TTM, Chasse K, Dupuis G, Kergoat MJ, Bherer L. Benefits of physical exercise training on cognition and quality of life in frail older adults. *J Gerontol B Psychol Sci Soc Sci.* 2012;68(3):400–4.
76. Minino AM, Heron MP, Murphy SL, Kochanek KD. *Deaths: Final Data for 2004.* Report No.: 55-19. U.S. Department of Health and Human Services; 2007.
77. Bish CL, Blanck HM, Serdula MK, Marcus M, Kohl HW, Khan LK. Diet and physical activity behaviors among Americans trying to lose weight: 2000 behavioral risk factor surveillance system. *Obes Res.* 2005;13:596–607.
78. Kohrt WM, Bloomfield SA, Little KD, Nelson ME, Yingling VR. Physical activity and bone health. *Med Sci Sports Exerc.* 2004;36(11):1985–96.
79. Mazzeo RS, Cavanagh PR, Evans WJ, et al. Exercise and physical activity for older adults. *Med Sci Sports Exerc.* 1998;30(6):992–1008.
80. Flegal KM, Carroll MD, Ogden CL, Curtin LR. Prevalence and trends in obesity among US adults, 1999-2008. *JAMA.* 2010;303(3):235–41.
81. Flegal KM, Williamson DF, Pamuk ER, Rosenberg HM. Estimating deaths attributable to obesity in the United States. *Am J Public Health.* 2004;94(9):1486–9.
82. Heron MP. *Deaths: Leading Causes for 2010.* Report No.: 61. Atlanta (GA): Centers for Disease Control and Prevention; 2012.
83. Ogden CL, Carroll MD, Kit BK, Flegal KM. Prevalence of childhood and adult obesity in the United States, 2011-2012. *JAMA.* 2014;311(8):806–14.
84. Office of Disease Prevention and Health Promotion. *Healthy People 2030.* 2021. Available from: https://health.gov/healthypeople.
85. Myers RS, Ross DL. Perceived benefits of and barriers to exercise and stage of exercise adoption in young adults. *Health Psychol.* 1997;16(3):277–83.
86. Russell SJ, Craig CL. *Physical Activity and Lifestyles in Canada.* Report No. 113. Ontario (ON): Canadian Fitness and Lifestyle Research Institute; 1996.
87. Sallis JF, Hovell MF, Hofstetter CR, et al. Distance between homes and exercise facilities related to frequency of exercise among San Diego residents. *Public Health Rep.* 1990;105:179–85.
88. Finkelstein EA, Brown DS, Brown DR, Buchner DM. A randomized study of financial incentives to increase physical activity among sedentary older adults. *Prev Med.* 2008;47(2):182–7.
89. Mitchell MS, Goodman JM, Alter DA, et al. Financial incentives for exercise adherence in adults: systematic review and meta-analysis. *Am J Prev Med.* 2013;45(5):658–67.

90. Ribeiro MA, Martins MA, Carvalho CRF. Interventions to increase physical activity in middle-age women at the workplace: a randomized controlled trial. *Med Sci Sports Exerc*. 2014;46(5):1008–15.
91. McAuley E, Mullen SP, Szabo AN, et al. Self-regulatory processes and exercise adherence in older adults: executive function and self-efficacy effects. *Am J Prev Med*. 2011;41(3):284–90.
92. Culos-Reed SN, Gyurcsik NC, Brawley LR. Using theories of motivated behavior to understand physical activity. In: Singer RN, Hausenblas HA, Janelle CM, editors. *Handbook of Sport Psychology*. 2nd ed. New York (NY): Wiley; 2001. p. 695–717.
93. Redding CA, Rossi JS, Rossi SR, Velicer WF, Prochaska JO. Health behavior models. *J Health Educ*. 2000; 3:180–93.
94. Rhodes RE, Nigg CR. Advancing physical activity theory: a review and future directions. *Exerc Sport Sci Rev*. 2011;39(3):113–9.
95. Rosenstock IM, Strecher VJ, Becker MH. Social learning theory and the health belief model. *Health Educ Behav*. 1988;15(2):175–83.
96. Ajzen I. The theory of planned behavior. *Organ Behav Hum Decis Process*. 1991;50:179–211.
97. Bandura A. *Self-Efficacy: The Exercise of Control*. New York (NY): Freeman; 1997.
98. Glanz K, Rimer BK, Lewis FM. *Health Behavior and Health Education. Theory, Research and Practice*. San Francisco (CA): Wiley & Sons; 2002.
99. Marcus BH, Dubbert PM, Forsyth LH, et al. Physical activity behavior change: issues in adoption and maintenance. *Health Psychol*. 2000;19(1):32–41.
100. Sallis JF, Hovell MF. Determinants of exercise behavior. In: Pandolf KB, Holloszy JO, editors. *Exercise and Sport Science Reviews*. Baltimore (MD): Williams & Wilkins; 1990. p. 307–30.
101. Dishman RK, Buckworth J. Adherence to physical activity. In: Morgan WP, editor. *Physical Activity and Mental Health*. Philadelphia (PA): Taylor & Francis; 1997. p. 63–80.
102. Dishman RK, Sallis JF. Determinants and interventions for physical activity and exercise. In: Bouchard C, Shephard RJ, Stephens T, editors. *Physical Activity, Fitness, and Health*. Champaign (IL): Human Kinetics; 1994. p. 214–38.
103. Rebar AL, Elavsky S, Maher JP, Doerksen SE, Conroy DE. Habits predict physical activity on days when intentions are weak. *J Sport Exerc Psychol*. 2014;36:157–65.
104. Cooper LB, Mentz RJ, Sun JL, et al. Psychosocial factors, exercise adherence, and outcomes in heart failure patients insights from heart failure: a controlled trial investigating outcomes of exercise training (HF-ACTION). *Circ Heart Fail*. 2015;8(6):1044–51.
105. Wong JN, McAuley E, Trinh L. Physical activity programming and counseling preferences among cancer survivors: a systematic review. *Int J Behav Nutr Phys Act*. 2018;15(1):1–22.
106. Kampshoff CS, van Mechelen W, Schep G, et al. Participation in and adherence to physical exercise after completion of primary cancer treatment. *Int J Behav Nutr Phys Act*. 2016;13(1):100.
107. Meeuwisse WH, Tyreman H, Hagel B, Emery C. A dynamic model of etiology in sport injury: the recursive nature of risk and causation. *Clin J Sport Med*. 2007;17(3):215–9.
108. Herring SA, Kibler WB, Putukian M. Psychological issues related to illness and injury in athletes and the team physician: a consensus statement — 2016 update. *Med Sci Sports Exerc*. 2017;49(5):1043–54.
109. Gennarelli SM, Brown SM, Mulcahey MK. Psychosocial interventions help facilitate recovery following musculoskeletal sports injuries: a systematic review. *Phys Sportsmed*. 2020;48:370–7.
110. Slimani M, Bragazzi NL, Znazen H, Paravlic A, Azaiez F, Tod D. Psychosocial predictors and psychological prevention of soccer injuries: a systematic review and meta-analysis of the literature. *Phys Ther Sport*. 2018;32:293–300.
111. Hildingsson M, Fitzgerald UT, Alricsson M. Perceived motivational factors for female football players during rehabilitation after sports injury — a qualitative interview study. *J Exerc Rehabil*. 2018;14(2):199–206.
112. Covassin T, Beidler E, Ostrowski J, Wallace J. Psychosocial aspects of rehabilitation in sports. *Clin Sports Med*. 2015;34(2):199–212.
113. Timpka T, Bargoria V, Halje K, Jacobsson J. Infographic: elite athletes' anxiety over illness ups risk of injury in competition. *Br J Sports Med*. 2018;52(15):955.
114. Wolanin A, Gross M, Hong E. Depression in athletes: prevalence and risk factors. *Curr Sports Med Rep*. 2015;14(1):56–60.
115. Montenegro SO. Disordered eating in athletes. *Athl Ther Today*. 2006;11(1):60–2.

116. McDuff D, Stull T, Castaldelli-Maia JM, Hitchcock ME, Hainline B, Reardon CL. Recreational and ergogenic substance use and substance use disorders in elite athletes: a narrative review. *Br J Sports Med.* 2019;53(12):754–60.
117. Schinke RJ, Stambulova NB, Si GY, Moore Z. International Society of Sport Psychology position stand: athletes' mental health, performance, and development. *Int J Sport Exerc Psychol.* 2018;16(6):622–39.
118. Putukian M. The psychological response to injury in student athletes: a narrative review with a focus on mental health. *Br J Sports Med.* 2016;50(3):145–8.
119. Arthur-Cameselle JN, Baltzell A. Learning from collegiate athletes who have recovered from eating disorders: advice to coaches, parents, and other athletes with eating disorders. *J Appl Sport Psychol.* 2012;24(1):1–9.
120. Donohue B, Pitts M, Gavrilova Y, Ayarza A, Cintron KI. A culturally sensitive approach to treating substance abuse in athletes using evidence-supported methods. *J Clin Sport Psychol.* 2013;7(2):98–119.
121. Williams SE. Comparing movement imagery and action observation as techniques to increase imagery ability. *Psychol Sport Exerc.* 2019;44:99–106.
122. Cumming J, Williams SE. Introducing the revised applied model of deliberate imagery use for sport, dance, exercise, and rehabilitation. *Mov Sport Sci.* 2013;82:69–81.
123. Hall CR, Mack DE, Paivio A, Hausenblas HA. Imagery use by athletes: development of the sport imagery questionnaire. *Int J Sport Psychol.* 1998;29(1):73–89.
124. Jenny O, Ely FO, Magalas S. It's all about timing: an imagery intervention examining multiple image speed combinations. *J Appl Sport Psychol.* 2020;32(3):256–76.
125. Munroe-Chandler KJ, Hall CR. Imagery. In: Schinke RJ, McGannon KR, Smith B, editors. *The Routledge International Handbook of Sport Psychology.* London (UK): Routledge; 2016. p. 357–68.
126. Quinton ML, Cumming J, Williams SE. Investigating the mediating role of positive and negative mastery imagery ability. *Psychol Sport Exerc.* 2018;35:1–9.
127. Filgueiras A, Quintas Conde EF, Hall CR. The neural basis of kinesthetic and visual imagery in sports: an ALE meta-analysis. *Brain Imaging Behav.* 2018;12(5):1513–23.
128. Slimani M, Bragazzi NL, Tod D, et al. Do cognitive training strategies improve motor and positive psychological skills development in soccer players? Insights from a systematic review. *J Sports Sci.* 2016;34(24):2338–49.
129. Brobst B, Ward P. Effects of public posting, goal setting, and oral feedback on the skills of female soccer players. *J Appl Behav Anal.* 2002;35(3):247–57.
130. Prochaska JO, Johnson SS, Lee P. The transtheoretical model of behavior change. In: Schron E, Ockene J, Schumaker S, Exum WM, editors. *The Handbook of Behavioral Change.* 2nd ed. New York (NY): Springer; 1998. p. 159–84.
131. Becker MH, Maiman LA. Sociobehavioral determinants of compliance with health care and medical care recommendations. *Med Care.* 1975;13:10–24.
132. Ajzen I, Madden TJ. Prediction of goal-directed behavior: attitudes, intentions, and perceived behavioral control. *J Exp Soc Psychol.* 1986;22:453–74.

CHAPTER 9

Motor Behavior

After completing this chapter, you will be able to:

1. Define motor behavior and provide examples of how motor development, motor learning, and motor control contribute to the understanding of movement as it relates to physical activity, exercise, sport, and athletic performance.
2. Identify the important historic events in the development of motor behavior as a scientific discipline.
3. Describe some of the important topics in the fields of motor development, motor learning, and motor control.
4. Discuss the important areas of study in motor development, motor control, and motor learning.

The basic and applied knowledge found in the discipline of **motor behavior** impacts exercise science professionals in many important ways. Motor behavior is an umbrella term that describes the study of the interactions between many of the physiological and psychological processes of the body. Motor behavior knowledge helps provide exercise science and other allied health professionals with an understanding of how the body develops, controls, and learns movement skills that individuals use in everyday activities as well as physical activity, exercise, sport, and athletic competition. Exercise science professionals use the knowledge gained from the study of motor behavior to improve physical activity and exercise performance and enhance success in sport and athletic competition. Motor behavior is comprised of three related areas of study: motor development, motor learning, and motor control (1).

Motor development is the study of change in motor behavior over a lifespan and the various processes that underlie these changes (2). Motor development is an area of study that examines the alterations in motor behavior that result from the maturation of the individual, rather than those alterations that occur owing to practice or experience. Motor development is concerned with how individuals learn and control movement as physical and mental changes occur over the lifespan. Originally, motor development involved the study of developmental changes from infancy to adulthood, but it now also includes changes that occur throughout the entire lifespan and into old age (1,3).

Motor learning is the study of how individuals learn skilled movements from experience or practice (4). When individuals learn how to move in efficient motor patterns, there is often a permanent change in the neural control of muscle actions. Motor learning has evolved primarily from the disciplines of psychology and education. The application of its principles, however, is found in all areas of exercise science and allied health professions (1).

Motor control is the study of the neurological, physiological, and behavioral aspects of movement (4). The neuromuscular system commands complex and coordinated movements by individuals, and motor control is concerned with how our brain and spinal cord plan and perform those movements. Motor control includes the study of how the central and peripheral nervous systems control the body prior to as well as during movement (1).

The areas comprising the discipline of motor behavior are interrelated and all have application to the performance of physical activity, exercise, sport, and athletic competition. The development of knowledge and the study of motor development, learning, and

Motor behavior An umbrella term that includes the disciplines of motor control, motor learning, and motor development.

Motor development The study of motor performance throughout the lifespan from birth through old age.

Motor learning The study of the acquisition of basic and advanced movement skills that are used in everyday activities.

Motor control The study of the understanding of the mechanisms by which the nervous system activates the muscles to coordinate body movements.

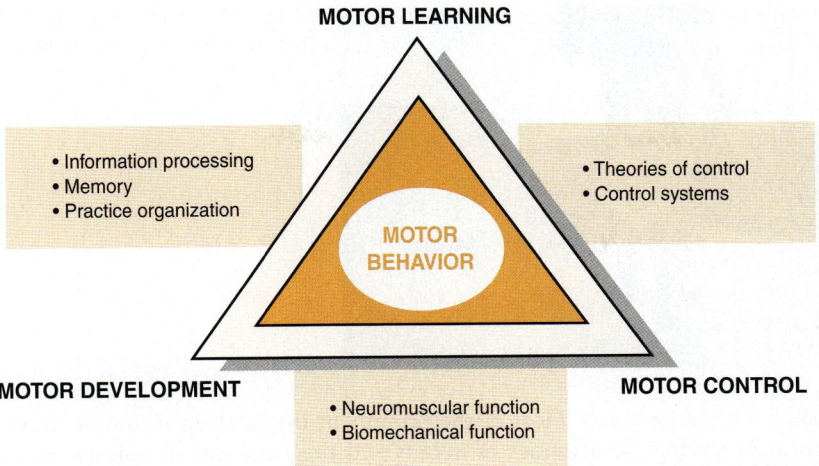

FIGURE 9.1 A schematic of the general organizational structure of motor behavior (1). (From Brown SP. *Introduction to Exercise Science*. Baltimore (MD): Lippincott, Williams & Wilkins; 2000. 335 p.)

control are, however, generally performed separately. This is partially because the individuals studying these areas come from different fields of science and training. It is also important to understand that the broad application of motor behavior goes beyond the study of movement associated with physical activity, exercise, sport, and athletic competition. The principles of motor behavior are applied to a variety of other skills such as factory work, operating equipment and machinery, and performing complex movements associated with leisure activities. Figure 9.1 provides a schematic of the general organizational structure of motor behavior (1). Exercise science and allied health professionals working in a variety of employment settings use the principles of motor behavior to enhance performance in a variety of activities and sports, as well as promote recovery from injury or medical interventions.

HISTORY OF MOTOR BEHAVIOR

Similarly to many of the other areas of study in exercise science, the history of motor behavior begins with the writings of ancient scholars who provided the framework for many of the principles associated with nervous system control over muscle contraction. The historical development of motor behavior is derived from an interaction of the parent disciplines of biology, psychology, and education. Trying to provide an overview of the history of motor behavior is challenging because much of the historical development of the three areas is interrelated. Only recently have distinct separations of the areas of motor development, learning, and control occurred.

Early Influences on Motor Development

The area of motor development had its early foundational period from 1787 to 1928. During this time, contemporary developmental psychology and motor development had

FIGURE 9.2 Motor development begins at an early age. (From Shutterstock.)

established a basis for study. Most of the prominent information in motor development came from observing the activities of infants and their changes in reflexes, movements, and feeding behaviors on a day-to-day basis (2). The publication of the book *Infancy and Human Growth* (5), in 1928, marked a period of increased research and scholarly study in motor development. Throughout the early 20th century, important work in motor development was being conducted on both infants and children (Figure 9.2). For example, a classic study by Myrtle McGraw (1899–1988) provided a thorough description of the behavior sequences that occur in infants and young children (6). Significant work was also performed on the role of maturation on motor development and the ability of children to reach, sit, and stand (7,8). During the mid-20th century, much of the research and scholarly activity performed in motor development was conducted by physical educators who had an interest in the growth, strength development, and motor performance in children (8).

Recent Influences on Motor Development

Beginning in the mid-1960s, the focus of motor development shifted from developmental psychology and the understanding of the influence of maturation to physical education, specifically how to improve children's motor behavior (2). This allowed motor development to become part of the study of physical activity (8). During the late 1960s and early 1970s, four influential books were published by scholars that served to define motor development (9–12). During the 1980s and 1990s, there was a shift in major areas of research and study in motor development that resulted in an expansion of knowledge in the following subjects:

- Variations in performance associated with age and sex
- Variations in performance associated with maturation
- Physical activity as a factor in growth and maturation
- Racial/ethnic and social factors influencing motor performance
- Cognitive factors in children's skill acquisition (8).

Individuals who have been instrumental in the advancement of motor development include G. Lawrence Rarick, Jane Clark, Eleanor Gibson, Robert Malina, Mary Ann Roberton, Vern Seefeldt, Esther Thelen, Beverly Ulrich, Dale Ulrich, and Jerry Thomas (8). In an effort to promote a better understanding of motor development and learning, the American Alliance for Health, Physical Education, Recreation, and Dance (AAHPERD) created a special interest area for scholars and professionals (8). As a professional organization, AAHPERD changed its name to the Society of Health and Physical Educators (SHAPE) in 2014 in an effort to better support professionals in the discipline. SHAPE has numerous professional councils that support motor development, including the Physical Activity Council and the Research Council.

Early Influences on Motor Control and Motor Learning

Much of the early work in motor control and motor learning focused on understanding and explaining the control of muscle contraction by the nervous system. Claudius Galen (ca. 129–ca. 216 AD), a Roman physician, proposed that muscle contraction was controlled by a fluid that passed down the nerves to the muscles, causing them to inflate. Galen termed the fluid, the key component to this hydraulic system, as "animal spirits." This was the main theory explaining contraction until Rene Descartes (1596–1650) expanded on Galen's proposed hydraulic system in the mid-17th century. According to Descartes, a sensory signal caused the movement of "animal spirits" from the heart and arteries into the muscles responsible for movement. Even though many scientists provided evidence that contradicted the hydraulic system of muscle control, this model remained the dominant theory for the control of muscle contraction until the late 18th century, when it was replaced with a model that was centered on the concept of bioelectricity (13–15).

In the late 18th century, Luigi Galvani (1737–89) conducted a series of experiments that resulted in the formation of the concept of bioelectricity. When Galvani applied an electric stimulus to a nerve or muscle of a frog, it caused a contraction. Galvani also demonstrated that muscle contraction in a frog could be caused by lightning or by contacting the nerve of one frog with the nerve of another frog. Galvani was convinced that certain tissues could generate electricity, which in turn resulted in muscle contraction. Figure 9.3 illustrates Galvani's experiment with a frog that helped establish the neurophysiologic concept of bioelectricity, which provided the foundation for the development of motor learning and motor control (8).

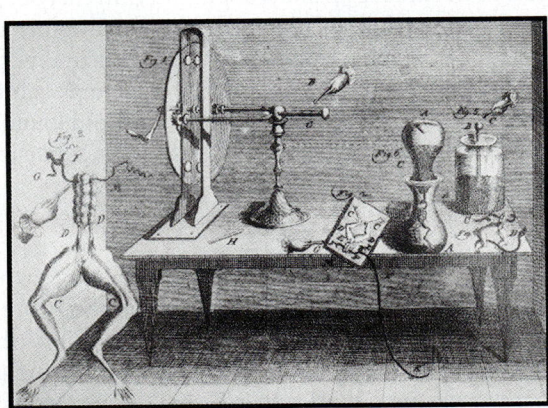

FIGURE 9.3 Galvani's experiment with a frog leads to the concept of bioelectricity.

FIGURE 9.4 The training of air force pilots led to the development of motor learning and motor control as disciplines of study. (From Shutterstock.)

The mid-19th century saw the birth of motor control and motor learning as they are presently understood. By the end of the 19th century, several significant research experiments started to shape the fields of motor control and motor learning. In the late 1890s, William Bryan and Noble Harter (16,17) reported on the existence of human learning curves and plateaus and different characteristics of novice and expert telegraph conductors (8). The study of motor control was further expanded in 1903 with the publication of the book *Le Mouvement* by R. S. Woodworth. This book was critical in establishing and expanding the field of motor skills research (8,18). The law of effect, which holds that stimuli that produce a pleasant or satisfying effect during a particular situation are likely to occur repeatedly during that situation (19), was developed during the late 1920s and significantly impacted motor learning research (8). In 1925, Coleman Griffith, the director of the Research in Athletics Laboratory at the University of Illinois, conducted significant research on the motor skills of athletes. As with many other areas of study in exercise science, World War II had a major influence on the development of motor control and motor learning as disciplines of study. In particular, the training of air force pilots (Figure 9.4) was instrumental in creating some of the early theories and foundational knowledge on memory, muscle control, movement, transfer of learning, and practice (8).

Recent Influences on Motor Control and Motor Learning

Franklin Henry played an instrumental role in the emergence of motor control and learning as areas of study in exercise science (8). In particular, his approach to memory (commonly referred to as the "memory drum" theory) spurred on our understanding of cognitive activity in motor control and learning (8,20). Henry is often known as the "father of motor skills research (21)." Other individuals who played a significant role in the development of motor control and learning included Alfred Hubbard (University of Illinois) and Arthur Slater-Hammel (Indiana University), who are well known for starting graduate programs in motor control and learning at their respective universities during the late 1960s. The *Journal of Motor Behavior*, the first journal devoted entirely to the publishing of scholarly research in motor learning and control, was first published in 1969. Other journals that have been instrumental in disseminating research and knowledge in motor control and learning include the *Journal of Human Movement Studies* (founded in 1970), *Research Quarterly for Exercise and Sport* (founded in 1930), *Human Movement Science* (founded in 1982), and *Motor Control* (founded in 1997) (8).

Several professional organizations have been instrumental in the development of motor control and motor learning. The primary organization that has served all areas of motor development, control, and learning has been the North American Society for Psychology of Sport and Physical Activity (NASPSPA). Although many individuals associate NASPSPA primarily with exercise and sport psychology, there are actually three major components of NASPSPA: motor control and learning, motor development, and sport psychology. NASPSPA publishes the *Journal of Motor Learning and Development*, which is designed to advance the understanding of movement skill attainment and expression across the lifespan. Additional information on the founding of NASPSPA is contained in Chapter 8, "Exercise and Sport Psychology." The Society for Neuroscience (founded in 1969), Society for Neural Control of Movement (founded 1990), and the International Society of Motor Control (founded 2002) all support motor control and learning by supporting development activities in the form of professional conferences and workshops (8). Throughout the 21st century there has been an increased emphasis on how motor control and learning is altered in the elderly, with specific emphasis on gait, balance, falls, and the ability to operate a motor vehicle (22,23).

Thinking Critically

In what ways has each of the areas of motor behavior (motor development, motor control, and motor learning) contributed to a broader understanding of health, physical fitness, and exercise?

In what ways has each of the areas of motor behavior (motor development, motor control, and motor learning) contributed to a broader understanding of how individuals can become successful in sport and athletic competition?

The development of knowledge that forms the basis of motor behavior has occurred through the interactions and efforts of individuals comprising numerous academic disciplines. Each of the areas that collectively constitute motor behavior has significant professional development opportunities and important application to the activities of exercise science and allied health care professionals. Table 9.1 provides some of the significant historic events in the development of motor behavior.

MOTOR DEVELOPMENT

Motor development is concerned with the study of motor performance and the factors that underlie changes throughout the lifespan. Motor development has several different aspects that distinguish it from motor learning and motor control. First, the origins of motor development arise primarily from the discipline of education, with contributions from the disciplines of educational psychology and physiology. Second, motor development has a closer association with the discipline of physiology than either motor learning or motor control. Because physical maturation and growth play an important role in an individual's

Table 9.1	Significant Events in the Historic Development of Motor Behavior
DATE	SIGNIFICANT EVENT
1903	Publication of the book *Le Mouvement*, which established the field of motor skills research
1925	Coleman Griffith established the Athletics Research Laboratory at the University of Illinois
1929	Publication of the book *Infancy and Human Growth*
1960	Franklin Henry proposed his "memory drum" theory
1967	Establishment of the NASPSPA
1969	Publication of the *Journal of Motor Behavior*
1970	Publication of the *Journal of Human Movement Studies*
1978	Formation of the Motor Development Academy in the AAHPERD
1997	Publication of the Journal *Motor Control*
2002	International Society of Motor Control founded
2014	AAHPERD changes its name to Society of Health and Physical Educators (SHAPE)

motor development, these factors require considerable attention. Finally, the research and scholarly methods employed in motor development are different from motor learning and control. In motor development, **longitudinal** and **cross-sectional** studies of individuals or groups of individuals are more prevalent than in the areas of motor learning and motor control (1). The changes that occur throughout the lifespan are a result of the requirements of the movement task, the biology of the individual, and the specific environmental conditions (24). If any of these three factors change, then movement also changes (3). Many exercise science professionals will encounter individuals from across the lifespan during their work, making the understanding of motor development important for ensuring the success of individuals involved in physical activity, exercise, sport, and athletic competition.

Lifespan Stages

Human development encompasses all aspects of behavior and can only artificially be separated into stages (24). The separation into stages, however, allows for a partitioning of motor development activities into the following: infancy, childhood, adolescence, adulthood, and older adult. Each of these stages is characterized by either developmental markers or a chronologic age; however, there is not always agreement between motor development experts as to when these stages

Longitudinal Following a sample of individuals over a period of time.
Cross-sectional Selection of a sample of subjects to represent the population as a whole.

start and finish. Some experts suggest adding an additional prenatal stage, splitting childhood into early childhood and later childhood stages, and splitting older adults to include an oldest old adult (1,25). This splitting would allow for the further differentiation of human development in those stages. The following sections provide a brief overview of each developmental stage.

Prenatal

Several prenatal factors affect motor development during infancy and throughout later years. Positive prenatal factors include appropriate nutritional intake, proper weight gain, and maintaining good physical fitness. There are also negative prenatal factors that can lead to birth defects or developmental abnormalities that can affect normal motor development in the later stages of growth. Some of the more common negative influences include malnutrition and the use of drugs, alcohol, and tobacco by the birth mother; hereditary factors that include chromosome-based and gene-based disorders; environmental factors, including radiation and chemical pollutants; and medical problems, including sexually transmitted diseases, maternal infection, and stress experienced by the birth mother during pregnancy (26).

Infancy

Much of the information about child infancy comes from describing what movements and activities infants engage in during the early period of development. Much of this information describes the primitive reflexes of infants, which are associated with the basic human needs of nourishment and security. Not all movements performed by infants, however, are reflexive in nature. For example, **maturational theorists** have identified and described several landmark activities that occur during an infant's early development of locomotion and manual control. Locomotion for infants includes crawling, creeping, and walking. Each of these forms of locomotion is important for the normal development of an infant. Manual control includes the movements of the hands and arms to hold and manipulate an object. The stages of manual control include reaching, grasping, and releasing behaviors. The development of locomotion and manual control progresses in an organized and orderly fashion during the normal growth and development of an infant (3,25–28).

Early Childhood and Later Childhood

Early childhood is considered the period from 2 to 6 years of age, whereas later childhood is the period from 6 to 10 years of age. Motor development in childhood involves the improvements in fundamental movement skills and the practice of these movements in everyday activities as well as for participation in physical activity, exercise, sport, and athletic competition. Common fundamental movement patterns during early and late childhood involve specific movements such as walking, running, jumping, and throwing (Figure 9.5). By 3 years of age, children should display some acceptable level of fundamental movement

Maturational theorists Individuals who believe that the chief principle of developmental change in an individual is maturation.

FIGURE 9.5 The development of fundamental movement patterns such as running (**A**) and throwing (**B**) occurs in early childhood. (From Shutterstock.)

patterns; however, confusion often exists in body direction, temporal, and spatial adjustments. Gross motor control is developing rapidly during this period. By 6 years of age, the child should refine many of these movement patterns to the degree that they would be classified as mature. Children should become aware of body shape, body size, and physical capacity at this point, and basic mechanical principles of movement are being developed during this time. This period marks a transition from refining fundamental movement abilities to the establishment of transitional movement skills to simple games and athletic skills (1,26).

Adolescence

As a child progresses into and through adolescence, significant improvements in motor skill performance occur as a result of substantial physical and physiological changes. Many of these changes are the result of body growth and changes in body structure. Many of the differences between males and females are the result of structural changes that give males several physical advantages over females. For example, as a result of sexual maturation and the increased production of anabolic hormones, males start to produce more muscle mass, and there are physical changes in the arms, hips, and shoulders that give a mechanical advantage in certain activities to males over females. These changes result in widening of the gap in motor performance between boys and girls during adolescence. Adolescents advance through different maturation levels and stages of learning new movement skills at different rates of development (25,27). The changes that occur during adolescence lead to physical differences later in life (Figure 9.6).

FIGURE 9.6 Developmental differences during adolescence (**A**) lead to physical differences later in life (**B**). (From Shutterstock.)

Adulthood

Early adulthood is the period when most individuals achieve their peak physical performances. Peak motor performance occurs around 22 to 25 years of age for women and around 29 years of age for men (25). Although very few changes in motor performance occur throughout early and middle adulthood, there is large individual variation in achieving peak performance. Time to peak physical performance and changes in movement patterns are a result of the type of motor skill or movement performed and the frequency of opportunities to perform the motor skill. The maintenance of motor skills through adulthood is a function of the motivation and opportunity to participate in physical activity, exercise, sport, and athletic competition (25).

Older Adulthood

As adults continue to age, the number of health-related problems increases. This is caused partly by a decrease from peak performance of various physical and physiologic functions (1). Decreases in performance commonly begin to occur in cardiorespiratory function, muscular function, and psychomotor function at around 50 to 60 years of age. Older adults tend to decrease more rapidly in physiologic function as they age, with some of the changes occurring differently between genders. Many factors can affect the rate of decline such as genetic predisposition, level of physical activity, participation in regular exercise, fitness level, and nutritional intake (1,25). It has been demonstrated that the older adult responds favorably to the intervention and practice of motor skills (Figure 9.7). Table 9.2 shows the primary changes in the cardiovascular, musculoskeletal, and psychomotor functions that occur with aging.

Of particular interest to the motor development changes that occur with aging is the alteration that transpires in **psychomotor function** or the ability to integrate cognition with motor abilities (1). Two types of intelligence play an important role in psychomotor function: crystal intelligence and fluid intelligence. **Crystal intelligence** is derived largely from educational experiences and knowledge. Crystal intelligence is primarily about storing information and can increase until an individual reaches about 60 years of age. **Fluid intelligence** is primarily about reasoning and abstract thought. Learning is considered a mechanism of

FIGURE 9.7 Older adults respond favorably to exercise and motor skill development. (From Shutterstock.)

Table 9.2	Primary Changes in Cardiovascular, Musculoskeletal, and Psychomotor Function with Aging (1,25)
SYSTEM	PRIMARY CHANGES
Cardiovascular	Decreased cardiac output, heart rate, myocardial muscle mass, peripheral blood flow, and maximal oxygen consumption
	Increased cardiac mass and time to return to resting heart rate
Musculoskeletal	Decreased lean body weight, bone mass, muscle mass, strength, muscle fibers, and motor neurons
	Increased body fat and muscle collagen
Psychomotor	Decreased attention to motor tasks
	Decreased motor unit recruitment
	Increased reaction time

Psychomotor function The ability to integrate cognition with motor abilities.
Crystal intelligence The ability to store information in the brain.
Fluid intelligence The ability to perform reasoning and abstract thought.

fluid intelligence. Essentially, fluid intelligence is a measure of the state of the brain because it is a measure of an individual's ability to make new and unique connections. Crystal intelligence is a state of the mind based on education, because it is a measure of well-established pathways in the brain, not the formation of new ones. Fluid intelligence starts to decrease when an individual enters the fourth decade of life and continues to decline as the individual ages. Older individuals do not learn as quickly as younger adults but can and do learn new motor skills and movements. The rate at which an individual loses fluid intelligence is related to the amount that an individual uses their fluid intelligence. The more an individual uses his or her mind, the slower the decline in psychomotor function (1,29).

Use of Motor Development Knowledge

The knowledge gained through the study of motor development can assist exercise science and allied health professionals working across a broad age range of individuals. For example, professionals working with young children must be aware of the various developmental progressions so that age-appropriate activities and games can be used to enhance individual fundamental movement skill development and acquisition (Figure 9.8). The development of fundamental movement skills is central to a child's physical, cognitive, and social development as well as essential for engaging in a physically active lifestyle. Performance of fundamental movement skills should be evaluated, and appropriate activities and games can be employed for development (30). Issues such as the rate of physical growth and readiness to learn are key components in movement skill development (31). For those professionals working with older children, knowledge of how physical and motor development characteristics are acquired is critical for ensuring that children refine the fundamental movement abilities in the areas of locomotion, manipulation, and stability (32). Skills such as running, skipping, jumping, throwing, catching, and balancing are critical for normal progression to advanced skills. The development of these skills further enhances the abilities of a child to participate in age-appropriate games and activities as well as developing normal psychological and social skills. As children move to the adolescent period of growth and development, numerous physiological and psychological changes occur that influence the acquisition of mature fundamental movement skills. It is during this period of development that gender differences begin to significantly appear. Exercise science and allied health professionals working with adolescents must ensure the mastery of the mature fundamental movement

FIGURE 9.8 Motor development knowledge can be used to establish age-appropriate activities for children. (From Shutterstock.)

skills such as skipping, swinging, and movement combinations before the introduction of specialized movement skills occurs (33). As individuals move through adulthood, individual characteristics, the demands of the movement skill, and the environmental circumstances are major factors that determine the level of success experienced by the adult in the performance of a motor task (34). As an individual moves into old age, the changes to the various systems of the body can have a significant impact on the ability to perform motor skills. In much the same manner that children cannot do many adult activities, many older adults cannot do many of the motor skill activities of a younger person (34). It is important for the exercise science professional to be aware of those motor development changes that occur across the lifespan so that appropriate measures can be taken to ensure safe, successful, and enjoyable participation in physical activity, exercise, sport, and athletic competition.

Thinking Critically

What specific information about motor development is necessary for an individual attempting to improve a motor skill associated with exercise?

MOTOR LEARNING

Motor learning is the study of how we become skilled at basic and advanced movements that are used in everyday life, including those involved with physical activity, exercise, sport, and athletic competition. Knowledge and content study in motor learning require an understanding of the process of learning, processing information, organizing practice to make learning efficient, and the process of memory. Exercise science and allied health professionals must understand the principles of motor learning to enhance the success of individuals beginning an exercise program or participating in a rehabilitation program, as well as for putting athletes in the best position to be successful during sport and athletic competition. There are several important components to understanding motor learning, including information processing, memory, and practice organization and learning.

Information Processing

The study of information processing comes from the parent academic discipline of **cognitive psychology**. Information is constantly presented to individuals during everyday activities and must be processed in a manner that often requires a response or action. The focus of the material in this section is on how information is processed efficiently and effectively during participation in physical activity, exercise, sport, and athletic competition. Individuals must determine what information is critical for movement and then organize the information so that the muscles can respond in a coordinated and efficient movement pattern. For example, an individual who is running outside must process information about the external environment and then make decisions about what action or movement the body should take. If during the run, the individual encounters a hill, information about the slope

Cognitive psychology Branch of psychology that studies the mental processes involved in perception, learning, memory, and reasoning.

and the length of the hill must be processed so that appropriate body lean, leg speed, and stride length can be achieved and the individual can run to the top of the hill. Information processing is customarily organized into three stages (1,4,21):

1. **Stimulus recognition** — collecting information from the environment, which is then identified or recognized as a pattern
2. **Response selection** — deciding what response to make with the information, including determining the stimulus–response compatibility
3. **Response programming** — organizing and initiating an action after a stimulus has been identified and a response has been selected

Information processing occurs in the brain and cannot be directly observed or easily studied. An assortment of indirect observation methods are therefore used to create hypotheses about brain activity during information processing. A commonly used method is to record how quickly an individual responds to a stimulus. This method requires an individual to receive an external cue from the environment and then respond to that cue. The time required to initiate the response is referred to as **reaction time** and is an indirect measurement of how long it takes for an individual to process the information involved in making a decision and to begin responding. Using reaction time to analyze a person's mental processing is called the **chronometric method** (1,4,27).

Traditionally, information processing has been viewed as occurring in series or one stage at a time, a position that has continued to define experimentation and study within the field. To accurately study, this method of information processing involves changing and controlling each of the stages individually to see how a particular stage affects overall reaction time. The manipulation of processes in one stage allows researchers and scholars to hypothesize on the role of that individual stage in processing information for a particular movement skill (1,21,27).

The stimulus recognition stage requires an individual to recognize that something has changed in the environment and that some stimulus has appeared. This is followed by the individual deciding on an appropriate response to that stimulus. The brain organizes a specific set of instructions for muscle actions to send out to the body to cause a particular movement to occur. These instructions give specific commands about **body stabilization**

Stimulus recognition When an individual collects information from the environment and recognizes it.
Response selection When an individual decides what to do after collecting and processing information.
Response programming When an individual initiates an action after a response has been selected.
Reaction time The time it takes to receive and respond to a stimulus.
Chronometric method Using reaction time to measure an individual's response to a stimulus.
Body stabilization The process of holding the body in a desired position.

and positioning. The recognition and identification of a stimulus is strongly affected by the intensity and clarity of the stimulus. For a stimulus to be identified, the brain must be aroused to the point at which it contacts memory, which makes an association between the stimulus and something meaningful (1,21,27). For example, Figure 9.9 shows how a runner approaching a hill must recognize the hill, including identifying its grade or slope and its distance, and then process the information so that appropriate muscle actions may occur and the runner can successfully run up the hill. When required to run up a steep incline, individuals will change their movement patterns to successfully engage the incline (35,36). Changes with uphill running include no aerial phase, faster stride frequency, and shorter foot-ground contact time (37).

The response selection stage of information processing requires an individual to decide on a suitable response to the stimulus. Factors such as the number of choices (called stimulus–response alternatives) and how natural the relationship between the stimulus and appropriate response is (called stimulus–response compatibility) affect the response selection. Practice may affect both of these factors. Generally, the more natural the relationship between the stimulus and appropriate response is, the quicker the individual decides on a response (1,27,38). Using the preceding example, a runner can select the appropriate body position, stride length, and stride speed based on the information derived from the stimulus recognition stage. If the runner has encountered the same type of hill during a previous run, then the practice of running up that hill will help guide the appropriate response selection of body position, stride length, and leg speed for successfully running up the hill. If

FIGURE 9.9 To be successful, a runner must cognitively process information about running up a hill.

the runner encounters a new hill, then the response will be selected using a combination of previous experiences.

In the response programming stage, the commands to the muscles are organized, and the response is initiated by the brain. The complexity and duration of the response affect response programming. As the complexity of the response increases, the reaction time of the individual also increases (39). The amount of time an individual has to organize muscular commands also affects response programming (27,39,40). If a runner encounters a hill after a long stretch of flat running during, which the hill can be seen, there is time for the runner to make the appropriate programming response. If, however, the runner encounters the hill after coming around a curve that blocked the view of the hill, then the response programming can be affected and perhaps the runner has to start up the hill before the appropriate body position, stride length, and stride speed are selected.

Information processing can be influenced by various factors that cause reaction time to increase or decrease. During sport and athletic competitions, correctly anticipating a movement by an opponent can decrease the time required by an athlete to respond to a movement by an opponent. For example, if a basketball player anticipates a specific move by an opponent during a game, the player can adjust their position to the opponent and effectively decrease the time required to respond to the movement. In sport and athletic competitions, an individual or team can gain an advantage by using strategies to influence each of the three components of information processing (1,27):

1. Stimulus identification — conducting different types of plays from the same formation
2. Response selection — increasing the number of options to which an opponent must respond
3. Response programming — forcing the opponent to perform a more complicated task

Each of these strategies can significantly influence the stages of information processing. When individuals are challenged at each stage of information processing, there are increases to the overall reaction time. Some disagreement exists as to whether the stages of information processing are separate and different or whether information is processed in a sequential order so that individuals are able to perform the processing of more than one stage concurrently (27,41,42).

Memory

Memory is important for retaining and recalling facts, events, and impressions and for remembering or recognizing previous experiences. A commonly used model to explain memory is called the **multistore memory model** (43). This model has three stores: short-term sensory store, short-term memory, and long-term memory. Although this model is

Multistore memory model The most widely used model to explain memory storage in humans.

oversimplified, it is useful in understanding how information is processed. Information going from short-term store (also known as working memory) to long-term memory signifies information going into the long-term memory. This process is known as **encoding**. Going from long-term memory to short-term memory signifies information moving from the permanent memory to the working memory. This process is called **decoding** (27).

Each memory store has a specific function and capacity for keeping information. The capacity of a memory store is the amount of information it can effectively hold. The duration is the length of time that the information remains in the memory store. The short-term sensory store collects information from the environment through the senses. The short-term sensory store has an unlimited capacity for storing information but a very short storage duration. This means individuals can hold a lot of information from the senses but for a short time, perhaps less than 1 second. The short-term sensory store holds information, while a decision is made on the importance of the information. This decision is made by a process called **selective attention,** a process that requires an individual to actively choose one unit of information to pay attention to at a time. The information that is attended to is sent to the short-term memory store (4,21,27).

There has been considerable interest in the role of regular physical activity and exercise in psychological function, principally, the influence that cardiovascular fitness and aerobic exercise have on various markers of cognitive function, including memory, attention, reaction time, and crystallized and fluid intelligence (Figure 9.10). The working theory has been that age-associated reductions in cardiovascular function lead to lower oxygen levels in the brain and that high levels of cardiovascular fitness and regular aerobic exercise can slow or decrease cognitive declines because of an increased oxygen delivery to the brain (44). Several observational and longitudinal studies (45,46) suggest that physical activity

FIGURE 9.10 Regular exercise can lead to improvements in cognitive function in older adults. (From Shutterstock.)

Encoding The process of moving information from short-term store to long-term memory store.

Decoding The process of moving information from long-term memory store to short-term memory store.

Selective attention A process that requires an individual to actively choose one unit of information to pay attention to at a time.

reduces the rate of cognitive decline in healthy individuals (47); however, more experimental evidence is still required before a complete understanding of the process can be achieved (47,48). Recent evidence also suggests that modest improvements in cognitive function can be observed in those individuals with cognitive impairment. Participation in a 6-month physical activity intervention results in improvements in cognitive function in Alzheimer disease patients (48), and regular exercise is associated with a delay in the onset of dementia and Alzheimer disease (49).

Short-term memory store is an individual's conscious or working memory (50). Units of information are collected from either the short-term sensory store or from the long-term memory and stored for short periods of time. The presentation of information and the organization of practice can affect how an individual organizes information in the short-term memory store. Units of information can be remembered more easily if grouped together in some systematic way. The duration of short-term memory is considered to be 1 to 60 seconds, if the thought process of the individual is uninterrupted (27,50,51). Repeatedly practicing a movement skill with feedback on performance can help an individual remember how to do the skill. If a novice engages in the act of practicing a golf shot, the movement skills and patterns will be more easily remembered if the individual continues to repeatedly practice the skill (Figure 9.11). To keep the movement skill in memory, however, an individual must move it to the long-term memory for storage.

Information deemed important enough to store permanently is sent from short-term memory to long-term memory, which is believed to have an unlimited storage capacity and storage duration. Even though capacity and duration are seemingly limitless, individuals can forget information that they used to remember. The information is not lost in the memory, but the individual has simply failed to retrieve the information from long-term memory. Memory can also be thought of as retaining information, which, by definition, is learning. To maximize successful performance when teaching an individual a new movement skill, exercise science professionals, including coaches and allied health care professionals, must have an understanding of how memory works. Motor learning scholars are very interested in understanding how memory functions and identifying ways that practice can be organized to optimize the efficiency with which individuals learn (27,52).

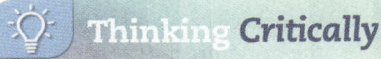

Thinking Critically

What specific information about motor learning is necessary for an individual attempting to improve performance against an opponent during a sport or athletic competition?

FIGURE 9.11 Repeatedly practicing a movement skill will commit the movement pattern to memory. (From Shutterstock.)

Practice Organization and Learning

To fully understand how individuals learn, it is important to clarify the distinction between practice and learning. Performance is typically defined as observable or measurable behavior. Learning is defined as a relatively permanent change in behavior that results from practice or experience. As learning results from a change in an individual's memory and cannot be directly observed, the assessment of how much learning has occurred must be inferred from performance, typically via a physical retention test (1,4,21).

Understanding the distinction between practice and learning is critical because an individual's performance during practice is not necessarily an indicator of learning (52). A practicing-learning paradox exists because certain variables affect practice performance and retention performance in an opposite manner. For example, practicing variations of a movement, instead of practicing the same movement repeatedly, hinders practice performance but enhances learning. For example, asking an individual to repeatedly shoot a basketball from the same spot on the floor will enhance the individual's ability to successfully make that particular shot in practice. However, asking the individual to repeatedly take shots from different spots on the floor will enhance the learning of the shooting skill but will likely result in fewer successful shots being made during the practice. A practice session that does not sufficiently challenge the individual enhances practice performance, but restricts learning of the skill. Individuals must strike a balance between practicing skills repeatedly to enhance performance of the skill and practicing different skills to enhance learning (1,53).

Contextual Interference

Contextual interference describes the interference that results from practicing several different tasks within the context of a single practice session. A high degree of contextual interference can be established by having the individual practice several different skills during the same practice session (54). A low degree of contextual interference can be established by having the individual practice only one skill during a practice session. Low contextual interference practice (relative to high contextual interference) leads to superior practice performance, but much inferior learning (55). For example, a softball infielder who is practicing catching ground balls hit directly at them by a coach will become skilled at catching ground balls (Figure 9.12A) but does not improve the motor skills required to catch ground balls that are hit to their left or right. A coach could increase the contextual interference by hitting ground balls to the left and then to the right of the athlete and requiring them to move and field the ball (Figure 9.12B). A further increase in contextual interference could be accomplished by not informing the athlete where the ball will be hit, making the athlete learn to field the ball at a higher level. This practice-retention paradox may be dependent on the experience of the individual. In those individuals with limited experience of a particular motor skill or activity, high levels of contextual interference during practice of the skill does not enhance learning more than low levels of contextual interference. In fact,

Contextual interference The interference that results from practicing a number of different tasks within the context of a single practice session.

FIGURE 9.12 Skill at fielding ground balls (**A**) can be improved through increasing the contextual interference by fielding ground balls hit to the right or left of the player (**B**).

until the individual is experienced at a movement task or skill, high levels of contextual interference may be detrimental to efficient learning. After some degree of proficiency is reached, however, high contextual interference is beneficial for learning. From a practical standpoint, decreasing extraneous interference is desired during the beginning stages of learning a motor skill, but as the individual becomes more proficient, greater extraneous interference enhances learning of the skill (1,4).

Variability of Practice

Repeatedly practicing the same movement or motor skill has been shown to impede practice performance but enhance the learning of the skill (4). Similarly to that observed with contextual interference, there is substantial variability that appears to be dependent on the individual engaged in the movement or motor skill. Furthermore, practice variability affects children differently than adults and affects men differently than women, suggesting that an individual's ability and previous knowledge, relative to the practice organization, influences learning (1,4,56).

Knowledge of Results

Providing individual feedback after completion of a motor activity is an extremely important factor affecting both performance and learning. The relationship between performance feedback and learning of motor skills is often studied using knowledge of results techniques. Knowledge of results is defined as information about the success of a movement with respect to its goal given to the individual after the completion of the motor skill or movement. For example, a physical therapist might provide knowledge of results to a patient with a back injury who is performing a rehabilitation exercise designed to improve lower back and hip flexibility. In this instance, the physical therapist will inform the patient about correct and incorrect movements. The two most commonly used feedback strategies are the summary knowledge of results and fading knowledge of results (1,4,56).

Summary Knowledge of Results

The summary knowledge of results requires an individual to complete several trials of a single skill or movement without receiving any information about their performance. After completion of the trials, knowledge of results about those trials is provided to the individual. For example, a basketball player may shoot a total of 10 shots. After completing all shots, a coach will provide feedback to the athlete about various aspects of the technique and movement. Summary knowledge of results can be strongly detrimental to practice performance as compared to giving knowledge of results immediately after each trial (52), but it may lead to better learning of a motor skill (57).

Some level of knowledge of results is important for motor learning. The appropriate amount of time between when a skill is performed and when the knowledge of results is provided depends on **task complexity** (4,21). Motor skills and movements of greater complexity require more help to solve the motor problem than a simple motor task. If an individual is to learn a complex task, more immediate knowledge of results should be given than if the same individual was learning a simple motor skill or movement. As an individual completes a practice session and improves performance of the movement or motor skill, the optimal time between completion of the movement or skill and the delivery of knowledge of results increases. Inexperienced individuals perform better during both practice and competition if the summary knowledge of results is short; as the individual becomes more skilled, a short summary may lead to better performance during practice but produces less learning (1,4).

Fading Knowledge of Results

The fading knowledge of the results process involves a systematic reduction in the amount of knowledge of results given to an individual during a practice session. This technique benefits learning by helping an individual solve the motor skill problem early in practice. For example, an individual shooting a basketball might receive knowledge of results from the coach after shooting five shots at the beginning of practice. However, as practice progresses, the coach may use the fading knowledge of results strategy and provide feedback only after shooting 25 shots. Reducing the knowledge of results as skill proficiency increases results in a highly effective practice schedule (1).

Part-Whole Practice

The part-whole practice method is commonly used to teach complex motor skills and movements. Motor skills high in complexity (*i.e.*, having a high number of components

Task complexity The level of difficulty required to complete a motor task.

to the motor skill) and low in organization (*i.e.*, a low dependent relationship between components) usually benefit from part-whole practice. Conversely, motor skills low in complexity (*i.e.*, having a low number of components to the skill) and high in organization (*i.e.*, a high dependent relationship between components) benefit from practicing the skill as a whole. As a result, a progressive-part method of practice has emerged for teaching skills and movements. The progressive-part method requires parts of skill to be practiced independently but ordered according to the sequence in which each part occurs in the skill. As components of the skill are learned, they are progressively linked together until the skill is practiced as a whole component. For example, a strength and conditioning coach might teach an athlete the power clean weightlifting movement by breaking the complete movement into two or three skill components. These three components of the power clean, commonly called the beginning phase, clean phase, and jerk phase, are practiced in parts until the athlete is sufficiently skilled to begin to link the phases together to perform the complete power clean. The progressive-part practice organization requires limits or boundaries to be established during the initial practice of a skill, when the individual components are practiced independently, but as the skill becomes better learned the boundaries are removed (1).

Learning Difficult Skills

Learning difficult motor skills is obviously important for successful participation in exercise, sport, and athletic competition. Task difficulty is defined as the complexity of the motor problem an individual must resolve to successfully complete a task. Task difficulty is often used to describe the process of learning difficult motor skills. There is a relationship between practice performance and task difficulty that is shown in Figure 9.13 (1). As illustrated in the figure, when the motor task difficulty increases, practice performance of the skill decreases. As the movement skill becomes more challenging, an individual's performance deteriorates. Figure 9.14 shows the hypothetical relationship between learning and task difficulty and how the quality of performance is affected by task difficulty.

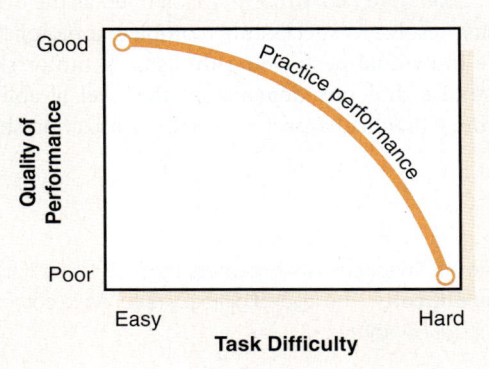

FIGURE 9.13 The relationship between task difficulty and practice performance (paradox principle). (From Brown SP. *Introduction to Exercise Science*. Baltimore (MD): Lippincott, Williams & Wilkins; 2000. 343 p.)

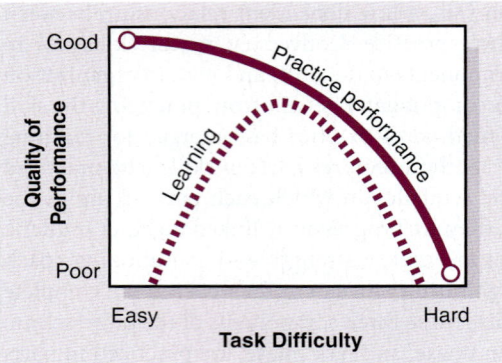

FIGURE 9.14 A hypothetical relationship between learning and task difficulty (paradox principle). (From Brown SP. *Introduction to Exercise Science*. Baltimore (MD): Lippincott, Williams & Wilkins; 2000. 344 p.)

Motor learning increases with increasing task difficulty to what some experts call the **challenge point**. At this point, the individual is being optimally challenged to enhance learning of the skill or movement. Interestingly, at the challenge point, practice performance is not optimal but learning is optimal. Increasing the task difficulty beyond this point continues to inhibit practice performance and also begins to inhibit learning (1).

As practice organization becomes more complex and challenging, it leads to poorer practice performance but superior retention performance (52,55). The performance of motor skills and movements in practice does not necessarily indicate effective learning. Furthermore, too much or too little relative difficulty of the motor skill may actually hinder learning. As expected, optimal task difficulty depends on the level of the performer and the complexity of the skill or movement. For example, when learning to hit a baseball or softball the initial practice session might involve hitting the ball off a standing tee. After acquiring a certain level of performance, the individual might progress to hitting a tossed or thrown ball. This level of practice is much more difficult because it requires an individual to track the ball and time the bat swing to hit the ball successfully. This higher level of practice organization will initially lead to poorer practice performance but provide a higher level of retention performance that ultimately leads to successful acquisition of the motor skill. If the ball is thrown too fast for the skill level of the individual, then the higher level of difficulty associated with hitting a ball thrown fast may hinder the ability of the individual to hit the thrown ball (1).

Relative task difficulty is defined as the difficulty of the motor problem an individual must resolve to successfully complete a motor task relative to the performance abilities of the individual performing the task. If motor skills or movements are constant, the relative task difficulty depends on the level of ability of the individual performing the task. From a practical aspect, practice organization should be adjusted as the performer's level of

Challenge point That point in the learning process where optimal learning is occurring.
Relative task difficulty The level of difficulty required to complete a motor task relative to an individual's level of ability.

performance changes, meaning that the difficulty of the required task should increase as the performer becomes more proficient at the movement skill (1,4).

Exercise science professionals must have a sound knowledge of the principles and concepts of motor learning to provide the individuals they work with the best opportunities for success in skill acquisition and performance. Regardless of the age or skill level of the individual or the type of activity being performed, the use of knowledge about memory and practice will help ensure that the experience encountered during physical activity, exercise, sport, or athletic competition is a safe and effective one.

Thinking Critically

What motor behavior information about a sport technique would you need to help an athlete improve performance of a complex skill such as throwing a ball or performing a dive off a diving board?

MOTOR CONTROL

Motor control is concerned with understanding the mechanisms by which the nervous system activates the muscular system to coordinate body movements. Different physical movements and motor skills require different programs of selected muscle actions. The successful performance of physical activity, exercise, sport, and athletic competition depends on the ability to execute the movements required to complete the movement or motor skill. For example, older adults tend to exhibit slower hesitant movements compared to the faster fluid movements that occur in younger adults. This type of muscle action is believed to result from a decline in motor coordination rather than any developed strategy for caution in the movement (58). The study of neuroanatomy and motor control theories is fundamental to the understanding of motor control and its relationship to movements and performance.

Neuroanatomy

Neuroanatomy is the study of the structure and function of the central (including the brain and spinal cord) and peripheral (outside the central nervous system) nervous systems. Figure 9.15 illustrates the components of the nervous system, and Chapter 3 provides an explanation of the functions of the nervous system. The primary functional components of the nervous system are the neurons (*i.e.*, nerve cells). In the brain, neurons transport information among the specialized areas of the brain that allows an individual to do all of the things that make humans unique (*e.g.*, think, react, move). Major motor areas of the central nervous system include the cerebellum, basal ganglia, and motor cortex, which is comprised of the primary motor, premotor, and supplementary motor cortices.

 Neuroanatomy The study of the structure and function of the central and peripheral nervous systems.

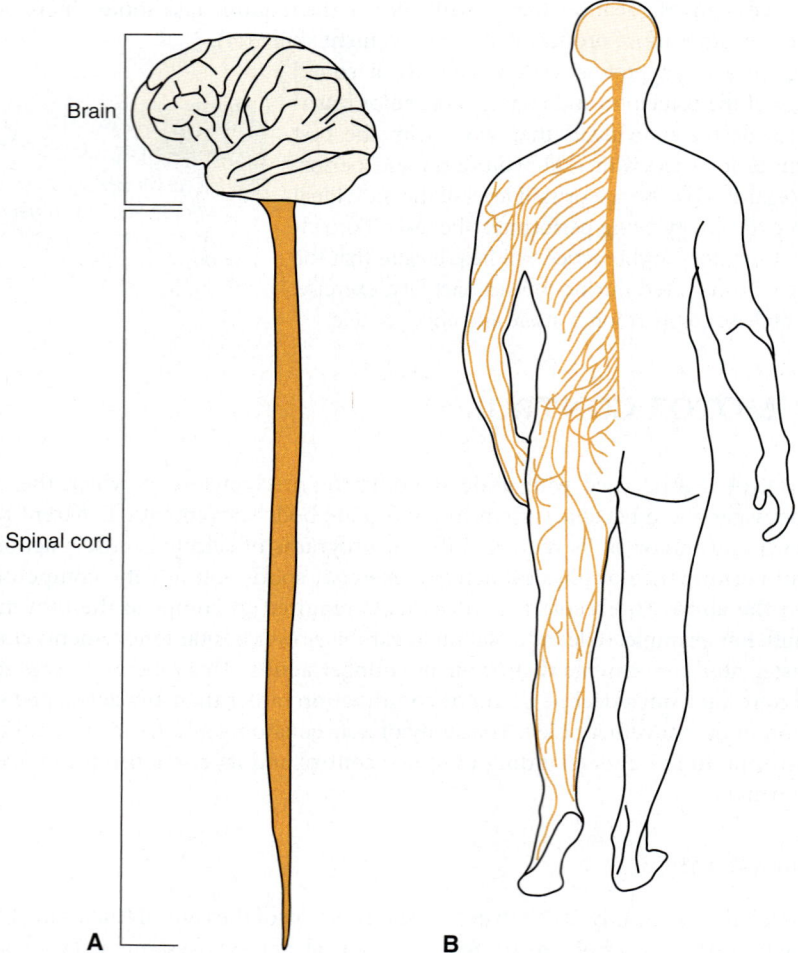

FIGURE 9.15 Components of the nervous system. **A**: The central nervous system includes the brain and the spinal cord. **B**: The peripheral nervous system consists of cranial nerves and spinal nerves, through which the central nervous system transmits commands to and receives information from the end organs. (From Bhatnagar S. *Neuroscience for the Study of Communicative Disorders.* 3rd ed. Baltimore (MD): Lippincott, Williams & Wilkins; 2008.)

Cerebellum

The cerebellum is the area of the central nervous system in humans that serves to coordinate complex voluntary movements, posture, and balance. It is located in the back of and below the cerebrum and consists of two lateral lobes and a central lobe. The cerebellum is

Cerebellum Area of the brain that serves to coordinate complex voluntary movements, posture, and balance in humans.

important for the performance of physical activity, exercise, sport, and athletic completion because it plays a central role in the integration of sensory perception, coordination, and motor control. The cerebellum fine-tunes body movements and motor skills using information derived from the sensory neurons (59).

Basal Ganglia

The **basal ganglia** are large masses of gray brain matter located at the base of the cerebral hemisphere. The basal ganglia constitute a set of interconnected structures in the forebrain. They serve several functions, including movement organization, scale and amplitude of movement, movement modulation, and perceptual-motor integration. The basal ganglia are believed to help in choosing and organizing how individuals perform basic and complex movement. They work in close association with the cerebral cortex and corticospinal system. Common diseases associated with improper functioning of the basal ganglia include Huntington disease and Parkinson disease (59).

Motor Cortex

The **motor cortex** is a region within the cerebral cortex involved in the planning, control, and execution of voluntary motor functions. The motor cortex is typically divided into three separate areas: primary motor cortex, premotor cortex, and supplementary motor cortex. The motor cortex generates a plan for movement and then executes that plan. It is one of the last areas of the brain to be active before movement of the body begins. The motor cortex also receives feedback from all areas of the body, fine-tunes the motor response to the desired movement, and then sends signals to the skeletal muscles to contract or relax and create the desired movement (1,59).

Primary Motor Cortex

The **primary motor cortex** is located in the dorsal portion of the frontal lobe of the brain. It contributes to generating neural impulses that move through the spinal cord and control the execution of body movement (59).

Supplementary Motor Cortex

The **supplementary motor cortex** is a part of the sensorimotor cerebral cortex, and it is important for planning movements of the body. It collects and processes information from

Basal ganglia Structures in the central nervous system that are responsible for movement modulation, movement organization, scale and amplitude of movement, and perceptual-motor integration.

Motor cortex A region of the cerebral cortex involved in the planning, control, and execution of voluntary motor functions.

Primary motor cortex A region of the cerebral cortex that generates neural impulses that pass through the spinal cord and control the execution of body movement.

Supplementary motor cortex Area of the brain that collects and processes information from other areas of the brain and initiates an organized movement.

other areas of the brain, such as the basal ganglia and cerebellum, and initiates an organized movement. Information from the supplementary motor cortex is sent to the premotor cortex and primary motor cortex for further processing before and during movement (59).

Premotor Cortex

The **premotor cortex** is an area of the motor cortex in the frontal lobe of the brain. The premotor cortex sends nervous signals to the musculature close to the body midline, such as the shoulders and arms, so that the hands become properly oriented to perform specific movements and tasks. The premotor cortex also receives sensory information from neurons that helps with the orientation of the body in space. The premotor cortex evaluates initial conditions of the body and then initiates a plan of action, making sure to stabilize the body or body part before movement begins. The premotor cortex works with the basal ganglia, the thalamus, and the primary motor cortex to control many of the body's more complex patterns of coordinated muscle activity such as those associated with exercise, sport, and athletic competition (1,59).

Peripheral Motor System

The **peripheral motor system** consists of the nerves of the peripheral nervous system and muscles, which those nerves innervate. The primary functional unit of the peripheral motor system is the motor unit, which is comprised of a motor neuron and all the muscle fibers that the motor neuron innervates. The recruitment of motor units during movement is responsible for the motor patterns that are executed and the force generated by a muscle during contraction (59). Different levels of force can be generated by recruiting different numbers and types of motor units within a whole muscle or group of muscles (60).

General Motor Control Systems

Motor control involves both the sending of information to the muscles to contract and receiving information through afferent neurons from the body's **proprioceptors** in the joints and muscles. The study of motor control during movement must consider both nervous system control of muscle contraction and the influence of the sensory information coming back from the peripheral tissues of the body (4). An individual must be able to coordinate muscle contraction and relaxation to successfully execute a movement or skill. The closed-loop and open-loop systems of motor control are used as explanations of how the brain, nervous system, and muscles coordinate body movement. **Dynamic systems theory**

Premotor cortex Works to control many of the body's more complex patterns of coordinated muscle activity.

Peripheral motor system Component of the nervous system that is responsible for controlling the motor patterns that are executed and the force generated by a muscle during contraction.

Proprioceptors Nervous structures in the body that are responsible for sensing body position.

Dynamic systems theory A theory used to describe the complex interaction of systems that can change in response to a stimulus.

has also been suggested as an additional means by which cognitive development and motor control occurs in humans. The advantage of using dynamic systems theory is that it brings order to a complex interaction of multiple systems, which commonly occurs during sport and athletic competition. Dynamic systems theory suggests that all the possible factors that may be in operation at any given developmental moment must be considered at that specific moment in time (61).

Closed-Loop System

The closed-loop system of motor control asserts that sensory information necessary to control motor performance is received by the nervous system during the movement. By using this information, muscle activity can be altered during the performance of the movement to correct for changes that need to be made to successfully perform the movement or to respond to something in the external environment (62,63). Figure 9.16 shows an expanded conceptual model of motor performance (21). Corrections and alterations in

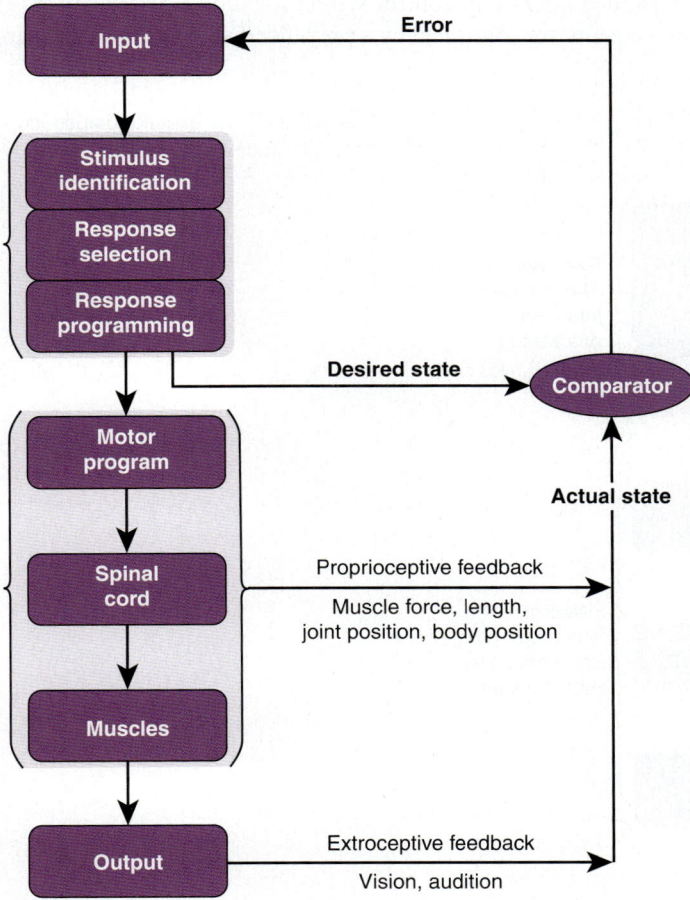

FIGURE 9.16 An expanded conceptual model of motor performance (21).

motor performance can occur because as the movements are being performed, feedback is sent to those areas in the brain responsible for correcting and fine-tuning the motor pattern. The closed-loop system of motor control enhances the accuracy of muscle actions because movements of the body can be controlled and adjusted as they are occurring. A disadvantage of the closed-loop system is lack of speed by which corrections to muscle actions can be made by the nervous system during the motor performance (1).

Thinking Critically

How might information in the area of motor control provide for improvements in preventive and rehabilitative health care for those individuals affected by a neuromuscular disease?

Open-Loop System

The control of motor performance is also explained using an open-loop system. This theory suggests that individuals do not receive feedback from the joints, proprioceptors, and muscles of the body during movement (64). The open-loop system suggests that body movements are completely preplanned prior to the initiation of the movement. Figure 9.17 illustrates an expanded open-loop control system for motor performance (4). In this instance, body movements are controlled by a predefined set of motor commands that once

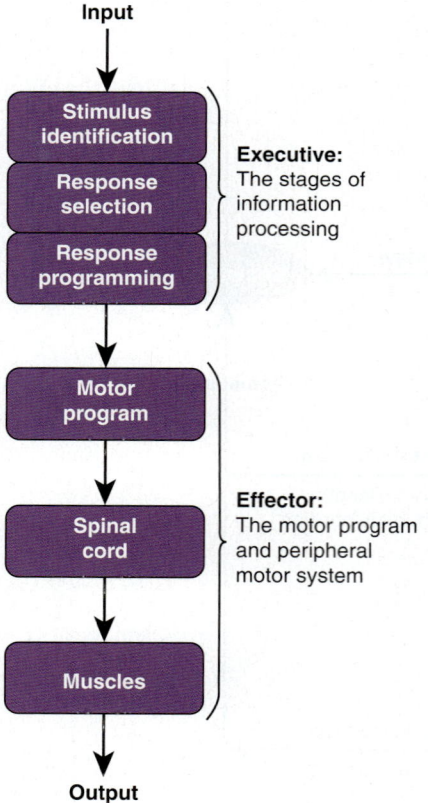

FIGURE 9.17 An expanded open-loop control system for motor performance (21).

sent to the muscles complete a movement without the involvement of feedback from the muscles. This type of system results in faster movements than those in the closed-loop system because the time it takes to provide feedback is eliminated from the process. A disadvantage of the open-loop system is that modifications to muscle movements cannot be made to correct for errors in the movements being performed (1).

AREAS OF STUDY IN MOTOR BEHAVIOR

The study and application of knowledge derived from motor development, learning, and control is evident in a wide range of activities and situations for both healthy and diseased individuals. Exercise science professionals should be fundamentally familiar with each area of motor behavior. Understanding the motor development of individuals throughout the lifespan is critical for applying the proper physical activity or exercise intervention to enhance health and reduce disease risk and for modifying sport and athletic competitions to ensure safety and success. Activities that are not developmentally appropriate for a child or an adult may lead to ineffective results and discouragement of the participant. Correct use of motor learning strategies is critical for an athletic coach to make effective use of practice time to enhance sport and athletic competition. The progression of basic and advanced motor skills in a developmentally appropriate sequence is critical for enhancing motor skill acquisition and promoting successful performance of a movement or activity. Understanding how individuals control motor movements is important for ensuring positive patient-based outcomes during the rehabilitation process of those individuals with neurologic disease. Effective rehabilitation strategies must incorporate an understanding of how muscles and movements are controlled by the nervous system. Several examples of how motor behavior are studied and applied by exercise science professionals are included in the following sections. The information in the following sections is not meant to be an exhaustive list, but merely examples of how the components of motor behavior are studied and used in a variety of situations.

Modification of Equipment, Games, and Playing Environment

To ensure safe and successful participation in exercise, sport, and athletic activities for individuals of all ages and body sizes, consideration should be given to modification of the equipment and the exercise or sport environment. Knowledge of how individuals develop skills across the lifespan can assist exercise science and coaching professionals in making appropriate changes to the equipment, the exercise, and sport environment. This process is called scaling (4). For example, many years ago females were expected to participate in basketball using a ball that was of the same size as the men's "regulation" ball even though the average size of the hand of a female is much smaller than that of a male. Adoption of a ball that was 1 inch smaller in circumference and 2½ oz lighter than the "regulation" ball used by males occurred in 1984 and allowed for improved motor skills and performance. This type of equipment scaling in basketball is now also used with children.

Scaling has been shown to improve accuracy when performing hitting motions in tennis (65,66), shooting in basketball (67,68), and hitting the ball in cricket (69). Simplifying a motor skill via equipment or playing environment scaling is believed to minimize working memory involvement during motor learning (70), which is particularly important for children given that working memory capacity is still developing during childhood (67,71).

The use of scaled strategies for children and adolescents ensures an enhanced development of motor skills for successfully striking an object. Recreational baseball and softball leagues employ tee-ball games for young children to enhance skill development (Figure 9.18) and then subsequently allowed older children to bat using pitching machines (Figure 9.19) before progressing to hitting a pitch thrown by an individual player. These types of modifications to the game environment are designed to increase the safety and success of young children playing baseball and softball. The body size scaling of soccer goals and height for basketball goals allows for improvements in developmentally appropriate skill and motor movement of young children, improving the opportunity for successful performance.

Along with an increased emphasis on improving fitness and health in children has come exercise equipment scaled to body size for younger children. Numerous Young Men's Christian Associations (YMCAs), physical therapy facilities, public and private school districts, youth fitness training facilities, special needs facilities, hospitals, health clubs, and wellness centers have all begun to provide developmentally age-appropriate exercise equipment for use. For older adults, the modification of playing equipment and surfaces can increase participation and enjoyment, improve fitness levels, and increase safety. The use of tennis rackets and golf clubs with larger hitting surfaces provides an increased probability that the tennis ball or golf ball will be struck with accuracy. Increasing the success of the older individual when playing the game will likely lead to a more enjoyable and worthwhile experience (4).

FIGURE 9.18 The use of a batting tee enhances success in baseball and softball games for young children. (From Shutterstock.)

FIGURE 9.19 Progression to a pitching machine improves the chances of further skill development in baseball and softball games for young children.

Understanding Parkinson Disease

Neurological disorders can affect individuals at any age but are usually more common with aging. The specific cause of a neurological disorder can vary considerably. Common neurological disorders include Alzheimer disease, muscular dystrophy, and amyotrophic lateral sclerosis. Parkinson disease is a progressive neurologic disorder characterized by a host of alterations to motor and nonmotor characteristics that can impact normal physical function. The disease is caused by a decrease in the synthesis of the neurotransmitter dopamine resulting from the death of dopaminergic cells in the brain. The decreased concentration of dopamine results in the following clinical features: muscle tremor at rest, muscle rigidity, slow movement (called bradykinesia), and postural instability. In addition, flexed posture and freezing (called motor blocks) have been included among the classic features of Parkinson disease (72,73).

Parkinson disease affects many aspects of motor control and movement (Figure 9.20). Muscle tremors are evident both at rest and with movement. Muscle rigidity often begins in the neck and shoulders and spreads to the trunk and extremities, making body movement difficult. The ability to move the fingers, hands, arms, or legs rapidly is significantly reduced, and motor control to rise from a chair is diminished. Standing posture is characterized by increased hunched shoulders (kyphosis) and fixed knees and elbows, as well as rounded (adducted) shoulders. Individuals with Parkinson disease have a slow and shuffling walking pattern, with quick and shortened steps (called festination), decreased arm swing, and

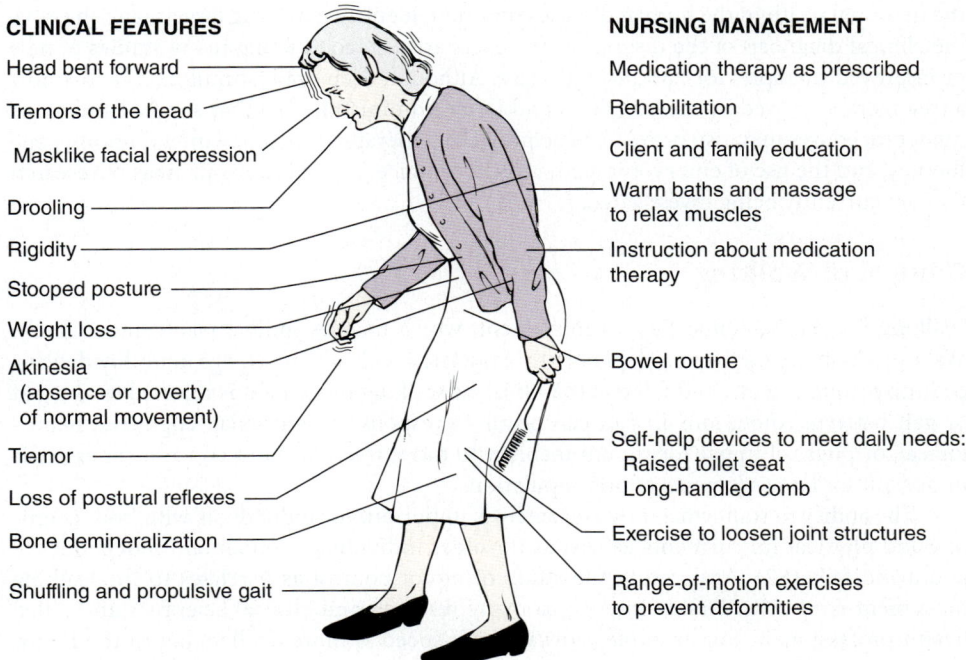

FIGURE 9.20 Changes in motor control and movement in Parkinson disease. (LifeArt image © 2010 Lippincott, Williams & Wilkins. All rights reserved.)

difficulty initiating a step. Postural control and righting reflexes are compromised and lost, and, as a result, falls can become a recurring problem. Episodes of decreased movement or freezing become more frequent during walking. Individuals with Parkinson disease often have problems with the volume and clarity of their speech (73).

The treatment of Parkinson disease involves making adjustments to an individual's lifestyle, participating in regular exercise and physical therapy, eating a healthy diet, and using medications. Exercise is important for general health, but especially for maintaining good physical function in Parkinson disease. Resistance exercise training can improve muscular endurance and strength (74–76). Cardiorespiratory fitness can also be improved in individuals with Parkinson disease (74,77), although possibly not to the same extent as muscular fitness (74). Physical therapy may be advisable and can help improve mobility, joint range of motion, and muscular tone. Although specific exercises cannot stop the progression of the disease condition, improving muscle strength can help an individual feel more confident and capable in their movements. Progressive exercise and physical therapy can help an individual improve their gait and balance (78,79). Working with a speech therapist or speech pathologist can improve problems associated with speaking and swallowing. Medications can help manage problems with walking, movement, and tremor by increasing the brain's supply of dopamine (73).

Recent research in the area of Parkinson disease has focused on halting the progression of the disease, restoring lost function, and even preventing the disease. Studying the genes responsible for the disease and identifying gene defects (80) can help researchers understand how Parkinson disease occurs, help develop animal models that accurately mimic the neuronal death in the human disease condition, identify new drug targets, and improve the clinical diagnosis of the disease. Researchers are also conducting many studies of new or improved therapies for Parkinson disease. Although deep brain stimulation (81) is now a treatment approved by the U.S. Food and Drug Administration, other surgical and treatment procedures may also prove to be helpful. Drug therapy, nutritional interventions, gene therapy, and the use of embryonic stem cells to produce dopamine are all areas of research that are currently being investigated.

Control of Walking

Walking is a fundamental human movement, which enables individuals to move freely. Walking disability has been independently associated with advancing age, mobility deficits, cognitive impairments, and fall risk (82–85). The walking movement is referred to as a gait or gait pattern. Alterations in gait can occur in response to muscular, cognitive, neurological, or physical impairments. An inability to move freely without concern for safety is important for large segments of our population.

The ability to walk efficiently and safely is important for individuals with both cognitive and physical impairments as well as for older individuals to maintain independence and avoid falls (82). As a result, the study of motor control as it relates to the walking movement is of considerable interest to many research and clinical scientists and allied health professionals. For example, slow walking speed is more challenging to the motor control of the gait pattern and may be more sensitive to age-related declines in gait than normal or faster walking speeds (86). Individuals with Huntington or Parkinson disease are at a high risk for falling, and the understanding of how patients control movement is important for ensuring their safety during walking (87,88). Individuals with physical disabilities (89) or recovering from replacement surgery must learn to control muscle movement patterns in order to have pain-free movement during walking (90). Cognitive impairment has been shown to predict falls risk in older adults, especially in challenging conditions such as uneven or slippery surfaces. As a result, many exercise science professionals study the relationships between executive function and gait in older adults (91,92).

> **Thinking Critically**
>
> How might coursework in motor development, motor control, and motor learning help prepare an individual for a career as an allied health professional, strength and conditioning coach, or fitness specialist?

Interview

Minoru "Shino" Shinohara, PhD, FACSM
School of Biological Sciences, Georgia Institute of Technology

Brief Introduction – I was born, raised, and educated in Tokyo, Japan. I received my Bachelor's and Master's degrees in Physical Education/Exercise Science and my PhD in Biomechanics/Exercise Physiology, all from the University of Tokyo. After working as an Assistant Professor at the University of Tokyo for several years, I moved to the United States, had research positions at the Pennsylvania State University and the University of Colorado at Boulder, and took a faculty position at the Georgia Institute of Technology. I direct the Human Neuromuscular Physiology Laboratory, in which we study the physiological mechanisms underlying human motor behavior/control.

Q: What are your most significant career experiences?

As an assistant professor at the University of Tokyo, I was so busy with teaching and service that I could find my research time only after hours. When I had a 1-year visit to the University of Colorado at Boulder for focusing on research, I was so impressed by the research quality and protected research time in research universities in the United States. I decided to quit the tenured faculty position in Japan and moved to the United States to build a career more focused on research. In this transition, I had to learn different ways of preparing, conducting, and reporting research and for educating students in the United States. This transition from a permanent position with research activity performed in spare time to a nonpermanent position with research activity as the main job in another country is one of my most significant career experiences. On the application side, I have applied my exercise science knowledge to training elite rhythmic gymnasts after hours. While I have no experience as an athlete in this sport, my science-based training strategies worked efficiently to produce and improve a national team gymnast with short training hours. This achievement as an exercise scientist and national-level coach helped to get the trust of the gymnastics federation, and they provided me opportunities to share exercise science knowledge with gymnastics coaches across the country.

Q: Why did you choose to become an "exercise scientist"?

The first exposure to exercise science happened to me at the time when I was a college student training for long-distance triathlon competitions (*e.g.* Hawaii Ironman Race). I was so excited to find this field of study in a book that could help me train efficiently for better performance. I was then attracted by the breadth and depth of knowledge on exercise science that I learned through the coursework at the University of Tokyo. I conducted an undergraduate research project in which I examined the hematological profiles of subjects who performed a long-distance triathlon race using a predetermined strategy for work intensity and fluid replacement. I was fascinated by making new findings through experimental research on what I was interested in and decided to become an exercise scientist. Later, my research interest transitioned to motor behavior/control because it deals with issues directly related to motor performance.

Q: Why it is important for exercise science students to have an understanding of "motor behavior/control"?

Motor behavior/control is a field of knowledge that describes the characteristics and mechanisms for voluntary and nonvoluntary movement. It includes psychological, neurophysiological, and biomechanical aspects. Repetitive movement of body parts in exercise is the accumulation of movements that may be described in motor behavior/control. Various factors in motor behavior/control during motor practice influence motor learning and motor rehabilitation. Hence, motor control/behavior comprises a fundamental part of exercise science, and that is why it is important for exercise science students to understand motor behavior/control.

Q: What advice would you have for a student exploring a career in any exercise science profession?

In exploring a career, students are advised to look for a profession that they would be passionate about and enjoy, rather than one in which they can be well paid. The possession of deeper, wider, and updated knowledge in exercise science and other related fields would make the enjoyment and contribution in the profession greater. Each organ and system has its specific functions but does not exist independently in our human body. So it is important to have both specific and holistic standpoints in understanding exercise science. Interaction between human sensorimotor control and technologies is evolving every day. Students are encouraged to learn both established knowledge in textbooks in exercise science and other related fields and new knowledge in journal papers to foster their capability in any exercise science profession.

Interview

Lisa Griffin, PhD, FACSM

Department of Kinesiology and Health Education, The University of Texas at Austin

Brief Introduction – I grew up in Canada and completed a Bachelor's degree in Human Kinetics with a minor in Biochemistry at the University of Guelph, Guelph, ON, Canada. At the University of Western Ontario in London, ON, Canada, I completed both a Master's degree and a PhD in Neuroscience. I then completed a postdoctorate at the Miami Project to Cure Paralysis, Miami, FL, and another postdoctorate at York University, Toronto, ON, Canada.

Q: What are your most significant career experiences?

My early career experiences involved watching my mentors for my graduate and postdoctoral work, Jayne Garland, Christine Thomas, and Enzo Cafarelli. I learned their styles of running their labs, advising their graduate students, publishing, applying for grants, and balancing work and family. My most significant career experiences involved the development of my research interests, which began with a pure interest in neuromuscular physiology and expanded into the desire to help others through my research. As a young scientist, it was a powerful experience to work at the Miami Project to Cure Paralysis and be a part of a group of scientists working together to improve the lives of patients with paralysis from spinal cord injury. It was also very exciting when I first started my own lab at the University of Texas at Austin. I felt that I had the complete freedom and all the resources needed to research whatever I wanted. I have always enjoyed helping graduate students develop their research ideas, and I also enjoy reviewing articles for scientific journals and grant applications for research proposals. I enjoy teaching and have expanded my research to include investigating neuromuscular control mechanisms for rehabilitation for musculoskeletal injuries and stroke. I value collaborations with engineers who design robotic exoskeletons and who work on brain-machine interfaces for rehabilitation.

 Q: Why did you choose to become an "exercise scientist"?

As an undergraduate student, I developed a passion for basic neuromuscular physiology. I wanted to become a scientist because I desired intellectual stimulation and the freedom to think about whatever I wanted and to manage my own schedule. I liked the idea of contributing to our basic understanding of neuromuscular physiology and motor control as well as helping others gain freedom from immobility and pain.

 Q: Why is it important for exercise science students to understand motor development/control/learning?

Motor control is the fundamental understanding of how humans move. It is an essential part of life and is fascinating to study. It is also important for clinical rehabilitation post injury. The effects of damage to the central nervous system at the level of the brain or spinal cord can have a devastating impact on quality of life. We are in the very early stages of understanding how to treat neurological injury, and the field is very challenging but also very important.

 Q: What advice would you have for a student exploring a career in any exercise science profession?

Exercise science is a field in which the fundamental knowledge of the function and movement of the human body is established. Pursuing an undergraduate degree in exercise science would benefit anyone who is interested in working in the context of personal training, athletic performance, sports and coaching, rehabilitation, physical therapy, occupational therapy, medicine, or research. Undergraduate students should remember that they are learning the foundations of anatomy, physiology, biomechanics, motor control, and motor learning and that this will serve as the foundation upon which they will add more specific clinical or performance-based knowledge for their future careers.

 # SUMMARY

- Motor behavior is the study of the interactions between the physiological and psychological processes of the body and includes the disciplines of motor development, motor control, and motor learning.
- The principles of motor behavior are applied to a variety of everyday skills, including those involved with work and leisure time activities, physical activity, exercise, sport, athletic competition, and rehabilitation.
- Motor development includes the study of changes throughout the lifespan and how motor performance is affected by those changes.
- Motor learning is the study of the acquisition of basic and advanced movement skills that are used in everyday activities, including those involved with physical activity, exercise, sport, and athletic competition.
- Motor control is the study of the mechanisms by which the nervous and muscular systems coordinate body movements.

FOR REVIEW

1. Define the following terms:
   a. Motor development
   b. Motor control
   c. Motor learning
2. Name the primary professional organizations in the areas of motor control, motor learning, and motor development.
3. What are the three stages of information processing?
4. Describe the following components of the multistore memory model:
   a. Short-term sensory store
   b. Short-term memory
   c. Long-term memory
5. What is the difference between practice and learning?
6. How are summary knowledge of results and fading knowledge of results used to improve motor skill performance?
7. Describe the relationship between practice performance and task difficulty.
8. What are the primary functions of the following neural structures?
   a. Cerebellum
   b. Basal ganglia
   c. Supplementary motor cortex
   d. Premotor cortex
   e. Motor cortex
   f. Primary motor cortex
9. What is the difference between the closed-loop and open-loop systems of motor control?
10. List the primary stages of motor development.
11. Describe how crystal intelligence and fluid intelligence interact with aging to influence psychomotor function.

Project-Based Learning

1. Identify an individual who does not participate in regular physical activity or exercise and suffers from instability while standing and walking. Prepare a presentation that includes at least five key recommendations that you would give to that individual in an effort to improve their balance during physical activity and exercise. What are those key points, and how does the motor behavior literature support your recommendations?
2. Identify a sport or athletic competition that you believe could be enhanced for younger or disabled athletes. Prepare a presentation that includes at least five points that you would give to another group of exercise science professionals who are trying to modify the equipment or competition for the young or disabled athletes. What are those key points, and how does the motor behavior literature support your recommendations?

REFERENCES

1. Guadagnoli MA. Motor behavior. In: Brown SJ, editor. *Introduction to Exercise Science*. 1st ed. Philadelphia (PA): Lippincott, Williams & Wilkins; 2001. p. 334–58.
2. Clark JE, Whitall J. What is motor development?: The lesson of history. *Quest*. 1989;41:183–202.
3. Haywood KM, Getchell N. *Life Span Motor Development*. 6th ed. Champaign (IL): Human Kinetics; 2014.
4. Schmidt RA, Lee T. *Motor Control and Learning: A Behavioral Emphasis*. 5th ed. Champaign (IL): Human Kinetics; 2011.
5. Gesell A. *Infancy and Human Growth*. New York (NY): Macmillan; 1928.
6. McGraw MB. *Growth, A Study of Johnny and Jimmy*. New York (NY): Appleton-Century-Crofts; 1935.
7. Dennis W. The effect of restricted practice upon the reaching, sitting, and standing of two infants. *J Genet Psychol*. 1935;47:17–32.
8. Thomas JR. Motor behavior. In: Massengale JD, Swanson RA, editors. *The History of Exercise and Sport*. 1st ed. Champaign (IL): Human Kinetics; 2003. p. 203–92.
9. Connolly KJ. *Mechanisms of Motor Skill Development*. New York (NY): Academic Press; 1970.
10. Espenshade A, Eckert HM. *Motor Development*. Columbus (OH): Merrill; 1967.
11. Rarick GL. *Physical Activity: Human Growth and Development*. New York (NY): Academic Press; 1973.
12. Wickstrom R. *Fundamental Movement Patterns*. Philadelphia (PA): Lea & Febiger; 1970.
13. Jeannerod M. *The Brain Machine*. Cambridge (MA): Harvard University Press; 1985.
14. Sabbatini REM. The discovery of bioelectricity. *Brain Mind*. 1998;2:1–4.
15. Sherwood DE. Motor control and motor learning. In: Housh TJ, Housh DJ, Johnson GO, editors. *Introduction to Exercise Science*. 3rd ed. Scottsdale (AZ): Holcomb Hataway; 2008. p. 234–52.
16. Bryan WL, Harter N. Studies in the physiology and psychology of the telegraphic language. *Psychol Rev*. 1897;4:27–53.
17. Bryan WL, Harter N. Studies on the telegraphic language: the acquisition of a hierarchy of habits. *Psychol Rev*. 1899;6:345–75.
18. Woodworth RS. *Le mouvement*. Paris (France): Doin; 1903.
19. Thorndike EL. The law of effect. *Am J Psychol*. 1927;39:212–22.
20. Henry FM, Rodgers DE. Increased response latency for complicated movements and a "memory drum" theory of neuromotor reaction. *Res Q*. 1960;31:448–58.
21. Schmidt RA, Wrisberg CA. *Motor Learning and Performance: A Situation-Based Learning Approach*. 4th ed. Champaign (IL): Human Kinetics; 2008.
22. Kannan L, Vora J, Bhatt T, Hughes SL. Cognitive-motor exergaming for reducing fall risk in people with chronic stroke: a randomized controlled trial. *NeuroRehabilitation*. 2019;44(4):493–510.
23. Payne VG, Isaacs LD. *Human Motor Development: A Lifespan Approach*. 9th ed. New York (NT): Routledge; 2017.
24. Gallahue DL, Ozmun JC, Goodway JD. Understanding motor development: an overview. In: *Understanding Motor Development: Infants, Children, Adolescents, Adults*. 7th ed. New York (NY): McGraw-Hill; 2012. p. 2–22.
25. Gabbard CP. *Lifelong Motor Development*. 6th ed. Dubuque (IA): Pearson Education; 2012.
26. Gallahue DL, Ozmun JC, Goodway JD. Prenatal factors affecting development. In: *Understanding Motor Development: Infants, Children, Adolescents, Adults*. 7th ed. New York (NY): McGraw-Hill; 2012. p. 82–105.
27. Anastasi A. *Psychological Testing*. 1st ed. New York (NY): Macmillan; 1988.
28. Gibson EJ. Introductory essay: what does infant perception tell us about theories of perception? *J Exp Psychol*. 1987;13(4):515–23.
29. Cattell RB. *Abilities: Their Structure, Growth and Action*. New York (NY): Houghton Mifflin; 1971.
30. Cools W, De Martelaer K, Samaey C, Andries C. Movement skill assessment of typically developing preschool children: a review of seven movement skill assessment tools. *J Sports Sci Med*. 2009;8(2):154–68.
31. Gallahue DL, Ozmun JC, Goodway JD. Childhood growth and development. In: *Understanding Motor Development: Infants, Children, Adolescents, Adults*. 7th ed. New York (NY): McGraw-Hill; 2012. p. 164–79.
32. Gallahue DL, Ozmun JC, Goodway JD. Fundamental movement abilities. In: *Understanding Motor Development: Infants, Children, Adolescents, Adults*. 7th ed. New York (NY): McGraw-Hill; 2012. p. 180–236.
33. Gallahue DL, Ozmun JC, Goodway JD. Specialized movement abilities. In: *Understanding Motor Development: Infants, Children, Adolescents, Adults*. 7th ed. New York (NY): McGraw-Hill; 2012. p. 304–22.

34. Gallahue DL, Ozmun JC, Goodway JD. Physiological changes in adults. In: *Understanding Motor Development: Infants, Children, Adolescents, Adults.* 7th ed. New York (NY): McGraw-Hill; 2012. p. 358–80.
35. Giovanelli N, Ortiz ALR, Henninger K, Kram R. Energetics of vertical kilometer foot races; is steeper cheaper? *J Appl Physiol.* 2016;120(3):370–5.
36. Ortiz ALR, Giovanelli N, Kram R. The metabolic costs of walking and running up a 30-degree incline: implications for vertical kilometer foot races. *Eur J Appl Physiol.* 2017;117(9):1869–76.
37. Whiting CS, Allen SP, Brill JW, Kram R. Steep (30°) uphill walking vs. running: COM movements, stride kinematics, and leg muscle excitations. *Eur J Appl Physiol.* 2020;120(10):2147–57.
38. Proctor RW, Reeve TG. *Stimulus-Response Compatibility: An Integrated Perspective.* Amsterdam (The Netherlands): Elsevier; 1990.
39. Klapp ST. Reaction time analysis of central motor control. In: Zelaznick HN, editor. *Advances in Motor Learning and Control.* Champaign (IL): Human Kinetics; 1996. p. 13–35.
40. Guadagnoli MA, Reeve TG. Movement complexity and foreperiod effects on response latencies for aimed movements. *J Hum Mov Stud.* 1992;23:29–39.
41. Sanders AF. Issues and trends in the debate on discrete vs. continuous processing of information. *Acta Psychol.* 1990;74:123–67.
42. Sternberg S. The discovery of processing stages: extensions of Donder's method. *Acta Psychol.* 1969;30:270–315.
43. Atkinson RC, Shiffrin RM. Human memory: a proposed system and its control processes. In: Spence KW, Spence JT, editors. *The Psychology of Learning and Motivation.* Vol 8. London (UK): Academic Press; 1968.
44. Mazzeo RS, Cavanagh PR, Evans WJ, et al. Exercise and physical activity for older adults. *Med Sci Sports Exerc.* 1998;30(6):992–1008.
45. Yaffe K, Barnes D, Nevitt M, Lui L-Y, Covinsky K. A prospective study of physical activity and cognitive decline in elderly women: women who walk. *Arch Intern Med.* 2001;161(14):1703–8.
46. Sofi F, Valecchi D, Bacci D, et al. Physical activity and risk of cognitive decline: a meta-analysis of prospective studies. *J Intern Med.* 2011;269(1):107–17.
47. Bherer L, Erickson KI, Liu-Ambrose T. A review of the effects of physical activity and exercise on cognitive and brain functions in older adults. *J Aging Res.* 2013;2013:1–8.
48. Lautenschlager NT, Cox KL, Flicker L, et al. Effect of physical activity on cognitive function in older adults at risk for Alzheimer disease: a randomized trial. *JAMA.* 2008;300(9):1027–37.
49. Larson EB, Wang L, Bowen JD, et al. Exercise is associated with reduced risk for incident dementia among persons 65 years of age and older. *Ann Intern Med.* 2006;144(2):73–81.
50. Baddley AD, Hitch G. Working memory. In: Bower GH, editor. *Psychology of Learning and Motivation.* New York (NY): Academic Press; 1974.
51. Ericcson KA, Chase WG, Faloon S. Acquisition of a memory skill. *Science.* 1980;208:1181–2.
52. Guadagnoli MA, Dornier LA, Tandy R. Optimal length of summary knowledge of results: the influence of task related experience and complexity. *J Exerc Sport Psychol.* 1996;67:239–48.
53. Guadagnoli MA, Holcomb WR, Weber T. The relationship between contextual interference effects and performer experience on the learning of a putting task. *J Hum Mov Stud.* 1999;37:19–36.
54. Porter JM, Magill RA. Systematically increasing contextual interference is beneficial for learning sport skills. *J Sport Sci.* 2010;28(12):1277–85.
55. Shea CH, Kohl RM, Indermil C. Contextual interference: contributions of practice. *Acta Psychol.* 1990;73:145–57.
56. Wrisberg CA, Ragsdale MR. Further tests of Schmidt's schema theory: development of a schema rule for a coincident timing task. *J Mot Behav.* 1979;11:159–66.
57. Lavery JJ. Retention of simple motor skills as a function of type of knowledge of results. *Can J Psychol.* 1962;16:300–11.
58. Morgan M, Phillips JG, Bradshaw JL, Mattingly JB, Iansek R, Bradshaw JA. Age-related motor slowness: simply strategic? *J Gerontol.* 1994;49(3):M133–M139.
59. Guyton AC, Hall JE. *Textbook of Medical Physiology.* 13th ed. Oxford (UK): Elsevier; 2016.
60. Ferguson RA, Aagaard P, Ball D, Sargeant AJ, Bangsbo J. Total power output generated during dynamic knee extensor exercise at different contraction frequencies. *J Appl Physiol.* 2000;89(1):1912–8.
61. Spencer JP, Austin A, Schutte AR. Contributions of dynamic systems theory to cognitive development. *Cogn Dev.* 2012;27:401–18.
62. Adams JA. A closed loop theory of motor learning. *J Mot Behav.* 1971;3:111–49.

63. Keele SW, Posner MI. Processing visual feedback in rapid movement. *J Exp Psychol.* 1968;77:155–8.
64. Schmidt RA. A schema theory of discrete motor skill learning. *Psychol Rev.* 1975;82:225–60.
65. Buszard T, Garofolini A, Reid M, Farrow D, Oppici L, Whiteside D. Scaling sports equipment for children promotes functional movement variability. *Sci Rep.* 2020;10(1):1–8.
66. Timmerman E, De Water J, Kachel K, Reid M, Farrow D, Savelsbergh G. The effect of equipment scaling on children's sport performance: the case for tennis. *J Sports Sci.* 2014;33(10):1093–100.
67. Buszard T, Reid M, Masters R, Farrow D. Scaling the equipment and play area in children's sport to improve motor skill acquisition: a systematic review. *Sports Med.* 2016;46(6):829–43.
68. Gorman AD, Headrick J, Renshaw I, McCormack CJ, Topp KM. A principled approach to equipment scaling for children's sport: a case study in basketball. *Int J Sports Sci Coach.* 2020:1–8.
69. Dancy PAJ, Murphy CP. The effect of equipment modification on the performance of novice junior cricket batters. *J Sports Sci.* 2020:1–8.
70. Buszard T, Farrow D, Reid M, Masters RSW. Scaling sporting equipment for children promotes implicit processes during performance. *Conscious Cogn.* 2014;30:247–55.
71. Alloway TP, Gathercole SE, Pickering SJ. Verbal and visuospatial short-term and working memory in children: are they separable? *Child Dev.* 2006;77(6):1698–716.
72. Jankovic J. Parkinson's disease: clinical features and diagnosis. *J Neurol Neurosurg Psychiatry.* 2008;79(4):368–76.
73. Protas EJ, Stanley RK, Jankovic J. Parkinson's disease. In: Durstine JL, Moore GE, Painter PL, Roberts SO, editors. *ACSM's Exercise Management of Person's with Chronic Diseases and Disabilities.* 3rd ed. Champaign (IL): Human Kinetics; 2009. p. 350–6.
74. Uhrbrand A, Stenager E, Pedersen MS, Dalgas U. Parkinson's disease and intensive exercise therapy — a systematic review and meta-analysis of randomized controlled trials. *J Neurol Sci.* 2015;353(1–2):9–19.
75. Paul SS, Canning CG, Song J, Fung VS, Sherrington C. Leg muscle power is enhanced by training in people with Parkinson's disease: a randomized controlled trial. *Clin Rehabil.* 2014;28(3):275–88.
76. Corcos DM, Robichaud JA, David FJ, et al. A two-year randomized controlled trial of progressive resistance exercise for Parkinson's disease. *Mov Disord.* 2013;28(9):1230–40.
77. Katzel LI, Sorkin JD, Macko RF, Smith B, Ivey FM, Shulman LM. Repeatability of aerobic capacity measurements in Parkinson disease. *Med Sci Sports Exerc.* 2011;43(12):2381–7.
78. Keus SHJ, Bloem BR, Hendriks EJM, Bredero-Cohen AB, Munneke M; Practice Recommendations Development Group. Evidence-based analysis of physical therapy in Parkinson's disease with recommendations for practice and research. *Mov Disord.* 2007;22(4):451–60.
79. Elleuch MH, Mallek A, Fakhfakh R, Yahia A, Ghroubi S. Physical neurorehabilitation therapy in Parkinson's disease "OFF" period. *Ann Phys Rehabil Med.* 2017;60:e68–e69.
80. Le W, Appel SH. Mutant genes responsible for Parkinson's disease. *Curr Opin Pharmacol.* 2004;4(1):79–84.
81. Weaver FM, Follett KA, Stern M, et al. Randomized trial of deep brain stimulation for Parkinson disease: thirty-six-month outcomes. *Neurology.* 2012;79(1):55–65.
82. Callisaya ML, Blizzard L, Schmidt MD, McGinley JL, Srikanth VK. Ageing and gait variability — a population-based study of older people. *Age Ageing.* 2010;39(2):191–7.
83. Hausdorff JM, Cudkowicz ME, Firtion R, Wei JY, Goldberger AL. Gait variability and basal ganglia disorders: stride-to-stride variations of gait cycle timing in Parkinson's disease and Huntington's disease. *Mov Disord.* 1998;13(3):428–37.
84. Herman T, Mirelman A, Giladi N, Schweiger A, Hausdorff JM. Executive control deficits as a prodrome to falls in healthy older adults: a prospective study linking thinking, walking, and falling. *J Gerontol A Biol Sci Med Sci.* 2010;65A(10):1086–92.
85. Martin KL, Blizzard L, Wood AG, et al. Cognitive function, gait, and gait variability in older people: a population-based study. *J Gerontol Ser Biol Sci Med Sci.* 2013;68(6):726–32.
86. Almarwani M, VanSwearingen JM, Perera S, Sparto PJ, Brach JS. Challenging the motor control of walking: gait variability during slower and faster pace walking conditions in younger and older adults. *Arch Gerontol Geriatr.* 2016;66:54–61.
87. Morris ME, Huxham F, McGinley J, Dodd K, Iansek R. The biomechanics and motor control of gait in Parkinson disease. *Clin Biomech.* 2001;16(6):459–70.
88. Yogev G, Giladi N, Peretz C, Springer S, Simon ES, Hausdorff JM. Dual tasking, gait rhythmicity, and Parkinson's disease: which aspects of gait are attention demanding? *Eur J Neurosci.* 2005;22(5):1248–56.

89. Alkjaer T, Raffalt PC, Dalsgaard H, et al. Gait variability and motor control in people with knee osteoarthritis. *Gait Posture*. 2015;42(4):479–84.
90. Zeni J, Pozzi F, Abujaber S, Miller L. Relationship between physical impairments and movement patterns during gait in patients with end-stage hip osteoarthritis. *J Orthop Res*. 2015;33(3):382–9.
91. Holtzer R, Wang C, Verghese J. The relationship between attention and gait in aging: facts and fallacies. *Motor Control*. 2012;16:64–80.
92. Persad CC, Jones JL, Ashton-Miller JA, Alexander NB, Giordani B. Executive function and gait in older adults with cognitive impairment. *J Gerontol A Biol Sci Med Sci*. 2008;63(12):1350–5.

CHAPTER 10

Clinical and Sport Biomechanics

After completing this chapter, you will be able to:

1. Define biomechanics and provide examples of the relationship between biomechanics and exercise science.
2. Identify the important historic events in the development of biomechanics.
3. Describe the important concepts of kinematics and kinetics.
4. Describe the importance of loading on body tissues.
5. Explain the differences between clinical biomechanics and sport biomechanics.
6. Discuss some of the important topics of study in biomechanics.

Biomechanics is the study of the human body at rest and in motion using principles and concepts derived from physics, mechanics, and engineering (1). The study of biomechanics involves examining forces acting on and within a biologic structure and the effects produced by such forces (2,3). Biomechanics can be subdivided into statics and dynamics. **Statics** is the branch of mechanics dealing with systems in a state of constant motion, including a system at rest (no movement) or moving with a constant speed and direction. **Dynamics** is the branch of mechanics dealing with systems when they are speeding up or slowing down (2). In order to further help the exercise scientist with the analysis of movement, biomechanics can also be subdivided into the areas of **kinematics** and **kinetics** (2). Kinematics is the study of motion, including the patterns and speed of movement of the body segments, without consideration of the forces acting on it. Kinetics is concerned with the forces acting on the body, especially forces that do not originate within the body itself. Anthropometric factors, such as the size, shape, length, and weight of the body segments, are principal considerations in a kinetic analysis (2,4).

The examination and understanding of the structural and functional mechanisms underlying human movement are critical to biomechanics. The analyses of biomechanical movement range from fundamental motor skills to complex sport activities (2). For exercise science and related professions, the relationship between humans and mechanical movement is typically studied in two primary environments: the clinical setting (called clinical biomechanics) and the sport setting (called sport biomechanics). **Clinical biomechanics** focuses on improving the ability of an injured or disabled individual to perform activities of daily living including work and leisure activities, physical activity, or exercise. For example, exercise science professionals such as athletic trainers and physical therapists use biomechanical principles and techniques to facilitate the recovery of an injured individual. **Sport biomechanics** applies the laws and principles of physics and mechanics to enhance sport performance and prevent injury through the improvement in movement techniques or the development or enhancement of equipment. For example, athletic coaches and other exercise science professionals may use the knowledge of biomechanical techniques to enhance the skilled movement of athletes and make them perform better during practice and competition, whereas manufacturers of sports equipment make alterations to a piece of

Static biomechanics The study of bodies and masses at rest or in a constant state of motion.
Dynamic biomechanics The branch of mechanics dealing with systems when they are speeding up or slowing down.
Kinematics The study of motion, including the patterns and speed of movement of the body segments, without consideration given to the mass of the body or the forces acting on it.
Kinetics The study of the forces acting on a body or system of bodies, especially of forces that do not originate within the system itself.
Clinical biomechanics A branch of biomechanics centered on improving the ability of an injured or disabled individual to perform activities of daily living, including work and leisure activities, physical activity, or exercise.
Sport biomechanics A branch of biomechanics centered on improving sport performance by athletes through the improvement in movement techniques or the development of equipment.

equipment to improve safety or enhance the performance of the athlete (2). Biomechanics and the associated principles of biomechanics are used by exercise science professionals in a variety of settings and have been around for a long time.

HISTORY OF BIOMECHANICS

Although biomechanics is a relatively young area of scientific study, many of the basic principles that form the foundation of biomechanics can be traced back thousands of years. Similarly to many of the other areas of study in exercise science, the developmental history of biomechanics starts with the ancient Greeks and Romans. The writings of many of the Greek scholars provided the framework for the guiding principles of modern biomechanics and the development of both clinical and sport biomechanics. Today, exercise science professionals and other professionals from the disciplines of physics, mechanics, and engineering continue to expand the knowledge base and discipline of biomechanics.

Early Influences

The development of mechanical, mathematical, and anatomical models and the first attempt to examine the human body biomechanically were key contributions from academics and scientists between 700 BC and 200 AD (5). Scholars such as Aristotle (384–322 BC) and Archimedes (287–212 BC) were instrumental in writing about walking, running, and movement in water. One of the early influential books was written by Aristotle and called "De Motu Animalium" or *On the Motion of Animals*. Aristotle viewed the bodies of animals as mechanical systems and pursued answers to questions such as "What is the physiological difference between imagining performing an action and actually doing it" (6). Archimedes' essay, titled *Floating Bodies,* described the principle of water displacement by physical structures, and this became the basis for determining the density of an object (5). This principle is used to form the basis for determining the density of a human body and, ultimately, body composition. Galen, a physician to the Roman emperor Marcus Aurelius, wrote *On the Function of the Parts* (meaning the parts of the human body), which was used as the world's standard medical text for over 1,400 years. This text included anatomic descriptions and terminology still used in certain areas of biologic science (5).

Throughout the Renaissance period of the 14th to 17th centuries, the science of biomechanics had its foundation further developed by some of the greatest scientists of all time. For example, Leonardo da Vinci examined the structure and function of the human body, analyzed muscle forces as acting along lines connecting origins and insertions, and studied joint function. Andreas Vesalius (1514–64), a Flemish physician, furthered the foundational development of biomechanics by publishing his brilliantly illustrated text *On the Structure of the Human Body* (Figure 10.1) (5,6). Galileo Galilei (1564–1642), an Italian physicist and mathematician, studied the action of falling bodies, the mechanical aspects of bone, and the mechanical analysis of movement. Adding to the work of previous scholars, Giovanni Alphonso Borelli (1608–79) examined various relationships between muscular movement and mechanical principles (Figure 10.2). Borelli's work *De Moti Animalium* demonstrated how geometry could be used to describe complex human and animal movements such as jumping, running, flying, and swimming. For this work, Borelli

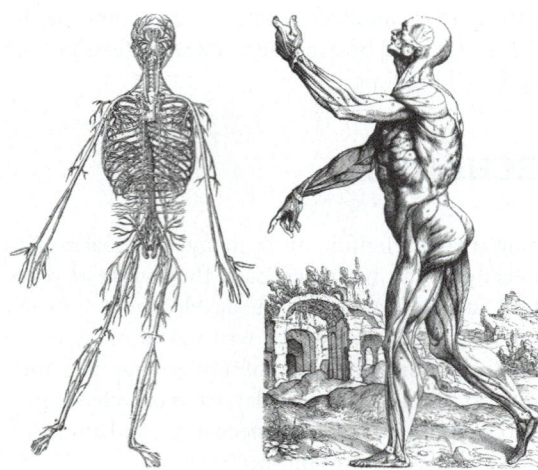

FIGURE 10.1 Anatomic structure of the body by Andreas Vesalius.

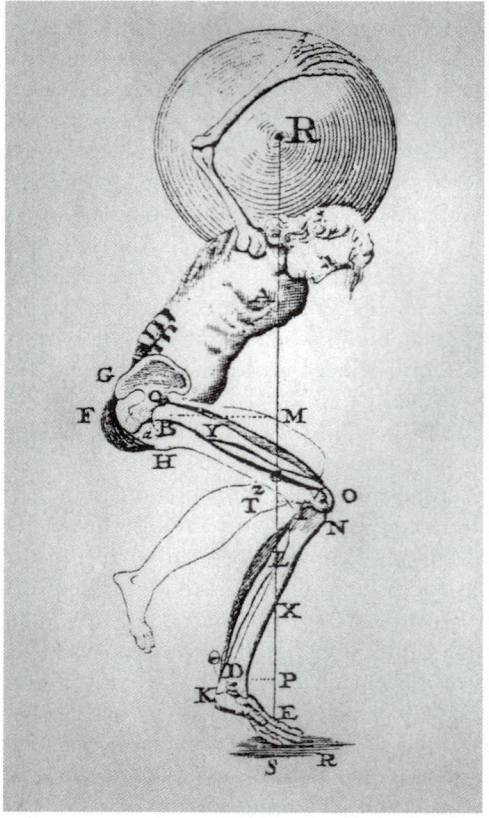

FIGURE 10.2 Giovanni Borelli's use of geometry to describe movement. (Image as seen in Provencher MT, Abdu WA. Historical perspective: Giovanni Alfonso Borelli: "Father of Spinal Biomechanics." *Spine*. 2000;25:131.)

is often referred to as "the father of biomechanics" (5). Borelli was the first to propose that levers of the musculoskeletal system magnify motion rather than force so that muscles must produce much greater forces than those forces resisting the motion (5). Sir Isaac Newton published his basic laws in *Philosophia Naturalis Principia Mathematica* in 1686. Newton's contribution to biomechanics during this period provides us with a theory for mechanical analysis and an improvement in science through the development of the process of theory and experimentation (5).

During the 19th century, the field of biomechanics expanded greatly through several key discoveries, including the development of **electromyography**, the development of measuring techniques to examine the kinematics and kinetics of movement, and the beginning of the use of engineering principles in biomechanical analysis (5). During the 1800s, scholars and scientists studied the path of the center of gravity during movement and the influence of gravity on limb movements in walking and running. Several individuals used photographic and cinemagraphic techniques to study the movements of animals and humans. For example, Eadweard Muybridge used different cameras to capture the movements of animals and humans (Figure 10.3) (7). Further study of movement resulted in a better understanding of gait and led to the development of prosthetic devices for assisting with movement. These developments have led some scholars to refer to the 19th century as the "gait century" of biomechanics (5).

FIGURE 10.3 Eadweard Muybridge's use of pictures to capture human movement. (Photos by Time & Life Pictures/Bernard Hoffman/Getty Images.)

 Electromyography A technique for evaluating and recording the electrical activity of a muscle or muscle group.

Twentieth-Century Influences

Throughout the 20th century, biomechanics further evolved into a science-based discipline for the study of animal and human movements. With the expansion of industrial technology came the need to examine the physical and physiologic aspects of industrial work. Jules Amar's book *The Human Motor*, published in 1920, focused on the efficiency of human movement and helped establish standards for human engineering in the United States and Europe. Nicholas Bernstein (1896–1966) studied the coordination and regulation of movement in both children and adults. This work provided the basis for his theories on motor control and coordination (5). Foundations for the use of biomechanics to examine the efficiency and energy cost of human movement were established by Archibald Hill (1886–1977), in the Harvard Fatigue Laboratory, and by Wallace Fenn (1893–1971), who published the first cinematographic analysis of sprint running in humans (5,8). As the interest in using mechanical and engineering principles to study movement during exercise and sport increased, biomechanical laboratories developed around the United States, including those founded by Richard Nelson at the Pennsylvania State University, Charles Dillman at the University of Illinois, and Barry Bates at the University of Oregon. These individuals were instrumental in promoting and developing biomechanics into a scientific discipline for the study of human movement in both clinical and sport environments (9).

From the mid-1960s through the end of the 20th century, the study of biomechanics was fueled by several new professional societies and academic journals, along with national and international conferences in the discipline. Five main professional societies have been instrumental in promoting biomechanics: the International Society of Electrophysiology and Kinesiology (founded in 1965), International Society of Biomechanics (founded in 1973), the American Society of Biomechanics (founded in 1977), the European Society of Biomechanics (founded in 1976), and the International Society for Biomechanics in Sport (founded in 1982). Dissemination of knowledge through peer-reviewed journals began in 1968 with the publication of the *Journal of Biomechanics*. The *International Journal of Sports Biomechanics* (now titled *Journal of Applied Biomechanics*) was first published in 1985, and the International Society of Biomechanics first published *Clinical Biomechanics* in 1986. These professional organizations and peer-reviewed journals have helped further the development of biomechanics for use in the clinical and sport settings.

Twenty-First-Century Influences

Progress in numerous areas of biomechanical study continue in the 21st century. Scholars in the area of orthopedic biomechanics continue to study the biomechanics of bone, articular cartilage, soft tissues, upper extremities, and spine. Advanced equipment and analyses, mathematical modeling, and improved engineering are being employed to create a better understanding of joint kinematics and tissue function during walking, running, and other movement activities. Biomedical and biomechanical engineers continue to work in collaboration with orthopaedic surgeons to study clinically relevant problems, improving patient treatments and outcomes (10). Notable developments and advancements in wireless technology and high-speed tracking have allowed standard biomechanical equipment to be used for complex and sophisticated assessments of movement in healthy, disabled, and injured individuals. This has created novel methods for developing cutting-edge equipment for rehabilitation and sport.

Table 10.1 Historic Events in the Development of Biomechanics

DATE	HISTORIC EVENT
ca. 384–322 BC	Aristotle publishes "On the Motion of Animals."
ca. 130–200	Galen publishes "On the Function of the Parts."
1543	Andreas Vesalius publishes "On the Structure of the Human Body."
1679	Giovanni Alphonso Borelli publishes "*De Moti Animalium*."
1920	Jules Amar publishes "The Human Motor."
1929	W.O. Fenn publishes the first cinematographic analysis of sprint running in humans.
1968	*Journal of Biomechanics* is first published.
1973	International Society of Biomechanics is founded.
1975	American Society of Biomechanics is founded.
1982	International Society for Biomechanics in Sport is founded.
1985	*Journal of Applied Biomechanics* is first published.
1986	*Clinical Biomechanics* is first published.
2000–10	Major advancements in the use of wireless and digital technology.

Professional organizations remain a critical aspect of advancement in the discipline of biomechanics. For example, the International Society for Biomechanics in Sport continues to be a leader in the development of coaching, teaching and training strategies, as well as patented sports, exercise, and rehabilitation equipment (11). For additional information on the history of biomechanics, please see the reviews provided by Wilkerson (12) and Nigg (5). Table 10.1 provides a list of some of the significant historical events in the development of the discipline of biomechanics.

Thinking Critically

In what ways has the study of biomechanics contributed to a broader understanding of physical fitness and health?

STUDY OF BIOMECHANICS

The study and application of biomechanics involves the use of various mechanical and computational principles from the academic disciplines of physics, mathematics, and engineering. The effective analysis of human movement requires an understanding of the concepts of kinematics and kinetics. Most human movement involves a complex combination of linear and angular motion components, where body segments and limbs are moving in different directions at different speeds. Motion of the body or body parts can be classified as linear, angular, or general (13). Exercise science professionals need a solid understanding of these mechanical and computational principles to appropriately use biomechanics to assist individuals in improving the execution of physical activity, exercise, sport, and athletic performance.

Types of Body Motion

True **linear motion** occurs when all points of the body are moving in the same direction at the same speed. Linear motion may also be thought of as movement along a line of travel. There are two forms of linear motion: rectilinear translation and curvilinear translation. **Rectilinear translation** (sometimes thought of as linear motion) occurs when all points on a body move in a straight line the same distance, with no change in direction. Examples of rectilinear translation might include a downhill skier in a tuck position (Figure 10.4) or a cyclist coasting during a ride. These two examples show how all points of the body can be moving forward in a straight line at the same time. **Curvilinear translation** occurs when all points on a body move in a parallel line the same distance, but the paths followed by the points on the object are curved and there is no change in body orientation. An example of curvilinear translation might be the upper body movement, which occurs during jogging or running. In curvilinear translation, the direction of the motion of the object is constantly changing, even though the orientation of the object does not (13). Figure 10.5 illustrates rectilinear and curvilinear translations.

Angular motion, also referred to as rotary motion or rotation, is movement around a central imaginary point known as the axis of rotation, which is oriented perpendicular to

FIGURE 10.4 A downhill skier in a tuck position provides an example of rectilinear translation. (From Shutterstock.)

Linear motion When all points of the body are moving in the same direction at the same speed.
Rectilinear translation When all points on a body move in a straight line, the same distance, and with no change in direction.
Curvilinear translation Occurs when all points on a body move the same distance but the paths followed by the points on the object are curved.
Angular motion The motion of a body about a fixed point or fixed axis.

FIGURE 10.5 Examples of rectilinear (**A**) and curvilinear translation (**B**) and whole body rotation (**C**). (From Dorling Kindersley RF/Getty Images.)

the plane in which the rotation occurs. Most voluntary human movement involves the rotation of a body part around an imaginary axis of rotation that passes through the center of the joint to which the body part attaches. Examples of angular motion include the shoulder press and the seated knee extension exercise, which involve movements of a ball and socket joint and a hinge joint, respectively. These movements involve a rotating action, whereby moving portions of the body are constantly in motion relative to other areas of the body (2,13). Figure 10.5C provides an illustration of whole body rotation.

General motion occurs when translation and angular movements are combined. Human movement usually consists of general motion rather than linear or angular motion. Classifying human movement as linear, angular, or general motion simplifies the biomechanical analysis of movement. If a movement can be broken down into the linear and angular components, those components can use the mechanical laws that govern linear and angular motions. The linear and angular analyses are then combined to provide a better understanding of the general motion of the individual or object (2,13). For example, while walking, the body performs both angular and translation motions. As steps are taken, the leg is performing angular motion owing to movement at the hip, knee, and ankle joints, whereas the upper torso is performing translational motion. When analyzing walking movement for the correct form, both types of motion must be considered in the analysis. Exercise science professionals should have a solid understanding of the principles of motion so that effective exercise and rehabilitation programs can be developed. This is especially true for athletic trainers, physical therapists, and occupational therapists.

Mechanical Systems

The complete biomechanical analysis of a movement requires that the system of interest be operationally defined. This helps improve the quality and usefulness of the analysis. In some analyses, the complete body is the system of interest, whereas in other instances only a body segment or an individual limb will be analyzed during the movement. For example, during

General motion Occurs when the translation and rotation movements are combined.

the analysis of walking patterns of an individual with osteoarthritis of the knee there may be interest in how the person shifts weight from one leg to the other during the movement. This might require the entire body being designated as the system of interest. If only the movements of the hip and knee are of interest, then only the affected leg might be designated as the mechanical system of interest (2).

Standard Reference Terminology

The analysis of human movement requires the use of common and specific terminology that precisely identifies body positions and directions of movement. By using consistent terminology, exercise science and allied health care professionals can precisely understand the movements describing the actions of the human body (2). This is particularly important for health care professionals, such as physical and occupational therapists, who must evaluate body movement and function to help individuals during rehabilitation from injury. Table 10.2 provides a list of terms and definitions commonly used in the biomechanical analysis of human movement.

Joint Movement Terminology

The most efficient and effective use of a biomechanical analysis occurs when exercise science professionals use consistent joint terminology to describe the movements of bones and joints of the body. For example, when the human body is in the anatomical reference position, all body segments are considered to be positioned at zero degrees. The subsequent movement of a body segment can then occur in one of the three planes: sagittal, frontal, or transverse. Figure 10.6 illustrates the planes of the body in an anatomical position.

The movement of a body segment away from the anatomical position is described according to the direction of motion and is measured as the angle between the body segment's position and the anatomic position (2). The various skeletal muscles and bones of the body are responsible for creating the movement of the body segments and limbs (2). Table 10.3 provides the terminologies of the various planes and movements of the body.

Table 10.2 Terms and Definitions Used in Biomechanical Analysis

TERM	DEFINITION
Anatomic reference position	An erect standing position with the feet slightly separated, the arms hanging relaxed at the sides, and the palms of the hands facing forward.
Directional terms	Used to describe the relationship of body parts or the location of an external object with respect to the body.
Anatomic reference planes	The division of the body by three imaginary cardinal planes into three dimensions: sagittal, frontal, and transverse.
Anatomic reference axes	The use of three reference axes for describing the rotation of the human body: mediolateral, anteroposterior, and longitudinal.

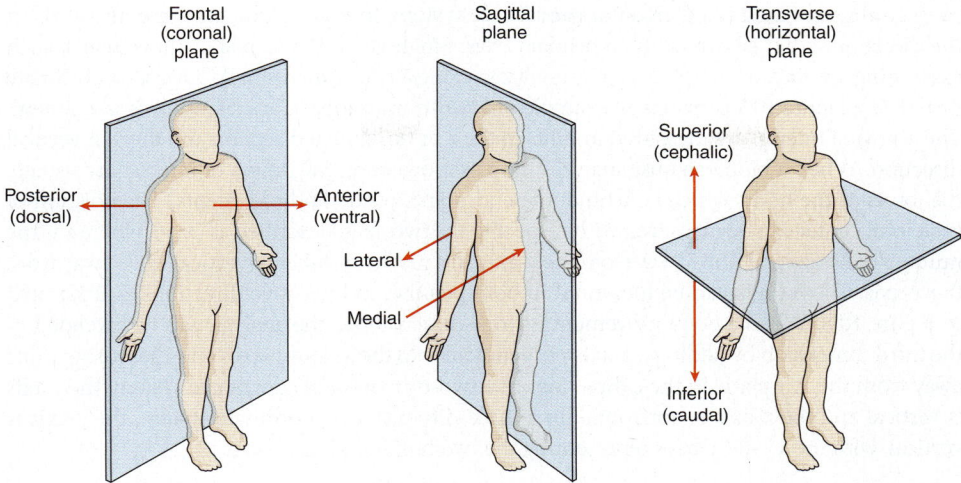

FIGURE 10.6 Illustration of the planes of the body in an anatomic position. (Adapted from Cohen BJ. *Memmler's the Human Body in Health and Disease*. 10th ed. Baltimore (MD): Lippincott Williams & Wilkins; 2005.)

Table 10.3 Terminologies of the Various Movements of the Body (2)

PLANES	BODY MOVEMENT
Sagittal	Flexion, extension, and hyperextension
Frontal	Abduction, adduction, lateral flexion, elevation and depression, deviation, eversion, and inversion
Transverse	Rotation, supination, pronation, abduction, and adduction
Other	Circumduction

Spatial Reference Systems

An understanding of biomechanics and the associated principles requires knowledge of spatial reference systems. For example, during body movement, the three coordinal planes and their associated axes of rotation also move. Therefore, to more easily perform a biomechanical analysis of a body movement or sport skill, it is often useful to employ a fixed system of reference. Clinical and sport biomechanics professionals quantitatively describe the movement of humans using a spatial reference system to standardize the measurements collected. The reference system most commonly used is a **Cartesian coordinate system**. In this system, units are measured in the direction of either two or three primary axes. Single direction or planar movements, such as running, cycling, or jumping, can be analyzed using a two-dimensional Cartesian coordinate system (2). Figure 10.7 provides an example of a two-dimensional Cartesian coordinate system. The points of interest are measured in units in the x or horizontal direction and the y or vertical direction. When a biomechanist analyzes human movement, the points of interest are usually the joints of the body, which constitute the end points of the body segments. The position of each joint center can be measured with respect to the two axes, described as (x, y), where x is the number of horizontal units away from the y-axis and y is the number of vertical units away from the x-axis. These units can be measured in both positive and negative directions as illustrated in Figure 10.8. When a body movement is three dimensional, the analysis can be extended to the third dimension by adding a z-axis perpendicular to the x- and y-axes and measuring units away from the x, y plane in the z direction. In a two-dimensional coordinate system, the y axis is vertical and the x-axis is horizontal. In a three-dimensional coordinate system, the z-axis is vertical, with the x- and y-axes representing the two horizontal directions (2).

Qualitative Analysis of Human Movement

Often, the analysis of human movement takes a **qualitative** form. The ability of exercise science and allied health care professionals, as well as sport and athletic coaches, to qualitatively assess human movement requires knowledge of the movement characteristics desired

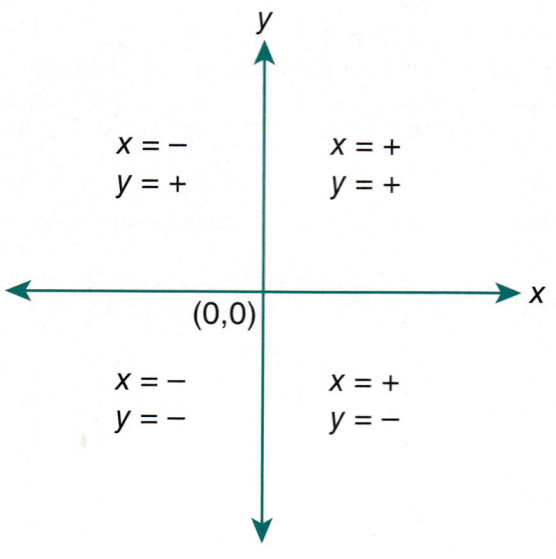

FIGURE 10.7 A two-dimensional Cartesian coordinate system.

Cartesian coordinate system A system in which the location of a point is given by coordinates that represent its distances from perpendicular lines that intersect at a point called the origin.

Qualitative The use of subjective and descriptive terms to evaluate movement and performance.

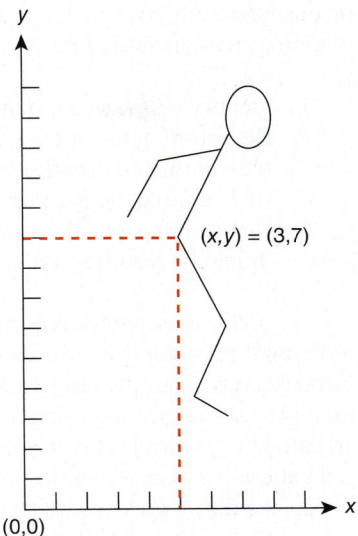

FIGURE 10.8 Positive and negative positions in a two-dimensional Cartesian coordinate system.

and the ability to observe and analyze whether a given performance incorporates these characteristics (2,13). Visual observation is the most commonly used approach to qualitatively analyze the mechanics of human movement, although many exercise science and allied health care professionals are employing the use of video recordings (See Chapter 11) to assist in the process. Using information gained from watching, in real time or on video, an athlete perform a skill or a patient's movement pattern, coaches and clinical allied health care professionals make judgments and recommendations based on movement patterns. Qualitative analyses can be further broken down into quality analysis and anatomical analysis. In the quality analysis, the movement may be evaluated in terms of the ranking of movement quality, such as "very poor," "poor," "fair," "good," and "very good." In an anatomical analysis, the movement may be evaluated in terms of joint action, primary muscles, and types of contraction. Qualitative analyses must be carefully planned and conducted with the knowledge of the biomechanics of the movement or motor skill (2,13).

Knowledge for a Qualitative Analysis

The analysis of a movement skill is a vital component of study in biomechanics. Two important factors to consider in the qualitative analysis of a movement are the techniques exhibited by the performer and the performance outcome. Effective skill analysis requires the person performing the analysis to understand the specific purpose or outcome of the motor skill being studied. For example, in the clinical environment, the successful performance outcome of rehabilitation from anterior cruciate ligament (ACL) surgery of the knee is to have a normal walking gait that does not put excessive stress on the joints of the lower extremity. In a sport environment, the successful performance outcome of a placekicker in American football is to kick the ball through the uprights of the goalpost to score points. A knowledge of this outcome is critical for understanding how to begin an examination of the movements and forces required to perform the kicking skill (2). Knowledge of relevant biomechanical principles is important for identifying the factors that contribute to successful or unsuccessful performance. An analysis of a movement skill requires several important planning steps, including the following (2):

1. Identifying the major question or questions of interest.
2. Determining the optimal perspective(s) from which to view the movement.
3. Identifying the distance from which to view the movement.
4. Determining the number of trials of the movement needed to formulate an analysis.
5. Determining whether visual observation alone is acceptable or the movement should be recorded with a motion capture system.

A **qualitative analysis** requires the progressive identification of the aspects critical to the movement or motor skill. Movements affecting the outcome of the motor skill are identified through a systematic process that often requires the individual performing the analysis to view multiple trials of the motor skill from different viewpoints. For a physical therapist, observing an individual from all sides during walking allows for the determination of whether a normal gait pattern has been established during the rehabilitation program. An athletic coach trying to improve the performance of the placekicker in American football might observe the athlete from each side during the performance of several repetitions of a kick. This allows multiple viewpoints from which the coach might make corrections to the movement being performed and therefore improve the chances of success (*i.e.*, kicking the ball through the uprights of the goalpost). It is also important to remember that the performance of a movement skill is affected by the physical, developmental, and psychological characteristics of the performer (2). The qualitative analysis of a movement skill can often result in the refinement of the movement and then possibly a new analysis of the revised skill. Figure 10.9 shows how a qualitative analysis of a movement skill can result in a cyclical process to improve performance (2).

BASIC CONCEPTS RELATED TO MOVEMENT

The term kinetics describes the study of forces on the motion or movement of the body or body part. When sufficient forces are produced by muscles, body movement and the manipulation of objects can occur. During the course of daily activities, exercise, and participation in sports, the human body generates and responds to forces. For example, recovery from injury requires the systematic increase of forces by muscles and muscle groups during rehabilitation. During the early phase of rehabilitation, small forces are generated during muscular contraction so that the injured body part is not damaged any further. As the rehabilitation program proceeds, there is a progressive increase in force production during muscular contraction. Often, the decision of an individual to return to work or competition is based on how much force can be generated during an assessment of muscle and joint movement. Sport participation requires muscles to apply forces to a variety of objects, including balls, bats, and racquets, and also the absorption of forces from impacts with ball, the playing surface, and opponents in contact sports (2). Individuals studying and using

Qualitative analysis Requires the progressive identification of the aspects critical to the movement through a systematic process that often requires the analyst to view multiple trials from different viewpoints.

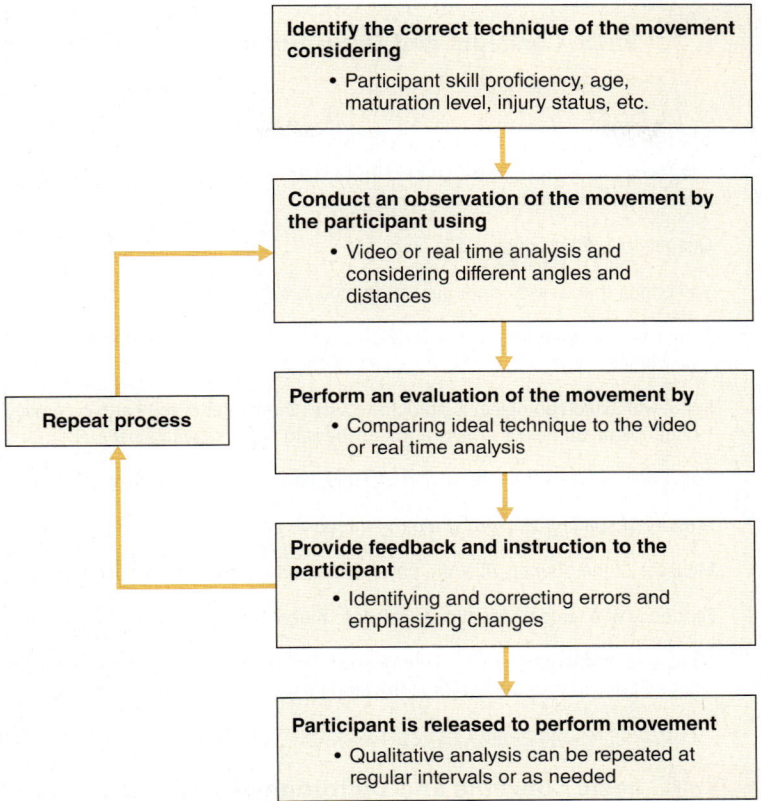

FIGURE 10.9 Process for a qualitative analysis of a movement to improve performance. (Based on data from 2,3.)

biomechanics to improve human performance must understand the basic concepts related to kinetics that are displayed in Table 10.4 (2).

Mechanical Loads on the Human Body

External forces acting on an object impose a **mechanical load** on the object. Forces from gravity and muscles and forces external to the body all affect the human body differently. The effect of a given force or forces on the body or an object depends on the direction, duration, and magnitude of the force. The types of mechanical loads on the human body are defined in Table 10.5 (2,13). The action of forces on the human body can be affected by the way in which the force is distributed on the body. There are two terms that are key to understanding the action of forces on the body: pressure and stress. Pressure represents the distribution of force that is applied externally to a body. Stress represents the resulting force distribution inside a body when an external force acts on the body (2,13).

Mechanical loads Forces that act on a body or object, including those from gravity, the muscles, and external to the body.

Table 10.4	Basic Concepts and Definitions Related to Kinetics (2)
BASIC CONCEPT	DEFINITION
Inertia	Tendency of a body to maintain its current state of motion, whether motionless or moving with a constant velocity.
Mass	Quantity of matter contained in an object.
Force	Something that causes a change in the motion of a body.
Center of gravity	Point around which the body's weight is equally balanced, no matter how the body is positioned.
Weight	Force with which an object is attracted toward the center of the earth by gravity; weight depends on an object's mass and the strength of the gravitational pull.
Pressure	Force per unit area that one region of a gas, liquid, or solid exerts on another region.
Volume	Amount of space occupied by a three-dimensional object or region of space.
Density	Measure of the quantity of some physical property expressed as mass per unit volume.
Torque	Tendency of a force applied to an object to make the object rotate about an axis.
Impulse	Change of momentum of a body or physical system over a time interval; equal to the force applied times the length of the time interval over which it is applied.

Table 10.5	Basic Concepts and Definitions Related to Mechanical Loads (2)
BASIC CONCEPT	DEFINITION
Compressive force (compression)	Force that tends to shorten or squeeze something, decreasing its volume.
Tensile force (tension)	Force that tends to stretch or elongate something.
Shear force	Force acting on a substance in a direction perpendicular to the extension of the substance.

Tension and Combined Loads

Pure compression and tension are directed along the longitudinal axis of the body and are called axial forces. When an **eccentric** (called nonaxial) force is applied to a structure, the structure bends, creating compressive stress on one side and tensile stress on the opposite

Eccentric A movement that results in a lengthening of the muscle under force.

side. For example, when an athlete performs the pole vault the end of the pole is placed in the vaulting box. The forces exerted on the pole cause a compressive stress on one side of the pole and tensile stress on the other side of the pole (Figure 10.10). **Torsion** occurs when a structure is caused to twist around its longitudinal axis, typically when one end of the structure is fixed. For example, many movements in ballet dance require the upper body to twist in one direction while the feet remain planted on the floor. In this instance, torsion is applied to the upper body. The most common type of loading on the body is combined loading, which is the presence of more than one form of force on the body (2).

Effects of Loading

There are two potential outcomes when a force acts on an object: acceleration and deformation. Acceleration is the rate of change in velocity of an object or the human body. When a baseball player starts the mechanical process of throwing a ball, the force imparted on the ball causes the ball contained in the hand and traveling at a velocity of 0 meters per second to leave the hand at an increased velocity and travel through space. Increased force production by the muscles of the body can increase the velocity of objects on which the body is acting (2). For example, professional baseball pitchers can generate enough force to throw a baseball over 100 miles per hour.

Deformation occurs when the external force causes a change in the shape or structure of an object or body component. For example, forces applied by an individual running down the track and placing the pole into the vaulting box will cause the pole to deform or bend. The bending of the pole stores elastic energy that can propel the individual upward and over the

FIGURE 10.10 An example of compression and tension during a pole vault. (From Shutterstock.)

Torsion The production of force at one end of a body that results in a twisting motion whereas the other end of the body remains fixed or moves in the opposite direction.

top of the bar. However, if too much force is applied to the pole during the jump, the physical structure of the pole would not be strong enough to withstand the forces, and therefore the pole would break. When an external force is applied to the human body, the structures of the body must withstand the external force. If too much force is applied, there is potential for injury to a body part. The magnitude and direction of the force and the area over which the force is distributed are important factors in determining the potential for tissue injury. The material properties of the loaded body tissues are also determinants of the risk of injury. If the amount of force applied to the body causes the deformation of the body tissues to exceed the point at which change to the structure occurs, some amount of deformation becomes permanent. In the human body, deformations exceeding the crucial failure point produce mechanical failure of the structure (called an injury), resulting in a fracturing of bone or rupturing of soft tissues (2).

Acute versus Chronic Loads

During participation in physical activity, exercise, sport and athletic competition, the body is subjected to acute (also called single) loads and chronic (also called repetitive) loads. The distinction between acute loading and chronic loading is important for understanding the response of the body tissues to physical activity and exercise. Understanding these concepts helps define the potential for the body tissues to experience a positive adaptation and define the potential for injury risk. When external loads of appropriate magnitude are applied to the body during physical activity and exercise, positive adaptations occur to the tissues and systems of the body. For example, during chronic resistance exercise training, performing several sets of bench press exercise can result in positive adaptations such as an increase in bone density and muscular strength. These adaptations are referred to as a training effect. When a single force large enough to cause an injury acts on body tissues, the injury is termed an acute injury. The force causing the injury is termed macrotrauma. A physical injury can also result from the repeated action of relatively small forces acting on tissue over a period of time. In this instance, the injury is called a chronic injury or a stress injury, and the causative mechanism is called microtrauma. For example, long distance running may contribute to stress fractures of the leg as a result of repeated microtrauma to the femur, tibia, and fibula (2,14).

Thinking Critically

What biomechanical information is helpful for an athlete attempting to improve movement technique following the rehabilitation of a sport-related injury?

COMPLEX MOVEMENT CONCEPTS

Movement skills that are very complex involve intricate coordination patterns among the nervous system and the musculoskeletal system. For example, throwing a ball to a specific location with great accuracy requires an individual to generate force by the body, often while moving, and then transfer that force to the ball being thrown in a manner that results in an accurate throw. Two biomechanical concepts that can help exercise science professionals in the understanding of complex movements and the principles associated with those movements include projectiles and kinetic link principles.

Projectiles

Many exercise and sport activities and athletic competitions involve the throwing of an object through the air and the hitting or kicking of an object that is traveling through the air. Objects that fly through the air free of external forces (with the exceptions of gravity and air friction) are considered projectiles. Flying objects are undergoing displacement over time and are considered to be in a free-fall state, such that gravity and air friction are the only forces that affect the flight. The instant any other external force is applied to an object, the object is no longer considered a projectile because its free-fall state has been disrupted. If the external force is removed and the two objects separate, the original object will begin another free fall. The original object will, however, have different kinematic qualities than it had before the external force was applied.

Projectiles moving in any direction can be better understood using a three-axis Cartesian coordinate system. The prediction and quantification of numerous aspects of a projectile's flight can occur if the assumption is made that no air resistance affects the object while it is traveling through its arc. This does not occur in a real-world situation because wind direction and velocity can significantly affect the projectile. If one assumes no air resistance, then the only external force that must be accounted for is gravity. Gravity affects only the vertical motion. Numerous predictions regarding the flight of the object can be made if one assumes a parabolic flight and the symmetry of the resulting arc created during flight. Success in throwing, hitting, and catching projectiles ultimately depends on the projectile's release velocity, angle of projection, and the height of the release. Changes in the throwing technique or force production by the muscles of an individual will alter the path of the projectile (4). The principles of biomechanics can be used to help an individual improve the success when tracking a projectile for hitting or kicking. This is typically done by altering movement patterns and training programs of the individual.

Kinetic Link Principle

Successful performance in exercise and sport activities and athletic competition depends on a coordination of muscle contraction for skilled movements. Highly skilled athletes often make very complex sports skills look simple and easy. Qualitative terms, such as good timing, smooth movement, effortless motion, and great coordination, are used to indicate that the nervous system appropriately controls the musculature, causing it to contract with the appropriate intensity or to relax at just the right time to produce the necessary movements for successful performance. Sport activities can be very challenging to perform because many requirements are necessary for success, including force production, velocity production, specific pattern of body motions and/or positions achieved, and conservation of energy while moving at a relatively fast velocity. Highly skilled athletes take advantage of the body's kinetic link system and create well-timed movements through coordinated muscle contractions. Two basic principles guide the body's kinetic link system: sequential movements and simultaneous movements of body segments (4).

The **sequential kinetic link principle** (also called sequential motion) means that segments of the body and joint movements occur in a specific sequence. This coordinated

Sequential kinetic link principle When segments of the body and joint rotations occur in a specific sequence or order.

movement typically leads to a high velocity generated during the last part of the performance. Sports skills that require the sequential kinetic link for success have the energy or momentum flowing from one body segment to another. The creation of momentum in the bigger slower segments of the body leads to effective transferal of momentum to the smaller, faster moving segments (4). An example would be pitching a baseball. To throw with maximum velocity, the pitcher must generate force by the body and then transfer that force to the ball. To successfully do this, a baseball pitcher must generate force using the legs and hips, then transfer that force to the shoulder and elbow. This sequential motion process will lead to the force being imparted to the ball on release from the throwing hand. This powerful and highly skilled sequential motion must be performed with great accuracy when pitching a baseball.

The **simultaneous kinetic link principle** means that major motor movements of the body occur at the same time (simultaneously) so that no observable difference exists between the contributions of the different body segments to the performance. Movements employing the simultaneous kinetic link principle are engaged when the athlete is required to move his or her body, an object, or another opponent, all of which offer varying degrees of resistance. An example would be performing a supine bench press. During the execution of the bench press, various muscle groups become active while the bar is lowered and raised. Simultaneous contraction of the pectoralis major muscles as well as other supporting muscles, including the anterior deltoids, serratus anterior, coracobrachialis, and the triceps occurs so that force can be generated by the muscles to move the weight.

Many exercise and sports activities require both the sequential and simultaneous kinetic link principles to occur during performance of the movement. Additionally, some activities require both the production of great force to move a massive object and the production of high velocity during the movement (4). For example, during the performance of a power clean lift, the muscle of the body must generate significant force to lift the bar off the floor and at the same time have the bar move at a fast velocity so that the weight can be moved quickly to the finishing position. Only when the muscles involved in the movements contract simultaneously can the power clean lift be completed successfully.

Thinking Critically

In what ways has the study of biomechanics contributed to a broader understanding of factors important for improving performance in sport and athletic competition?

 AREAS OF STUDY IN BIOMECHANICS

Biomechanics and its associated principles of study are used in a wide range of employment and professional activities. Working as a clinician in an allied health profession, ergonomist, personal trainer, or strength and conditioning coach requires a sound knowledge of basic and advanced biomechanical principles. Biomechanics contributes to a better understanding of numerous issues related to the safe and successful performance of individuals in physical activity, exercise, sport, and athletic competition through a variety of areas. The

Simultaneous kinetic link principle When major movements of the body occur at the same time.

areas of study discussed in the following sections provide a sampling of some of the primary interest areas in clinical and sport biomechanics and is not meant to be an exhaustive list.

Clinical Biomechanics

Clinical biomechanics focuses on the mechanics of injury and the principles of prevention, evaluation, and treatment of musculoskeletal problems. Clinical biomechanics professionals rely extensively on the fundamental knowledge and principles of anatomy, engineering, mathematics, physics, and psychology. Examples of the primary areas of interest in clinical biomechanics include designing individualized rehabilitation techniques, wheelchair design, tissue repair, surgical techniques, and bone and tissue design. With advanced study and preparation, clinical biomechanists can design environments that allow disabled individuals to live a satisfying and safe lifestyle and participate in recreational and sporting activities (2,13).

It is important for clinical biomechanists to know how the body moves and responds in normal healthy situations. This information is then used to set goals for recovery for injured and disabled individuals and help prevent injuries. Understanding normal patterns of movement and their variations for healthy individuals is critical so that an injured or disabled individual can be rehabilitated toward a normal and efficient movement pattern with a low risk of injury. Individuals using the principles of clinical biomechanics must be able to evaluate movement patterns in an injured or disabled individual to determine whether movement can return to normal. If normal movement cannot be attained, then adjustments in movement patterns must be made toward the most efficient and safe pattern for that patient. Clinical biomechanists work with other health care providers, such as physicians and physical and occupational therapists, to help individuals return to normal function as quickly as possible (15). Examples of areas where clinical biomechanists and the application of clinical biomechanical principles help improve physical function include osteoarthritis, gait patterns (*e.g.*, Parkinson disease), and rehabilitation from knee injuries.

Osteoarthritis is a dynamic, progressive disease that causes the loss of joint function and significant disability. In osteoarthritis, a wearing away of the cartilage that covers and acts as a cushion inside joints causes a low-grade inflammation, which ultimately results in pain in the joints (10). For example, in osteoarthritis of the knee, the bone surfaces become less well protected by the cartilage, and the individual experiences pain in weight-bearing activities, including walking and standing. Osteoarthritis of the knee is two to three times more prevalent in females than in males (16–18), and females also have a twofold higher risk of developing bilateral knee osteoarthritis (17,19). Intrinsic differences between males and females in muscle strength (20), quadriceps angle (21), joint laxity (22), and muscle activation patterns (23) may cause biomechanical differences and contribute to the differentiated risk for developing knee osteoarthritis.

For allied health care professionals, a key factor in understanding how to create rehabilitation programs for injured or disabled individuals is understanding the normal movement patterns of healthy individuals. Nonosteoarthritic individuals display similar joint kinetics at the knee (24), and similar knee joint kinematics and stride characteristics have been observed when comparing gait characteristics between females and males (25). Healthy and osteoarthritic populations have different gait patterns (Figure 10.11). Both male and female patients with osteoarthritis walk with an increased knee adduction movement (24), decreased knee flexion movement (26), and decreased knee flexion angle (27) and at a slower velocity than healthy individuals (28). Osteoarthritic females also display certain gait biomechanics that do not occur in males (29). It remains to be determined

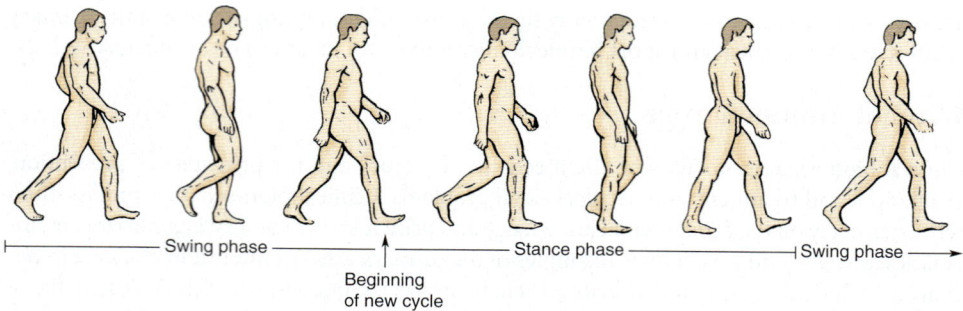

FIGURE 10.11 A normal gait cycle. (From Moore KL, Dalley AF II. *Clinical Oriented Anatomy*. 4th ed. Baltimore (MD): Lippincott Williams & Wilkins; 1999.)

whether the differences in gait biomechanics between males and females are the result of having osteoarthritis or a contributing factor that causes females to have higher rates of osteoarthritis.

Determining whether differences exist in the biomechanics of gait movement in males and females has significant implications for the treatment of osteoarthritis. Females with osteoarthritis may walk differently than healthy females or males, and treatment interventions should hence be designed based on gender and the differences that are present between females and males. Gender-specific design of biomechanical interventions to slow the progression of osteoarthritis should be examined to determine whether these interventions significantly influence the disease condition (29).

Individuals with osteoarthritis of the knee typically suffer from pain, stiffness, loss of joint range of motion and limitation of their daily activities (30). The movement of external knee adduction is a strong contributor to knee pain in those individuals with osteoarthritis (31). The use of orthotic devices is a nonsurgical and nonpharmacological way to treat knee osteoarthritis that is also less costly to the individual and contributes to improved treatment outcomes by modifying the biomechanical movements of the knee. There is support for ankle, foot, and/or knee orthoses to produce greater decreases in the external knee adduction, which is a movement that contributes to abnormal gait and increased pain (30,32). Of particular interest is the effect of shoe-worn insoles on biomechanical changes experienced at the knee joint. Placing small lateral wedges and arch support within the insoles produces small reductions in knee adduction angles, external moments, and ankle eversion and contributes to an improvement in pain relief and functional ability (32).

Total knee replacements are becoming an increasingly popular way to treat late stage osteoarthritis of the knee joint (33). Individuals with knee osteoarthritis often walk with a higher knee adduction movement when compared to individuals without the condition (34). For every 1% increase in knee movement above baseline, the risk of progression of osteoarthritis increases over sixfold (35). As is common with surgical procedures, an outcomes assessment is critical to understanding the success of the medical procedure. One way to assess outcomes is to evaluate the gait mechanics of the individual prior to and following surgery or by making comparisons to a healthy control group. For example, following surgery, the peak knee extension moment generated by the quadriceps muscles was weaker than in control subjects, whereas the knee range of motion, while improved, was still smaller than in controls (36). Both of these factors appear to contribute to a slower gait speed and suboptimal improvement of knee biomechanics (36). The biomechanical analysis of movement patterns after knee replacement surgery allows for the appropriate therapy after surgery to focus on retaining the reduction in knee adduction moment in the replaced knee and preventing a decline in the loading patterns of the nonsurgical knee (37).

Thinking Critically

What kinematic and kinetic information is helpful for an individual beginning a rehabilitation program to recover from an injury?

Ergonomics

Ergonomics is the study of the interaction between humans, the objects they use, and the environments in which they function (38). Professional preparation in ergonomics is similar to clinical biomechanics, with ergonomists taking additional coursework in engineering, anatomy, physiology, psychology, and statistics. The terms ergonomics and human factors are often used interchangeably, ergonomics referring to the work station and human factors referring to the wider system in which the individual works and operates. Ergonomists work to improve workplace efficiency, prevent workplace injuries, and improve the capacity of the individual to return to work after an injury has occurred. This includes the modification of the workplace environment and the working techniques employed by the individual worker (39). Ergonomists also design equipment and modify work and living conditions for special populations such as the elderly and disabled. Ergonomists work to design and implement changes to improve environments such as the home, recreational sites and facilities, motor vehicles, schools, clinics, workplace, and other human-built environments (15,38).

Several physical factors can cause, maintain, or worsen musculoskeletal pain and injury. Excessively forceful exertions, awkward postures, localized contact loads, and repetitive motion may lead to musculoskeletal disorders caused by **mechanical fatigue**. Ergonomists work to reduce the influence of these factors in causing pain and injury. Changing human movement patterns, redesigning work areas, and developing protective equipment are examples of how ergonomists might improve work performance and decrease the risk of pain and injury. Table 10.6 lists some of those factors that may increase the risk of pain and injury and that therefore need to be addressed by an ergonomist (15,38).

The health care environment is highly complex (40,41) and characterized by dynamic, hazardous, and distributed social systems, large space problems, and various other factors that contribute to an employee and patient environment that is uncertain at

Mechanical fatigue When the expected force production of a muscle, tendon, ligament, cartilage, or bone cannot be achieved or maintained.

Table 10.6	Factors Increasing the Risk of Injury and Pain (15,38)
RISK FACTOR	CONTRIBUTION TO INJURY AND PAIN
Forceful exertion	Used in activities that require a large magnitude of force to perform a task may easily lead to injury.
Awkward postures	Using improper technique to perform a task can produce disabling injuries to active and supporting tissues and muscles.
Localized constant loads	Occur between body tissues and objects in the environment and can cause increased compression and shearing on the tissues.
Repetitious motions	Occur during repeated performance of tasks during the day and may cause injury through inflammation and repetitive stress.

times (42). Within this environment, there exists considerable risk for employee and patient injury. Health care ergonomics is critical for providing a better understanding of the health professional and patient environment so that the work environment is improved and patient safety increased (42–44). For example, falls by patients within a health care environment are common and facilitated by a loss of physical and cognitive function by the patient and unfamiliarity with the environment (45). Ergonomists work to create a health care environment that reduces the risk of falls and increases the opportunity for safe mobility (45). Designing appropriate nonskid flooring, arranging patient care equipment and furniture to create a safe movement environment, and minimizing patient distractions are some of the interventions often employed to minimize the risk of falls by patients (45).

Sport Biomechanics

Sport biomechanics includes the examination of factors of human movement associated with exercise training and sport for the purpose of improving performance and preventing injury. Information provided by sport biomechanics professionals through quantitative investigations is often useful for coaches trying to provide a quantitative analysis of an athlete performing a motor skill or strength and a conditioning coach teaching a movement pattern during training. Sport biomechanics combines the study of applied human anatomy with mechanical physics to describe how and why the human body moves the way it does and why individuals perform at varying levels of success in sports activities (4). Athletic coaches work with sport biomechanists to use neuromuscular and mechanical factors associated with human movement to describe the requirements necessary for an athlete to perform at an optimal level. Detailed biomechanical descriptions of movement performance help coaches and athletes to refine their knowledge and approaches to training, as well as consider new and innovative techniques for improving sport performance. This knowledge may also provide information about the mechanical causes of sport-related injuries, potentially leading to safer sport participation (2,13).

Technique Improvement

Improving an athlete's technique is one of the most common methods for enhancing performance in sport and athletic competition (13). Using biomechanics to improve an athlete's technique can occur in two primary ways. First, coaches may use their knowledge of biomechanics to correct an athlete's technique to improve the execution of a movement skill. In this instance, coaches use qualitative biomechanical analysis methods to affect changes in the technique of the athlete. Second, research in biomechanics may discover a new and more effective technique for performing a sports skill. In this instance, a biomechanics researcher uses quantitative biomechanical analysis methods to discover new techniques, which will then be communicated to the coaches and athletes who will implement the new techniques (2,13).

Numerous examples exist for how sport biomechanics has helped improve athletic performance by changing techniques. Prior to the 1968 Olympics, most athletes performed the high jump by using techniques known as the Western roll, the straddle technique, or the scissors kick (Figure 10.12A). In the 1960s, Dick Fosbury developed a high-jump technique that allowed for increased height following the approach to the bar and takeoff from the ground. This technique, originally called the Fosbury Flop, allows the center of gravity to be lowered and a rotation of the body to occur immediately prior to the jump (Figure 10.12B). As a result, there is more force being created to move the body up and over the bar. Another example of how a change in technique improves sport performance would be placekicking in American football. Prior to the late 1960s, all placekickers in professional football employed a straight-on (called conventional) approach to kicking the ball. In this technique, the leg served as a pendulum to impart force onto the ball and propel it toward the goal posts. Pete Gogolak was the first placekicker to approach the ball at an angle and kicked it with his instep. This style (called the soccer style) allows for greater forces to be imparted on the ball. Virtually all American football kickers now use this soccer style approach. Table 10.7 provides brief examples of how biomechanics may be used to improve the performance of an athlete during sport or athletic competition (2,13).

FIGURE 10.12 A: The Western roll high-jump technique. (Photo by Hulton Archive/Evening Standard/Getty Images.) **B**: The Fosbury flop high-jump technique. (Photo by Bob Thomas Sports Photography/Getty Images.)

Table 10.7 — Examples of How Biomechanics May Be Used to Improve the Technique and Performance of an Athlete (2,13)

	PERFORMANCE TECHNIQUE	PERFORMANCE ANALYSIS	CHANGE IN TECHNIQUE	CHANGE IN PERFORMANCE
Coach improving performance	A baseball pitcher experiences a decrease in throwing velocity.	The coach observes the pitcher from different positions around the pitcher's mound.	The coach suggests opening up the front foot when stepping toward home plate during the pitch delivery.	This allows the pitcher's hips to open sooner, creating more force with the body and greater throwing velocity.
Biomechanist improving performance	A swimmer has a slow start off the stand, resulting in a poor entry into the water.	The biomechanist films a group of swimmers using different starting techniques and analyzes the video to determine which technique results in the fastest entry into the water.	The biomechanist recommends a change in foot placement on the starting block.	This allows the swimmers to generate more force during entry into the water, resulting in a faster entry into the water.

Equipment Improvement

Biomechanics also contributes to performance enhancement by improving designs for the shoes, apparel, and equipment used in various sports (13). For example, significant changes in shoe design and construction since the 1980s (Figure 10.13A) have most likely contributed to improved performance by athletes in all types of sports. Athletes can now choose from a wide range of shoe options that are specific to the structural features of the foot and body, the playing surface, or the environmental conditions (Figure 10.13B). Alterations in clothing design have resulted in reduced friction through the air and water and improvements in body temperature regulation. For example, competitive runners can wear full body suits in an effort to reduce air resistance on the body when sprinting. Equipment worn by athletes may have an effect on performance, either directly or through injury prevention. Improvements in helmet design in football (Figure 10.14A and B), ice hockey, and lacrosse have reduced the force impact felt by athletes competing in contact sports (46). Lighter and better designed equipment have contributed to improved performances by competitive athletes and recreational participants as well. For example, professional and amateur baseball players often carve out the end of their wooden bat in an effort to make the bat lighter and increase the velocity of the bat when swung at a pitch (13).

FIGURE 10.13 A: Original basketball sneakers (photo courtesy of Converse). **B**: Biomechanically improved basketball sneakers.

Road cycling is one sport where numerous equipment changes have resulted in an improvement in race performance. Cyclists can benefit from using formfitting clothing, aerodynamic helmets, lighter weight and aerodynamic bicycles, and cladded wheels during training and competition. Each of these equipment changes effectively reduces the amount of force required by the muscles to move the bicycle over the road surface. Changes in helmet styles have been shown to reduce drag and improve time-trial performance (47,48). Structural changes to bicycles include drop handlebars, which allow cyclists to create an aerodynamic position that improves speed and reduces the energy required for a given amount of work (49,50), and cladded wheels, which allow for a reduction in drag, have been shown to improve performance (51). Figure 10.15 shows some of the improvements in cycling equipment that are a result of biomechanical research.

What qualitative information about a sport technique would you need to help an athlete improve performance of a complex movement skill?

Training Improvement

The principles of biomechanics can be used to modify training and improve performance in many ways. A biomechanical analysis of skill performances may identify deficiencies

FIGURE 10.14 A: Old-time American football helmet. (Photo by Photodisc/Alexander Nicholson/Getty Images.) **B**: Modern American football helmet. (From Shutterstock.)

FIGURE 10.15 Improvements in cycling equipment have arisen through biomechanical research. (From Shutterstock.)

in technique that can be improved by altering training. For example, if an athlete participating in the high jump is having difficulty in determining the correct distance to run during the approach to the bar they may spend more time in practice working on the approach to the bar. If an athlete is limited by the strength or endurance of certain muscle groups, a biomechanical analysis may be helpful in determining which muscle groups are limiting performance. The training program of the athlete may then be altered to focus on improving the strength of the muscle group, which in turn may then improve the performance technique (4,13). For example, if a biomechanical analysis reveals that poor jumping technique is the result of a weakness in the strength of the quadriceps muscles, then a training program can be devised to improve the strength of that muscle group.

Plyometric exercise training is an example of how muscle biomechanics can be used to enhance force and power production in the muscles. This type of exercise training involves practicing motor movements to strengthen tissues and train nerve cells to stimulate a specific pattern of muscle contraction. The theoretical basis of plyometric training is in creating a prestretch of the muscle so that it generates as much force as possible during the contraction. Plyometric exercises involve an initial rapid muscle lengthening movement, followed by a short resting phase, and then an explosive muscle contraction movement. The combination of these movements enables the muscles involved in the contraction to generate maximal force. Plyometric exercise training engages the myostatic-reflex, which is the automatic contraction of muscles when their muscle spindle receptors are stimulated. Plyometric exercises use explosive movements to generate a large amount of force quickly, thereby improving muscle power. Plyometric exercise training acts on the nerves, muscles, and tendons of the body to increase an athlete's power output without necessarily increasing their maximum strength (52). Plyometric exercise training has been shown to improve muscular performance in athletes from a variety of sports, including male and female basketball players (53,54), soccer (55), and, most recently, older adults (56,57).

Injury Prevention

The use of biomechanical analyses helps athletic trainers and other sports medicine professionals identify factors that have caused an injury and shows them how to prevent the injury from recurring (or occurring in the first place) and what activities and exercises may assist with rehabilitation from the injury. Biomechanics can also provide the basis for alterations in technique, equipment, or training to prevent or rehabilitate injuries (4,13). For example, a kinematic analysis of an athlete running on a treadmill may reveal improper foot placement that is contributing to hip pain. Making adjustments in running mechanics may allow for a different foot placement that alleviates the pain in the hip caused by improper running mechanics. Biomechanical analyses can reveal factors that may contribute to ACL injuries (58–60) as well as assist with evaluating the recovery process from ACL surgery (61,62).

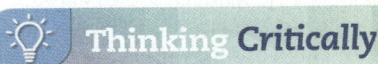

Thinking Critically

How might coursework in biomechanics and the associated disciplines prepare an individual for a career as a clinical biomechanist, a physical therapist, and an athletic coach?

Interview

Mark A. Heidebrecht, MSE, ACSM-EP, CPE/CHFP
Owner of ErgoMethods, LLC and Partner in Ergonomics International, LLC

Brief Introduction – Ergonomics or human factors is the scientific discipline that is concerned with designing tasks, jobs, products, environments, and systems that fall within the physical and cognitive abilities of an individual. I graduated with a BS in Exercise Science from the University of Kansas in 1992 and graduated with honors with my Master's degree in Exercise Physiology and Biomechanics in 1994. I feel my undergraduate and graduate education prepared me very well for my work in the fields of forensic biomechanics and ergonomics/human factors. As a Board-certified ergonomist/human factors professional, I am not only interested in the physical demands and risks associated with jobs, but also the cognitive and information processing abilities and/or limitations of individuals. I draw on my education in psychology, sociology, and statistics more than I ever anticipated I would when taking those classes. As an expert witness, I rely heavily on my ability to interpret epidemiological studies in the development of my opinions regarding the causation of traumatic and cumulative musculoskeletal injuries. Biomechanical, physiological, and cognitive fatigue are a precursor to many traumatic and cumulative musculoskeletal injuries. Understanding and applying concepts of biomechanical, physiological, and cognitive fatigue are critical in identifying risks that could lead to a workplace injury.

Q: What are your most significant career experiences?

In 1998, I started my own ergonomic consulting company, ErgoMethods, LLC and Ergo-Online, LLC, which provided live and online seminars and training. In 2015, my business partner and I formed Ergonomics International, LLC, which acquired Ergo-Online. Ergonomics International continues to provide online education but has expanded to become a leader in ergonomic and occupational health and safety software. Our software provides an epidemiological approach to understanding the development of musculoskeletal injury.

In 2000, I was asked to testify at the Occupational Safety and Health Administration (OSHA) hearing regarding the development of the Ergonomic Standard. Later that year, I was asked to present to the National Academy of Science in Washington, DC regarding the effectiveness of ergonomic interventions and the early identification of musculoskeletal disorders. I am listed as a contributor to the National Academy of Science publication, Musculoskeletal Disorders and the Workplace. Testifying at the OSHA hearing and presenting to the National Academy of Science are two experiences that have opened many doors professionally.

As an expert witness, I have provided expert opinion for very large, Fortune 500 companies, state and Federal Court. Many Board-certified ergonomists/human factors professionals have an advanced degree in psychology or industrial engineering. However, I feel my exercise physiology and biomechanics degree has provided me with a unique skill set that has separated me from many of my peers. Board-certified ergonomist/human factors professionals are in high demand, which has allowed me to present and work throughout the United States, Europe, Canada, Australia, and New Zealand.

Q: Why did you choose your professional career?

After completing several internships in the field of physical therapy, I became more interested in applying my knowledge of physiology and biomechanics in the prevention and causation of injury rather than rehabilitation. For me, ergonomics and human factors was the perfect fit. This field allows me to apply my knowledge of biomechanics and exercise physiology to the work environment. Ergonomics and human factors allow me to identify ways for companies to increase productivity, increase quality, and decrease the risk of injury.

Q: Why is it important for an exercise science student to have an understanding of biomechanics and exercise science?

My understanding of biomechanics provides me with the knowledge to identify postures and motions that may be contributing to the development of fatigue or a musculoskeletal injury. This knowledge base also allows me to develop more efficient or safer movement patterns. For instance, having a very good understanding of the rotator cuff and the muscles involved allows me to identify and reduce the exposure of ergonomic risks that may otherwise be overlooked. My understanding of muscle fatigue and the impact of environmental factors on the human body provides me with the knowledge to identify when job requirements may exceed the physical capabilities of the worker. A complete understanding of biomechanics and exercise science is critical when looking at cause and prevention of both traumatic and cumulative musculoskeletal injury.

Q: What advice would you have for an undergraduate student beginning to explore a career in exercise science?

Think outside of the box! An exercise science degree gives you a skill set that can be applied to many different traditional and nontraditional occupational settings. Take advantage of internships and independent studies when in school. The experience and contacts you make are just as important as the knowledge you gain. Find professionals who are established in your chosen field and ask if they will act as a mentor for you. Look at others in the field as resources rather than competitors. Get a broad exercise science foundation with as much practical hands-on experience as you can.

Interview

Rafael Escamilla, PhD, PT, CSCS, FACSM

Professor of Physical Therapy; Co-Director, Biomechanics Laboratory
Department of Physical Therapy, California State University, Sacramento

Brief Introduction – I grew up in the Pacific Northwest in Washington State, Idaho, and Oregon. I received an AA degree from Walla Walla Community College, a BS degree in Mathematics and Physical Education at Linfield College, an MS degree in Biomechanics from Washington State University, an MPT degree in Physical Therapy at Elon University, and a PhD degree in Biomechanics at Auburn University. I am currently a professor of physical therapy at California State University and a research coordinator at Results Physical Therapy and Training Center. I also work as an outpatient licensed physical therapist. I previously served as Director of Research at the Andrews Institute, I was Professor of Orthopaedic Surgery at Duke University Medical Center, Director of the Michael W. Krzyzewski Human Performance Laboratory at Duke University, and I taught biomechanics and anatomy courses in the Doctor of Physical Therapy program at Duke University. I also worked at the American Sports Medicine Institute in Birmingham, AL, with world-renowned sports medicine orthopedic surgeon Dr. James Andrews. I am a certified strength and conditioning specialist (CSCS) and have taught numerous CSCS classes and workshops. My area of expertise is in biomechanics and patho-mechanics of the overhead throwing shoulder and elbow, and knee biomechanics during rehabilitation exercise. I have published over 200 peer-reviewed scientific articles, abstracts, and book chapters in biomechanics, physical therapy, and sports medicine–related journals and textbooks and have given over 250 professional presentations at scientific conferences, primarily in the areas of overhead throwing biomechanics, knee biomechanics during exercise, and strength and conditioning. In 2016, I was chosen the outstanding senior faculty researcher at California State University, Sacramento. I have won three national and international powerlifting championships and was a collegiate athlete in football, baseball, and track.

Q: What are your most significant career experiences?

My most significant career experiences have been working as an educator in both kinesiology and physical therapy for the past 35 years and working as a sports medicine researcher during that same time frame. I have worked with numerous high-profile professional athletes focusing on training and rehabilitation, athletic performance, and injury prevention. My focus was to maximize human performance in sport and minimize injuries during training and competition. I conducted biomechanical research in baseball and tennis (studied baseball pitching and the tennis serve) on medical research grants at the 1996 Atlanta Summer Olympic Games and 2000 Sydney Summer Olympic Games, examining shoulder and elbow biomechanics in addition to total body biomechanics. Contributing to the science of sports medicine by publishing in scientific journals has been a career highlight, as well as presenting my research all over the world at scientific conferences.

Q: Why did you choose to become an "exercise scientist"?

I was a college athlete and also a competitive powerlifter for 15 years after college. Exercise science enhanced my knowledge on how to maximize performance. Eventually, I also got interested in injury prevention and rehabilitation, which is also under the umbrella of an exercise scientist.

Q: What advice would you have for a student exploring a career in any exercise science profession?

First determine what you are interested in and then assess how those interests can lead to a lifelong career in exercise science. Understand all subdisciplines under the umbrella of exercise science and determine which ones best help you accomplish your short- and long-term goals. Assess your interests in teaching, research, clinical work, and human performance enhancement to determine what you will be.

SUMMARY

- Biomechanics is the study of the human body at rest and in motion using basic and advanced principles from the academic disciplines of mathematics, physics, and engineering.
- The use of kinematics and kinetic techniques and principles provides insight into how the human body moves and responds to forces when healthy, injured, or performing a sport or movement activity.
- Biomechanics has important applications in the clinical environment for promoting recovery from injury and the industrial setting for improving work performance and reducing the risk of job-related injuries.
- Biomechanical principles are also used to enhance sport and athletic performance through improvements in technique, equipment, training methods, and injury prevention.
- Advanced biomechanical concepts are used to understand complex movements and sport skills to promote successful performance.

FOR REVIEW

1. Explain the difference between static and dynamic biomechanics.
2. What significant events happened during the Renaissance period that contributed to the development of biomechanics?
3. What are the two types of linear motion?
4. What is a biomechanical system of interest?
5. How are the three planes of the body (sagittal, frontal, or transverse) used to describe movements of the body?
6. How is a Cartesian coordinate system used in a biomechanical analysis of movement?
7. Describe the steps involved in a qualitative analysis of a soccer kick.
8. Define the three types of mechanical loads: compressive force, tensile force, and shear force.
9. How does mechanical loading cause the deformation of an object such as bone?
10. Why must a clinical biomechanist understand how the body responds to a normal situation or movement?
11. What is the difference between a clinical biomechanist and an ergonomist?
12. Describe how a biomechanist can improve performance by changing technique, equipment, or training techniques.
13. What is the difference between the sequential movements and simultaneous movements of body segments?

Project-Based Learning

1. Identify an individual who participates in regular physical activity or exercise but suffers from poor biomechanical form when walking and running. Prepare a presentation that includes at least five key recommendations that you would give to that individual in an effort to improve their balance during physical activity and exercise. What are those key points and how does the biomechanics literature support your recommendations?
2. Identify an individual participating in a sport or athletic competition whose performance you believe could be enhanced by some changes in their biomechanical movement patterns. Prepare a presentation that includes at least five key points that you would give to another group of exercise science professionals who are evaluating that athlete's movement patterns. What are those key points and how does the biomechanics literature support your recommendations?

REFERENCES

1. Bates BT. The need for an interdisciplinary curriculum. In: *Third National Symposium on Teaching Kinesiology and Biomechanics in Sports Proceedings*; 1991: Ames, IA.
2. Hall SJ. *Basic Biomechanics*. 7th ed. Dubuque (IA): McGraw-Hill; 2015.
3. Hay JG. *Biomechanics of Sports Techniques*. 4th ed. Englewood Cliffs (NJ): Prentice-Hall; 1993.

4. Johnson BF. Sports biomechanics. In: Brown SP, editor. *Introduction to Exercise Science*. 1st ed. Philadelphia (PA): Lippincott, Williams & Wilkins; 2001. p. 264–88.
5. Nigg BM. Introduction. In: Nigg BM, Herzog W, editors. *Biomechanics of the Musculo-skeletal System*. 3rd ed. Chichester, England: John Wiley & Sons, Ltd.; 2007. p. 1–48.
6. Martin RB, editor. *A Genealogy of Biomechanics*. Pittsburgh (PA): American Society of Biomechanics; 1999.
7. Haas RB. *Muybridge: Man in Motion*. Berkley (CA): University of California Press; 1976.
8. Fenn WO. Mechanical energy expenditure in sprint running as measured in moving pictures. *Am J Physiol*. 1929;90:343–4.
9. Stergiou N, Blanke DJ, Chen SJ, Siu KC. Biomechanics. In: Housh TJ, Housh DJ, Johnson GO, editors. *Introduction to Exercise Science*. 3rd ed. San Francisco (CA): Pearson Education, Inc.; 2008. p. 207–31.
10. Innocenti B. Biomechanics: a fundamental tool with a long history (and even longer future!). *Muscles Ligaments Tendons J*. 2018;7(4):491–2.
11. International Society for Biomechanics in Sport 2020 Web site [cited 2020]. Available from: https://isbs.org/.
12. Wilkerson JD. Biomechanics. In: Massengale JD, Swanson RA, editors. *The History of Exercise and Sport Science*. 1st ed. Champaign (IL): Human Kinetics; 1997. p. 321–66.
13. McGinnis PM. *Biomechanics of Sport and Exercise*. 3rd ed. Champaign (IL): Human Kinetics; 2013.
14. Edwards WB, Gillette JC, Thomas JM, Derrick TR. Internal femoral forces and moments during running: implications for stress fracture development. *Clin Biomech*. 2008;23(10):1269–78.
15. Leveau BF. Clinical biomechanics. In: Brown SP, editor. *Introduction to Exercise Science*. 1st ed. Philadelphia (PA): Lippincott, Williams & Wilkins; 2001. p. 236–63.
16. Buckwalter JA, Lappin DR. The disproportionate impact of chronic arthralgia and arthritis among women. *Clin Orthop Relat Res*. 2000;458:159–68.
17. Helmich CG, Felson DT, Lawrence RC, et al. Estimates of the prevalence of arthritis and other rheumatic conditions in the United States: part I. *Arthritis Rheum*. 2008;58(1):15–25.
18. Lawrence RC, Felson DT, Helmich CG, et al. Estimates of the prevalence of arthritis and other rheumatic conditions in the United States: part II. *Arthritis Rheum*. 2008;58(1):26–35.
19. March LM, Bagga H. Epidemiology of osteoarthritis in Australia. *Med J Aust*. 2004;180:S6–10.
20. Cureton KJ, Collins MA, Hill DW, McElhannon FM. Muscle hypertrophy in men and women. *Med Sci Sports Exerc*. 1988;20:338–44.
21. Horton MG, Hall TL. Quadriceps femoris muscle angle: normal values and relationships with gender and selected skeletal measures. *Phys Ther*. 1989;69:897–901.
22. Bridges AJ, Smith E, Reid J. Joint hypermobility in adults referred to rheumatology clinics. *Ann Rheum Dis*. 1992;51:793–6.
23. White KK, Lee SS, Cutuk A, Hargens AR, Pedowitz RA. EMG power spectra of intercollegiate athletes and anterior cruciate ligament injury risk in females. *Med Sci Sports Exerc*. 2003;35:371–6.
24. Hurwitz DE, Sumner DR, Andriacchi TP, Sugar DA. Dynamic knee loads during gait predict proximal tibial bone distribution. *J Biomech*. 1998;31:423–30.
25. Resnicow K, Lazarus Yaroch A, Davis A, et al. GO GIRLS!: results from a nutrition and physical activity program for low-income, overweight African American adolescent females. *Health Educ Behav*. 2000;27(5):616–31.
26. Martinez JA. Obesity in young Europeans: genetic and environmental influences. *Eur J Clin Nutr*. 2000;54:S56–S60.
27. Childs JD, Sparto PJ, Fitzgerald GK, Bizzini M, Irrgang JJ. Alterations in lower extremity movement and muscle activation patterns in individuals with knee osteoarthritis. *Clin Biomech*. 2004;19:44–9.
28. Kaufman KR, Hughes C, Morrey BF, Morrey M, An K. Gait characteristics of patients with knee osteoarthritis. *J Biomech*. 2001;34:907–15.
29. McKean KA, Landry SC, Hubley-Kozey CL, Dunbar MJ, Stanish WD, Deluzio KJ. Gender differences exist in osteoarthritic gait. *Clin Biomech*. 2007;22:400–9.
30. Baghaei Roodsari R, Esteki A, Aminian G, et al. The effect of orthotic devices on knee adduction moment, pain and function in medial compartment knee osteoarthritis: a literature review. *Disabil Rehabil Assist Technol*. 2017;12(5):441–9.
31. Moyer R, Birmingham T, Dombroski C, Walsh R, Giffin JR. Combined versus individual effects of a valgus knee brace and lateral wedge foot orthotic during stair use in patients with knee osteoarthritis. *Gait Posture*. 2017;54:160–6.
32. Shaw KE, Charlton JM, Perry CKL, et al. The effects of shoe-worn insoles on gait biomechanics in people with knee osteoarthritis: a systematic review and meta-analysis. *Br J Sports Med*. 2018;52(4):238–53.

33. Ackerman INP, Bohensky MAP, de Steiger RM, et al. Substantial rise in the lifetime risk of primary total knee replacement surgery for osteoarthritis from 2003-2013: an international, population-level analysis. *Osteoarthritis Cartilage*. 2016;25(4):455–61.
34. Baliunas AJ, Hurwitz DE, Ryals AB, et al. Increased knee joint loads during walking are present in subjects with knee osteoarthritis. *Osteoarthritis Cartilage*. 2002;10(7):573–9.
35. Mandeville D, Osternig LR, Lantz BA, Mohler CG, Chou L-S. The effect of total knee replacement on the knee varus angle and moment during walking and stair ascent. *Clin Biomech*. 2008;23(8):1053–8.
36. Ro DH, Ro DH, Han H-S, et al. Slow gait speed after bilateral total knee arthroplasty is associated with suboptimal improvement of knee biomechanics. *Knee Surg Sports Traumatol Arthrosc*. 2018;26(6):1671–80.
37. Debbi EM, Bernfeld B, Herman A, et al. Frontal plane biomechanics of the operated and non-operated knees before and after unilateral total knee arthroplasty. *Clin Biomech*. 2015;30(9):889–94.
38. Pulat BM. *Fundamentals of Industrial Erogonomics*. 2nd ed. Prospect Heights (IL): Waveland Press; 1997.
39. Chaffin DB, Andersson GB. *Occupational Biomechanics*. 2nd ed. New York (NY): Wiley; 1991.
40. Carayon P. Human factors of complex sociotechnical systems. *Appl Ergon*. 2006;37(4):525–35.
41. Carayon P, Bass E, Bellandi T, Gurses A, Hallbeck S, Mollo V. Socio-technical systems analysis in health care: a research agenda. *IIE Trans Healthc Syst Eng*. 2011;1(1):145–60.
42. Valdez RS, McGuire KM, Rivera AJ. Qualitative ergonomics/human factors research in health care: current state and future directions. *Appl Ergon*. 2017;62:43–71.
43. Holden RJ, Carayon P, Gurses AP, et al. SEIPS 2.0: a human factors framework for studying and improving the work of healthcare professionals and patients. *Ergonomics*. 2013;56(11):1669–86.
44. Valdez RS, Holden RJ, Novak LL, Veinot TC. Transforming consumer health informatics through a patient work framework: connecting patients to context. *J Am Med Inform Assoc*. 2014;22(1):2–10.
45. Hignett S, Wolf L. Reducing inpatient falls: human factors & ergonomics offers a novel solution by designing safety from the patients' perspective. *Int J Nurs Stud*. 2016;59:A1–A3.
46. Pellman EJ, Viano DC, Withnall C, Shewchenko N, Bir CA, Halstead PD. Concussion in professional football: helmet testing to assess impact performance — part 11. *Neurosurgery*. 2006;58(1):78–96.
47. Beaumont F, Taiar R, Polidori G, Trenchard H, Grappe F. Aerodynamic study of time-trial helmets in cycling racing using CFD analysis. *J Biomech*. 2018;67:1–8.
48. Alam F, Chowdhury H, Wei HZ, Mustary I, Zimmer G. Aerodynamics of ribbed bicycle racing helmets. *Proc Eng*. 2014;72:691–6.
49. Chowdhury H, Alam F. Bicycle aerodynamics: an experimental evaluation methodology. *Sports Eng*. 2012;15(2):73–80.
50. Lukes RA, Chin SB, Haake SJ. The understanding and development of cycling aerodynamics. *Sports Eng*. 2005;8(2):59–74.
51. Arora BB, Bhattacharjee S, Kashyap V, Khan MN, Tlili I. Aerodynamic effect of bicycle wheel cladding — a CFD study. *Energy Rep*. 2019;5:1626–37.
52. Brooks GA, Fahey TD, Baldwin KM. *Exercise Physiology: Human Bioenergetics and Its Applications*. 4th ed. Mountain View (CA): Mayfield; 2004.
53. Asadi A, Ramirez-Campillo R, Meylan C, Nakamura FY, Canas-Jamett R, Izquierdo M. Effects of volume-based overload plyometric training on maximal-intensity exercise adaptations in young basketball players. *J Sports Med Phys Fitness*. 2017;57(12):1557–63.
54. McCormick BT, Hannon JC, Newton M, Shultz B, Detling N, Young WB. The effects of frontal- and sagittal-plane plyometrics on change-of-direction speed and power in adolescent female basketball players. *Int J Sports Physiol Perform*. 2016;11(1):102–7.
55. Ramírez-Campillo R, Vergara-Pedreros M, Henríquez-Olguín C, et al. Effects of plyometric training on maximal-intensity exercise and endurance in male and female soccer players. *J Sports Sci*. 2015;34(8):687–93.
56. Van Roie E, Walker S, Van Driessche S, Delabastita T, Vanwanseele B, Delecluse C. An age-adapted plyometric exercise program improves dynamic strength, jump performance and functional capacity in older men either similarly or more than traditional resistance training. *PLoS One*. 2020;15(8):1–6.
57. Vetrovsky T, Steffl M, Stastny P, Tufano JJ. The efficacy and safety of lower-limb plyometric training in older adults: a systematic review. *Sports Med (Auckland)*. 2018;49(1):113–31.
58. Cibin F, Pavan D, Trevisanato G, et al. Biomechanical analysis of the side cut in basketball athletes as non-contact ACL injury screening. *Gait Posture*. 2019;74:9–10.
59. Numata H, Numata H, Nakase J, et al. Two-dimensional motion analysis of dynamic knee valgus identifies female high school athletes at risk of non-contact anterior cruciate ligament injury. *Knee Surg Sports Traumatol Arthrosc*. 2018;26(2):442–7.

60. Thompson JA, Tran AA, Gatewood CT, et al. Biomechanical effects of an injury prevention program in preadolescent female soccer athletes. *Am J Sports Med*. 2017;45(2):294–301.
61. King E, Richter C, Franklyn-Miller A, et al. Biomechanical but not timed performance asymmetries persist between limbs 9 months after ACL reconstruction during planned and unplanned change of direction. *J Biomech*. 2018;81:93–103.
62. Hadizadeh M, Amri SB, Mohafez H, Ahmad SR, Mokhtar AHB. Gait analysis of national athletes after anterior cruciate ligament reconstruction following three stages of rehabilitation program: symmetrical perspective. *Gait Posture*. 2016;48:152–8.

CHAPTER

11

Assessment and Equipment in Exercise Science

After completing this chapter, you will be able to:

1. Explain the important issues associated with pretesting guidelines and procedures.
2. Describe the different types of equipment used in the assessment of cardiovascular and pulmonary function.
3. Describe the different types of equipment used in musculoskeletal assessments.
4. Explain the different types of equipment and instruments used in weight management and body composition assessments.
5. Describe the different types of equipment used in clinical assessment and rehabilitation.
6. Describe the different types of equipment used in motor performance assessment.
7. Explain the different types of instruments used in exercise behavior and sport psychology assessments.

The performance of individuals participating in physical activity, exercise, sport, and athletic competition can be affected by many factors, including, but not limited to, genetic endowment, health and injury status, nutrient intake, physiological status, psychological status, and biomechanical factors (1–6). Exercise science and allied health professionals must accurately assess and evaluate those dietary, physical, physiological, biomechanical, motor control, and psychological attributes that can provide insight into an individual's health status, the risk of certain diseases and illnesses, and the individual responses to rehabilitation and exercise training programs. Furthermore, the assessment and evaluation of numerous factors related to success is important to athletes working to enhance performance in sport and athletic competition (7).

Evaluation and assessment procedures can be expensive and time consuming, so the most appropriate measurements using the best available equipment and instruments is critical to ensuring valid and reliable measures. Regardless of whether the assessments are made on an individual participating in a regular physical activity or exercise program or on a highly trained athlete, the benefits derived from evaluation and assessment include (1,7):

- Identifying the individual's strengths and weaknesses from the measured variables
- Providing important feedback to the individual being tested, the exercise science or allied health care professional, or the coach on the assessment and evaluation results
- Identifying current health status, the risk of certain diseases, and the progress made on recovery from an injury or illness
- Providing educational information to the individual being tested on the assessment and evaluation process and the results of the testing

Much of the assessment and evaluation performed by exercise science professionals can be broadly categorized into **fitness and functional capacity testing** and **diagnostic testing**. Fitness and functional capacity testing is used to help assess the fitness and performance capabilities of an individual to do work- or job-related activity, physical activity, exercise, perform a sport, or participate in an athletic competition. Diagnostic testing is performed to help identify the presence of a disease condition, risk factors for a disease condition, or an existing injury within an individual. The type of assessments and evaluations performed will likely differ between a nonathlete and an athlete as the purpose and use of the testing information is different for the two types of individuals. Whether performing fitness and functional capacity or diagnostic testing, there are guidelines and procedures that should be followed to ensure accurate and consistent results.

PRETESTING GUIDELINES AND PROCEDURES

To ensure that the most accurate and consistent information is obtained during assessments and evaluations, the exercise science and allied health care professional must establish and

Fitness and functional capacity testing Used to provide an objective measure of an individual's safe functional abilities.

Diagnostic testing Used to determine a specific condition or possible illness.

follow specific pretesting guidelines and procedures. Two of the most important issues to consider before testing include making sure that the assessments and evaluations that are selected are **valid** and **reliable** and that the data being collected are specific and relevant to the information that is required of the testing. An assessment or evaluation of a particular aspect of fitness or physical or psychological function is valid when it actually measures what it claims to measure, and it is reliable when the results from the assessment or evaluation are consistent and reproducible. To ensure validity and reliability, the instruments used in testing must be calibrated and used according to the instructions of the manufacturer or according to established procedures. For example, the calibration of many metabolic measurement carts and blood analyzers require the use of known reference standards in the calibration process. These known reference standards are certified by the producers of the product to be of the exact value. When the known reference standards are used in the calibration of an instrument, the validity and reliability of the equipment are significantly enhanced (7).

When making an assessment or evaluation of a particular aspect of fitness, physical or psychological function, and performance, it is important to make sure that the measurement protocols, techniques, equipment, and instruments are specific to and designed for the characteristic being measured. For the test results to have optimal significance, the testing must be specific to the desired question of interest. As testing procedures move further away from the actual physical or psychological function measured in the assessment and evaluation, the validity declines despite the fact that the results may be reliable. For example, if an exercise science professional wanted to know which muscles were being used during the performance of a job-related task, it would be best to perform the assessment in the workplace environment. However, because specialized equipment is often needed to make this type of assessment, the measurements might have to be performed in a laboratory or clinical environment. This may necessitate constructing a test that would simulate the job-related task as closely as possible. If this is not possible, then the validity of the assessment might be compromised (7).

After identification of the specific assessments and evaluations is made, it is important to familiarize the individual with the testing environment and the procedures that they will undergo and closely control the test administration procedures to minimize the influence of extraneous factors on the results obtained (1,7). Important issues to consider and control during the assessment and evaluation process include the following:

- Ensure that standardized and clear instructions are given to subjects, patients, clients, or athletes.
- Ensure that the participant is wearing clothing and footwear appropriate for the testing to be conducted.
- Provide sufficient practice or warm-up procedures prior to the assessment and evaluation.
- Make certain that the order and number of the assessment and evaluation items does not influence the results.

Valid Providing an accurate measurement.
Reliable Providing a consistent measurement.

- Allow sufficient recovery time between assessment and evaluation items.
- Control the environmental conditions where the tests are being conducted.

An inability to sufficiently control or account for these factors has the potential to significantly affect the results obtained during testing, making them invalid and unreliable.

Individual Issues to Control

There are also numerous other issues to consider and possibly control for about the subjects, patients, clients, or athletes so that accurate assessment and evaluation results are obtained (1). Some of these issues include:

- Determining whether and for how long an individual has been participating in a physical activity, exercise, or sport and athletic training program
- Considering whether the time of the most recent physical activity, exercise, training session, or competition has any influence on the tests being conducted
- Considering if the time of day in which the assessments and evaluations are made have any relation to any previous assessments that have been conducted
- Evaluating the nutritional status of the individual to determine whether this has an impact on the assessments being made
- Ensuring that the amount of sleep or rest the individual has received is sufficient for testing
- Identifying whether the individual has an injury or illness and deciding if this will affect the assessments
- Ensuring that the individual is properly hydrated
- Identifying any drugs or medications used by the individual and considering whether these will affect the testing results
- Identifying the psychological status of the individual (*e.g.*, anxious or nervousness) and considering the impact of the status on the testing results

Laboratory and Field Testing

Certain types of testing assessments and measurements must be made in a laboratory or clinical environment, whereas other tests can be made outside the laboratory and are usually called a field test. Laboratory and clinical tests and assessments are conducted in a controlled environment and use protocols and equipment that simulate part or all of the physical activity, exercise, sport, or athletic performance. Measurement procedures for field tests are conducted while the individual is performing in a simulated physical activity, exercise, or competitive sport situation. Results obtained from field tests are often not as reliable as those obtained from laboratory tests but can sometimes be more valid because of their greater specificity to the activity or sport. Performance varies more in the field setting than in the laboratory environment because it is difficult to control variables such as wind velocity, temperature, humidity, and other environmental factors. Additionally, portable data collection systems necessary for field testing may not be as accurate as those used in the laboratory; however, recent advancements in technology have improved the validity and reliability of much of the equipment and instrumentation used in field testing.

The information received from fitness and functional capacity testing and diagnostic testing can be used to enhance health and improve physical performance. Many of the assessments of health and physical performance are conducted by most exercise science and allied health professionals and are not limited to a particular field or discipline. Therefore, the information contained in this chapter will attempt to provide a description of the current and most commonly used equipment according to the following broad-based categories:

- Cardiovascular and pulmonary function
- Musculoskeletal function
- Energy balance assessment
- Body composition measurement
- Blood collection
- Injury rehabilitation
- Motor performance
- Behavioral and psychological function

CARDIOVASCULAR AND PULMONARY FUNCTION ASSESSMENT

The health and functional capacity of the cardiovascular system and the pulmonary system play an important role in an individual's risk for developing cardiovascular and pulmonary diseases (8) and has considerable application in the clinical assessment of patients (9,10). Additionally, cardiovascular and pulmonary fitness can have considerable influence on the potential for successful performance in a variety of sports and athletic competitions that involve cycling, running, swimming, and other movement activities (11). The assessment of cardiovascular and pulmonary function can range from inexpensive, easily administered field tests to very expensive time-consuming measurements performed in a controlled laboratory environment or clinical facility.

Treadmills and Ergometers

When assessing the health and functional capacity of the cardiovascular and pulmonary systems, the intensity and amount of exercise performed by the individual must be precisely controlled and accurately measured. Furthermore, the administration of safe and effective exercise and training prescriptions requires an accurate control of the exercise intensity. The motor-driven treadmill and cycle ergometer are the most commonly used pieces of equipment in both diagnostic and functional capacity exercise testing. The selection of equipment for testing should be based on the individual's primary mode of physical activity and exercise. The treadmill is the preferred equipment for most individuals because of the familiarity with walking and running movements. Upright or **recombinant cycle ergometers** can be

 Recombinant cycle ergometer A cycle ergometer that allows the rider to be seated with the legs supine and the back supported.

used for those individuals, such as the overweight/obese, injured, or disabled, who require support during exercise testing or whose primary physical activity and exercise activity are cycling (12).

During treadmill exercise, the intensity is controlled by manipulating the speed of the treadmill belt and the grade (*i.e.*, slope) of the treadmill. This allows for a regulated and incremental increase in the workload administered to the individual walking or running on the treadmill. The intensity of exercise is controlled on the cycle ergometer by increasing the resistance against which an individual must pedal. Many laboratory cycle ergometers (Figure 11.1) administer resistance by the use of a friction belt against the rotating flywheel. In friction belt cycle ergometers, the individual pedaling rate will affect the workload and thus the intensity of the exercise. For example, if two individuals are pedaling against the same resistance but one is pedaling at a faster rate, then that individual is doing more work at a higher exercise intensity. This limitation of the friction belt cycle ergometers can be overcome by using an electronically braked cycle ergometer that employs an electromagnet to impart resistance during exercise. An electronically braked cycle ergometer automatically adjusts the intensity of exercise to compensate for changes in pedaling rate. During diagnostic and functional capacity testing of the cardiovascular and pulmonary systems, the exercise begins at a low intensity and increases until a predetermined workload or intensity is reached or until the subject reaches their maximal exercise capacity (7,12).

Other equipment used in cardiovascular and pulmonary testing and exercise training includes arm cycle ergometers (Figure 11.2A), stairclimbers, step boxes, elliptical machines, rowing machines (Figure 11.2B), simulated skiing machines (Figure 11.2C), and swimming flumes (Figure 11.2D). For example, arm ergometers can be used to test cardiovascular function in those individuals who cannot use their legs for physical activity or exercise. An arm cycle ergometer is similar to a cycle ergometer, except that arm pedals replace the foot pedals and the device is placed on a tabletop. Although stairclimbers and elliptic machines can

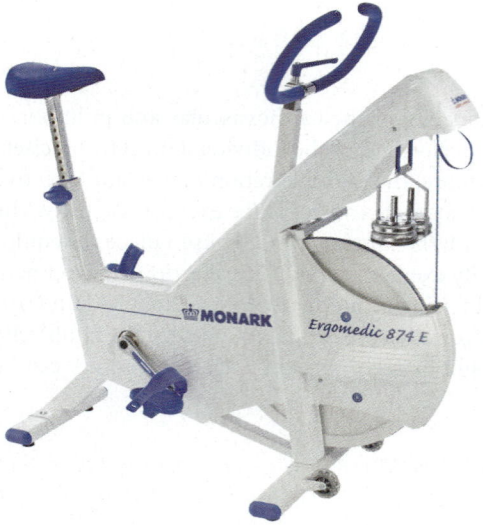

FIGURE 11.1 Laboratory cycle ergometer. (Photo courtesy of Monark.)

FIGURE 11.2 A: Arm ergometer. (Photo courtesy of Monark.) **B:** Rowing machine. (Photo courtesy of Hydrow.) **C:** Simulated skiing machine. (Photo courtesy of Nordic Track.) **D:** Swimming flume. (Photo courtesy of Endless Pools.)

be employed for exercise testing, their use is not widespread because many individuals do not commonly perform the movements required of those devices and it can be difficult to accurately control and measure the exercise intensity. Rowing machines, simulated skiing machines, and swimming flumes are frequently used to test athletes in their specific sport, and this often provides a more accurate assessment of cardiovascular and pulmonary fitness. The use of this type of equipment allows an athlete to perform the activity they are engaged in during competition, resulting in a more accurate assessment of cardiovascular and pulmonary function (7).

Although treadmills and ergometers are most commonly used in diagnostic and functional capacity testing of the cardiovascular and pulmonary systems, they can also be used for other purposes. For example, a biomechanical analysis of certain movement skills can be performed in controlled conditions using treadmills, cycle ergometers, simulated skiing machines, and swimming flumes. The use of ergometers allow assessments of an individual's readiness to return to work, physical activity, or exercise to be more accurately determined following the participation in a rehabilitation program. Finally, measurements of energy expenditure for weight management can be best made during controlled exercise on a motor-driven treadmill or cycle ergometer where the exercise intensity and volume can be accurately controlled and measured (13).

Metabolic Measurement Equipment

The assessment of cardiovascular and pulmonary function is commonly performed while using equipment that measures the volumes of air inhaled and exhaled, the amount of oxygen consumed, and the amount of carbon dioxide produced during rest and exercise. This equipment, commonly referred to as a metabolic cart (aka **metabolic measurement cart**), includes highly sensitive instruments for the measurement of air volume and the oxygen and carbon dioxide concentrations within the expired air (Figure 11.3). A metabolic cart collects and measures the volume of expired air from an individual's mouth and then analyzes the air for oxygen and carbon dioxide concentrations, comparing them to environmental concentrations to produce assessments of the amount of oxygen consumed and the type of fuel source (carbohydrates or fats) used to produce energy during rest and exercise.

During diagnostic and functional capacity testing, the levels of oxygen consumption attained at different workloads or at maximal effort exercise can be used to evaluate fitness levels, health, and performance and develop individualized exercise prescriptions. The level of oxygen consumption at maximal effort exercise (called VO_2max) is a measure of cardiorespiratory fitness and is a strong predictor of the risk of certain diseases (14) and performance during aerobic endurance sports and competitions (11). Most metabolic carts are large instruments designated for laboratory or clinical facility use. A portable metabolic device allows for the assessment of oxygen consumption and carbon dioxide production during activities outside a controlled laboratory testing facility (Figure 11.4). The freedom of movement in a portable metabolic device allows for measurements during work, physical activity, exercise, and athletic and sporting activities.

The information collected by a metabolic cart can also be used to determine the amount of energy expended during physical activity and exercise and the relative amounts of energy derived from carbohydrate and fat fuel sources. This is referred to as **indirect calorimetry**.

Metabolic measurement cart An instrument that measures the volumes of oxygen consumed and carbon dioxide produced.

Indirect calorimetry The assessment of energy expenditure through the measurement of oxygen consumption and carbon dioxide production.

FIGURE 11.3 Metabolic measurement cart for measuring oxygen consumption. (Photo courtesy of Parvo Medics.)

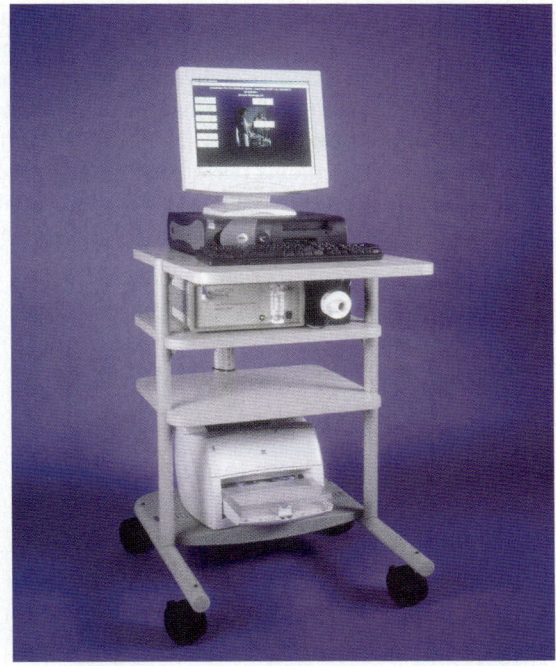

FIGURE 11.4 A portable metabolic measurement system for measuring oxygen consumption. (Photo courtesy of COSMED.)

Information from indirect calorimetry measurements is often used during the development of individualized exercise prescriptions, especially in weight management programs.

Pulmonary Function Equipment

Pulmonary function assessments include a broad range of tests that measure how well an individual inhales and exhales air from the lungs and how efficiently the lungs transfer oxygen into the blood and remove carbon dioxide from the blood. The measurement of the inhalation and exhalation rates and volumes and the amount of oxygen that diffuses into the blood can assist in the diagnosis of restrictive and obstructive lung disease. **Restrictive lung disease** results when an individual cannot inhale a normal volume of air and may be caused by inflammation or scarring of the lung tissue or by abnormalities of the muscles or skeleton of the chest wall (12,15). Restrictive lung disease can be diagnosed by several lung volume measurements performed in one of two ways. The most accurate way is for an individual to sit inside a **whole body plethysmograph** (Figure 11.5). This instrument is a sealed, transparent box. When inside the whole body plethysmograph, the individual

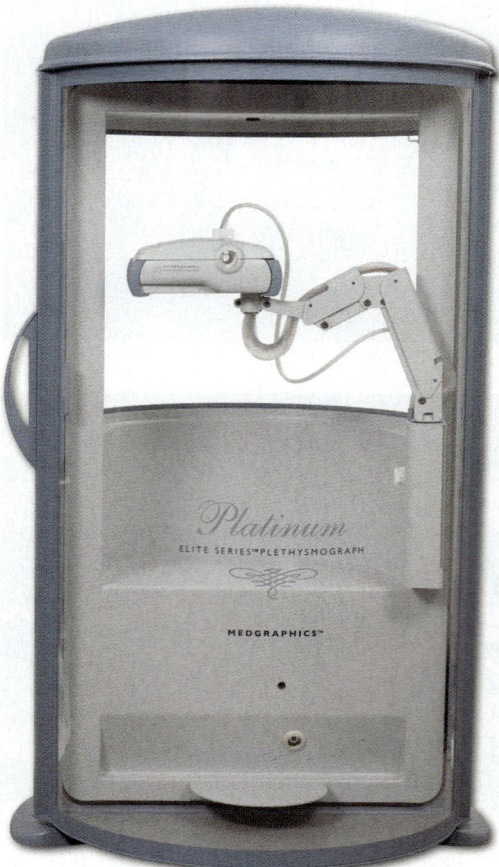

FIGURE 11.5 A whole body plethysmograph for measuring pulmonary function. (Photo courtesy of Medical Graphics Corporation.)

Restrictive lung disease Disease characterized by a reduced lung volume.
Whole body plethysmograph An instrument that allows for the measurement of lung volumes.

breathes normally into and out of a mouthpiece. Changes in pressure inside the plethysmograph allow for the determination of different lung volumes such as **residual lung volume** and **total lung capacity**. Lung volume can also be measured by having an individual breath nitrogen or helium gas through a tube for a specified period of time. The concentration of the gas in a sealed chamber attached to the tube is measured, allowing an estimation of the individual's lung volume (15).

The measurement of how well the lungs exhale air can be made through a spirometer (Figure 11.6). The spirometer is a device that records the amount and the rate of air that is inhaled and exhaled over a specified time through the use of a pneumotach or turbine. The assessments are usually performed during normal, quiet breathing and during forced inhalation or exhalation after a deep breath. The information gathered during this test is useful in diagnosing certain types of lung disorders, especially when making evaluations for **obstructive lung diseases** (*e.g.*, asthma and chronic obstructive pulmonary disease) (15).

Testing the diffusion capacity of a gas permits an estimate of how efficiently the lungs transfer oxygen from the air into the bloodstream. The diffusion capacity is measured when a person breathes carbon monoxide for a very short time, often one breath. The concentration of carbon monoxide in exhaled air is then measured, often by a metabolic cart. The

FIGURE 11.6 An incentive spirometer for measuring pulmonary function. (From Willis MC. *Medical Terminology: A Programmed Learning Approach to the Language of Health Care*. Baltimore (MD): Lippincott Williams & Wilkins; 2002.)

Residual lung volume The volume of air left in the lungs after a maximal exhalation.
Total lung capacity The volume of air in the lungs after a maximal inhalation.
Obstructive lung disease A condition that results in the narrowing of the passageways of the lungs making the exhalation of air difficult.

Thinking Critically

How might a comprehensive cardiopulmonary testing program improve the health and physical performance of an athlete and nonathlete?

difference in the amount of carbon monoxide inhaled and the amount exhaled allows an estimation of how quickly gas can travel from the lungs into the blood (15). Impaired diffusion capacity limits oxygen diffusion from the lungs into red blood cells and may limit a person's ability to be physically active or exercise.

Electrocardiograph Equipment

One of the primary instruments used during functional capacity and diagnostic testing of the cardiovascular system is the electrocardiograph (ECG) machine (Figure 11.7). This instrument detects and records the electric impulses generated by the heart during and between contractions (Figure 11.8). When individuals are suspected of having heart disease or a cardiovascular abnormality, an ECG test may also be included as part of a comprehensive physical examination (Figure 11.9). During functional capacity and diagnostic testing, the ECG is used to record the electric activity of the heart at rest and in response to graded or incremental exercise. An ECG uses 10 electrodes (also called leads) that are attached to an individual's chest. The electrical activity of the heart is recorded from 12 different views, which the ECG displays as 12 separate readings. In addition to the standard recording of electrical activity during rest and exercise, an ECG can assist in monitoring the heart during other diagnostic procedures such as **thallium testing**, and for therapeutic purposes in a cardiac rehabilitation program (12). Portable ECG instruments are often called Halter monitors. These monitors are used in patients with heart disease. Halter monitors are worn continuously so that the electric activity of the heart can be recorded for prolonged periods of time (*e.g.*, 24–48 hours) as the individual performs normal daily activities.

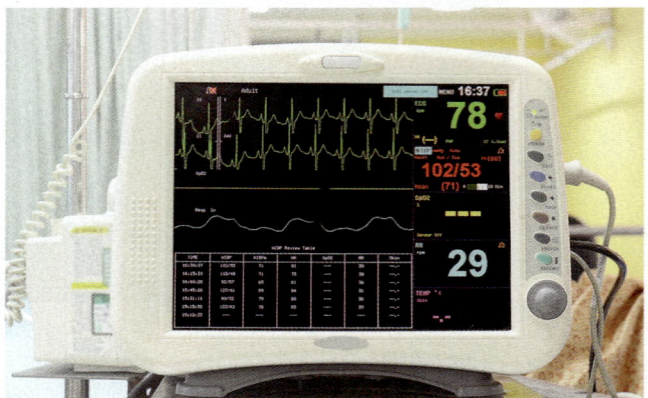

FIGURE 11.7 An ECG machine monitor for measuring electrical activity of the heart. (Photo courtesy of Shutterstock.)

Thallium testing A nuclear imaging test that shows how well blood flows into the heart while at rest or during exercise.

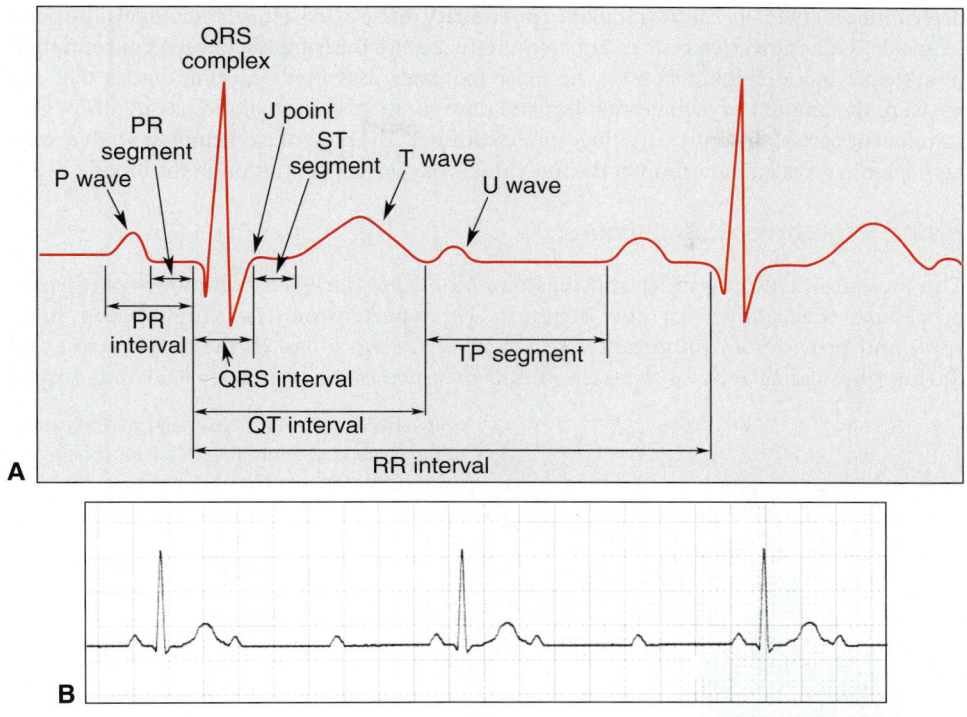

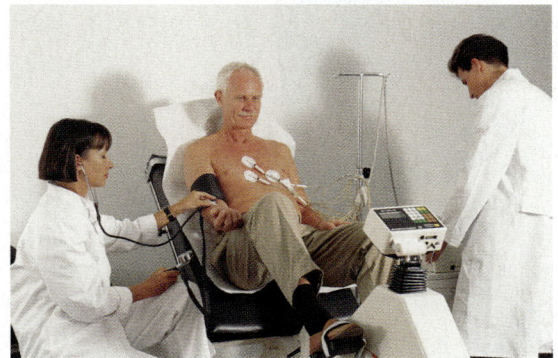

FIGURE 11.8 A normal ECG waveform (**A**). Waveforms are shown on printouts from an ECG machine (**B**). (**A:** From Ehrman JK, deJong A, Sanderson B, Swain D, Swank A, Womack C, editors. *ACSM's Resource Manual for Guidelines for Exercise Testing and Prescription.* 6th ed. Baltimore (MD): Lippincott Williams & Wilkins; 2009. **B:** From Dunbar CC, Saul B. *ECG Interpretation for the Clinical Exercise Physiologist.* Baltimore (MD): Lippincott Williams & Wilkins; 2009.)

FIGURE 11.9 An ECG machine and blood pressure measurement during a graded exercise test. (Shutterstock.)

Pulse Oximeter

The cardiovascular and pulmonary systems are responsible for exchanging oxygen and carbon dioxide between the lungs and the blood and then delivering the blood to the tissues of the body. Often, it is important to measure the oxygen concentration in the blood to help determine whether any cardiovascular or pulmonary diseases exist in an individual. The pulse oximeter is a noninvasive instrument commonly used to measure the oxygen concentration in systemic blood (Figure 11.10). The pulse oximeter uses light-emitting diodes that can measure the amount of **oxyhemoglobin** and **deoxyhemoglobin** in blood passing through a translucent part of the body (*e.g.*, fingertip or earlobe). The ratio of oxyhemoglobin to deoxyhemoglobin gives an indication of the individual's oxygen concentration in the blood.

Blood Pressure Assessment

The measurement of systolic and diastolic blood pressure is a commonly performed procedure that allows for the diagnosis of hypertension (*i.e.*, high blood pressure) and provides an important assessment of the workload of the heart at rest and during physical activity or exercise. Blood pressure is the product of **cardiac output**

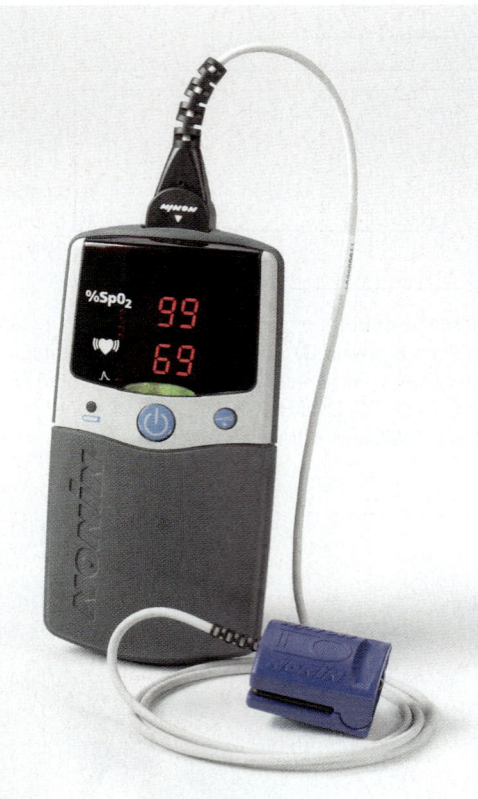

FIGURE 11.10 Pulse oximeter for measuring the oxygen concentration in blood. (Photo courtesy of Nonin Medical, Inc.)

Oxyhemoglobin When the hemoglobin protein in red blood cells is bound with oxygen.
Deoxyhemoglobin When the hemoglobin protein in red blood cells is not bound with oxygen.
Cardiac output The volume of blood pumped from the heart in a specified time, usually 1 minute.

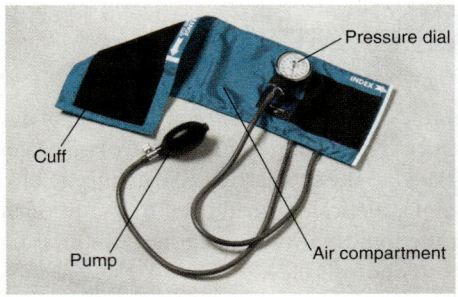 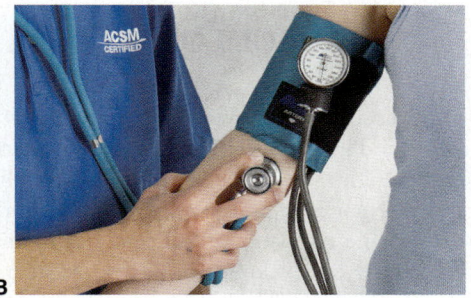

FIGURE 11.11 A: Blood pressure cuff and stethoscope. **B**: Manual assessment of blood pressure. (From Thompson WR, Bushman BA, Desch J, Kravitz L, editors. *ACSM's Resources for the Personal Trainer*. 3rd ed. Baltimore (MD): Lippincott Williams & Wilkins; 2009.)

and **total vascular peripheral resistance**, making the use of blood pressure measurements an indirect assessment of the health of the cardiovascular system. Hypertension is a common disease condition that can lead to other health problems if not diagnosed and properly treated (12). The assessment of blood pressure during a graded exercise test or during the performance of physical activity or exercise provides information about the health of the heart and cardiovascular system. Blood pressure can be measured using a sphygmomanometer, also known as a blood pressure cuff (Figure 11.11), and a stethoscope. The manual method of blood pressure assessment requires a trained individual to use a stethoscope and listen for sounds of blood flow after the blood pressure cuff has been inflated and the pressure is being released. Automated blood pressure machines (Figure 11.12) eliminate the need for a manual assessment of blood pressure and are a popular method of blood pressure assessment, especially in a clinical environment.

Thinking Critically

What types of measurement or assessment would you use to evaluate an individual's risk level for cardiovascular disease?

FIGURE 11.12 Automated blood pressure cuff. (Photo courtesy of Omron.)

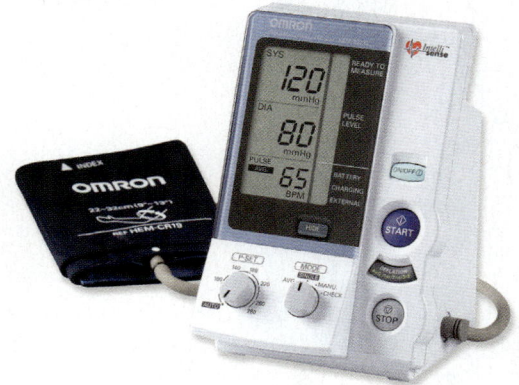

Total vascular peripheral resistance The resistance to blood flow provided by the blood vessels of the body.

MUSCULOSKELETAL ASSESSMENT

The musculoskeletal system plays an important role in the performance of activities of daily living and in physical activity, exercise, sport, and athletic competition endeavors. The generation of force by skeletal muscle is strongly related to the successful performance of physically demanding movement and most sports and athletic competitions (16). Additionally, the specific assessment of skeletal health and function can provide important information on the risk of developing bone disease such as osteoporosis (17).

Electromyography Equipment

Electromyography (EMG) is an assessment technique that allows for the measurement and recording of the electric activity of skeletal muscles at rest and during contraction. An EMG machine (Figure 11.13) is used to record the electric activity within the muscles when the muscles are stimulated by an internal signal from a nerve or by an external signal from an electric stimulation instrument. Muscles respond to both internal and external stimuli with the generation of electric signals in the individual muscle fibers, and the recording of this electric activity can occur through the use of an electromyograph. There are two forms of EMG: intramuscular and surface. Intramuscular EMG requires the insertion of a needle electrode containing a fine wire into a specific area, usually the belly of the muscle. The electric activity of the muscle at rest and during contraction is recorded, and trained personnel can make evaluations about the condition and contractile properties of the muscle. Surface EMG can also be used to assess the condition of the muscle

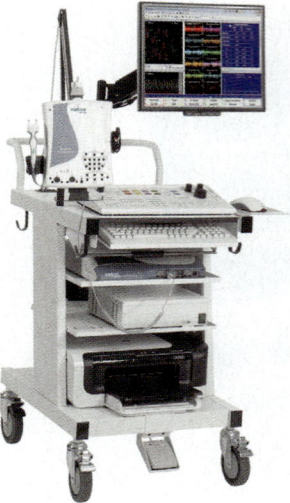

FIGURE 11.13 An EMG is used for the measurement of muscle electrical activity. (Photo courtesy of Natus Medical Incorporated.)

or muscle groups. Surface EMG is typically used when an individual is performing a movement, when information about a large muscle's or muscle group's action is needed, or when the insertion of a needle electrode is considered too invasive or unwarranted for the movement being studied (18).

Force Platforms

Force platforms are used to provide a measure of force production by a muscle or muscle group. A force platform provides voltage signals proportional to the forces exerted on the platform's surface in the vertical, horizontal, and lateral directions (Figure 11.14). The signals are recorded on a computer, and analyses of pressure and muscle force and power output can be made. Force platforms are often used to analyze **ground reaction forces** while standing, walking, running, and jumping (16). This information can be used to make adjustments to motor movements or skills that will promote more efficient or more powerful movements. The results obtained from a force platform can also be used to assist in the correction of movements by injured or disabled individuals and in the creation of devices or instruments that may improve movement performance.

Pressure-Sensitive Insoles

The measurement of pressure distribution in the foot and the vertical force generated during movement can be obtained using pressure-sensitive insoles inserted into the footwear of the individual being assessed. Using principles similar in nature to the force platform, the pressure-sensitive insoles allow for a more sophisticated and continuous measurement of pressure and force during ambulation, whether the movement is walking, running, jumping, or some type of sport movement. These measurements are important in the evaluation of abnormalities of walking and running for diagnostic and rehabilitation purposes and in the evaluation of sport and athletic performance.

FIGURE 11.14 A force plate for the measurement of ground reaction forces. (Photo courtesy of Kistler Instrument Corporation.)

Ground reaction force The force produced by a body part when in contact with the ground.

Handgrip Dynamometer

A handgrip dynamometer is an instrument that measures force production of the lower arm. As the individual applies force to the dynamometer with their hand, a calibrated scale within the instrument records the amount of force produced. This type of strength measurement is easy to administer but is limited to the muscle groups of the forearm and wrist. The handgrip dynamometer is considered good for evaluating overall upper body strength, but it does not measure dynamic strength.

Isokinetic Dynamometers

Isokinetic dynamometers are used to measure force during isometric and isokinetic movements of muscles and various body parts. When skeletal muscles contract, force is generated and transmitted, usually to an object external to the body. Isokinetic dynamometers measure muscular force (also called torque) at a constant velocity (Figure 11.15). Isokinetic dynamometers have computerized monitors that continuously alter the resistance of the dynamometer so that the movement velocity is held constant. Movement velocities can range from 0 to 450 degrees$\cdot$s^{-1}. Typically, a joint is selected for assessment or evaluation, and the maximum force or torque output is calculated for the movement by the muscles of the joint. Isokinetic dynamometers are used to measure static and dynamic muscle movements, perform muscle balance assessments, assist with pathology diagnosis, and are employed in occupational and physical rehabilitation programs. Isokinetic dynamometers allow for assessments of force production from almost all of the muscles and joints of the body as special attachments can be used to test numerous functional activities and movements (16).

Magnetic Resonance Imaging

Magnetic resonance imaging (MRI) is a radiology technique that uses a strong magnetic field, radio waves, and a computer to produce images of body structures. The MRI equipment includes a scanner tube surrounded by a giant circular magnet (Figure 11.16). During testing, the individual is placed on a movable bed that is inserted into the

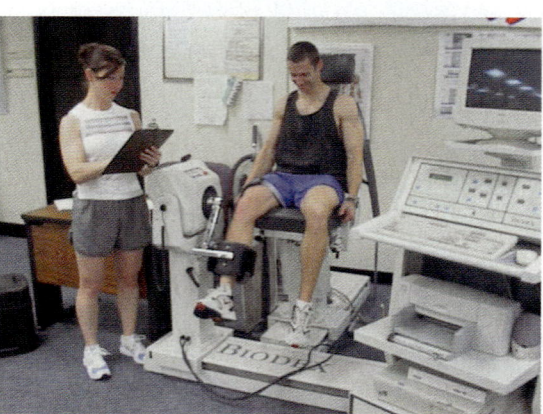

FIGURE 11.15 An isokinetic dynamometer for the measurement of muscle force and torque.

FIGURE 11.16 MRI equipment used to generate pictures of the internal organs and tissues of the body. (Photo courtesy of GE Healthcare.)

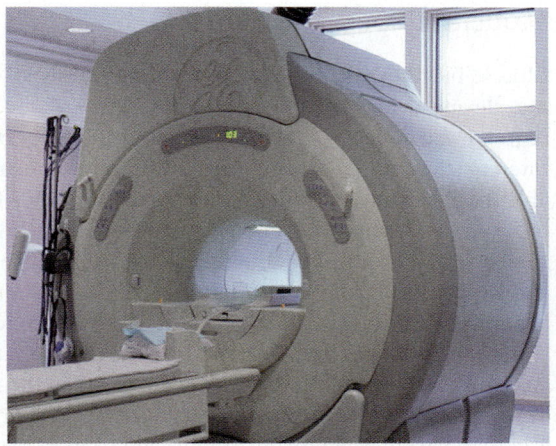

magnet. The magnet creates a strong magnetic field that aligns the protons of hydrogen atoms within the tissues of the body. The atoms are then exposed to a beam of radio waves from a radiofrequency coil. This spins the various protons of the body, which produce a weak signal that is detected by the receiver component of the MRI scanner. The receiver information is processed by a computer, and a detailed digital image is created. The image and resolution produced by the MRI can allow for the detection of tiny changes of structures within the body. To increase the accuracy of the images, contrast agents can be used. An MRI scan can be used as an extremely accurate method of detecting abnormalities of the tissues of the body, including glands, organs, joints, soft tissues, and bones (19).

Magnetic Resonance Spectroscopy

Magnetic resonance spectroscopy (MRS) can be used to measure the levels of different metabolites in body tissues. The principle of MRS is similar to that of MRI, except that in the MRS technique the noninvasive measurement of tissue substrates and metabolites can occur. The principle of MRS is based on the spin of an atom. Any nucleus with an odd atomic number or an odd atomic weight will produce a net spin like a spinning ball of energy. When a charged atom moves, it creates a magnetic field. When the magnet creates a strong magnetic field and the atoms are exposed to a beam of radio waves, they will create a magnetic field. Some biologically important nuclei that have spin include ^{1}H, ^{13}C, ^{17}O, ^{23}Na, and ^{31}P. Most MRS assessments of metabolites in body tissues use ^{1}H, ^{13}C, and ^{31}P. MRS can be used to assess disease status in various tissues, including metabolic disorders, tumor growth, changes in tissue metabolism following exercise, and noninvasive identification of muscle fiber type (19).

Metabolite A substance produced during a chemical reaction.

Muscle Biopsy Equipment

Muscle biopsies can be used to assess the level of substrates and metabolites in skeletal muscle and determine the types of fibers in specific muscles. The muscle biopsy procedure requires the collection of a tissue sample, usually from the belly of the skeletal muscle of interest (Figure 11.17). The procedure requires a small incision on the surface of the skin and the insertion of a specialized needle through the muscle fascia into the largest area of the muscle. On removal of the needle (Figure 11.18), the tissue sample is immediately frozen (usually in isopentane cooled by liquid nitrogen). This snap freezing stops all metabolic processes from continuing to occur in the muscle sample. Various biochemical and analytical techniques can then be used to determine the concentrations of substrates such as glucose, glycogen, and fatty acids in the muscle sample and the concentrations of various enzymes that control the metabolic processes in the muscle cells. Muscle fiber type can also be determined using a muscle biopsy sample (20). Serial slices of the biopsy tissue are made and then various analytic techniques are used to determine the fiber type within the tissue sample. The techniques used in fiber typing include histochemical, gel electrophoresis, and immunohistochemical. Muscle biopsies are often collected in response to acute exercise

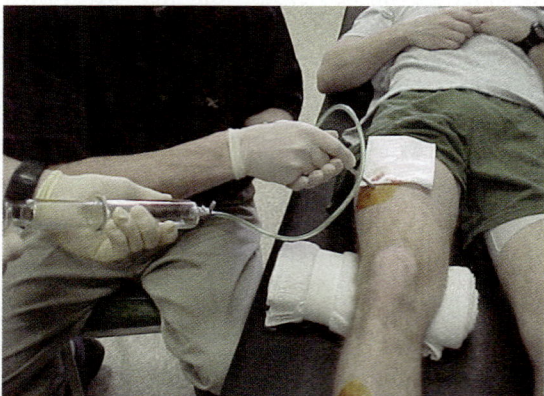

FIGURE 11.17 Muscle biopsy collection from the vastus lateralis.

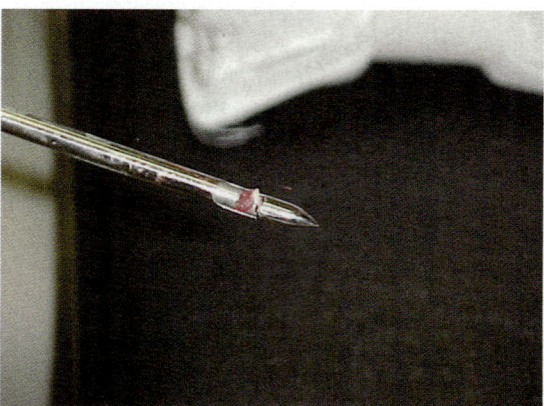

FIGURE 11.18 Muscle tissue sample from the vastus lateralis.

and to examine the changes that occur following chronic exercise or sport training. Muscle biopsy samples may also be used to help predict the risk of certain metabolic and neuromuscular diseases.

Computed Tomography

Computed tomography (CT) is a diagnostic tool that provides scans or views of internal body structures using X-rays. The principle behind the CT scan involves the use of digital geometry to allow a computer to generate a three-dimensional image of an object from a series of two-dimensional images taken around a single axis of rotation. During a CT scan, the individual lies supine on a movable bed that is inserted into the center of the scanner (Figure 11.19). The X-ray beam then rotates around the individual to collect the images. Precise locations of the body can be viewed as cross-sectional scans allowing for various densities of tissues to be easily distinguished. CT scans allow for the viewing of abnormal structures and organs, assessment of tumor growth in various areas of the body, strokes or lesions in the brain, and differentiation of tissue structure in the body. CT scans are commonly performed in a hospital or outpatient imaging center (21).

Dual Energy X-Ray Absorptiometry

Dual energy X-ray absorptiometry (DXA or DEXA) is used to measure bone density, which is altered in diseases such as **osteopenia** and **osteoporosis**. Special equipment called a bone densitometer is used to measure the bone mineral content (Figure 11.20). Two X-ray beams with differing energy levels are used to scan the entire body or a specific body region. Different tissues of the body absorb and reflect the X-rays to varying degrees, and using this

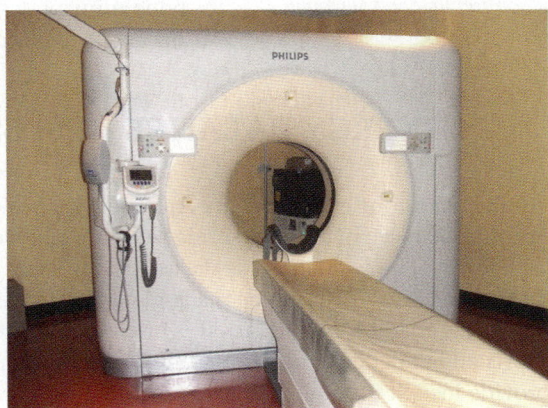

FIGURE 11.19 A CT scanner used to generate pictures of the internal organs and tissues of the body.

Osteopenia A decrease in the density of bone.
Osteoporosis A disease that results in low bone mineral density and an increased risk of bone fractures.

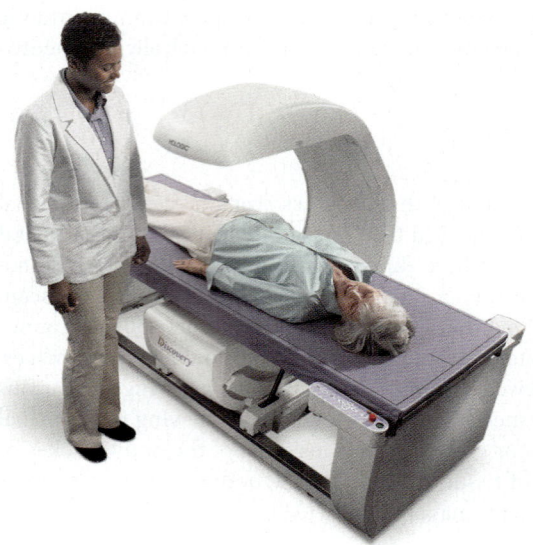

FIGURE 11.20 Dual energy X-ray absorptiometer for the measurement of skeletal mass and body composition. (Photo courtesy of Hologic Inc.)

> **Thinking Critically**
>
> What assessments would you make and why would you make them, for an athlete attempting to improve performance in the sports of marathon running and Olympic weightlifting?

principle allows for the measurement of lean mass, fat mass, and skeletal mass. When lean mass and fat mass are removed from the scan, the bone mineral density can be determined from the absorption of each beam by the bone. DXA is the most widely used technology to study bone mineral density and make a diagnosis of osteopenia and osteoporosis (21).

ENERGY BALANCE ASSESSMENT

The regulation of energy intake and energy expenditure is important for maintaining a healthy body weight and promoting weight loss or weight gain (13). Excess body weight and high levels of body fat increase the risk of developing cardiovascular disease, hypertension, diabetes mellitus, hyperlipidemia, and certain forms of cancer (22–24). The acquisition of body weight and lean body mass is often important for successful performance of certain sports and athletic competition (25). Assessing energy balance and body composition can provide insight into the risk level associated with certain diseases and the effectiveness of weight loss or weight gain intervention programs.

Measuring Energy Intake

The assessment of energy intake can be valuable for determining the nutritional needs of individuals involved in weight management programs or for those individuals required to consume specialty diets (*e.g.*, low sodium or low fat) for the purpose of disease risk reduction (*e.g.*, hypertension) or personal choice (*e.g.*, **vegetarian**). Individuals attempting to lose weight require the knowledge of dietary intake for the purpose of creating a negative energy balance for weight loss (13). For some individuals, weight gain, or often, more specifically, the gain of lean muscle tissue is critical for enhancing performance in certain sports and athletic competitions. In this instance, the knowledge of total calorie intake and protein consumption is beneficial for creating a positive energy balance and creating an **anabolic state** of the body (26). The two most common methods for measuring nutritional intake are the dietary recall and dietary record (27). Descriptions of these two methods for measuring energy intake are contained in Chapter 7 "Exercise and Sport Nutrition."

Measuring Energy Expenditure

The assessment of energy expenditure during resting and nonresting conditions can be easily performed using a stationary or portable metabolic cart, as described earlier in this chapter. The measurement of the carbon dioxide production and oxygen consumption provides a calculated value called the respiratory exchange ratio (RER). The RER is determined by the formula CO_2 production/O_2 consumption. It is a valid measure of fat and carbohydrate oxidation rates during steady state metabolism. During rest or submaximal exercise, the RER is normally between 0.70 (100% of the energy expenditure from fat) and 1.0 (100% of the energy expenditure from carbohydrate). When the RER is used in conjunction with the volume of oxygen consumed, a measure of total energy expenditure can be determined. This is because fat and carbohydrate molecules require different amounts of oxygen to be completely metabolized to provide energy. Figure 11.21 shows the measurement of resting

FIGURE 11.21
Measurement of resting energy expenditure. (Photo courtesy of COSMED.)

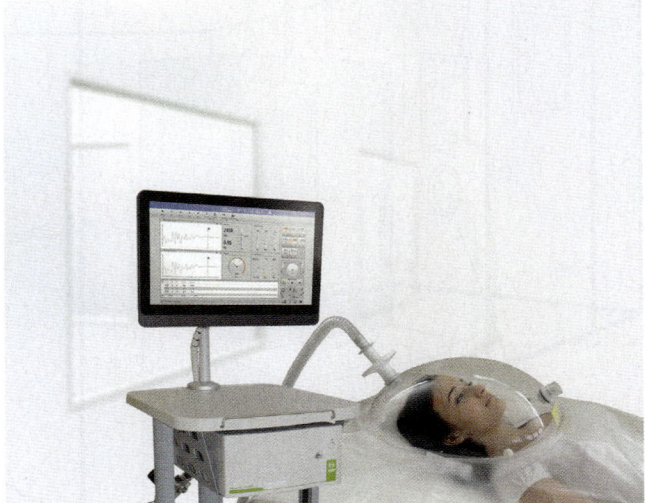

Vegetarian The consumption of vegetables, fruits, grains, and nuts only in the diet.
Anabolic state The process where the body is building larger molecules or compounds from smaller molecules or compounds.

energy expenditure and **substrate** utilization. In addition to using metabolic measurement equipment, other methods and instruments can be used to measure the energy expenditure of the body during different metabolic conditions. The most technically advanced methods include the use of whole room calorimeters and doubly labeled water. Other methods include using pedometers, accelerometers, and physical activity questionnaires.

Whole Room Indirect Calorimeter

A whole room indirect calorimeter requires individuals to live in a small room (approximately 10′ × 10′) for a specific period, usually 24 to 48 hours (Figure 11.22). Airflow into and out of the room is controlled, allowing for the determination of oxygen consumption and carbon dioxide production by special gas analyzers. Food is delivered through a sealed passageway. Human waste is collected and analyzed for metabolic by-products. If needed, blood samples can be collected through a special portal. Over the designated period, accurate assessments can be made for total energy expenditure; the relative energy production from carbohydrate, fat, and protein; and changes in various body metabolites. Whole

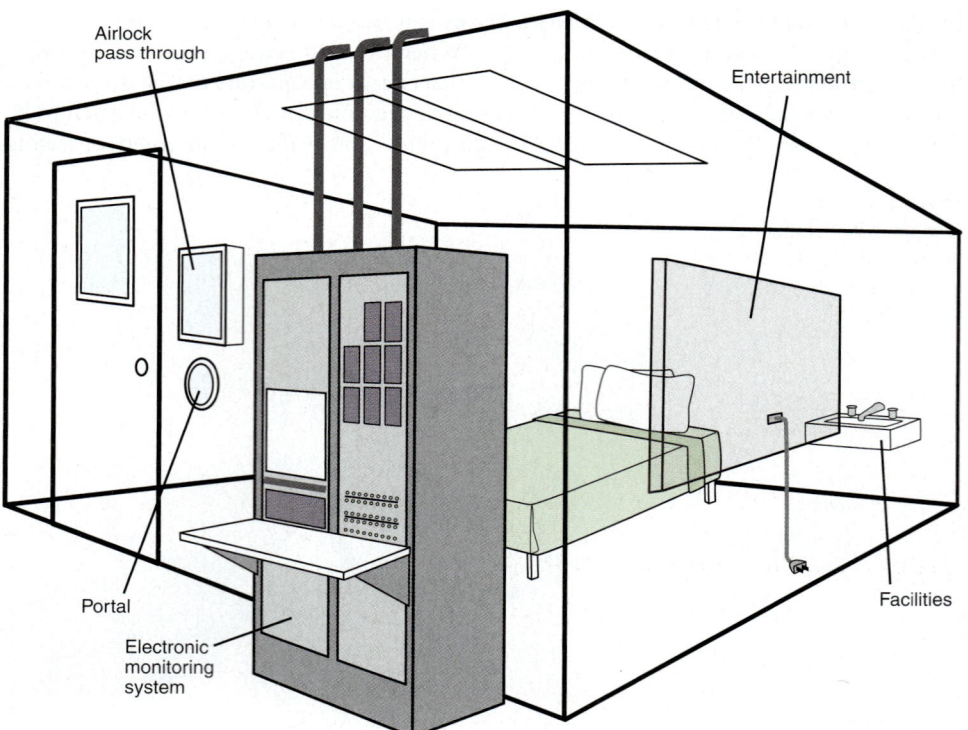

FIGURE 11.22 Schematic of whole room calorimeter.

Substrate Substances in the body that are acted on by enzymes in a chemical reaction.

room indirect calorimeters can also be used to determine the effects of different exercise regimens, nutritional programs, and pharmacological interventions on whole body metabolism (28). Although the assessment of metabolism can be accurately measured in a whole room indirect calorimeter, a limitation of this technique is that it does not reflect a free-living environment where the individual is able to perform activities of daily living.

Doubly Labeled Water

The use of the doubly labeled water technique allows for the assessment of energy expenditure in a free-living condition, unlike the whole room calorimeter, and without the need to be connected to portable metabolic measurement equipment. The doubly labeled water technique requires an individual to drink an **isotope**-based solution of **enriched water**. The fundamental theory of the doubly labeled water technique is that oxygen and hydrogen turnover in the body over time can provide a measure of oxygen consumption and carbon dioxide production. By collecting samples of body fluids (*e.g.*, saliva or urine) and then analyzing the fluids for the concentrations of oxygen and hydrogen, a measure of energy expenditure can be made. The more active an individual, the faster the turnover of enriched oxygen and hydrogen and the greater the measured energy expenditure. Measurement periods are usually 7 to 14 days in length (28). Although this technique provides an accurate measure of free-living energy expenditure, it is rather expensive to use, and the analysis of the collected body fluids is technically challenging.

Other Assessment Instruments

Heart rate monitors, pedometers, accelerometers, and physical activity questionnaires are commonly used instruments for the determination of the volume of physical activity performed, which can then be converted to energy expenditure using specific equations or algorithms. Heart rate monitors are straps worn around the chest, often with a watchlike receiver worn around the wrist that can measure and record an individual's heart rate over a designated period of time (Figure 11.23). The heart rate information can then be downloaded to a computer or handheld device and an estimate of energy expenditure is calculated based on the measured relationship between heart rate and oxygen consumption during rest and exercise (7,29).

FIGURE 11.23 Heart rate monitor used to measure heart rate during rest and movement. (Photo courtesy of Polar Electro Inc.)

Isotope One of the two or more atoms having the same atomic number but different mass numbers.

Enriched water Isotopic water that has the same atomic number as water but a different mass.

Pedometers measure the number of steps taken during a specific period, usually a 12- or 24-hour period (Figure 11.24). These small instruments are worn on the waist or the wrist and allow for free movement during physical activity or exercise. Pedometers, however, measure movement only when the foot strikes the ground at a sufficient force to create a recording of the movement. Pedometers do not make recordings of movement during other forms of activity such as cycling, swimming, or resistance exercise.

The use of accelerometers can increase the accuracy of physical activity and exercise measurement. **Accelerometers** can measure body movement in three planes allowing for the detection of movement during activities such as cycling, resistance exercise, machine stairclimbing, yoga, and pilates (Figure 11.25). Accelerometers provide counts of movement that can be converted to energy expenditure using specific equations or algorithms (7,29).

Physical activity questionnaires are typically used when attempting to obtain information from large numbers of individuals. These instruments require individuals to answer questions about daily activities and exercise over a certain period, typically 7 days. The physical activity questionnaires can provide information about total activity levels, but it is often very difficult to convert the information obtained to energy expenditure because of differences in body size among individuals and lack of precise information about the intensity of the activity (28).

Thinking Critically

What battery of assessments would you prescribe and why would you prescribe them for an individual attempting to achieve significant weight loss and maintain a healthy body weight?

FIGURE 11.24 Pedometer for the measurement of physical activity. (Photo courtesy of Omron.)

Accelerometer A device that measures and records the movement of the human body.

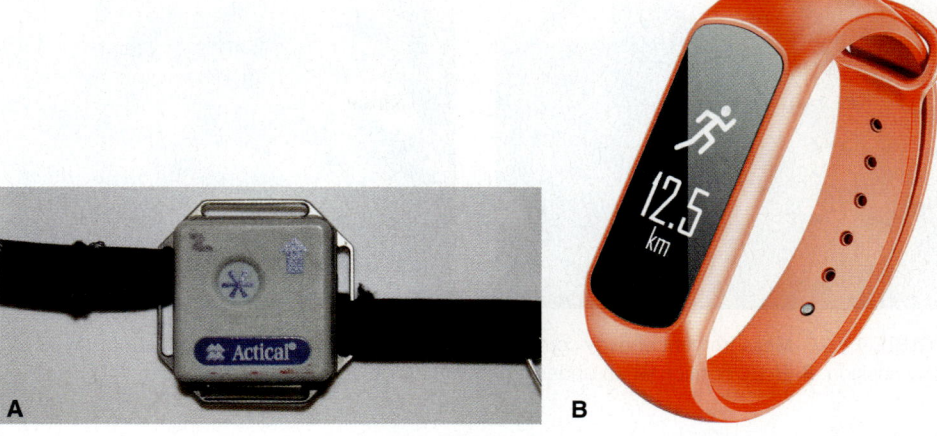

FIGURE 11.25 Accelerometer **(A)** and activity tracker **(B)** for the measurement of physical activity.

BODY COMPOSITION ASSESSMENT

An accurate assessment of body composition is necessary to correctly estimate an individual's risk for certain health conditions such as cardiovascular disease, hypertension, diabetes mellitus, hyperlipidemia, and certain forms of cancer (30). Additionally, the relative and sometimes total amounts of body fat mass and lean body mass can influence performance in certain types of sports and athletic competitions (31). Furthermore, the periodic measurement of body composition can be useful in determining the effectiveness of nutritional and exercise training interventions for both athletic and nonathletic individuals. There are several valid and reliable methods for assessing body composition to determine the amounts of fat mass, fat-free mass, and bone mass within an individual (32).

Densitometry

The assessment of body composition can be determined by **densitometry**, which is a process that provides a measure of body density. To calculate body density, a measurement of body mass and body volume is required. Body mass is easily determined from a calibrated scale. Body volume can be determined from hydrostatic weighing (Figure 11.26) or air displacement plethysmography (Figure 11.27).

Hydrostatic weighing requires an individual to be weighed underwater when expiring all of the air from the lungs and achieving residual lung volume. The underwater weight is then used with the land weight (*i.e.*, body mass) to calculate body density.

Densitometry Process that determines the density of a body mass.

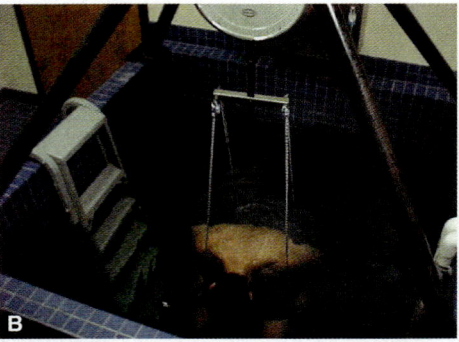

FIGURE 11.26 Measurement of body composition using hydrostatic weighing procedure. **A:** Initial body position. **B:** Body position when underwater weight is determined.

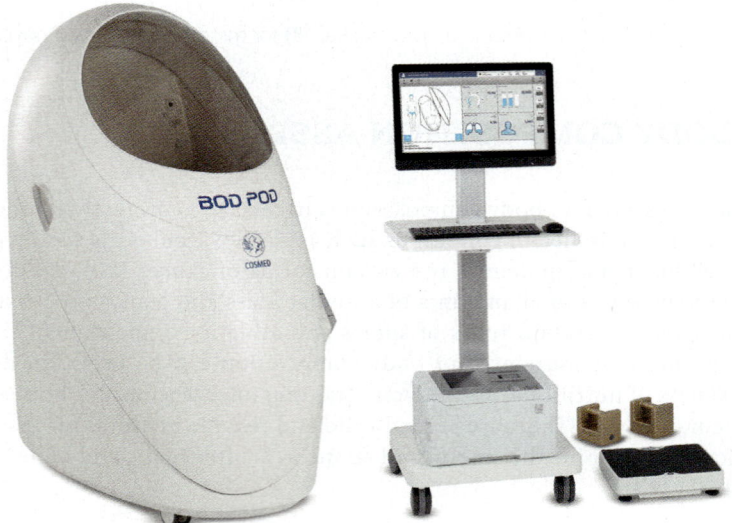

FIGURE 11.27 A Bod Pod air displacement plethysmography unit for assessing body composition. (Photo courtesy of Life Measurement, Inc.)

Air displacement plethysmography requires an individual to sit inside a whole body plethysmograph while body volume is measured. The body volume is then used with the body mass value to calculate body density. In both hydrostatic weighing and air displacement plethysmography, the body density is then converted into a percentage body fat using mathematical **regression equations** (21).

Air displacement plethysmography A process for measuring body composition that utilizes the inverse relationship between pressure and volume to measure body volume directly.

Regression equation A mathematical equation that measures the relationship between two different variables.

Dual Energy X-Ray Absorptiometry

In addition to measuring bone mineral density, DXA can also be used to measure body composition. As mentioned previously, the different tissues of the body absorb and reflect the X-rays from the DXA machine to varying degrees. When a complete body scan is performed, it is possible to determine the total amounts of lean mass and fat mass. The DXA can also partition the regional areas of lean mass and fat mass. These values, when combined with total body mass, can be used to determine the percentage of body fat within an individual (21).

Bioelectric Impedance

Bioelectric impedance analyzers can be used to measure body composition by sending a safe low-voltage level electric current through the body (Figure 11.28). Electric currents travel faster in body tissues that have a higher water and electrolyte content than in those tissues with a lower water and electrolyte content. Lean body tissue, which has a high water content (approximately 50%–70% water) allows the electric current to travel faster than in fat tissue, which has a low water content (approximately 15%–20%). The speed at which the current passes through the body can be used to determine the percentages of lean tissue and fat tissue in the body. By using an individual's height and weight, it is possible to calculate the percentage of body composition that is fat mass and fat-free mass (21).

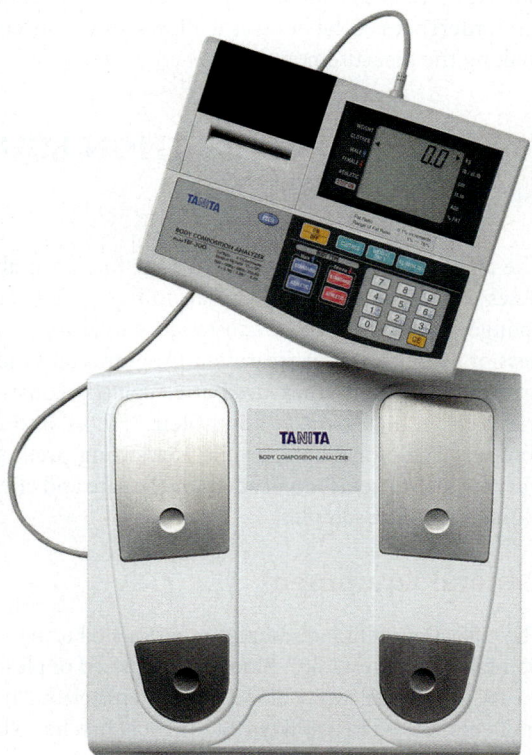

FIGURE 11.28 Bioelectric impedance analyzer for assessment of body composition. (Photo courtesy of Tanita Corporation.)

Skinfold Assessments

The measurement of the skinfold thickness at various sites around the body can allow for the calculation of body density and the determination of body composition. The use of skinfold thickness measurements for determining body composition is based on the principle that a certain percentage of total body fat lies directly beneath the surface of the skin. Skinfold calipers are calibrated instruments that allow for the accurate determination of the thickness of fat immediately below the skin's surface (Figure 11.29). When predetermined sites are measured using standard measurement techniques, the resulting information can be used in a regression equation to calculate body density, which is then used to provide an estimate of body composition (21).

Anthropometric Measurements

Body composition can be estimated by measuring the size and proportion of the human body and its various segments. Measures of height using a stadiometer and weight using a calibrated scale can be used to calculate the body mass index (BMI). The BMI (body mass [kg]/height^2 [m]) can be used to provide an estimate of body fatness (33,34) and help determine the risk of certain disease conditions and health outcomes (35). Circumference measurements of various body segments can also be used to assess the risk of disease conditions and mortality (36). In particular, a circumference measure of the waist at the level of the navel can be used to assess the level of risk for cardiovascular and other disease conditions (37). A spring-loaded tape measure is the most commonly used device for measuring circumferences of the body as it allows for consistent tension to be applied to the tape when making the measurements.

 BLOOD COLLECTION EQUIPMENT AND ASSESSMENT

The collection of blood from the body for the analysis of various compounds and metabolites can provide insight into an individual's health and risk of disease conditions. The changes observed in these substances can also provide information regarding the effectiveness of intervention or treatment programs and medications. For example, the blood lipid profile is frequently measured and typically consists of determining the total cholesterol, triglyceride, high-density lipoprotein (HDL), and low-density lipoprotein (LDL) concentrations. The blood lipid profile is a strong predictor of cardiovascular disease risk (38). Various instruments are needed for the safe and effective collection, storage, and analysis of blood and tissue samples.

General Equipment

The collection of blood samples is performed using needles, syringes, vacutainers, and collection tubes (Figure 11.30). Blood is comprised of plasma, which is the watery component, and the red blood cells, white blood cells, and platelets. An analysis of blood in the laboratory is generally performed in two ways. Blood must first be collected and then permitted to clot. When a

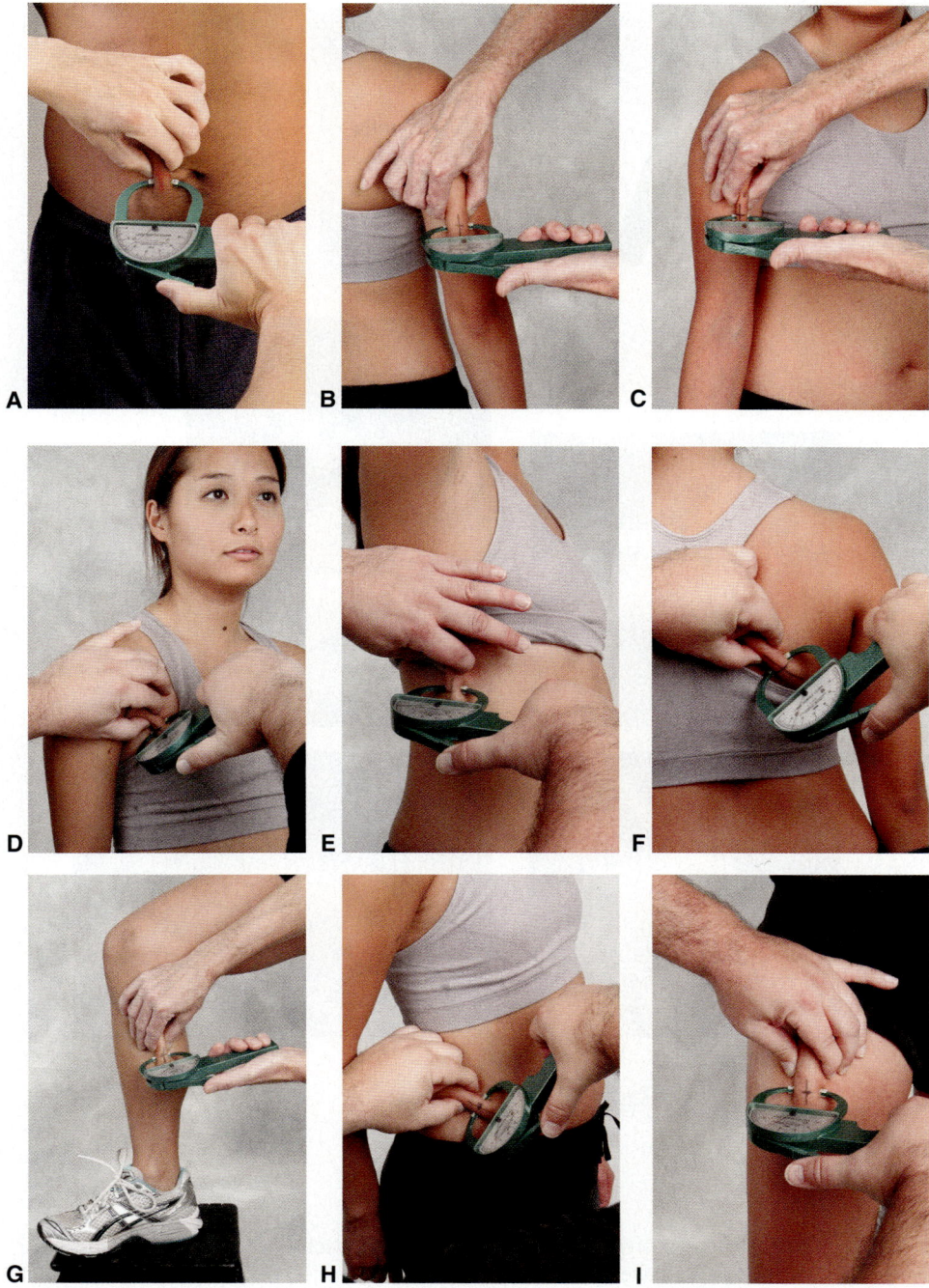

FIGURE 11.29 Assessment of body composition at various sites (**A-I**) using skinfold calipers. (From Thompson WR, Bushman BA, Desch J, Kravitz L, editors. *ACSM's Resources for the Personal Trainer*. 3rd ed. Baltimore (MD): Lippincott Williams & Wilkins; 2009.)

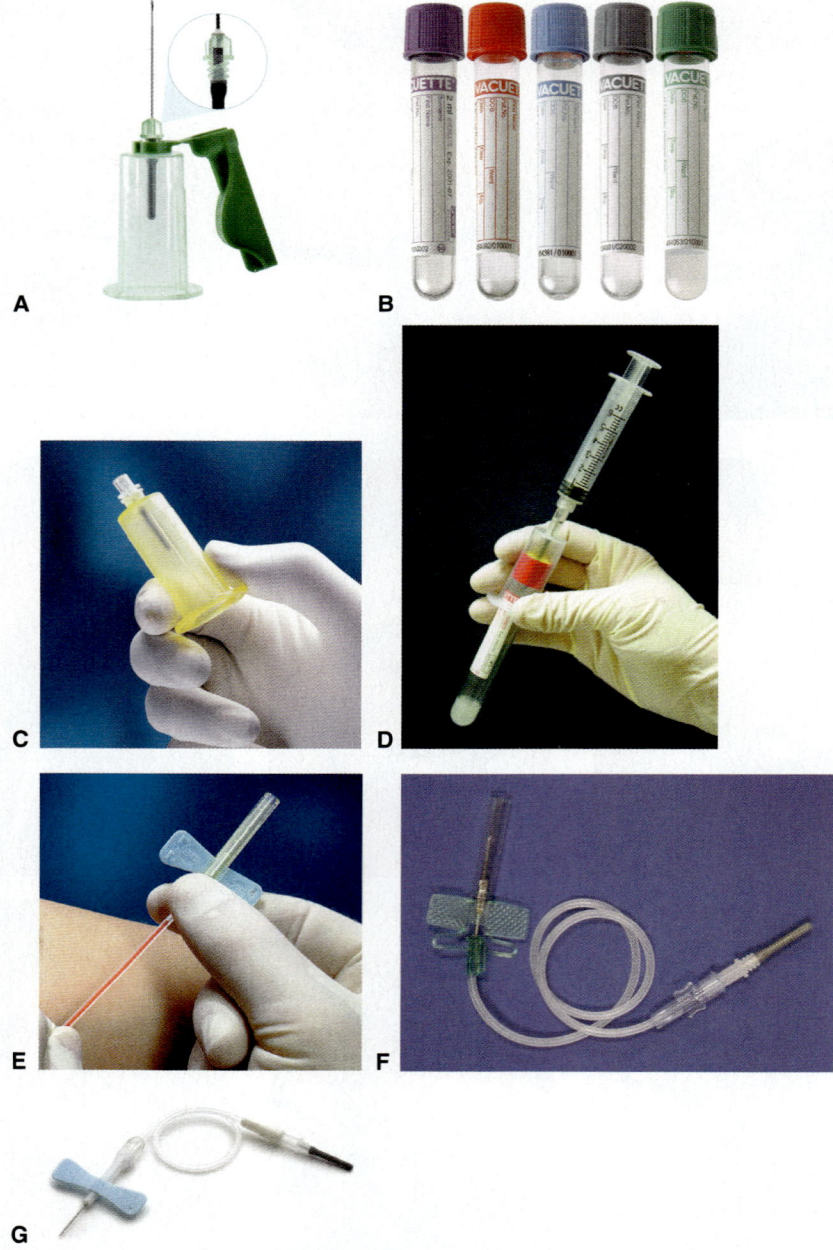

FIGURE 11.30 Blood collection equipment: evacuated tubes (Vacutainer Plus Plastic brand evacuated tubes [**A**] and Vacuette evacuated tubes [**B**]); syringe safety devices (a BD transfer device [**C**] and a Greiner transfer device attached to a syringe [**D**]); and examples of safety winged infusion sets (SAFETY-LOK [**E**], Monoject Angel Wing [**F**], and the Vacuette [**G**] safety butterfly blood collection systems). (From McCall RE, Tankersley CM. *Phlebotomy Essentials*. 4th ed. Baltimore (MD): Lippincott Williams & Wilkins; 2008.)

clot forms, a liquid, called serum, escapes from the clotted cells, and the serum is used in many tests. Anticoagulants are used to prevent the coagulation of the blood specimen. The anticoagulant used should not alter blood components as this may affect the results of the analysis performed. The most commonly used anticoagulant in **hematology** is ethylenediaminetetraacetic acid (EDTA). The proper anticoagulant must be used for specific test procedures. Blood collected with one anticoagulant may be suitable for one test or a group of tests but not for others.

Once the blood sample has been collected from the individual, the collection tube is placed into a centrifuge for the separation of serum or plasma, and the blood cells. The serum or plasma sample can then be analyzed using special equipment, or the sample can be stored in a deep freezer, usually at −80 °C until analysis. The nature of the specific compound will determine the type of chemical analysis required. This, in turn, will dictate the specific equipment that will be used. Often, an automated analyzer will be used to determine the concentration of the compound in the collection sample. In other instances, specialized equipment will be used in the analysis. Examples of common equipment used by exercise science and allied health care professions to measure compounds of interest include the **spectrophotometer**, the **fluorometer**, the **high-pressure liquid chromatograph**, and the **gas chromatograph/mass spectrometer**.

Common Blood Measures

Some of the more common measures of blood samples in exercise science, allied health professions, and related fields include hematocrit, hemoglobin, glucose, lactic acid, and blood lipids. The hematocrit is an indirect measure of the red blood cell number. Red blood cells are responsible for carrying almost all of the oxygen in the blood to the tissues of the body. A decrease in the red blood cell number (called anemia) may compromise oxygen delivery to the body. Increases in the red blood cell number and/or hemoglobin concentration can result in an increased oxygen delivery to the tissues of the body. Glucose is the primary energy source for nervous tissues in the body and an important energy source for skeletal muscles during exercise. An inability of the body to regulate blood glucose concentration may lead to insulin resistance and diabetes mellitus. Lactic acid is a metabolic by-product of carbohydrate metabolism in the body. Increased levels of lactic acid can contribute to muscular fatigue and a reduction in force production during muscular contraction. The primary blood lipids that increase the risk of cardiovascular disease include cholesterol, triglycerides, and LDL cholesterol. Conversely, high levels of HDLs reduce the risk of

Hematology The study of the nature, function, and diseases of the blood and of blood-forming organs.

Spectrophotometer An instrument used to determine the intensity of a variety of wavelengths in a spectrum of light.

Fluorometer An instrument used to measure the fluorescence being emitted from a substance.

High-pressure liquid chromatograph An instrument used to separate, identify, and quantify compounds.

Gas chromatograph/mass spectrometer An instrument used to detect the presence of a substance in a test sample.

cardiovascular disease. Periodic monitoring of these blood lipids is important for reducing the risk of cardiovascular disease (8).

REHABILITATION EQUIPMENT AND ASSESSMENT

Various types of instrumentation and equipment are used to assist in the rehabilitation of an injured body part owing to a disease condition or participation in physical activity, exercise, sport, or athletic competition. This equipment is generally designed to enhance the recovery process and improve muscle and joint functions. Activities such as stretching and resistance exercise are commonly used by exercise science and allied health professionals and sports medicine personnel during the rehabilitative process. Commonly used equipment in rehabilitation includes resistance bands, stability devices, resistance exercise equipment, devices for thermotherapy, and transcutaneous electric stimulation units.

Resistance Devices and Exercise Equipment

Various devices are used to overload the nervous system, skeletal muscle, and joints during rehabilitation assessment and exercise activities. Assessments of force production during rehabilitation and exercise training are important for initial assessments of muscle and joint performance and for determining the effectiveness of the intervention. A description of that equipment occurs earlier in this chapter. Elastic bands, medicine balls, stability devices, parachutes, weighted vests, and other types of equipment are used to increase the resistance against the muscles and joints involved during muscle contractions. By providing an overload on the muscle, improvements in force-generating capacity can be obtained. More traditional resistance exercise equipment includes free weights (dumbbells and barbells with weights) and resistance exercise machines. Resistance devices and exercise equipment are also used to promote muscular strength and endurance for improving health and fitness and for enhancing performance in sport and athletic competition (39,40).

Thermotherapy

Thermotherapy is the use of cold or heat to assist in the rehabilitation and recovery from injury or surgery. **Cryotherapy** includes the use of ice massage, cold or ice water immersion, ice packs, and vapocoolant sprays. The purpose of cryotherapy is to reduce the temperature of the injured tissue. The reduction of temperature causes a decrease in tissue metabolism, inflammation, pain, muscle spasm, and blood flow, and when the cold is removed there is an increase in blood flow to the injured tissue. Heat therapy includes the use of a hot cloth, hot water, ultrasound, and heating pad. The purpose of heat therapy is to aid in recovery from injury by increasing blood flow to the injured area and decreasing muscle and joint stiffness and soreness (41).

Thermotherapy The use of heat or cold in rehabilitation therapy.
Cryotherapy The use of cold in rehabilitation therapy.

Whirlpool Tubs

Whirlpool tubs are instruments used to circulate hot or cold water around an injured body part. The tubs are large enough to fit an entire person; however, there are smaller units that can be used for the limbs of the body. A small electronic motor is used to circulate water in the tub, creating an enhanced thermal gradient between the water and the body part submerged in the water (Figure 11.31). The circulation of water assists with the rehabilitation and healing processes.

Therapeutic Ultrasound

Therapeutic ultrasound is a rehabilitation technique that uses high-frequency sound waves to enhance the healing process of an injured joint or muscle. During therapeutic ultrasound, the body tissue at the site of application is heated, creating an increase in blood flow, which assists with the healing process. At the same time, the therapeutic ultrasound unit creates cellular vibration of the tissues that may assist in the repair of damaged tissue (41). Figure 11.32 shows a therapeutic ultrasound unit.

Transcutaneous Electric Nerve Stimulation

A transcutaneous electric nerve stimulator (TENS) is a device that produces electric signals that stimulate nerves through the skin. A TENS unit consists of two or more electrodes

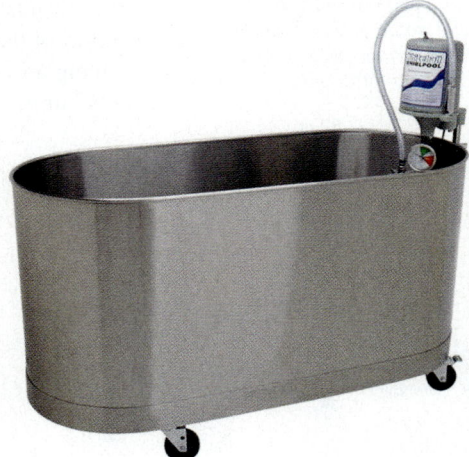

FIGURE 11.31 Whirlpool tub for thermotherapy rehabilitation. (Photo courtesy of Whitehall Manufacturing.)

Therapeutic ultrasound An instrument that causes vibrations and a deep heating of the tissue it is used on.

Transcutaneous electric nerve stimulator An instrument that causes relief of pain or an enhanced healing of injured tissue.

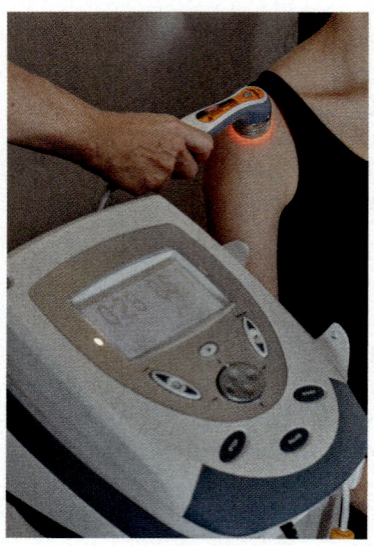

FIGURE 11.32 A therapeutic ultrasound unit for muscle rehabilitation. (Photo courtesy of Shutterstock.)

> 💡 **Thinking Critically**
>
> What assessments and tests would you recommend, and why would you recommend them for a recreational runner participating in a rehabilitation program for an injury to the lower leg?

that are placed over an area of the body that is designated to receive the stimulation. The frequency and intensity of the stimulation can be adjusted to assist with the alleviation of pain or to assist with muscular reeducation following an injury or surgery (41). Figure 11.33 shows a TENS unit.

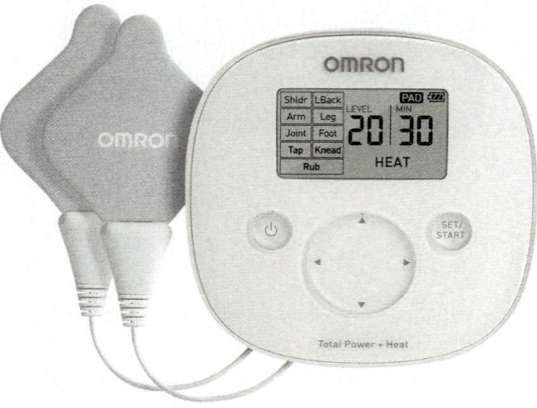

FIGURE 11.33 A TENS unit for muscle and nerve rehabilitation. (Photo courtesy of Omron.)

MOTOR PERFORMANCE ASSESSMENT

The assessment and evaluation of motor performance employs instruments and equipment that can be used to track movements and responses to stimuli. The knowledge gained is important for making corrections to motor movements to enhance performance in daily activities, as well as improving performance in sport and athletic competition (42). Movements of body parts can be directly measured using a goniometer and potentiometer or through imaging techniques such as computerized cinematography. Other methods used to assess motor performance include EMG, eye-tracking instruments, and equipment for measuring brain activity.

Goniometers and Potentiometers

Goniometers are hinged devices that can be attached to a body part to record movement and the range of motion. Goniometers move with the body joint and provide information about changes in the angle of movement (Figure 11.34). Many goniometers are connected to a potentiometer, which sends a voltage signal to a computer that accumulates and analyzes the information. **Potentiometers** can also be used in conjunction with many other devices, including specialized equipment that is physically moved by an individual and recorded by a computer (42).

FIGURE 11.34 Goniometer for measurement of joint range of movement. (Photo courtesy of Shutterstock.)

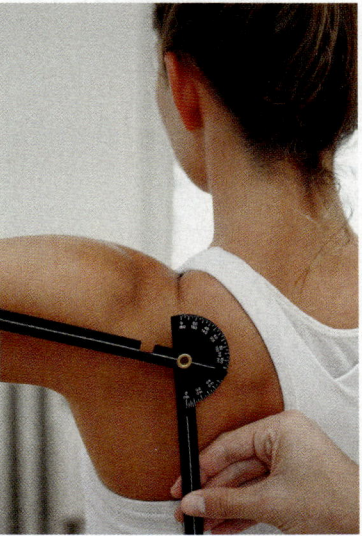

 Potentiometer An instrument that measures an unknown voltage to a standard known voltage.

Motion Capture Systems

The recording of physical movements is commonly used to assess motor performance. The digital recordings of body and limb movements can be analyzed using sophisticated computer software. During the assessment of a movement, the individual being recorded usually wears reflective tape or fluorescent markers that can easily be tracked during the analysis of the movement. Advanced computer software programs are able to track body movements and make a very detailed analysis of motor performance (42). Figure 11.35 shows the use of a high-speed motion capture system for recording body movement.

Electromyography

In addition to using EMG to evaluate musculoskeletal function, EMG recordings can be used to provide knowledge about the control of motor performance by the central nervous system. When combined with other **kinematic** information, the EMG activity obtained from skeletal muscle can provide additional insight into the execution of a motor skill (42).

Eye-Tracking Instruments

The use of eye movement recording systems allows for the determination of what visual factors influence attention during motor tasks. Vision tracking is used to determine where

FIGURE 11.35 Motion capture system for recording body movements during physical activity or exercise. (Photo courtesy of VICON.)

Kinematic The study of motion of a body without consideration given to its mass or the forces acting upon the body.

the head and various other body parts are directed. Tracking eye movements requires the use of two video imaging techniques. One camera is mounted directly on the head or on a helmet that the individual is wearing. Information from this camera provides an image of the individual's line of vision. A second camera measures the movements of the eyes by means of corneal reflection. The combination of the two recording devices provides a measure of where the individual's eyes are directed during the performance of a motor task or movement (42).

Information Processing Measurement

The ability to identify a stimulus, select a response, and then program and initiate the response is very important for performing activities of daily living and responding to situations encountered during participation in exercise, sport, and athletic competition. An individual's response to a stimulus is referred to as reaction time. To accurately measure reaction time, devices that present a stimulus and then measure the time it takes an individual to respond to the stimulus are used (Figure 11.36). These reaction time devices have the ability to measure a response to a stimulus in hundredths and thousandths of a second. Changes in reaction time are often measured in response to training programs or environmental distractions.

Measuring Brain Activity

The equipment and assessment techniques used to measure brain activity allows for fast and flexible assessments of brain activity during motor performance. **Electroencephalography** (EEG) is the measurement of electric activity produced by the brain and recorded by a computer through the use of electrodes placed on the head. The primary advantage of using EEG is the almost instantaneous feedback that is provided during motor performance. Figure 11.37 shows the electrode placement and the output from an EEG

FIGURE 11.36 Reaction time measuring instrument. (Photo courtesy of Lafayette Instrument Co.)

Electroencephalography The measurement and recording of the electric activity of the brain.

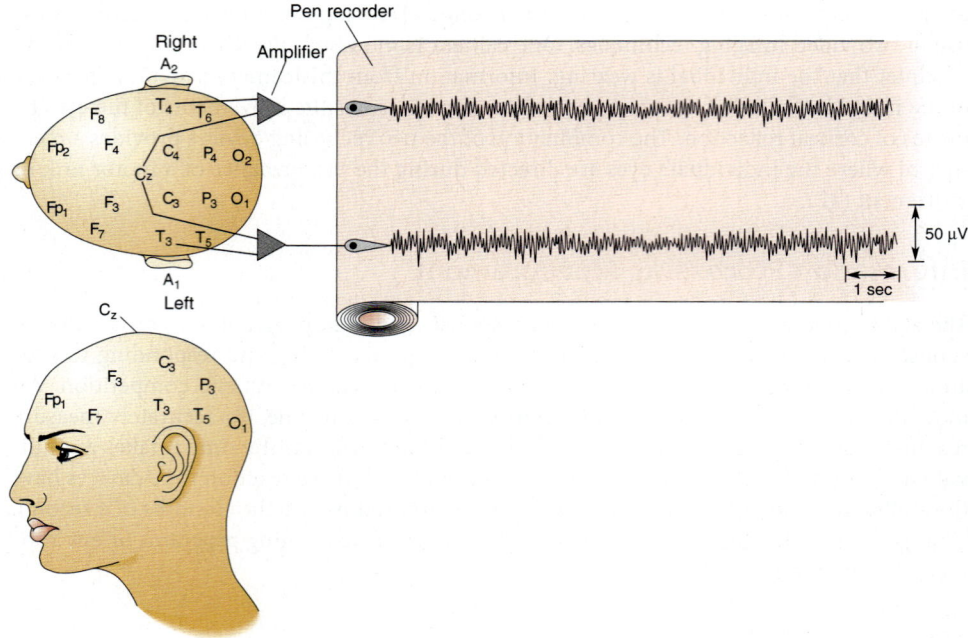

FIGURE 11.37 The standard placement positions of electroencephalograph (EEG) electrodes. A, auricle (or ear); C, central; Cz, vertex; F, frontal; Fp, frontal pole; O, occipital; P, parietal; T, temporal. Wires from pairs of electrodes are fed to amplifiers, and these drive pen recorders. (From Bear MF, Connors BW, Parasido MA. *Neuroscience: Exploring the Brain.* 2nd ed. Philadelphia (PA): Lippincott Williams & Wilkins; 2001.)

instrument. Magnetoencephalography (MEG) is an imaging technique used to measure the magnetic fields produced by the electric activity in the brain using extremely sensitive recording devices. These measurements are used to determine the function of various parts of the brain. **Positron emission tomography** is a nuclear medicine medical imaging technique that can be used to produce a three-dimensional image of brain structure and activity. Another form of neuroimaging, called **functional magnetic resonance imaging**, allows for the measurement of blood flow related to neural activity in the brain or spinal cord (42). Each of the brain activity measurement techniques has advantages and disadvantages that must be considered when determining what equipment to use during testing (42).

Positron emission tomography A computer-generated image of metabolic or physiologic activity in the body generated through the detection of gamma rays.

Functional magnetic resonance imaging An imaging technique used to examine relationships between physical changes in the brain and mental functioning.

BEHAVIORAL AND PSYCHOLOGICAL ASSESSMENT

Assessments of behavioral and psychological issues associated with physical activity, exercise, sport, and athletic completion include the use of both quantitative and qualitative instruments and methods. With appropriate training, an individual may use various quantitative instruments to assess variables of interest in exercise and sport psychology. These instruments are valid and reliable psychological tests or questionnaires. The use of a qualitative method requires a more individualized approach to the assessment of variables of interest.

Quantitative Assessments

Assessing psychological phenomena in exercise and sport psychology requires a variety of analytic methods often not used in other areas of exercise science. The most common method has been to use a **constructionist approach** in which significant consideration is given to the individual's subjective experience. The self-report, using standardized questionnaires or psychological inventories, is the predominant analytic strategy used in the constructionist approach. Some examples of commonly used questionnaires and inventories include (43):

- Spielberger State-Trait Anxiety Inventory
- Profile of Mood States
- Beck Depression Inventory
- Sport Competition Anxiety Test
- Competitive State Anxiety Inventory 2
- Test of Attentional and Interpersonal Style

The use of these questionnaires and others can provide valuable information on an individual's psychological state in situations such as acute and chronic physical activity and exercise or in response to practice or game situations in sport or athletic competition (44).

Qualitative Assessments

The use of qualitative assessments for understanding the psychosocial phenomena in exercise and sport psychology helps understand the personal significance that individuals have constructed within their natural environment. Qualitative assessments have been used to develop a deep understanding of the individual and personal factors that give meaning to why an individual believes and acts in a specific manner. Numerous qualitative approaches

Constructionist approach A sociologic theory that considers how a social phenomenon develops in a specific social context.

are used to help understand the meaning of what happens to an individual or group participating in an exercise or sport setting. Table 11.1 provides a description of the most commonly used types of qualitative research techniques (45). A mixed methods approach, whereby both quantitative and qualitative assessments are used, can draw on the strengths of both types of assessment methods and is commonly used to evaluate behavioral and psychological aspects of an individual participating in an exercise or sport setting.

Table 11.1	The Most Commonly Used Types of Qualitative Research Techniques (45)
QUALITATIVE RESEARCH TECHNIQUE	DESCRIPTION
Grounded theory	Attempts to derive a theory from the views of the participants.
Life histories	Derives information about the lives of one or more individuals providing stories about their lives.
Case study	Provides an intensive, holistic, and in-depth understanding of a single event, activity, program, process, or individual.
Phenomenological	Identification of the core meaning of an experience.
Ethnographic	Provides a description and interpretation of a cultural or social group.
Basic and generic	Identifies recurrent patterns in the form of themes and categories and provides a descriptive understanding and exploratory interpretation.

Interview

Sandra Hunter, PhD, FACSM

Professor of Exercise Science. Department of Physical Therapy, and Director of the Athletic and Human Performance Research Center at Marquette University, Milwaukee, Wisconsin

Brief Introduction – My formative education occurred in Australia. I received a Bachelor of Education (Physical Education) degree from the University of Sydney, and a graduate diploma in Human Movement Science from the University of Wollongong. After teaching physical education for several years, I completed a PhD from the University of Sydney with an emphasis on Neuromuscular Function, Strength Training, and Aging. In 1999, I came to the United States and began a postdoctoral research fellowship at the University of Colorado,

Boulder, under the mentorship of Roger Enoka. My primary research focus was the neural control of movement and included studying the age and sex differences in fatigability of limb muscles. Since 2003, I have been on the faculty at Marquette University. My current research examines the mechanisms of the sex differences and age difference in muscle function and fatigability in healthy and clinical populations such as diabetes and the protective effects of exercise training. I also conduct studies to determine the limits of human performance in men and women across all ages using big data sets from real-world performances in the marathon, ultramarathons, swimming, and rowing as examples.

Q: What were your most significant career experiences?

Two impactful career experiences were the following: A 4-year postdoctoral research fellowship under the mentorship of Roger Enoka. It was in this dynamic and rich academic environment of a thriving laboratory, where interdisciplinary views were welcome, that I learned the discipline and structure required for writing publications and grants, delivery of presentations, and mentoring students. These are the currency of success in academia. An American Physiological Society Research Career Enhancement Award (2004) allowed me to perform collaborative studies with Professors Janet Taylor and Simon Gandevia (Sydney, Australia), who are world leaders in the neural control of movement. Over an eight-week research-intensive collaboration, I learned how to use transcranial magnetic stimulation to understand the neural control of muscle activation during fatiguing contractions.

Q: Why did you choose to become an "exercise scientist"?

I have a passion to educate and empower people of all ages and abilities to optimize their performance and well-being through physical activity. I initially trained as a physical education teacher at a time when degrees in exercise science were not available in Australia. Shortly after I graduated, however, the field of exercise science exploded and quickly grew in the 1990s. I loved to teach, but I also wanted to understand more about the mechanisms of exercise training adaptations, especially in females. I realized I could teach and research within a university, so I pursued graduate studies in exercise science.

Q: Why is it important for exercise science students to have an understanding of equipment and assessment in the various fields of exercise science?

Appreciation of the available and latest cutting-edge techniques will allow students to pursue the best strategies and options for either addressing research questions or the possibilities for their clients as an exercise scientist. It is important to remember, however, that techniques are only tools. A deep understanding of the physiology and the response of the body during exercise using these tools is very important.

Q: What advice would you have for a student exploring a career in any exercise science profession?

Follow your passion and be inquisitive. Choose where you study and your training opportunities based on the quality of training and mentorship, rather than purely on the topic within a specific field. If you pursue research, do hypothesis-driven research rather than technique-driven research. This means, chase the answer to the significant questions, and don't be limited by techniques. Do something toward advancing your career and your profession every workday, even 15 minutes! Embrace diverse opinions and perspectives.

Interview

L. Bruce Gladden, PhD, FACSM

Professor of Exercise Physiology in the School of Kinesiology at Auburn University in Auburn, AL

Brief Introduction – I grew up in middle Tennessee and enjoyed playing on the basketball team in high school. I was particularly fortunate to attend an excellent high school that prepared me well for college. I planned to major in Mathematics at the University of Tennessee but quickly became interested in Biology. I graduated with a major in Zoology and a minor in Chemistry. During the last term of my undergraduate career, I took an exercise physiology course and I was hooked! From there, I had the good fortune to participate in a research project in Dr. Hugh Welch's Exercise Physiology Laboratory. This ultimately led to a PhD in Exercise Physiology at the University of Tennessee and postdoctoral experience in the Department of Physiology at the University of Florida.

Q: What are your most significant career experiences?

After the completion of 2 years of postdoctoral work, I spent 11 years at the University of Louisville, where I had the opportunity to perform both human and animal research. Then I was recruited to Auburn University, where I have remained for the past 31 years. Along the way I have had the pleasure of working in the American College of Sports Medicine (ACSM) and the American Physiological Society. The regional chapter of ACSM provided numerous opportunities to interact with both professors and students from other institutions. In recent years, my highest honors have been to receive the ACSM Citation Award, to be named Editor-in-Chief of ACSM's flagship journal, *Medicine & Science in Sports & Exercise*, and to be elected President-Elect of ACSM. As a researcher, I have focused on exercise bioenergetics, specifically lactate metabolism and oxygen uptake on-kinetics. As a result of my work, I have had the opportunity to write several review/synthesis articles in prominent publications.

Q: Why did you choose to become an "exercise scientist"?

I was "hooked" when I took my first undergraduate exercise physiology class. As an undergraduate, I was a "gym rat" who played basketball for several hours almost every day, but I also enjoyed biology. Exercise physiology was an absolutely perfect discipline for my overall interests. Since that time, my education and my career have been completely enjoyable. Every day when I come to work, I feel like I am being paid to have fun.

Q: Why is it important for exercise science students to have an understanding of equipment and assessment in the various fields of exercise science?

Reading textbooks and listening to lectures can provide a great deal of knowledge, but your depth of understanding is incredibly enhanced by hands-on experience. This includes the actual use of equipment in class laboratories and performing various types of testing. While laboratory classes are important, anyone who has an interest in study beyond the Bachelor's level should search for opportunities to participate in research with faculty members. I think it's especially important to understand the basic principles behind various types of tests and equipment. Just because a piece of equipment (*e.g.*, oxygen uptake system) costs many thousands of dollars doesn't mean it automatically gives the right numbers every time you turn it on!

Q: *What advice would you have for a student exploring a career in any exercise science profession?*

Take some classes and get involved in research and application of exercise science to see if you enjoy it. Being well educated is important for being a well-informed and active citizen. From a personal view, education is about choice. Always get enough education and experience so that you have choices in your career path. No matter what career you choose, be absolutely certain that you like it, maybe even love it! If that is the case, you're never just working; instead, you are having fun.

SUMMARY

- Professionals from all areas of exercise science use the equipment and assessment methods for evaluating physical and psychological performances and functions.
- The assessment of physiological, psychological, and biomechanical measures can provide useful information about an individual's risk of disease and illness.
- Periodical assessment of variables can provide insight into the acute responses to physical activity and exercise as well as practice and game situations in sport.
- The regular assessment of certain variables can provide information on the effectiveness of an individual exercise or team training program.

FOR REVIEW

1. Why should the type of ergometer (*i.e.*, treadmill or cycle) used during testing be matched to the exercise mode of the individual being tested?
2. A metabolic cart is used to make assessments of what physiological measure?
3. The measurement of lung function provides an assessment of what two broad categories of pulmonary disease?
4. An ECG machine measures what aspect of cardiovascular function?
5. An EMG machine measures what aspect of muscular function?
6. An isokinetic dynamometer controls the speed of muscle contraction and makes a measurement of what muscle function?
7. Describe the basic scientific principle associated with the use of MRI and MRS.
8. Dual energy X-ray absorptiometry can be used to provide a quantitative assessment of what three body components?
9. What are the three most commonly used instruments for the determination of free-living physical activity and energy expenditure?
10. What is the conceptual basis for using bioelectric impedance analyzers to measure body composition?
11. The principle that a certain percentage of total body fat lies directly beneath the surface of the skin allows for the use of which technique for the assessment of body composition?
12. What are the two types of thermotherapy, and how does each type work to enhance rehabilitation from injury.
13. High-speed motion capture systems are often used to assess what aspect of motor performance?

14. Electroencephalography (EEG) provides the assessment of what physiologic measure?
15. What is the difference between quantitative and qualitative instruments in the measurement of behavioral and psychological assessments of exercise and sport psychology?

Project-Based Learning

1. An individual who is 50 years of age wants to participate in a regular exercise program but has been diagnosed with metabolic syndrome. Prepare a presentation for that individual that includes the different types of assessments you would perform, and include the type of equipment you would use. Using research literature, provide justification for the decisions and choices you would make.
2. Identify an individual participating in a sport or athletic competition whose performance you believe could be enhanced by periodic assessment of characteristics critical to their success as an athlete. Prepare a presentation that you would give to the athlete and their coach that includes at least five performance-related assessments that you would make for that athlete. What are those five assessments, what equipment would you use, and how does the research literature support your recommendations?

REFERENCES

1. MacDougall JD, Wenger HA. The purpose of physiological testing. In: MacDougall JD, Wenger HA, Green HJ, editors. *Physiological Testing of the High-Performance Athlete*. 2nd ed. Champaign (IL): Human Kinetics; 1999. p. 1–6.
2. Mattsson CM, Wheeler MT, Waggott D, Caleshu C, Ashley EA. Sports genetics moving forward: lessons learned from medical research. *Physiol Genomics*. 2016;48(3):175–82.
3. Wang G, Tanaka M, Eynon N, et al. The future of genomic research in athletic performance and adaptation to training. *Med Sport Sci*. 2016;61:55–67.
4. Glazier PS. Towards a grand unified theory of sports performance. *Hum Mov Sci*. 2017;56(Pt A):139–56.
5. Maffulli N, Marigotti K, Longo UG, Loppini M, Fazio VM, Denaro V. The genetics of sports injuries and athletic performance. *Muscles Ligaments Tendons J*. 2019;3(3):179.
6. American College of Sports Medicine, Academy of Nutrition and Dietetics, Canadian Dietetic Association. Nutrition and athletic performance. *Med Sci Sports Exerc*. 2009;41(3):709–31.
7. Maud PJ, Foster C. *Physiological Assessment of Human Fitness*. 2nd ed. Champaign (IL): Human Kinetics; 2006.
8. American College of Sports Medicine. *ACSM's Guidelines for Exercise Testing and Prescription*. 11th ed. Philadelphia (PA): Lippincott Williams & Wilkins; 2021.
9. Guazzi M, Bandera F, Ozemek C, Systrom D, Arena R. Cardiopulmonary exercise testing what is its value? *J Am Coll Cardiol*. 2017;70(13):1618–36.
10. Balady GJ, Arena R, Sietsema K, et al. Clinician's Guide to cardiopulmonary exercise testing in adults: a scientific statement from the American Heart Association. *Circulation*. 2010;122(2):191–225.
11. Hagberg JM, Moore GE, Ferrell RE. Specific genetic markers of endurance performance and VO_2max. *Exer Sport Sci Rev*. 2001;29(1):15–9.
12. American College of Sports Medicine. *ACSM's Guidelines for Exercise Testing and Prescription*. 10th ed. Philadelphia (PA): Lippincott Williams & Wilkins; 2017.
13. Donnelly JE, Blair SN, Jakicic JM, et al. Appropriate physical activity intervention strategies for weight loss and prevention of weight regain for adults. *Med Sci Sports Exerc*. 2009;33(12):2145–56.
14. Wing RR, Jakicic JM, Neiberg R, et al. Fitness, fatness, and cardiovascular risk factors in Type 2 diabetes: look AHEAD study. *Med Sci Sports Exerc*. 2007;39(12):2107–16.

15. Levitzky MG. *Pulmonary Physiology*. 8th ed. New York (NY): McGraw-Hill; 2013.
16. Harman EA. The measurement of human mechanical power. In: Maud PJ, Foster C, editors. *Physiological Assessment of Human Fitness*. 2nd ed. Champaign (IL): Human Kinetics; 2006. p. 87–113.
17. American College of Sports Medicine. Osteoporosis and exercise. *Med Sci Sports Exerc*. 1995;27(4):i–vii.
18. Hall SJ. *Basic Biomechanics*. 7th ed. Dubuque (IA): McGraw-Hill; 2015.
19. Brooks GA, Fahey TD, Baldwin KM. *Exercise Physiology: Human Bioenergetics and Its Applications*. 4th ed. Mountain View (CA): Mayfield; 2004.
20. McGuigan MRM, Sharman MJ. Skeletal muscle structure and function. In: Maud PJ, Foster C, editors. *Physiological Assessment of Human Fitness*. 2nd ed. Champaign (IL): Human Kinetics; 2006.
21. Pollock ML, Kanaley JA, Garzarella L, Graves JE. Anthropometry and body composition measurement. In: Maud PJ, Foster C, editors. *Physiological Assessment of Human Fitness*. 2nd ed. Champaign (IL): Human Kinetics; 2006.
22. Lahey R, Khan SS. Trends in obesity and risk of cardiovascular disease. *Curr Epidemiol Rep*. 2018;5(3):243–51.
23. Clarke MA, Fetterman B, Cheung LC, et al. Epidemiologic evidence that excess body weight increases risk of cervical cancer by decreased detection of precancer. *J Clin Oncol*. 2018;36(12):1184–91.
24. Scherer PE, Hill JA. Obesity, diabetes, and cardiovascular diseases. *Circ Res*. 2016;118(11):1703–5.
25. Academy of Nutrition and Dietetics Web site [Internet; cited 2020]. Available from: www.eatright.org.
26. Kraemer WJ, Volek JS, Bush JA, Putukian M, Sebastianelli WJ. Hormonal responses to consecutive days of heavy-resistance exercise with or without nutritional supplementation. *J Appl Physiol*. 1998;85(4):1544–55.
27. Thompson FE, Subar AF. Dietary assessment methodology. In: Coulston AM, Rock CL, Monsen ER, editors. *Nutrition in the Prevention and Treatment of Disease*. San Diego (CA): Academic Press; 2001. p. 3–30.
28. Melby CL, Ho RC, Hill JO. Assessment of human energy expenditure. In: Bouchard C, editor. *Physical Activity and Obesity*. Champaign (IL): Human Kinetics; 2000. p. 103–31.
29. Lee SK, Sobal J. Socio-economic, dietary, activity, nutrition and body weight transitions in South Korea. *Public Health Nutr*. 2003;6(7):665–74.
30. Pi-Sunyer X. Changes in body composition and metabolic disease risk. *Eur J Clin Nutr*. 2019;73(2):231–5.
31. Roelofs EJ, Smith-Ryan AE, Trexler ET, Hirsch KR. Seasonal effects on body composition, muscle characteristics, and performance of collegiate swimmers and divers. *J Athl Train*. 2017;52(1):45–50.
32. Wagner DR, Heyward VH. Techniques of body composition assessment: a review of laboratory and field methods. *Res Q Exerc Sport*. 1999;70(2):135–49.
33. Dietz WH, Bellizzi MC. Introduction: the use of body mass index to assess obesity in children. *Am J Clin Nutr*. 1999;70:123S–125S.
34. Morabia A, Ross A, Curtin F, Pichard C, Slosman DO. Relation of BMI to a dual-energy X-ray absorptiometry measure of fatness. *Br J Nutr*. 1999;82:49–55.
35. Kirk S, Zeller M, Claytor R, Santangelo M, Khoury PR, Daniels SR. The relationship of health outcomes to improvement in BMI in children and adolescents. *Obes Res*. 2005;13:876–82.
36. Pischon T, Boeing H, Hoffmann K, et al. General and abdominal adiposity and risk of death in Europe. *N Engl J Med*. 2008;359(20):2105–20.
37. Seidell JC, Perusse L, Despres JP, Bouchard C. Waist and hip circumferences have independent and opposite effects on cardiovascular disease risk factors: the Quebec Family Study. *Am J Clin Nutr*. 2001;74(3):315–21.
38. Jeppesen J, Hein HO, Suadicani P, Gyntelberg F. Low triglycerides-high high-density lipoprotein cholesterol and risk of ischemic heart disease. *Arch Intern Med*. 2001;161:361–6.
39. Kohrt WM, Bloomfield SA, Little KD, Nelson ME, Yingling VR. Physical activity and bone health. *Med Sci Sports Exerc*. 2004;36(11):1985–96.
40. Pearson D, Faigenbaum AD, Conley MS, Kraemer WJ. The National Strength and Conditioning Association's basic guidelines for the resistance training of athletes. *Strength Cond J*. 2000;22(4):14–27.
41. Prentice WE. *Principles of Athletic Training: A Guide to Evidence-Based Clinical Practice*. 16th ed. New York (NY): McGraw-Hill Companies; 2017.
42. Schmidt RA, Lee T. *Motor Control and Learning: A Behavioral Emphasis*. 5th ed. Champaign (IL): Human Kinetics; 2011.
43. Petruzzello SJ. Exercise and sports psychology. In: Brown SP, editor. *Introduction to Exercise Science*. 1st ed. Philadelphia (PA): Lippincott Williams & Wilkins; 2001. p. 310–33.
44. Weinberg RS, Gould D. *Foundations of Sport and Exercise Psychology*. 6th ed. Champaign (IL): Human Kinetics; 2015.
45. Baumgartner TA, Hensley LD. *Conducting and Reading Research in Health and Human Performance*. 4th ed. New York (NY): McGraw-Hill Publishers; 2006.

CHAPTER 12

Careers and Professional Issues in Exercise Science

After completing this chapter, you will be able to:

1. Describe the differences among the credentialing titles of certification, licensure, and registration.
2. Identify various professional job opportunities available in exercise science and related areas.
3. Identify the role and mission that professional organizations play in promoting the advancement of exercise science and related areas.
4. Explain the role that governmental and international agencies and organizations play in advancing the professional growth of exercise science.

Graduates of programs in exercise science can gain professional employment and career development in a wide variety of job settings. Examples might include work as a personal trainer (Figure 12.1), health and wellness coordinator, or clinical exercise physiologist (Figure 12.2). In other instances, graduates can take advantage of opportunities in related professional fields where the knowledge base developed through a program of study in exercise science can be a valuable asset. This might include, for example, a career in an allied health care field as a physical therapist, occupational therapist, or a physician assistant. In all of these professions, the knowledge, skills, and abilities developed as a student in an undergraduate program of study in exercise science will provide a solid foundation for a successful professional career.

Educational qualifications for professional employment can range from a Bachelor's degree (*e.g.*, BS) to an advanced professional degree (*e.g.*, Doctor of Medicine [MD], Doctor of Chiropractic [DC], Doctor of Dentistry [DD], Doctor of Osteopathic Medicine [DO], Doctor of Physical Therapy [DPT]) or educational degree (*e.g.*, MS, EdD, PhD, DSci). It is also becoming increasingly common for employers to require potential employees to demonstrate achievement of certain skills and competencies prior to hiring. These may

FIGURE 12.1 Personal trainers work with clients in a variety of settings. (From Shutterstock.)

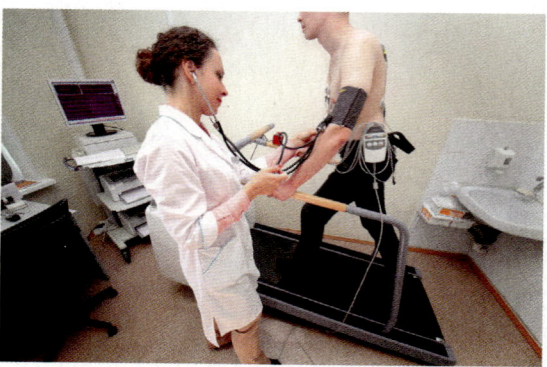

FIGURE 12.2 Clinical exercise physiologists can administer graded exercise tests to measure cardiorespiratory fitness. (From Shutterstock.)

include successfully attaining **certification** and/or **licensure** and being included in **registration** databases. Graduates of exercise science programs must pay close attention to the professional requirements for working with individuals in physical activity, exercise, sport, and athletic competition settings. Several professional organizations have developed certification and registration programs to help individuals demonstrate knowledge and proficiency in content, skills, and abilities of the specific areas for which they are intending to work. Professional organizations have also worked with the state and federal governmental agencies to develop licensure programs for individuals in those professions that require a higher level of professional skills and competency when approving an individual for professional practice. Professional organizations also provide membership benefits and services to assist in the professional development for a career in an exercise science or allied health profession. Furthermore, numerous governmental and private organizations promote and support initiatives and programs that are related to the various areas of study in exercise science (1). This chapter provides an overview of certification, licensure, and registration and the role that professional and governmental organizations play in the credentialing processes. Potential employment opportunities for graduates of exercise science programs are presented. Finally, information is provided about the various professional organizations and governmental agencies that support exercise science and its related areas of study and provide for career development.

CERTIFICATION, LICENSURE, AND REGISTRATION

Credentialing is defined as providing certified documents showing that an individual is professionally qualified and entitled to recognition or has a right to perform certain functions or actions. Credentialing is designed to ensure that, within a profession or service, standards of a safe and ethical practice are being maintained (1). By achieving a credential, an individual has demonstrated that they have achieved a set standard of knowledge, skills, and abilities as defined by the **credentialing organization** in a particular area of study or specialty. For many exercise science professionals, the most common form of credentialing

Certification A process whereby individuals demonstrate knowledge and proficiency in content, skills, and abilities of the specific areas for which they are intending to work.
Licensure Granting of permission by an official or legal authority (usually a government agency) to an individual or organization to engage in a practice or activity that would otherwise be illegal.
Registration Documentation of professional qualification information relevant to government licensing regulations.
Credentialing Providing certified documents showing that an individual is entitled to credit or has a right to perform certain functions or actions.
Credentialing organization A professional organization or governmental agency that oversees and administers examinations for certification, licensure, or registration of an individual or program.

is certification. Professional certification recognizes those individuals who possess the knowledge and competency to undertake a variety of responsibilities in health, fitness, wellness, coaching, training, rehabilitative, and sports medicine programs (1). Obtaining certification from a credentialing organization:

- Provides recognition of competency to work with healthy, athletic, injured, diseased, or disabled populations
- Demonstrates commitment to the profession and the standards that have been established
- Assists with job employment and advancement in professional development

Licensure is a credential that requires a higher level of professional competency than certification and is often required to gain employment and practice as an allied health care and medical professional (1).

Definitions of Certification, Licensure, and Registration

Within exercise science, graduates of degree and professional preparation programs can attain credentials from several professional organizations. There are distinctions among certification, licensure, and registration that allow individuals who possess a credential to perform specific duties within a defined area of professional practice. Higher education programs that receive certification or accreditation have demonstrated the ability to deliver to students a program or curriculum that meets standards acceptable to the credentialing organization. In some instances (*e.g.*, athletic training, dietetics and nutrition, physical therapy), students must graduate from an accredited program of study to qualify to take a national certification (athletic training) or licensure (dietetics and nutrition, physical therapy) examination (1).

Certification

The certification process requires individuals, educational programs, or institutions to be evaluated and recognized as meeting predetermined standards through the successful completion of a valid and reliable examination (*e.g.*, for individuals) and program review (*e.g.*, educational programs and colleges or universities) (1). Professional organizations usually administer certification, licensure, and registration examinations and reviews to individuals and programs that meet a defined level of curricular expectations. Individuals and programs voluntarily participate in the credentialing process; however, as professional expectations among employers and government agencies change, certification or licensure is often desired (*e.g.*, certified clinical exercise physiologist [CEP] and registered clinical exercise physiologist [RCEP]) and/or required (athletic trainer certified [ATC] and registered dietician and nutritionist [RDN]) in many exercise science areas.

The process for individuals seeking certification typically includes the following steps (1):

- Achievement of a core of knowledge, skills, and abilities through formal educational and/or professional experience

- Submission of an application form and candidacy credentials
- Review of prerequisite knowledge and skills through the use of study guides, resource manuals, review courses, videos, webinars, and/or workshops
- Taking and passing the certification examination

Certification examinations generally include both written and **interactive components**. The written component is primarily comprised of objective questions of content knowledge, skills, and abilities. The interactive component typically requires candidates to view a video excerpt and then provide an answer to questions about a specific task or activity in the video clip. An individual who achieves a passing score on both the written and interactive components is recognized by the credentialing organization as a certified member.

Certified individuals are usually required to obtain **continuing education units** (CEUs, aka continuing education credits in some countries) to retain certification in the credentialing organization. This can usually be accomplished by attending professional meetings and conferences, taking additional study courses, participating in online professional development programs, and/or becoming involved in the certification organization. Certification costs vary widely depending on the level of certification obtained and the fee structure of the organization administering the examination (1,2).

Certification can also be extended to include educational programs and facilities. In this type of certification, individual programs or facilities are required to meet specific criteria related to curriculum, staff, facilities, and learning environment. For instance, **program accreditation** is currently required for academic programs in athletic training and dietetics. Students in these areas of study must graduate with a degree from an accredited program before being allowed to take the examinations for certification (*e.g.*, athletic training) or licensure/registration (*e.g.*, dietetics). Programs of exercise science can also become accredited. For example, the **Commission on Accreditation of Allied Health Education Programs** (CAAHEP) oversees the **Committee on the Accreditation for the Exercise Sciences** (CoAES). The role of the CoAES is to establish standards and guidelines for academic programs that facilitate the preparation of students seeking employment in the health, fitness, and exercise industry. Programmatic accreditation through CAAHEP is specifically intended for exercise science programs or the academic departments in which the program is contained (*e.g.*, Kinesiology, Exercise Science, Movement Science) (3). The American

Interactive component A part of an examination whereby the candidate taking the examination must respond to a visual situation.

Continuing education units Additional professional education that is required to maintain certification, licensure, or registration.

Program accreditation The granting of an academic program the standing of meeting acceptable criteria for the preparation of students enrolled in the program.

Commission on Accreditation of Allied Health Education Programs The largest organization for program accreditation in the health and exercise sciences field.

Committee on the Accreditation for the Exercise Sciences An organization designed to establish standards and guidelines for academic programs that facilitate the preparation of students seeking employment in the health, fitness, and exercise industry.

Association of Cardiovascular and Pulmonary Rehabilitation (AACVPR) confers program certification for those cardiovascular and pulmonary rehabilitation programs in hospitals, medical centers, and outpatient facilities that meet specific guidelines of professional practice. The goal of program certification in this instance is to assure that programs meet the essential standards of patient care (1,4).

Licensure

Licensure is the granting of permission by an official or legal authority (usually a state government agency or board) to an individual or organization to engage in the legal practice of a professional activity that would otherwise be illegal. Licensure requirements and regulations vary among states, and it is important for the exercise science and allied health professional to be aware of the state requirements for professional practice. Individuals such as physicians, physician assistants, nurses, physical and occupational therapists, and dietitians are required to attain licensure to practice as a professional. Licensure is obtained after completing an approved educational curriculum from an accredited program in a college or university and passing a licensure examination, similar to the process for certification. Licensure is usually permanent, but periodic fees, competency examinations, and/or CEUs may be required to maintain active licensure. Individual states usually establish the requirements for licensure within their state, and this allows individuals to legally practice a regulated profession. Professional certification does not allow an individual to practice a regulated occupation unless the individual state recognizes the certification requirements as equivalent to the licensure requirements (1).

Registration

In certain areas of exercise science and allied health, registration is required to engage in professional practice. Registration is the documentation of professional qualification information relevant to government licensing requirements (1). Individuals who are registered with an organization or agency must also complete CEUs to maintain professional registration. For example, the Academy of Nutrition and Dietetics, through the Commission on Dietetic Registration, confers the title of RDN on those individuals who complete academic training and pass the Academy of Nutrition and Dietetics examination (5). Registration and licensure are similar in scope except that a licensed professional generally has a broader range and scope of professional practice. An individual who is registered as a professional typically has their name listed in the registry of the organization. This practice provides information to the general public and potential employers about the qualifications of the individuals listed in the registry. For example, an organization may provide the names of professionals within a particular geographic area who are ACSM Registered Clinical Exercise Physiologists (RCEPs). Table 12.1 provides a summary explanation of certification, licensure, and registration. The selection of an appropriate credential should be based on the professional requirements and the intended range of professional practice. Several professional organizations offer certifications, licensure, and registrations that have relevance to exercise science professionals. Table 12.2 lists some of the professional organizations and the credentialing that each organization provides to the members.

Thinking Critically

How do certification, licensure, and registration promote an increase in the confidence of the general public when working with exercise science and allied health professionals?

Table 12.1 A Description of Certification, Licensure, and Registration (1)

CATEGORY	DESCRIPTION
Certification	Individual, institution, or educational program is evaluated and recognized as meeting certain predetermined standards through successful completion of a valid and reliable examination
Licensure	Granting of permission by a competent authority (usually a government agency) to an organization or individual to engage in a practice or activity that would otherwise be illegal
Registration	Recording of professional qualification information relevant to government licensing regulations; similar to licensure, except that the scope of practice is usually more narrow than for a licensed professional

Table 12.2 Professional Organizations in Exercise Science and Related Areas and the Credentialing Each Provides

PROFESSIONAL ORGANIZATION	CREDENTIAL
American College of Sports Medicine (ACSM)	Health and Fitness Certification
	Personal Trainer
	Group Exercise Instructor
	Exercise Physiologist
	Clinical Certification
	Clinical Exercise Physiologist
	Specialty Credentials
	Exercise Is Medicine
	Cancer Exercise Trainer
	Inclusive Fitness Trainer
	Physical Activity in Public Health
National Strength and Conditioning Association (NSCA)	Certified Strength and Conditioning Specialist
	Certified Personal Trainer
	Certified Special Population Specialist
	Tactical Strength and Conditioning Facilitator
American Council on Exercise (ACE)	Personal Trainer
	Medical Exercise Specialist
	Group Fitness Instructor
	Health Coach

(continued)

Table 12.2	Professional Organizations in Exercise Science and Related Areas and the Credentialing Each Provides (continued)
PROFESSIONAL ORGANIZATION	CREDENTIAL
Academy of Nutrition and Dietetics (AND)	Registered Dietician
	Registered Dietician Nutritionist
	Dietetic Technician, Registered
	Nutrition and Dietetics Technician, Registered
	Specialist Certification
	Advanced Practice Certification
National Athletic Trainers' Association (NATA)	Certification through the NATA Board of Certification
YMCA	Personal Training Instructor

CAREER EMPLOYMENT AND PROFESSIONAL OPPORTUNITIES

One of the most frequently asked questions by undergraduate exercise science students is: "What type of job can I get with an exercise science degree?" Well, the possibilities for professional employment are numerous and vary widely. It is important that every undergraduate student take some time to seriously consider what they are willing to invest in professional education and career development and what they really want to do with the professional aspect of their life. Thoughtful reflection is critical because it is important to consider the many different components of your professional education and development as you prepare yourself for your professional career. Often professional careers require additional education in the form of a graduate degree in a specialized field (e.g., athletic training or physical therapy). Listed here are some questions every student should consider when deciding on a professional career in exercise science or related area (6).

- Do you participate in and enjoy physical activity, exercise, and sports?
- Have you enjoyed coursework in biology, chemistry, math, nutrition, and physiology?
- Are you willing to commit to the necessary investment in education, academic training, and professional education that are required of working exercise science and allied health professionals, including progression from an undergraduate degree to a graduate degree?
- Have you spoken to individuals who currently work in your particular field of interest?
- In what type of professional employment setting do you wish to work (e.g., hospital, clinic, school, fitness facility, industrial setting, corporate setting, outpatient clinic, college, university)?
- Do you enjoy working with all types of people, or are you only interested in certain populations (e.g., athletes, children, elderly, or patients with a chronic disease condition)?
- Do you want to work with people to prevent disease and injury or with patients desiring treatment and rehabilitation?
- Do you prefer to work in a research or an educational capacity?

Starting salaries for individuals with an undergraduate degree and no experience will vary widely. Factors such as previous work experience, geographic location, employment setting, and market demand combined with other factors such as whether you hold professional licensure or certification will influence your beginning salary. Speaking to a professional who currently works in your field of interest in your geographic location is a good way to obtain an approximation of an expected starting salary and benefits (6).

There are numerous reasons for the increased opportunities for professional careers in exercise science and allied health professions. The greater interest in overall health by the general public has resulted in an explosion of possible career opportunities in exercise science–related fields such as health coaches, wellness coaches, personal trainers, fitness coaches, fitness directors, exercise and sport nutritionists, ergonomists, and exercise psychologists. The rise in lifestyle-related diseases has resulted in the need for more highly trained medical doctors, physician assistants, dietitians, physical and occupational therapists, rehabilitation specialists, and researchers. Advances in sport technology and product development, along with increases in participation in sport and athletics, have resulted in the need for more highly trained persons as athletic trainers, strength and conditioning coaches, sport biomechanists, sport psychologists, and researchers. The U.S. Department of Labor, Bureau of Labor Statistics (www.bls.gov/) periodically projects the employment opportunities in all sectors of the economy and is a good place to identify employment trends in many of the professional jobs that exercise science graduates can expect to gain (7). The following sections provide a short description of the major types of professional employment and career opportunities for graduates of exercise science programs. In some instances, an undergraduate degree is sufficient for employment, whereas in other professions an advanced or professional degree is required to work in a particular field. In almost every instance, obtaining a credential provides a benefit for gaining employment or is a required component of the job.

Athletic Trainer

An athletic trainer is a sports medicine professional, with a professional credential of athletic trainer certified (ATC), who is involved in the prevention, treatment, and rehabilitation of injuries to physically active individuals and athletes (see Chapter 6). Many people often view athletic trainers as only working with athletes in a sport setting; however, athletic trainers also work closely with exercise science professionals, sports medicine physicians, and other allied health professionals to provide care to anyone who may have an injury caused by participation in physical activity, exercise, or sport (Figure 12.3). Only graduates of an athletic training program accredited by the **Commission on Accreditation of Athletic Training Education**

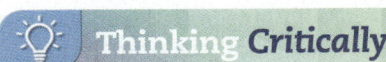

Thinking *Critically*

How do academic coursework and experiences in exercise science prepare future professionals for working in the allied health and exercise science professions?

Commission on Accreditation of Athletic Training Education The agency responsible for the accreditation of professional athletic training educational programs.

FIGURE 12.3 Clinical athletic trainers work to rehabilitate individuals injured during physical activity and exercise. (From Shutterstock.)

may take the National Board of Certification Examination and become a certified athletic trainer. In many states, athletic trainers must also become licensed by the state to practice athletic training. Athletic trainers are employed in a variety of work environments, including secondary schools, colleges and universities, sports medicine clinics, professional sports programs, industrial and other occupational settings, as well as other allied health care employment settings, including those as a physician extender (6,8).

Biomechanist

With proper academic training and professional experience, an individual can work as a clinical or sport biomechanist. In general, an exercise science student will need additional undergraduate coursework in physics or engineering or, possibly, a graduate degree (MS or PhD) to work as a professional biomechanist. Clinical biomechanics focuses on the mechanics of injury and the principles of prevention, evaluation, and treatment of musculoskeletal problems. Sport biomechanics examines factors of human movement associated with exercise and training for the purpose of improving sport and athletic performance. There is currently no credentialing available in the area of biomechanics, although certification in a related area may be helpful in gaining employment. Employment opportunities include working in colleges and universities, sport and athletic performance enhancement centers, sports medicine clinics, hospitals, and other allied health care environments (6,8).

Biomedical Engineer

With proper academic training and professional experience, an individual can work as a biomedical engineer. To gain employment as a biomedical engineer, an exercise science student will need undergraduate coursework in math, physics, engineering, chemistry, molecular biology, and genetics and a dual undergraduate degree or graduate degree (MS or PhD) in engineering. There is currently no credentialing available in the area of biomedical engineering, although certification in a related area may be helpful in gaining employment. Biomedical engineers apply the principles of biology and engineering to health care and medicine designing artificial organs, prosthetics, surgical instruments, and medical devices to support physiological functions. Employment opportunities include working in business and industry, rehabilitation centers, medical clinics, hospitals, and other allied health care environments.

Chiropractor

A doctor of chiropractic medicine is a health care professional who is involved in the prevention, treatment, and rehabilitation of musculoskeletal injuries by using spinal manipulations to improve spinal motion and the body's function. Individuals involved in physical activity, exercise, sport, and athletic competition may benefit from chiropractic adjustments that increase the range of motion of joints and muscles. Only graduates of an accredited chiropractic school may take an examination to become a licensed chiropractor. Chiropractors work in a variety of environments, including private practice, long-term care facilities, medical institutions, health departments, residential care facilities, hospitals, colleges and universities, and other health care settings (6).

Clinical Exercise Physiologist

CEPs work with healthy and diseased individuals in a variety of employment settings. Individuals who have chronic disease conditions, including cardiovascular, respiratory, and metabolic diseases, can benefit from regular participation in physical activity and exercise. CEPs are responsible for performing health and fitness assessments, developing and implementing exercise prescriptions, and monitoring the effectiveness of the interventions. Keeping appropriate records to determine the effectiveness of the physical activity or exercise intervention is often an important responsibility of the CEP. Frequently, advanced coursework in electrocardiography, pathophysiology, and specific populations (*e.g.*, children or the elderly) is required, and many clinical exercise physiologists have a graduate (MS) degree. Certification and registration are becoming increasingly important and in many instances a requirement for employment as a clinical exercise physiologist. Beginning in 2027, only individuals who graduate with a Bachelor's degree (or higher) in Exercise Science or Exercise Physiology from a regionally accredited college or university that is also programmatically accredited by the Commission on Accreditation of Allied Health Education Programs (CAAHEP) can take the ACSM Certified Exercise Physiologist® (ACSM-EP®) and the ACSM Certified Clinical Exercise Physiologist® (ACSM-CEP®) examinations. Some states also require licensure to practice. Employment opportunities exist primarily in hospitals, outpatient and allied health care centers, cardiac rehabilitation centers, and within wellness and fitness programs (6,8).

Dietitian/Sport Dietician

A dietitian is a licensed professional who assesses the nutritional needs of individuals and then develops and assists with the implementation of nutrition programs for those individuals (Figure 12.4). Dietitians may also advise patients and clients on several health- and disease-related conditions, including weight loss, diabetes, high blood pressure, and elevated cholesterol levels. Only individuals who graduate from an Academy of Nutrition and Dietetics accredited program complete an Academy of Nutrition and Dietetics–approved internship and pass the certification examination may become a licensed RDN. RDNs work in a variety of environments, including private practice, long-term care facilities, medical institutions, health departments, social service agencies, residential care facilities, hospitals, primary and secondary school systems, colleges and universities, and other allied health

FIGURE 12.4 Dieticians perform nutritional counseling with healthy and diseased children and adults, as well as athletes. (From Shutterstock.)

settings. RDNs may also practice sports nutrition by working with individuals to develop nutritional programs that will serve to improve sport and athletic performance (6).

Ergonomist

With proper academic training and professional experience, an individual can work as an ergonomist to evaluate and make changes at the human and work place interface. In general, an exercise science student will need additional undergraduate coursework in engineering, anatomy, physiology, psychology, and statistics. Ergonomics is the study of the interaction between humans, the objects they use, and the environments in which they function (see Chapter 10). While not required, certification may be helpful in gaining employment working in the areas of human factors, user experience, or general ergonomics. Employment opportunities include working in private business and industry, industrial ergonomic settings, sports medicine clinics, performance enhancement centers, hospitals, and other allied health care environments (9).

Exercise and Sport Psychologist

Exercise and sport psychology professionals work with healthy and diseased individuals, as well as athletes of all competition levels to enhance the psychological components related to the successful performance of physical activity, exercise, sport, and athletic competition. The principles of exercise and sport psychology are also used by exercise science and allied health professionals in a variety of employment settings, including the wellness and fitness industry, athletic training, coaching, clinical exercise and rehabilitation settings, and long-term care facilities. Individuals who obtain a graduate degree and become certified as consultants by the Association for Applied Sport Psychology (AASP) can seek consultant positions with individual athletes and sport and athletic teams. Advanced academic coursework resulting in licensure as a clinical psychologist can also be valuable for developing a professional career in exercise or sport psychology. Attaining an advanced degree (PhD or EdD) can lead to employment as an instructor of exercise and sport psychology at a college or university (6,8). Employment opportunities exist primarily in outpatient and allied health care centers, cardiac rehabilitation centers, wellness and fitness programs in colleges and universities, performance enhancement centers, private business, industrial settings, and sports medicine clinics.

Graduate School and Researcher

Individuals with an undergraduate degree in exercise science or a related area who wish to pursue an advanced graduate degree in a particular area of exercise science or become actively involved in research have several opportunities to choose from. A Master's degree (MS or MEd) generally requires 1 to 2 years of graduate coursework and related experiences such as clinical or field experiences. The culminating experience may be a Master's thesis, independent research project, or an internship. The completion of a Master's degree generally requires 30 to 40 credit hours beyond the undergraduate degree. A doctoral degree (PhD, DSci, or EdD) typically requires between 3 and 5 years of graduate coursework and research experiences. The traditional culminating experience is a doctoral dissertation demonstrating competency in conducting independent research and scholarship. Table 12.3

Table 12.3 Areas of Advanced Graduate Study in Exercise Science
Biomechanics
Cardiac rehabilitation
Environmental physiology
Epidemiology
Exercise and aging
Exercise behavior and psychology
Exercise biochemistry
Exercise physiology
Genetics
Integrative physiology
Motor behavior
Motor control
Motor development
Neuroscience/Neurophysiology
Nutrition for health promotion and performance enhancement
Occupational physiology
Pediatric exercise physiology
Rehabilitation for disease and injury
Sport psychology
Therapeutic exercise

provides some examples of areas of advanced study and research in exercise science. Advanced coursework and participation in research activities are the cornerstones of graduate school education. Coursework in content areas, research design, and statistical analysis is critical for pursuing a career in research. Individuals with graduate degrees work in a variety of environments, including pharmaceutical companies, food and beverage companies, technology companies, sports apparel/equipment companies, colleges and universities, hospitals, medical schools and institutions, governmental agencies, including state and local health departments, and private research foundations (6).

Medical Doctor

A medical doctor is a licensed professional who is involved in the prevention, treatment, and rehabilitation of illness and injuries to individuals. Medical doctors can be educated to practice medicine in the following areas: allopathic, chiropractic, osteopathic, podiatric, and ophthalmology. Each type of medical doctor has a defined scope of practice that can involve work with individuals participating in physical activity, exercise, sport, and athletic competition. Many people often view medical doctors as only working with ill or sick individuals in a health care setting. Medical doctors, however, work closely with other allied health professionals to provide preventive care to a wide range of healthy and diseased individuals. Individuals can also work as a sports medicine physician in clinical practice or with local high school, college, or university sports teams. Only graduates of an accredited medical school may become a licensed medical doctor. Medical doctors, also called physicians, provide medical coverage at amateur, collegiate, and professional sport and athletic competitions. Certification as a sports medicine specialist is also becoming increasingly important if an individual has a desire to work with athletes. Medical doctors work in a variety of environments, including private practice, long-term care facilities, medical institutions, health departments, residential care facilities, hospitals, colleges and universities, and other health care settings (6).

Occupational Therapist

Occupational therapists are licensed professionals who assist individuals with physically, mentally, emotionally, or developmentally limiting conditions to maintain or improve working skills and daily function. Often, occupational therapists teach individuals how to compensate for some temporary or permanent loss of motor function. Occupational therapists help individuals learn or regain the ability to perform activities of daily living, including dressing, preparing meals, and eating. To become a licensed occupational therapist, students must graduate from an occupational therapy program accredited by the Accreditation Council for Occupational Therapy Education, complete a fieldwork requirement, and pass a certification exam administered by the National Board for Certification in Occupational Therapy. A graduate degree at the Master's or doctoral level is required. Occupational therapists work in a variety of environments, including private practice, long-term health care facilities, medical institutions, community health centers, residential care facilities, hospitals, school systems, adult day care centers, and other health care settings (8,10).

Personal Trainer

Personal trainers work with individuals to assess functional capacity and then develop and implement exercise programs for enhancing physical fitness and health. Exercise sessions with personal trainers are generally individual or small group (4–8 individuals) sessions, and they typically occur in the client's home, the personal trainer's place of employment, or at a fitness facility. Personal trainers also conduct group exercise sessions in activities such as spinning, yoga, Pilates, kickboxing, Zumba, core strengthening, aerobics, water aerobics, and types of exercise sessions. Personal trainers benefit from having a strong academic background in exercise physiology, biomechanics, fitness assessment, exercise prescription, exercise psychology, and nutrition as many aspects of this job involve the development of individualized muscular strength and endurance training programs and sound nutritional practices, as well as providing goal setting and motivational strategies for success. In addition to obtaining an undergraduate degree in exercise science, gaining professional certification as a personal trainer or exercise leader from one of the organizations listed in Table 12.2 is highly recommended (6).

Physical Therapist

Physical therapists are licensed professionals who help individuals recover from an injury or disabling physical condition (Figure 12.5). Physical therapists develop structured treatment and rehabilitation programs designed to improve mobility, reduce pain, and prevent or limit permanent disability. Physical therapists conduct evaluations of muscular fitness, range of motion, and muscle and joint functions and then use that information to develop and implement individualized treatment programs for patients. Only graduates of physical therapy professional education programs accredited by the Commission on Accreditation in Physical Therapy Education (CAPTE) may take the national licensure examination that allows them to practice as a licensed physical therapist in the United States. Graduates of

FIGURE 12.5 Physical therapists work in both inpatient and outpatient facilities. (From Shutterstock.)

programs outside the United States may apply for a physical therapy license. CAPTE now requires all academic programs to offer the Doctor of Physical Therapy (DPT) as the minimal entry-level degree. Physical therapists work in a variety of employment environments, including private practice, long-term care facilities, medical schools and institutions, community health centers, residential care facilities, hospitals, school systems, adult daycare centers, and other allied health care facilities (6).

Physician Assistant

A physician assistant is a licensed professional who practices medicine under physician supervision in the prevention, diagnosis, treatment, and rehabilitation of illness and injuries to individuals (Figure 12.6). Physician assistants work closely with other allied health professionals to provide health care to a wide range of healthy and diseased individuals. Physician assistants can benefit from undergraduate coursework in fitness assessment, exercise physiology, biomechanics, nutrition, exercise psychology, and premedicine courses. Only graduates of an accredited program may take the Physician Assistant National Certification Examination to become a certified physician assistant. Once certified, an individual may apply for state licensure. Physician assistants may practice in all medical specialty areas. Physician assistants work in a variety of environments, including private practice, long-term care facilities, medical institutions, health departments, residential care facilities, surgical centers, hospitals, colleges and universities, and other allied health care settings.

Podiatrist

A doctor of podiatric medicine (often called a podiatrist) may diagnose and treat conditions of the leg, ankle, and foot. Podiatrists complete 4 years of training in a medical school and then complete 3 years of residency training. A podiatrist can specialize in sports medicine, diabetic care, wound care, pediatrics, and surgery. A podiatrist can earn board certification after completing advanced training, gaining clinical experience, and taking an examination provided by either the American Board of Foot and Ankle Surgery or the American Board of Podiatric Medicine. Podiatrists are employed in a variety of work environments, including private practice, long-term care facilities, medical institutions, residential care facilities, surgical centers, hospitals, and other allied health care settings.

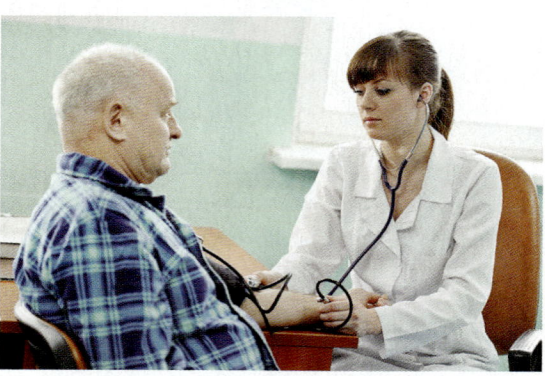

FIGURE 12.6 Physician assistants evaluate and treat patients in a variety of health care settings. (From Shutterstock.)

Public and Private School Teacher

Health and physical education teachers are licensed professionals who work to help children, adolescents, and young adults develop healthy behaviors, motor skills, and physical fitness within a school setting. With proper academic training, an individual can pursue a career as a teacher. Individuals interested in teaching health education and/or physical education must graduate from an accredited program and pass a state and national licensing exam. This is typically accomplished as an undergraduate teacher education major. However, many colleges and universities are offering graduate programs to prepare individuals to become licensed teachers. The cornerstone of teacher education programs is the practicum teaching experience, which must be completed before the licensure examination. The majority of teachers work in public and private school systems teaching kindergarten through grade 12. Coaching sport and athletic teams can often be a part of the additional responsibilities of a health education or physical education teacher. Individuals can also pursue teaching careers at the college and university levels. Employment at this level does not require passing a licensure examination but almost always requires a Master's or doctoral degree, usually in one of the areas of study listed previously in Table 12.3 (6).

Recreational Therapist

Recreational therapists use a structured process that employs recreational and other activity-based interventions to address the needs of individuals with illnesses and/or limiting conditions. Recreational therapy is a means to improve or maintain physical, cognitive, social, emotional, and spiritual functioning in order to facilitate full participation in life. Individuals who graduate from a program that meets the standards of the National Council for Therapeutic Recreation Certification are eligible to take the certification examination. Recreational therapists provide treatment services and recreation activities to individuals using arts and crafts, animals, physical activity, exercise, sports, games, dance, drama, music, and community events. Recreational therapists work in a variety of environments, including private practice, long-term care facilities, medical institutions, health departments, residential care facilities, hospitals, and other allied health care settings (11).

Strength and Conditioning Coach

Strength and conditioning coaches are involved in the development and implementation of specialized training programs for athletes (Figure 12.7). Strength and conditioning coaches work with a variety of individual and team sport athletes to increase muscular strength and endurance, cardiovascular fitness, flexibility, and movement skills in an effort to improve athletic performance. The evaluation and assessment of physical performance and training improvements are important responsibilities of the strength and conditioning coach. In addition to coursework in an exercise science curriculum, individuals wishing to pursue this career option should complete an internship or acquire volunteer training experience within an established strength and conditioning program. Individuals should strongly consider obtaining an appropriate certification credential and, possibly, a graduate degree in an exercise science–related area of study. Strength and conditioning coaches work primarily in secondary schools, colleges and universities, professional sports programs, sports medicine clinics, and commercial sports performance enhancement and development businesses (6).

FIGURE 12.7 Strength and conditioning coaches work with athletes of all ages. (Photo courtesy of Stockbyte/DKP/Getty Images.)

Wellness and Fitness Industry Professional

A wellness and **fitness industry** professional can expect to work with members of the general public to develop and implement physical activity and exercise programs to improve health, wellness, and fitness and reduce the risk for lifestyle-related diseases. Exercise science graduates can obtain employment in a variety of professional jobs within the wellness and fitness industry. A broad knowledge of biomechanics, fitness assessment, exercise physiology, and exercise psychology is critical for an exercise science graduate to establish and develop a professional career in the wellness and fitness industry. As more individuals use participation in physical activity and exercise as a means to stay healthy and physically fit and decrease the risk of developing lifestyle diseases, there will be an increased need for professional employees in the wellness and fitness industry (8). Table 12.4 provides examples of some employment opportunities in the wellness and fitness industry. Within each of the work settings, exercise science professionals have the opportunity to work as individual and group exercise leaders, fitness director, operations or facility manager, club manager, or general manager. Each of these positions requires a specific content knowledge base and in some instances years of experience and additional education (*e.g.*, club manager or general manager). In many instances, professional certification is expected and often required.

Fitness industry A global term used to describe components related to improving health and fitness of individuals through physical activity and exercise.

Table 12.4	Employment Opportunities in the Wellness and Fitness Industry
EMPLOYMENT OPPORTUNITY	DESCRIPTION
Club fitness programs	For-profit business operating to provide a service to members who join the club
Community programs	Operated by local communities and nonprofit organizations such as the YMCA and the YWCA
Corporate wellness programs	Operated by businesses and corporations as a means of providing employees with an opportunity to enhance health and wellness
Spa fitness programs	For-profit business operating to provide a variety of traditional and nontraditional exercise, health, and relaxation programs to guests

As a professional employed in one of the wellness and fitness industry settings, you can expect to work with individuals across the lifespan from preschool-aged children to the older adult. In many instances, the individuals will be healthy, but physically **deconditioned**. There is the possibility of undiagnosed disease, and often individuals who want to begin an exercise program have risk factors for a variety of disease conditions. That makes it very important to follow appropriate guidelines for exercise testing and exercise prescription (12). Participants in wellness and fitness programs will also have a wide range of knowledge, skills, and abilities related to physical activity and exercise. Within the wellness and fitness industry, exercise science professionals can be expected to be involved in health and fitness screening and assessment, fitness program development and implementation, and program evaluation and assessment (8).

PROFESSIONAL ORGANIZATIONS IN EXERCISE SCIENCE

The development of exercise science and the related areas of study has resulted in the establishment of numerous professional organizations that provide valuable services to professional members. Examples of these membership benefits and services include the distribution of newsletters and professional journals and providing conferences, webinars, professional development opportunities, including continuing education, and employment bulletin boards. Many of these professional organizations offer discounted membership fees and reduced conference registration fees to both undergraduate and graduate students. It is important to become actively involved in a professional organization as the benefits of membership can assist you in your professional development and future employment. The specific mission of an organization provides considerable insight into the scope of practice

Deconditioned A state of being unfit to perform physical activity or exercise.

that the members who belong to the organization participate in as professionals. This can help guide you in the selection of an appropriate professional organization. Many individuals often belong to more than one organization with a view to receiving the valuable benefits offered by each organization. Table 12.5 provides the mission statements of the primary exercise science professional organizations. The following section provides a summary of the primary professional organizations in exercise science and is not meant to be an inclusive list.

American Council on Exercise

The American Council on Exercise (ACE) is a nonprofit organization committed to enriching quality of life through safe and effective physical activity and exercise. Its purpose is to protect all segments of society against ineffective fitness products and promote programs and trends through its ongoing public education, outreach, advocacy, and research. ACE further protects the public by setting certification and continuing education standards for fitness professionals. ACE was founded in 1985 and serves to provide fitness certification, education, and training to its members. Membership benefits include career services, professional conferences, access to teleconferences, webinars, and blogs, and subscriptions to a variety of newsletters. Professional development opportunities are available through numerous continuing education programs and certifications. Additional information can be obtained by visiting the ACE home page at www.acefitness.org (1).

American Association of Cardiovascular and Pulmonary Rehabilitation

Founded in 1985, the American Association of Cardiovascular and Pulmonary Rehabilitation (AACVPR) is the premier professional organization dedicated to the development of its members who are involved in the profession of cardiovascular and pulmonary rehabilitation. The AACVPR provides educational and networking opportunities that inform individuals of advances in cardiovascular and pulmonary rehabilitation, current legislative and reimbursement initiatives, as well as member benefits to help improve the care and quality of life for patients with heart and lung diseases. Certification for professional cardiac and pulmonary rehabilitation programs is also a function provided by AACVRP. Membership benefits include career services, discounts on conference registration fees, access to teleconferences, podcasts, webcasts, and subscription to the *Journal of Cardiopulmonary Rehabilitation and Prevention*. Professional development opportunities are available through meetings and conferences. Additional information can be obtained by visiting the AACVPR home page at www.aacvpr.org (13).

American College of Sports Medicine

The American College of Sports Medicine (ACSM) was founded in 1954 for the purpose of bringing together individuals with an interest in exercise and health. The ACSM has a diverse membership of professionals from all disciplines of exercise science and sports medicine. The ACSM is the largest exercise science and sports medicine organization in the world, with more than 50,000 international, national, and regional chapter members and certified professionals. ACSM offers individual certification programs in health fitness,

Table 12.5 The Missions of Various Exercise Science Professional Organizations

PROFESSIONAL ORGANIZATION	MISSION
American Association for Cardiovascular and Pulmonary Rehabilitation (AACVPR)	Reducing morbidity, mortality, and disability from cardiovascular and pulmonary disease through education, prevention, rehabilitation, research, and disease management — central to the core mission is improvement in quality of life for patients and their families.
American Council on Exercise (ACE)	To get people moving.
American College of Sports Medicine (ACSM)	Advances and integrates scientific research to provide educational and practical applications of exercise science and sports medicine.
American Society of Biomechanics (ASB)	To foster the advancement, communication, and application of biomechanics to benefit society.
American Society of Exercise Physiologists (ASEP)	Represents and promotes the profession of exercise physiology and is committed to the professional development of exercise physiology, its advancement, and the credibility of exercise physiologists.
Association for Applied Sport Psychology (AASP)	Focused on human performance, holistic well-being and social functioning through education, research and practice, certification, and service to the profession of sport psychology.
International Society for Biomechanics in Sport (ISBS)	To provide a forum for the exchange of ideas for sports biomechanics researchers, coaches, and teachers; to bridge the gap between researchers and practitioners; to gather and disseminate information and materials on biomechanics in sports.
International Society of Biomechanics (ISB)	Promotes the study of the biomechanics of movement with a special emphasis on human beings; encouraging international contacts among scientists in this field, promoting knowledge of biomechanics on an international level, and cooperating with related organizations.
International Society of Motor Control (ISMC)	Stimulates and fosters education and open debate among scientists from all nations on basic and applied research in the area of the biological control of movement.
North American Society for the Psychology of Sport and Physical Activity (NASPSPA)	To develop and advance the scientific study of human behavior when individuals are engaged in sport and physical activity; facilitate the dissemination of information; improve the quality of research and teaching in the psychology of sport, motor development, and motor learning and control.
National Athletic Trainers' Association (NATA)	To represent, engage, and foster the continued growth and development of the athletic training profession and athletic trainers as unique health care providers.
National Strength and Conditioning Association (NSCA)	Advancing the strength and conditioning and related sport science professions around the world.
Society of Health and Physical Educators (SHAPE)	Advance professional practice and promote research related to health and physical education, physical activity, dance, and sport.

clinical, and specialty areas. Membership benefits include career services, interest group affiliations, discounts on conference registration fees, and access to the journals *Medicine and Science in Sports and Exercise, Exercise and Sport Science Reviews, The Translational Journal of ACSM, Current Sports Medicine Reports,* and *ACSM's Health & Fitness Journal.* Professional development opportunities are available through professional meetings, conferences, webinars, and podcasts. Continuing education credits and continuing medical education credits are available to members who participate in organizational activities. The ACSM also supports research activities through the ACSM Foundation and is active in advocacy for national and international issues related to exercise and health. Examples of advocacy programs include Exercise is Medicine® and the ACSM American Fitness Index™. Additional information can be obtained by visiting the ACSM home page at www.acsm.org (6).

American Physiological Society

The American Physiological Society (APS) was founded in 1887 to promote the advancement of physiology and facilitate interaction among American physiologists. The APS connects scientists and educators from around the world in a multidisciplinary community, driving collaboration and highlighting scientific breakthroughs in physiology and related disciplines. Membership benefits include career services, interest group affiliations, and access to journals and newsletters. Professional development opportunities are available through national meetings and events. The APS also provides awards and grants to professional and student members. Additional information can be obtained by visiting the APS home page at www.physiology.org (14).

American Society of Biomechanics

The American Society of Biomechanics (ASB) was founded in 1977. The ASB encourages and fosters the exchange of information and ideas among biomechanists working in various disciplines and fields of application for the improvement of health, well-being, and sport performance. Membership benefits include career services, interest group affiliations, and subscription to the *Journal of Biomechanics*. Professional development opportunities are available through national meetings and regional conferences. The ASB also provides awards and grants to professional and student members. Additional information can be obtained by visiting the ASB home page at www.asbweb.org (15).

American Society of Exercise Physiologists

The American Society of Exercise Physiologists (ASEP) was founded in 1997 for the purpose of bringing together exercise physiology professionals. The ASEP is committed to the professional development of exercise physiology, its advancement, and the enhanced credibility of exercise physiology as a health care profession. As part of its mission, ASEP offers an exercise physiology certified examination. Membership benefits include career services and subscriptions to the online resources such as the *Journal of Exercise Physiology*, the *Journal of Exercise Medicine*, and the ASEP newsletter. Professional development opportunities are available through professional meetings and conferences. Additional information can be obtained by visiting the ASEP home page at www.asep.org (16).

Association for Applied Sport Psychology

Founded in 1986, the Association for Applied Sport Psychology (AASP) promotes the ethical practice, science, and advocacy of sport and exercise psychology. AASP is an international, multidisciplinary, and professional organization and is the largest applied sport and exercise psychology organization in the world. AASP incorporates information and expertise from exercise and sport sciences and psychology into three interrelated focus areas: health and exercise psychology, performance enhancement/intervention, and social psychology. The association offers an AASP Certified Consultants program to its members. Membership benefits include career services, interest group affiliations, discounts on conference registration, webinars, special interest group affiliations, and subscriptions to the *Journal of Applied Sport Psychology* and the *Journal of Sport Psychology in Action*. Professional development opportunities are available through professional meetings and conferences, webinars, and continuing education programs. AASP also provides an online resource center for professionals and the general public. Additional information can be obtained by visiting the AASP home page at www.appliedsportpsych.org (17).

International Society of Biomechanics

The International Society of Biomechanics (ISB) was founded in 1973 to promote the study of all areas of biomechanics, although special emphasis is given to the biomechanics of human movement. The ISB promotes international contacts among scientists, promotes the dissemination of knowledge, and forms liaisons with national organizations. Membership in the ISB totals about 2000 and includes individuals from the disciplines of anatomy, physiology, engineering (mechanical, industrial, and aerospace), orthopedics, rehabilitation medicine, sport science and medicine, ergonomics, electrophysiological kinesiology, and others. The ISB promotes the organization of biennial international congresses, publication of congress proceedings and a biomechanics monograph series, distribution of a quarterly newsletter, sponsorship of scientific meetings related to biomechanics and affiliations with the *Journal of Biomechanics*, the *Journal of Applied Biomechanics, Clinical Biomechanics*, the *Journal of Electromyography and Kinesiology*, and *Gait and Posture*. Additional information can be obtained by visiting the ISB home page at www.isbweb.org (18).

International Society of Biomechanics in Sport

Established in 1982, the International Society of Biomechanics in Sport (ISBS) provides a forum for the exchange of ideas for sport biomechanics researchers, coaches, and teachers. The ISBS also works to bridge the gap between researchers and practitioners and gather and disseminate information and materials on biomechanics in sports. The ISBS offers annual symposia, a regular newsletter, and additional resources for members, including proceedings from the international conferences. Membership of the ISBS also includes a subscription to the journal *Sport Biomechanics*. Additional information can be obtained by visiting the ISBS home page at www.isbweb.org (18).

International Society of Motor Control

Founded in 2002, the International Society of Motor Control (ISMC) is dedicated to the development of its members who are interested in basic and applied research in the area of control of movements in biologic systems. The ISMC provides educational and networking opportunities that inform individuals of current activities in motor control. The ISMC membership benefits include career services and subscription to the journal *Motor Control*. Professional development opportunities are available through professional meetings and conferences, including the Progress in Motor Control conference. Additional information can be obtained by visiting the ISMC home page at www.i-s-m-c.org (19).

National Athletic Trainers' Association

The National Athletic Trainers' Association (NATA) was founded in 1950 and currently serves approximately 45,000 national and international members. The NATA develops and presents the most advanced information on injury prevention and rehabilitation. Athletic training is practiced by certified athletic trainers who are health care professionals and who collaborate with physicians to optimize the activity and participation of patients and clients. Athletic training encompasses the prevention, diagnosis, and intervention of emergency, acute, and chronic medical conditions involving impairment, functional limitations, and disabilities. The NATA offers the certification of athletic trainers through the Commission on Accreditation for Athletic Training Education. Membership benefits include discounts on conference registration fees, career services, interest group affiliations, a membership directory, and subscription to the *Journal of Athletic Training* and *NATA News*. Professional development opportunities are available through national, regional and state meetings and conferences, online CEUs, webinars, and continuing education programs. The NATA also provides grants and scholarships to student members through the NATA Research & Education Foundation. Additional information can be obtained by visiting the NATA home page at www.nata.org (20).

National Strength and Conditioning Association

The National Strength and Conditioning Association (NSCA) was founded by strength and conditioning professionals in 1978. The NSCA serves nearly 60,000 members throughout the world. It develops and makes available the most advanced information about strength training and conditioning practices, performance enhancement, injury prevention, and research findings. The NSCA is comprised of a diverse group of professionals from the sport and exercise science, athletic, allied health, and fitness industries working together to bridge the gap between the scientist in the laboratory and the practitioner in the field. The NSCA offers the Certified Strength and Conditioning Specialist, the NSCA-Certified Personal Trainer, the Certified Special Population Specialist, and the Tactical Strength and Conditioning Facilitator credentialing programs. Membership benefits include career services, interest group affiliations, access to electronic publications, webinars, and subscription to the *Journal of Strength and Conditioning Research* and the *Strength and Conditioning Journal*. Professional development opportunities are available through professional meetings and conferences and podcasts and videos and electronic downloads. The NSCA also grants student scholarships through the National Strength and Conditioning Association Foundation. Additional information can be obtained by visiting the NSCA home page at www.nsca.com (21).

North American Society for the Psychology of Sport and Physical Activity

The North American Society for the Psychology of Sport and Physical Activity (NASPSPA) was founded in 1967. It is a multidisciplinary association of scholars from the behavioral sciences and related professions that work together to develop and disseminate information about the psychology of physical activity, exercise, and sport. NASPSPA is also interested in improving the quality of research and teaching in sport psychology and motor behavior. Membership benefits include access to the resource center and a discounted subscription to the *Journal of Sport and Exercise Psychology* and *Journal of Motor Learning and Development*. Professional development opportunities are available through national and regional meetings and conferences and online resources. Additional information can be obtained by visiting the NASPSPA home page at www.naspspa.com (22).

Society for Health and Physical Educators

The Society for Health and Physical Educators (SHAPE) was originally titled the American Alliance for Health, Physical Education, Recreation and Dance (AAHPERD). SHAPE is the largest organization of professionals involved in physical education, leisure, fitness, dance, health promotion, sport and education, and all specialties related to achieving a healthy lifestyle. Its purpose is to provide members with wide-ranging and coordinated support, resources, and programs to help professionals improve their skills and abilities to help ensure that all children have the opportunity to lead healthy, physically active lives. Originally founded in 1885, the organization is an alliance of six district associations and four program councils, and serves more than 25,000 members. The program councils are School Health Council, Physical Activity Council, Physical Education Council, and Research Council. Membership benefits include discounts on conference registration fees and educational materials, access to professional development webinars, and subscription to journals such as *Research Quarterly for Exercise and Sport* and *Journal of Physical Education, Recreation, and Dance*. Professional development opportunities are available through meetings and conferences. SHAPE also provides grants and scholarships to student members. Additional information can be obtained by visiting the SHAPE home page at www.shapeamerica.org (23).

Sports, Cardiovascular, and Wellness Nutrition Dietetics Practice Group

The Sports, Cardiovascular, and Wellness Nutrition Dietetics Practice Group (SCAN) was established in 1981 and is one of the largest dietetic practice groups of the Academy of Nutrition and Dietetics. SCAN works to promote healthy active lifestyles through practice in sports, cardiovascular, and wellness nutrition and the prevention and treatment of disordered eating. RDNs may become certified as a Specialist in Sports Dietetics by the Commission on Dietetic Registration. SCAN offers professional development opportunities through professional meetings, electronic newsletters, e-library, podcasts, and conferences. Additional information can be obtained by visiting the SCAN home page at www.scandpg.org (24).

Thinking Critically

Why is it important to thoughtfully consider personal and professional goals when deciding on a career in an allied health or exercise science profession?

PROFESSIONAL ORGANIZATIONS RELATED TO EXERCISE SCIENCE

There are several other professional organizations that have an interest in exercise science–related activities. Each of these organizations has a specific mission, of which some aspect of physical activity or exercise may be a component. It is also common for exercise science professionals to be members of one or more of these professional organizations if membership provides a benefit in the professional development and employment of the individual. Table 12.6 provides the primary mission of those professional organizations with a secondary interest in exercise science.

Thinking Critically

How do professional organizations within exercise science work to fulfill their organizational mission and meet the needs of their members and the general public?

Table 12.6 Primary Mission of Professional Organizations with an Interest in Exercise Science

PROFESSIONAL ORGANIZATION	MISSION
American Cancer Society	To save lives, celebrate lives, and lead the fight for a world without cancer.
American College of Epidemiology	To serve the interests of the profession and its members by advocating for issues pertinent to epidemiology, a credential-based admission and promotion process, sponsorship of scientific meetings, publications and educational activities, and recognizing outstanding contributions to the field.
American Diabetes Association	To prevent and cure diabetes and to improve the lives of all people affected by diabetes.
Academy of Nutrition and Dietetics	To accelerate improvements in global health and well-being through food and nutrition.
American Heart Association	To be a relentless force for a world of longer, healthier lives.
American Medical Association	Dedicated to driving medicine toward a more equitable future, removing obstacles that interfere with patient care and confronting the nation's greatest public health crises.
American Nurses Association	To lead the profession to shape the future of nursing and health care.
American Physical Therapy Association	Building a community that advances the profession of physical therapy to improve the health of society.
American Physiological Society	To foster education, scientific research, and dissemination of information in the physiological sciences.
American Psychological Association	To promote the advancement, communication, and application of psychological science and knowledge to benefit society and improve lives.

U.S. GOVERNMENT AGENCIES WITH AN INTEREST IN EXERCISE SCIENCE

Participation in regular physical activity and exercise is recognized as a major factor in improving health and reducing the risk for disease. As a result, major initiatives by agencies of the federal government address various aspects of health, disease prevention and treatment, physical activity, and exercise. For example, agencies and offices of the federal government are responsible for developing and promoting the Healthy People initiatives (25), which originally began with the 1996 Surgeon General's report (26). The Health People initiatives are used to guide decision-making and policy development for enhancing the quality and years of healthy life and the elimination of health disparities. *Healthy People 2030* is the guiding document for assessing the major risks to health and wellness, changing public health priorities, and addressing the emerging issues related to our nation's health preparedness and prevention (25). There are several U.S. Government agencies that play an important role in advancing the public health agenda of our nation through the promotion of physical activity and proper nutrition.

U.S. Department of Health and Human Services

The Department of Health and Human Services (DHHS) is the federal government's principal agency for protecting the health of all Americans and providing essential human services, especially for those who are least able to help themselves. The DHHS's mission is to enhance the health and well-being of Americans by providing for effective health and human services and fostering strong, sustained advances in the sciences underlying medicine, public health, and social services. The DHHS provides administrative oversight through the Office of the Secretary of Health and Human Services. Figure 12.8 provides the current organizational chart of the U.S. DHHS (27).

Furthermore, the DHHS provides resource support for over 300 programs covering a wide range of activities (28). Table 12.7 provides some examples of activities administered by the DHHS. The DHHS programs are administered by 11 operating divisions, including 8 agencies in the U.S. Public Health Service and 3 human services agencies. In addition to the services they deliver, the DHHS programs provide for equitable treatment of beneficiaries nationwide and enable the collection of national health and disease data. Currently, the DHHS has five goals, each of which has multiple objectives that support and carry out its mission (28):

- *Goal 1* — Reform, Strengthen, and Modernize the Nation's Healthcare System
- *Goal 2* — Protect the Health of Americans Where They Live, Learn, Work, and Play
- *Goal 3* — Strengthen the Economic and Social Well-Being of Americans Across the Lifespan
- *Goal 4* — Foster Sound, Sustained Advances in the Sciences
- *Goal 5* — Promote Effective and Efficient Management and Stewardship

The DHHS had identified strategic objectives by which to accomplish each goal. Although all of the agencies of the DHHS are involved in the health care of the nation's population, some are especially important to exercise science and its related areas. For example, the

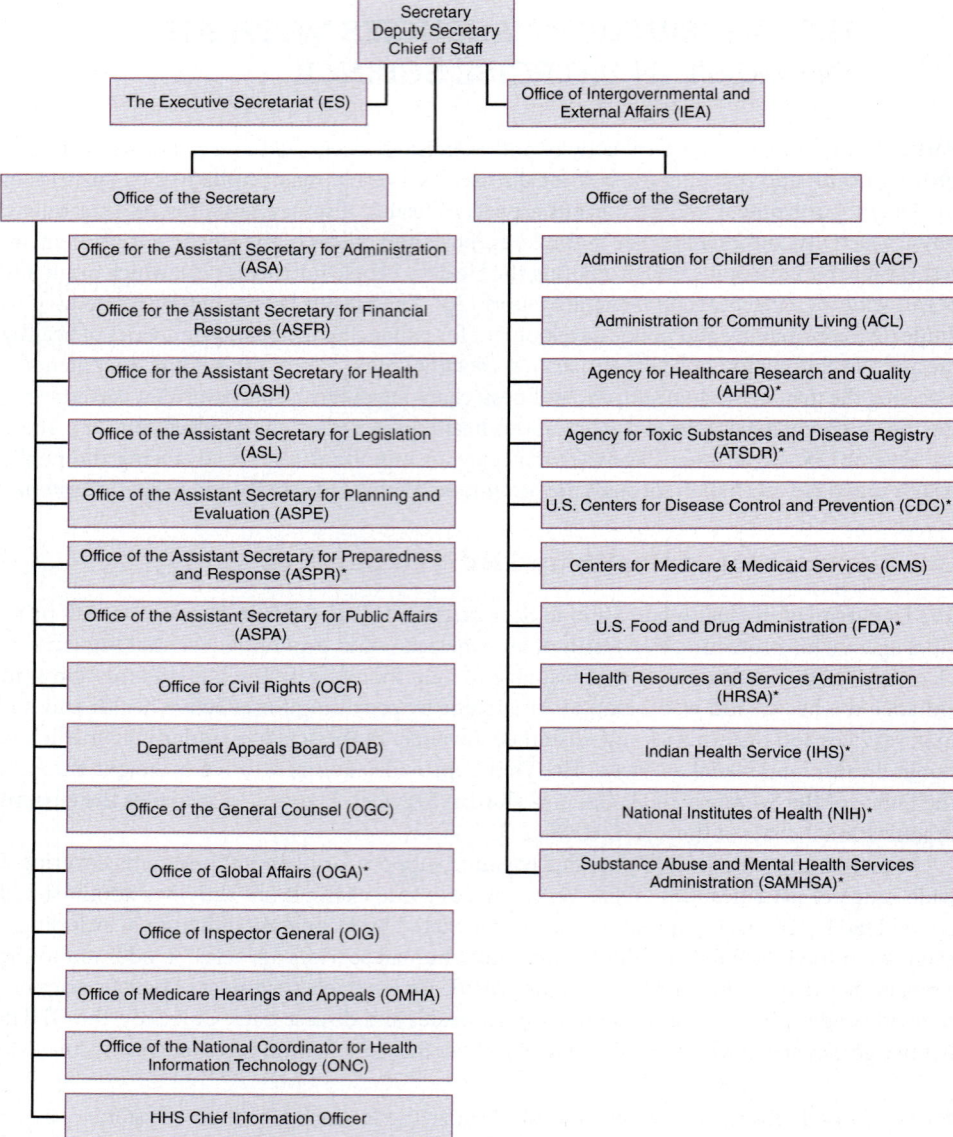

FIGURE 12.8 Organizational chart of the U.S. DHHS (29). (Available from www.hhs.gov/about/agencies/orgchart/index.html.)

Healthy People initiatives (*e.g.*, Healthy People 2010, 2020, and 2030) were established by the DHHS agency that represents the nation's disease prevention agenda. *Healthy People 2030* has five overarching goals (28):

1. Attain healthy, thriving lives and well-being free of preventable disease, disability, injury, and premature death.

2. Eliminate health disparities, achieve health equity, and attain health literacy to improve the health and well-being of all.
3. Create social, physical, and economic environments that promote attaining the full potential for health and well-being for all.
4. Promote healthy development, healthy behaviors, and well-being across all life stages.
5. Engage leadership, key constituents, and the public across multiple sectors to take action and design policies that improve the health and well-being of all.

It is the responsibility of the DHHS to work with other federal, state, and local government agencies, as well as additional public and private organizations to enhance the health and well-being of the nation's people. Additional information can be obtained by visiting the DHHS home page at www.hhs.gov (29) and the *Healthy People 2030* home page at www.health.gov/healthypeople (28).

National Institutes of Health

The National Institutes of Health (NIH) is the medical research branch of the DHHS. The NIH is the world's premier medical research organization, supporting close to 40,000 research projects nationwide in diseases including AIDS, Alzheimer's, arthritis, cancer, diabetes, and

Table 12.7 Examples of Activities Administered by the U.S. DHHS

Assuring food and drug safety
Comprehensive health services for Native Americans
Faith-based and community initiatives
Financial assistance and services for low-income families
Head start preschool education and services
Health Information technology
Improving maternal and infant health
Medical and social science research
Medical preparedness for emergencies, including potential terrorism
Medicaid — health insurance for low-income Americans
Medicare — health insurance for the elderly and disabled Americans
Preventing child abuse and domestic violence
Preventing disease, including immunization services
Services for older Americans, including home-delivered meals
Substance abuse prevention and treatment

Table 12.8	Health Institutes and Centers of the NIH
National Cancer Institute — Est. 1937
National Eye Institute — Est. 1968
National Heart, Lung, and Blood Institute — Est. 1948
National Human Genome Research Institute — Est. 1989
National Institute on Aging — Est. 1974
National Institute on Alcohol Abuse and Alcoholism — Est. 1970
National Institute of Allergy and Infectious Diseases — Est. 1948
National Institute of Arthritis and Musculoskeletal and Skin Diseases — Est. 1986
National Institute of Biomedical Imaging and Bioengineering — Est. 2000
Eunice Kennedy Shriver National Institute of Child Health and Human Development — Est. 1962
National Institute on Deafness and Other Communication Disorders — Est. 1988
National Institute of Dental and Craniofacial Research — Est. 1948
National Institute of Diabetes and Digestive and Kidney Diseases — Est. 1950
National Institute on Drug Abuse — Est. 1974
National Institute of Environmental Health Sciences — Est. 1969
National Institute of General Medical Sciences — Est. 1962
National Institute of Mental Health — Est. 1949
National Institute on Minority Health and Health Disparities — Est. 2010
National Institute of Neurological Disorders and Stroke — Est. 1950
National Institute of Nursing Research — Est. 1986
National Library of Medicine — Est. 1956
Center for Information Technology — Est. 1964
Center for Scientific Review — Est. 1946
Fogarty International Center — Est. 1968
National Center for Complementary and Integrative Health — Est. 1999
National Center for Advancing Translational Sciences — Est. 1962
NIH Clinical Center — Est. 1953

hypertension. Within the organizational structure of the NIH, there are 27 separate health institutes and centers (Table 12.8) (27). The budget of NIH, established by congress and the president of the United States, is used to support the mission of NIH, which is to promote science in pursuit of fundamental knowledge about the nature and behavior of living systems and the application of that knowledge to extend healthy life and reduce the burdens of illness and disability. The NIH uses much of its budget to fund grants and contracts supporting research and training throughout the United States and abroad. Many of the grants and contracts funded by NIH are connected to research in various areas of exercise science. Additional information can be obtained by visiting the NIH home page at www.nih.gov (29).

Centers for Disease Control and Prevention

The Centers for Disease Control and Prevention (CDC) are under the administrative oversight of the DHHS. The mission of the CDC is to promote the health and quality of life by preventing and controlling disease, injury, and disability. Figure 12.9 illustrates the factors that constitute the role of the CDC (29). Of the CDC's Coordinating Units, the National Center for Chronic Disease Prevention and Health Promotion has the closest link to exercise science. The mission of the National Center for Chronic Disease Prevention and Health Promotion is to provide national leadership in areas of health promotion and

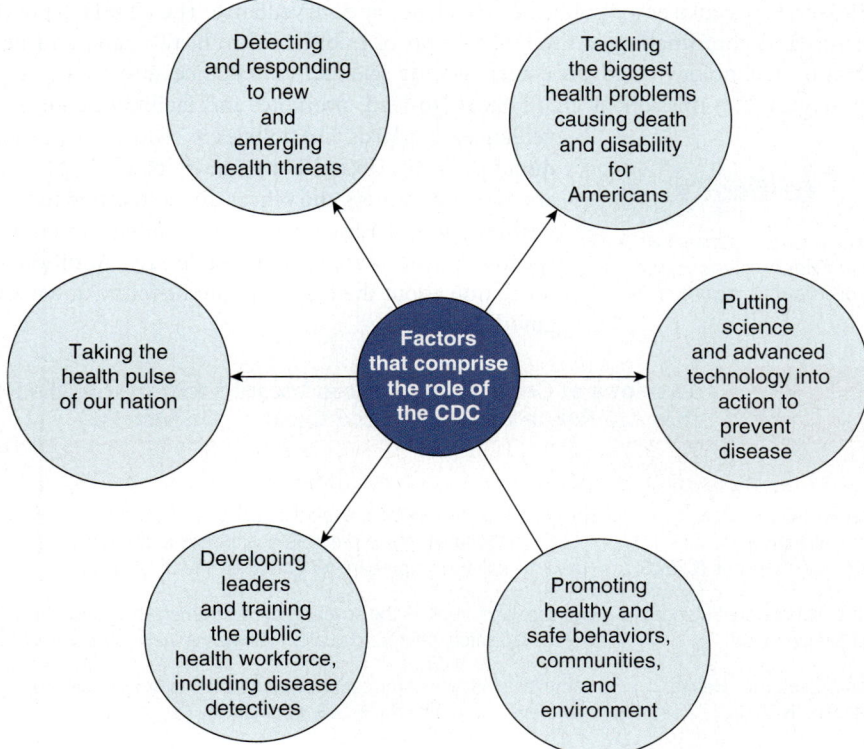

FIGURE 12.9 Mission of the Centers for Disease Control and Prevention (CDC).

chronic disease prevention by conducting public health surveillance, epidemiologic studies, and behavioral interventions; by disseminating guidelines and recommendations; and by assisting state health and education agencies to increase their capacity to prevent chronic diseases and promote healthful behaviors. These programs provide national leadership by offering guidelines and recommendations and by helping state health and education agencies promote healthy behaviors (27). There are eight divisions that oversee major programs:

- Cancer Prevention and Control
- Diabetes Translation
- Heart Disease and Stroke Prevention
- Nutrition, Physical Activity, and Obesity
- Oral Health
- Population Health
- Reproductive Health

Additional information about the Centers for Disease Control and Prevention can be found at www.cdc.gov (29).

Deputy Director of Public Health Science and Surveillance

The Deputy Director of Public Health Science and Surveillance (DDPHSS) serves as the DHHS Secretary's adviser for public health science and surveillance. The DDPHSS provides leadership and coordination at the intersection of public health, health care, and health information technology to advance agency-wide science, surveillance, and data priorities and strategies. The mission of DDPHSS is "to lead, promote, and facilitate science, surveillance, standards and policies to reduce the burden of diseases in the United States and globally" (29). There are four core centers and offices directly related to health, wellness, and exercise science. Those offices and their respective missions are listed in Table 12.9. Additional information about the DDPHSS can be found at www.cdc.gov/ddphss (30).

Thinking Critically

In what ways do governmental agencies work to accomplish their individual missions and provide for the improved public health of the nation?

Table 12.9	Missions of Centers and Offices Located within the DDPHSS that Are Related to Health and Exercise Science (30)
UNIT	MISSION
Center for Surveillance, Epidemiology, and Laboratory Services (CSELS)	Provides scientific service, expertise, skills, and tools in support of CDC's national efforts to promote health; prevent disease, injury and disability; and prepare for emerging health threats.
Office of Laboratory Science and Safety (OLSS)	Oversees and monitors the development, implementation, and evaluation of the laboratory safety and quality management programs across CDC.
National Center for Health Statistics (NCHS)	Compiles statistical information to guide actions and policies to improve the health of the U.S. population.
Office of Science (OS)	Provides CDC with scientific vision and leadership to advance the quality and integrity of CDC science.

ADDITIONAL ORGANIZATIONS AND AGENCIES IN EXERCISE SCIENCE

Numerous other organizations and agencies are involved in a variety of activities that are related to the various areas of study in exercise science. Some of these organizations contribute to enhancing the educational development of exercise science professionals and the general public, whereas other organizations attempt to provide guidance and support for addressing public health issues. Table 12.10 includes some of those organizations that are involved in exercise science–related activities.

Table 12.10 Additional Organizations that Are Involved in Various Aspects of Exercise Science and Related Areas

ORGANIZATION	RELATIONSHIP TO EXERCISE SCIENCE
World Health Organization	Supports projects, initiatives, activities, information products, and contacts in the following areas: health systems; promoting health through the life-course; noncommunicable diseases; communicable diseases; corporate services; preparedness, surveillance, and response.
Board of Certification in Professional Ergonomics	The certifying body for individuals whose education and experience indicate broad expertise in the practice of human factors/ergonomics. Provides ergonomics certification to protect the public, the profession, and its professionals by assuring standards of competency and advocating the value of certification.
Aerobics and Fitness Association of America	World's largest fitness educator delivering comprehensive cognitive and practical education for fitness professionals; grounded in industry research, using both traditional and innovative modalities.
The Cooper Institute	Dedicated to scientific research in the field of preventive medicine and public health and communicating the results of research to the scientific and medical communities as well as to the general public.

Interview

Trent A. Hargens, PhD, FACSM, CEP, EIM3

Associate Professor in the Department of Kinesiology, James Madison University, Harrisonburg, Virginia

Brief Introduction – I grew up in Iowa and received a Bachelor of Science degree in Health Promotion from the University of Iowa. I have a Master of Science degree in Clinical Exercise Physiology from Ball State University and a Doctor of Philosophy in Clinical Exercise Physiology from Virginia Polytechnic Institute and State University. Between my MS and PhD degrees, I worked for various hospital systems as a practicing clinical exercise physiologist in cardiac rehabilitation, pulmonary rehabilitation, and peripheral arterial disease rehabilitation. I also taught fitness classes for older populations and arthritis-based fitness classes. Currently, my research focuses on the cardiovascular, metabolic, and autonomic consequences of chronic disease, with a special emphasis on obstructive sleep apnea. Additionally, I am examining how poor sleep quality impacts physical activity and aerobic exercise performance.

Q: *What are two or three of your most significant career experiences?*

When I was a practicing CEP, one of my pulmonary rehabilitation patients was someone on the transplant list for two new lungs. My main goal as his CEP was to help keep him healthy enough to see the day that his name got to the top of the transplant list. During my time working with him, I watched as he went from no supplemental oxygen, to oxygen with exercise, to oxygen all the time. However, this person worked hard enough and stayed healthy enough that he received his new lungs. I also was privileged to witness his amazing transformation after that surgery, to see him with so much joy and energy without the need for any oxygen. It really drove home to me in a personal way how exercise can have a profound impact on people's lives.

A second experience that comes to mind was as a professor of Exercise Science. One of our students was nearing graduation and wasn't exactly sure what her next step was going to be. Through several candid, one-on-one chats, I was able to help her figure out what aspect of Exercise Science drove her passion. This former student has gone on for two additional degrees in the field and has literally thanked me for "inspiring her." It is those kinds of interactions with students that give us professors the inspiration to get out of bed each day. I will forever be grateful to her.

Q: *Why did you choose to become an "exercise scientist"?*

For me it was how exercise and physical activity can impact individuals with chronic disease; hence, my gravitation toward degrees in clinical exercise physiology. My grandfather had significant cardiovascular disease as well as diabetes, and I just happen to look exactly like him when he was younger. For me, it was learning about how exercise can impact me as a person to avoid the health issues of my grandfather, while also being able to help others through my work as a practicing clinical exercise physiologist, and now as a researcher.

Q: *Why is involvement in professional organizations important for career development?*

Between my MS and PhD degree work, I let my involvement in professional organizations fade. Looking back now, it was a very isolating time professionally and was likely a part of the reason I went on to obtain my PhD. I was enjoying the work that I was doing and was enjoying my various coworkers of differing backgrounds (nurses, dieticians, psychologists, respiratory therapists, etc.), but I was getting stagnant in my growth as a clinical exercise physiologist. Professional organizations allow you to network with other people of similar professional backgrounds. You get exposed to new and exciting research, keep up with new trends in your particular interest area, obtain needed continuing education, and you can also help shape the future of your profession as you become more involved in these organizations.

Q: *What advice would you have for a student exploring a career in any exercise science profession?*

There is still so much to be done in this field, from a practitioner standpoint or from a researcher standpoint. This field is growing in so many ways, and it's growing beyond the traditional job descriptions that people may think of (*e.g.*, personal trainer, cardiac rehabilitation). Explore those possibilities and find an area or a population of clientele that gives you passion, whether it's working with them directly or researching that population. We need more diversity in our field across all demographics, whether it's the exercise scientists themselves or the populations being served and/or researched by those exercise scientists. The future is bright for the field of exercise science!

Interview

Katie Hake, RDN, LD, ACSM CPT
Owner, Katie Hake Health & Fitness, LLC

Brief Introduction – I received a BS from Purdue University in Dietetics and Nutrition, Fitness, and Health. I am a Certified Intuitive Eating Counselor. I have worked over a decade in the fitness industry, in a variety of settings from clinical to commercial and boutique fitness to campus recreation and industry. I have worked one-on-one as a personal trainer, as well as in groups as a fitness instructor. I spent several years as a master trainer traveling the country to certify and mentor new and seasoned fitness instructors in predesigned workouts. While building my private practice, I worked as a dietitian in bariatrics, general surgery, and in a medical genetics department as a metabolic dietitian. In my private practice, I provide one-on-one nutrition and fitness counseling to women who struggle with diets and negative body image.

Q: What are two or three of your most significant career experiences?

I will never forget presenting on stage at the FIBO Global Fitness conference in Miami, Florida. I also presented at SCW Midwest Mania in Chicago, IL. Whether it's fitness or nutrition, I love to teach and empower others with the knowledge to change their life. Seeing my two passions for these different fields has been so rewarding when the client or fitness professional whom I am teaching has that lightbulb moment!

Q: Why did you choose to become an "exercise scientist"?

I chose to become an exercise scientist because I am fascinated by the human body — both the physical and mental aspects of it. I love being able to help clients move better, and feel better, because it can truly have a life-changing impact. I also love working on an industry level to collaborate and help our profession continue to evolve and improve.

Q: Why is involvement in professional organizations important for career development?

Professional organizations allow you to network and connect with other leaders. As the saying goes, if you are the smartest in the classroom, you are in the wrong classroom! Surrounding yourself with others who help you to level up as a professional is crucial to this field. Through ACSM professional organizations, I have been able to learn about unique job opportunities that I did not even know existed. By connecting with other professionals in these organizations I have been invited to work on special projects such as photoshoots, exam writing, and app testing developments. Even though I have earned my credentials, I always aim to stay up to date by staying involved with these organizations, by taking advantage of webinars and CEU opportunities, mentorship opportunities, and more.

Q: What advice would you have for a student exploring a career in any exercise science profession?

Talk to those in the profession, learn their experiences, hear their stories, and ask questions. You never know what doors and opportunities are out there unless you go seek them.

SUMMARY

- Credentialing is an important aspect of professional development that demonstrates individual competency in knowledge, skills, and abilities.
- Professional careers in exercise science and allied health require a core base of knowledge and additional preparation in specialty areas.
- Students should consider key interest areas when deciding on a professional career.
- Professional organizations supporting exercise science–related fields strive to enhance the development of their profession, their members, and their mission by providing member benefits and services.
- Departments and agencies within the Federal Government and other professional organizations work to enhance the health and wellness of Americans through policy initiatives and program development.

FOR REVIEW

1. Describe the differences among the following credentials:
   a. certification
   b. licensure
   c. registration
2. What are the primary benefits of certification?
3. What are the principal certifications offered by the primary exercise science professional organizations?
4. What are the primary employment opportunities in the fitness industry?
5. What is the difference between an occupational therapist and a physical therapist?
6. In what ways do a physician and a physician's assistant work together to promote health and reduce disease risk?
7. Why are professional certifications important for athletic trainers and clinical exercise physiologists?
8. Describe in your own words the mission of the following professional organizations:
   a. American College of Sports Medicine
   b. National Athletic Trainers Association
   c. American Alliance for Health, Physical Education, Recreation and Dance
   d. American Association of Cardiovascular and Pulmonary Rehabilitation
   e. North American Society for the Psychology of Sport and Physical Activity
   f. American Society of Exercise Physiologists
   g. International Society for Motor Control
   h. International Society of Biomechanics
9. What branches of the federal government are charged with enhancing the health and reducing the disease risk of the American public? What does each specific branch do to enhance the health of Americans?

Project-Based Learning

1. As an exercise science professional, your employer has provided you with $1000 to spend on professional membership for the purpose of keeping you informed of the most recent developments in your field. What professional organizations would you join, and what benefits would you receive that would give you the most information for developing your professional career?
2. You are charged by your employer with developing a comprehensive wellness and fitness program for the employees within the company. What resources would you use to develop the recommendations for the program, making sure to identify the health and physical activity and exercise guidelines for both healthy and high disease risk individuals in the program?

REFERENCES

1. Kreider RB, Cahill KM. Exercise science and fitness certifications. In: Brown SP, editor. *Introduction to Exercise Science*. 1st ed. Philadelphia (PA): Lippincott Williams & Wilkins; 2001. p. 67–81.
2. U.S. Department of Labor, Bureau of Labor Statistics Web site [Internet; cited 2020]. Available from: www.bls.gov.
3. Commission on Accreditation of Allied Health Education Programs Web site [Internet; cited 2020]. Available from: www.caahep.org.
4. Thompson WR, Brown SP. Professional issues. In: Brown SP, editor. *Introduction to Exercise Science*. 1st ed. Philadelphia (PA): Lippincott Williams & Wilkins; 2001. p. 117–29.
5. Academy of Nutrition and Dietetics Web site [Internet; cited 2021]. Available from: www.eatright.org.
6. American College of Sports Medicine Web site [Internet; cited 2021]. Available from: www.acsm.org.
7. U.S. Department of Labor, Bureau of Labor Statistics Web site [Internet; cited 2021]. Available from: www.bls.gov.
8. Kravitz L. Job activities and employment. In: Brown SP, editor. *Introduction to Exercise Science*. 1st ed. Philadelphia (PA): Lippincott Williams & Wilkins; 2001. p. 82–96.
9. Board of Certification in Professional Ergonomics 2021. Available from: www.bcpe.org.
10. The American Occupational Therapy Association Web site [Internet; cited 2021]. Available from: www.aota.org.
11. National Council for Therapeutic Recreation Certification Web site [Internet; cited 2021]. Available from: www.nctrc.org.
12. *ACSM's Guidelines for Exercise Testing and Prescription*. 11th ed. Philadelphia (PA): Lippincott Williams & Wilkins; 2021.
13. American Association of Cardiovascular and Pulmonary Rehabilitation Web site [Internet; cited 2021]. Available from: www.aacvpr.org.
14. American Physiological Society Web site [Internet; cited 2021]. Available from: www.physiology.org.
15. American Society of Biomechanics Web site [Internet; cited 2021]. Available from: www.asbweb.org.
16. American Society of Exercise Physiologists Web site [Internet; cited 2021]. Available from: www.asep.org.
17. Association for Applied Sport Psychology Web site [Internet; cited 2021]. Available from: www.appliedsportpsych.org.
18. International Society of Biomechanics in Sport Web site [Internet; cited 2021]. Available from: www.isbweb.org.
19. International Society of Motor Control Web site [Internet; cited 2021]. Available from: www.i-s-m-c.org.
20. National Athletic Trainers Association Web site [Internet; cited 2016]. Available from: www.nata.org.
21. National Strength and Conditioning Association Web site [Internet; cited 2021]. Available from: www.nsca.com.

22. North American Society for the Psychology of Sport and Physical Activity Web site [Internet; cited 2021]. Available from: www.naspspa.com.
23. Society for Health and Physical Educators Web site [Internet; cited 2021]. Available from: www.shapeamerica.org.
24. Sports, Cardiovascular, and Wellness Nutrition Dietetics Practice Group Web site [Internet; cited 2021]. Available from: www.scandpg.org.
25. Healthy People 2030 Web site [Internet; cited 2021]. Available from: https://health.gov/healthypeople.
26. U.S. Department of Health and Human Services, Centers for Disease Control and Prevention, National Center for Cardiovascular Disease Prevention and Health Promotion, The President's Council of Physical Fitness and Sports. *Physical Activity and Health: A Report of the Surgeon General.* 1995.
27. U.S. Department of Health and Human Services. HHS Organizational Chart Web site [Internet; cited 2021]. Available from: https://www.hhs.gov/about/agencies/orgchart/index.html.
28. U.S. Department of Health and Human Services Web site [Internet; cited 2008]. Available from: www.hhs.gov.
29. U.S. Department of Health and Human Services. National Institutes of Health Web site [Internet; cited 2021]. Available from: www.hhs.gov.
30. U.S. Department of Health and Human Services. Deputy Director for Public Health Science and Surveillance Web site [Internet; cited 2021]. Available from: https://www.cdc.gov/ddphss/.

CHAPTER 13

Exercise Science in the Twenty-First Century

After completing this chapter, you will be able to:

1. Describe how past research by exercise science professionals has influenced future trends in health promotion and disease prevention.

2. Identify major public and private initiatives for enhancing health promotion and disease prevention that include areas of study from exercise science.

3. Describe the impact of exercise science on sport and athletic performance enhancement.

4. Identify some future trends for exercise science in the area of sport and athletic performance enhancement.

Predicting what will happen in the future is a very difficult process. This is especially true for the areas of study that comprise exercise science where numerous factors can have a significant influence on the direction any particular profession or related allied health care field may move in. One only needs to examine the historical developments of exercise science to construct a vision of how the profession has been influenced by individuals, discovery, societal change, and needs for the future. Most recently, the COVID-19 pandemic serves as an all too real example of how an uncontrolled infectious disease can significantly alter the way we participate in physical activity, exercise, sport, and athletic competitions. The pandemic altered our physical activity and exercise habits, forcing the closure of gyms and fitness centers, expanding online exercise platforms, increasing sales of exercise equipment, and enhancing the participation of outdoor activities and exercise. Furthermore, the COVID-19 pandemic resulted in the cancellation of millions of sport and athletic events, changed the way athletes are recruited into college and university athletic programs, and altered how individuals socially engage at athletic venues. Our experiences with the highly infectious virus resulted in new protocols for the way we work with clients and treat patients, perform exercise, conduct training programs, and care for athletes.

Although one could probably construct a reasonably accurate short-term vision of the future for each area of study or discipline within exercise science, it is probably more realistic to take a longer view of the direction that exercise science and the related areas of study and professions will take. It is however, fairly safe to assume that many of the challenges facing exercise science and allied health care professionals and the opportunities derived from those challenges will be in the areas of health promotion, chronic disease prevention and disease risk reduction, and rehabilitation from injury or illness. For a variety of reasons, each of these has become a major area of focus for state and federal governments, politicians, nonprofit organizations, for-profit business, scholars, and health care providers and professionals. In addition, because many individuals of all ages participate in sport and athletic competition, exercise science professionals will continue to play important roles in the development of new training and nutritional strategies for enhancing performance; enhanced assessment methods for talent identification; better equipment for greater safety protection and improved performance; and more effective diagnostic, surgical, and rehabilitative programs for injured athletes.

EXERCISE SCIENCE AND HEALTH

Over the last 60 years, a considerable amount of research has strongly supported the role that physical activity, exercise, and physical fitness can play in reducing the risk of morbidity and early mortality from many lifestyle-related diseases. Much support for this view comes from research derived from **prospective studies**, **epidemiologic studies**,

Prospective studies Research investigations conducted to watch for outcomes, such as the development of a disease, during a period of time and relate this to other factors such as suspected risk or protection factor(s).

Epidemiologic studies Research investigations conducted to identify factors that affect the health and disease of populations.

Table 13.1 Risk Factors for CVD

Risk factors that cannot be changed	• Increasing age • Male sex • Family history of CVD • Race
Risk factors that can be changed or modified	• Smoking • High blood cholesterol • Uncontrolled high blood pressure • Physical inactivity • Obesity and overweight • Uncontrolled diabetes mellitus • High C-reactive protein • Uncontrolled stress or anger

clinical trials, and **clinical studies**. For example, extensive evidence about the importance of physical activity and exercise in decreasing the risk of cardiovascular disease (CVD) and the identification of factors that increase the risk of heart disease comes from data collected in clinical studies such as the Framingham Heart Study (1), the Harvard alumni study (2,3), Aerobics Center Longitudinal Study (2), and other research investigations. CVD has been the leading cause of death and serious illness in the United States for the past 70 years, and much of the research in this disease area has been directed toward gaining a greater understanding of the factors that contribute to the development of CVD and what lifestyle changes, nutritional modifications, and behavioral interventions can reduce the risk of disease development. Table 13.1 illustrates many of those risk factors for CVD that cannot be altered and those that can be affected by intervention strategies and behavior change (4).

One very important CVD risk factor that has been studied extensively by exercise science professionals is physical inactivity. As a result of much research, physical inactivity is now considered a primary risk factor for CVD, similar to smoking, high blood pressure, and high serum cholesterol levels (5–8). Physical inactivity also increases the risk of other disease conditions (9), including Parkinson disease (10), autoimmune rheumatic diseases (11), mental health (12), diabetes mellitus (13), obesity (14), and certain forms of cancer (15,16). Figure 13.1 shows the prevalence of self-reported physical inactivity among U.S. adults by state and territory for the years 2015–18 (from the Centers for Disease Control and Prevention). Conversely, higher levels of physical activity and regular exercise are beneficial for helping to maintain a healthy body weight, improving cardiovascular fitness, and strengthening muscles, bones, and joints (17–19). Physical activity and exercise can also enhance mental health (20), decrease the risk of cancer (21), and reduce the risk of falls in older adults (17,22).

Clinical trials Comparison test of a medical intervention versus a control condition, placebo condition, or the standard medical treatment for a patient's condition.

Clinical studies A research study involving human subjects (called volunteers or participants) that is meant to add to medical knowledge.

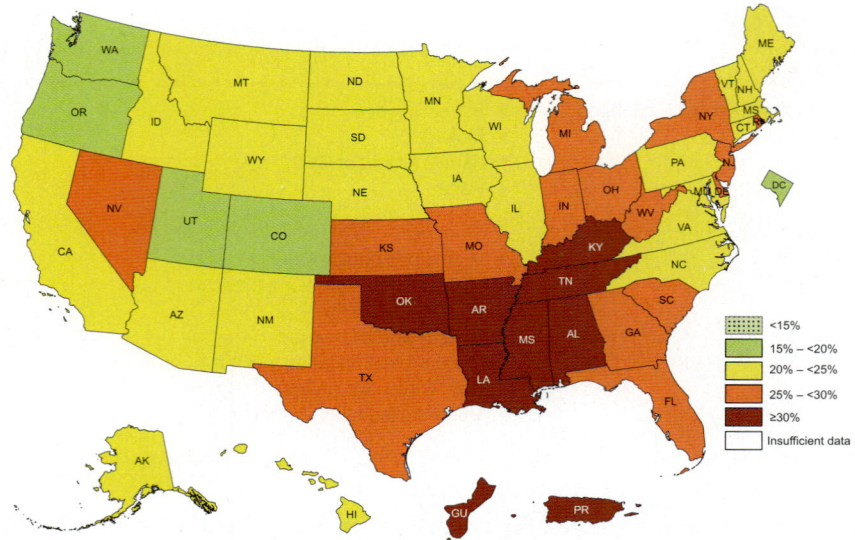

FIGURE 13.1 Prevalence of self-reported physical inactivity among U.S. adults by state and territory, BRFSS, 2015–2018. (From Centers for Disease Control and Prevention.)

Much of what we know about physical inactivity and increased disease risk and the role of physical activity and exercise in minimizing disease risk comes from epidemiologic studies. Information derived from these longitudinal research studies is used to better understand the disease risk from engaging in certain behaviors and also to guide future **health promotion** and disease prevention strategies by public and private organizations. An example of a health promotion strategy is the MyActivity Pyramid shown in Figure 13.2. Although not a foolproof method for predicting the future of exercise science and the allied health professions, an examination of some of these studies can give an indication as to the continued future direction of exercise science in helping to better understand the role of physical activity and exercise in promoting good health and reducing the risk of disease among individuals of all ages.

Why is research an integral component of the future of those areas that comprise exercise science?

 EPIDEMIOLOGY AND HEALTH PROMOTION

Epidemiology is the study of the causes, distribution, and control of disease in specifically defined populations. Epidemiology serves as the foundation for interventions made for the improvement of public health and the promotion of better health and medical care. It is a fundamental methodology of public health research and is highly regarded in

Health promotion The process of enabling people to increase their control over and improvements in their health.

Epidemiology The study of factors affecting the health and illness of populations.

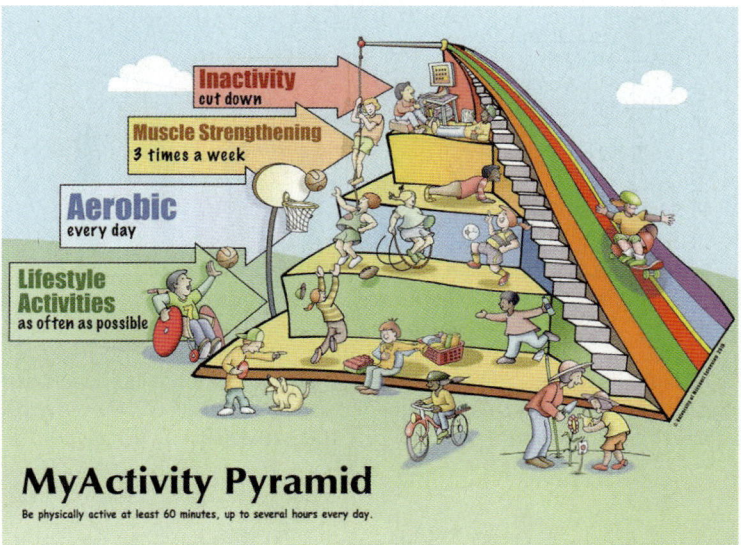

FIGURE 13.2 MyActivity Pyramid for children is an example of a health promotion strategy. (From University of Missouri Extension Office.)

evidence-based medicine for identifying risk factors for disease and determining optimal intervention and treatment approaches for use in clinical practice. Epidemiologic research is essential for examining the efficacy of treatments and preventive measures, and for elucidating their relationships among the various factors that affect the disease condition. Information about the frequency and prevalence of disease and health threats is essential for the development of effective prevention programs and treatment interventions for disease and for illuminating the environmental, behavioral, and biologic factors associated with good and poor health conditions. Information derived from epidemiologic research is used to inform and advance future research, develop public policy, and support initiatives designed to improve the health of large groups of people. Although countless epidemiologic investigations have been performed on all types of health issues and problems, several have provided significant information about public health that is relevant to exercise science and allied health professionals. Examples of some of the more noteworthy epidemiologic studies include the Framingham Heart Study, the Heritage Family Study, and the National Health and Nutrition Examination Survey (NHANES). It is the use of information from these types of studies that helps identify the future direction of health care and, indirectly, the future of many of those areas comprising exercise science and the allied health professions.

Framingham Heart Study

CVD has been the leading cause of death and serious illness in the United States since the early 20th century (23). In 1948, the Framingham Heart Study, under the direction of the National Heart Institute (now known as the National Heart, Lung, and Blood Institute),

Evidence-based medicine A process whereby the quality of evidence relative to the risks and benefits of treatment is determined.

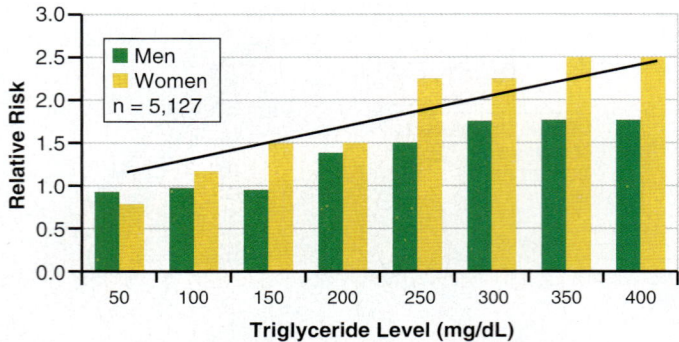

FIGURE 13.3 Relative risk of heart disease by triglyceride levels in those individuals participating in the Framingham Heart Study. (Adapted with permission from Castelli WP. Epidemiology of triglycerides: a view from Framingham. *Am J Cardiol*. 1992;70:3H–9H. Copyright 1992 Excerpta Medica.)

embarked on an ambitious project in cardiovascular health research. When the study began, little was known about the general causes of heart disease and stroke, but the death rates for CVD had been increasing steadily since the beginning of the 20th century and had become an epidemic in the United States. The primary objective of the Framingham Heart Study was to identify the common factors or characteristics that contribute to CVD by following its development over a long time in a large group of participants who had not yet developed overt symptoms of CVD or suffered a heart attack or stroke. The investigators recruited 5,209 men and women between 30 and 62 years of age from the town of Framingham, Massachusetts, and began the first round of extensive physical examinations and lifestyle interviews that would later be analyzed for common patterns related to CVD development (24). For additional detailed information about the Framingham Heart Study, please see the Web page at www.framinghamheartstudy.org.

Since its inception, the Framingham Heart Study has produced many major discoveries that have helped scientists, physicians, and allied health care professionals understand the development and progression of CVD and its risk factors (24). As an example, Figure 13.3 illustrates the extremely important relationship between blood triglyceride levels and heart disease. As can be observed in the figure, an increase in the triglyceride level also coincides with an increase in the relative risk of heart disease. Exercise science and health care professionals used this information to formulate studies on how exercise and nutritional interventions may be used to lower triglycerides, thus linking the findings of an epidemiologic study to the development of effective strategies to lower disease risk. These effective strategies include nutritional (25), exercise (26), and pharmacologic interventions (27). Some of the important milestones from the Framingham Heart Study are included in Table 13.2. More recently, the study of the role of genetic factors on the risk of CVD has expanded the understanding of how familial influences modify the risk of someone having an adverse cardiovascular event (28). The information and knowledge gained have helped generate new research questions for additional study; promoted drug development for the treatment of CVD; and facilitated behavioral, nutritional, and physical activity interventions for the reduction of disease risk and the promotion of recovery from CVD.

Table 13.2 Milestones in the Framingham Heart Study

YEAR	SIGNIFICANT MILESTONE
1948	Start of the Framingham Heart Study.
1960	Cigarette smoking found to increase the risk of heart disease.
1961	Cholesterol level, blood pressure, and electrocardiogram abnormalities found to increase the risk of heart disease.
1967	Physical activity found to reduce the risk of heart disease, and obesity found to increase the risk of heart disease.
1970	High blood pressure found to increase the risk of stroke.
1971	Framingham Offspring Study initiated to assess the familial and genetic factors as determinants of coronary heart disease.
1974	Overview of diabetes, its complications, and association with the development of CVD described.
1976	Menopause found to increase the risk of heart disease.
1978	Psychosocial factors found to affect heart disease.
1981	Major report issued on the relationship of diet and heart disease.
1987	High blood cholesterol levels found to correlate directly with the risk of death in young men; fibrinogen found to increase the risk of heart disease; estrogen replacement therapy found to reduce the risk of hip fractures in postmenopausal women.
1988	High levels of HDL cholesterol found to reduce risk of death; association of type "A" behavior with heart disease reported; isolated systolic hypertension found to increase the risk of heart disease; cigarette smoking found to increase risk of stroke.
1990	Homocysteine (an amino acid) found to be a possible risk factor for heart disease.
1994	Lipoprotein (a) found as a possible risk factor for heart disease; apolipoprotein E found to be possible risk factor for heart disease.
1995	OMNI Study of Minorities starts.
1997	Cumulative effects of smoking and high cholesterol on the risk for atherosclerosis reported.
1998	Work identifying a gene (angiotensin-converting enzyme deletion/insertion polymorphism) associated with hypertension in Framingham men.
2002	Excess body weight/obesity linked with an increased risk of heart failure; BMI (body mass index) is an independent risk factor for heart disease.
2002	Third Generation Study enrolls 3,900 grandchildren of the Framingham Heart Study's original enrollees; key goals are to identify new risk factors for heart, lung, and blood diseases; identify genes that contribute to good health and to the development of heart, lung, and blood disease; and to develop new imaging tests that can detect very early stages of coronary atherosclerosis in otherwise healthy adults.
2003	Likelihood of heart attack three times greater in individuals with common genetic variation in an estrogen receptor.
2004	Having a parent with a CVD history doubles the personal risk of the disease.
2005	An increase of up to 45% for risk of heart attack, stroke, or arterial disease may occur in middle-aged people with a sibling who suffered a similar cardiovascular event.
2010	Increased fat around the abdomen is associated with smaller, older brains in middle-aged adults.

(continued)

Table 13.2	Milestones in the Framingham Heart Study (Continued)
YEAR	SIGNIFICANT MILESTONE
2010	First definitive evidence that occurrence of stroke by age 65 years in a parent increased risk of stroke in offspring by threefold.
2015	Inclusion in a National Heart Lung and Blood Institute project to analyze the complete DNA sequences of Framingham Heart Study participants.
2018	Former smokers who quit smoking 25 or more years earlier still have three times higher risk of developing lung cancer compared to people who have never smoked.

From (24). Framingham Heart Study, a study of Boston University and the National Heart Lung and Blood Institute, funded under contract 75N92019D00031.

HERITAGE Family Study

The HERITAGE Family Study began in 1992 under the coordinated efforts of Claude Bouchard and other well-known scholars such as James K. Skinner, Arthur S. Leon, Jack H. Wilmore, and D. C. Rao. The justification for conducting the study included the following components:

- Regular aerobic exercise has favorable effects on the risk profile for CVD and Type 2 diabetes.
- There are considerable individual differences in the response to regular exercise.
- Genes are thought to play an important role in determining the general benefits derived from participating in regular physical activity. Thus, studying the influence of genetics on responses to exercise will increase our understanding of the relationship between exercise and health.

The primary objectives of the HERITAGE Family Study were to examine the role of the human **genotype** in the cardiovascular and metabolic responses to aerobic exercise training and the changes brought about by regular exercise for several CVD and diabetes risk factors (29). There have been three phases to the study. Phase I involved the testing and exercise training of 742 subjects. Responses of cardiovascular and metabolic variables to submaximal and maximal exercise were made prior to and following exercise training. Phase 2 included an investigation of **genetic epidemiology** issues pertaining to exercise **phenotypes** and CVD and Type 2 diabetes. Phase 3 included the expansion and further refinement of the search for genes and mutations affecting cardiorespiratory endurance and CVD and Type 2 diabetes risk factors as well as their response to regular exercise (24). For further information about The HERITAGE Family Study, please see the Web page at http://www.pbrc.edu/heritage.

To date, the results from the study have yielded some significant findings about the relationships between genetic factors, cardiorespiratory endurance, risk factors for CVD, and Type 2 diabetes. It has been determined that genetic factors explain about 40% of the individual variations in cardiorespiratory fitness (30). There was, however, large individual variation in the response to exercise training (often called responder and nonresponder). For example,

Genotype Genetic makeup of an organism is distinguished from its physical characteristics.
Genetic epidemiology The study of the role of genetic factors in determining health and disease in families and in populations.
Phenotype The physical appearance of an organism as distinguished from its genetic makeup and the interaction with the environment.

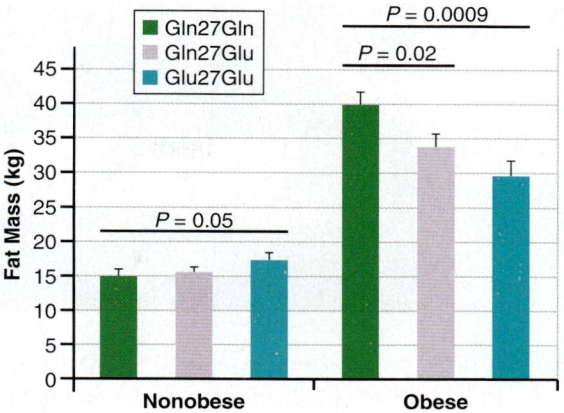

FIGURE 13.4 Association between the gene polymorphism and the total fat mass in nonobese and obese white men in the HERITAGE Family Study. (With permission from Garenc C, Pe'russe L, Chagnon YC, et al. Effects of 2-adrenergic receptor gene variants on adiposity: The HERITAGE family study. *Obes Res*. 2003;11:612–8.)

there was about 2.5 times more variance between families than within families for the improvement observed in cardiorespiratory fitness (30). Furthermore, it appears that one set of genes influence the initial level of cardiorespiratory fitness and another set of genes influence the response to exercise training (31–33). Also, it appears that familial/genetic factors are more important in determining the amount and distribution of subcutaneous body fat than in the response to exercise training (34). For example, Figure 13.4 shows the association between the gene polymorphism and the total fat mass in nonobese and obese white men in the HERITAGE Family Study. The information gained from this study has led to countless other investigations of the role of genetic profiles in health promotion and disease risk reduction. Unfortunately, the relatively easy assessment of genetic testing has led to a proliferation of direct to consumer testing that often reports false positives for various health conditions (35) and fails to consider all the critical implications for individuals having this information (36).

National Health and Nutrition Examination Survey

The goal of NHANES is to assess the health and nutritional status of adults and children in the United States. To achieve this goal, a complex series of statistical techniques are used to obtain a nationwide sample that is representative of the total population. The entire country is divided into geographic areas, called primary sampling units, which are then combined to form strata, and each stratum is then divided into a series of neighborhoods. From these neighborhoods, households are chosen at random, and inhabitants of the households are interviewed to determine eligibility for participation in the survey. Theoretically, each selected survey participant represents approximately 50,000 other U.S. residents (37).

Once a household has been identified through the sampling procedure, an interviewer conducts an initial in-house interview to determine study eligibility. Eligible participants are scheduled for an in-person appointment at a mobile examination center. Figure 13.5 shows a mobile examination center used for data collection in the NHANES. The mobile examination center consists of large trailers connected together that contain all of the equipment and personnel necessary to perform the following evaluations: (a) a physical examination; (b) a dental examination; (c) blood and urine specimen collection; and (d) personal interviews to collect information about nutrition, alcohol and tobacco use, sexual experience, mental illness, and assessments of cognitive development and learning achievement (37). For further information about the NHANES, please see the Web page at http://www.cdc.gov/nchs/nhanes.htm.

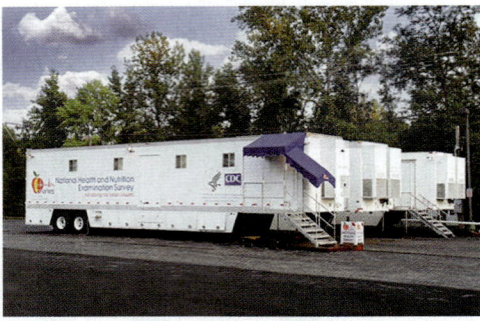

FIGURE 13.5 A mobile examination center used in the NHANES research activities. (Photo courtesy of NHANES.)

Since its inception, NHANES has produced an abundance of valuable information that has helped improve the overall health and health care of the U.S. population. Results of NHANES benefit the people of the United States in many ways. Facts about the distribution of health problems and risk factors in the population give researchers important clues to the causes of disease. Information collected from the current survey is compared with information from previous surveys. This allows public health professionals to detect the extent to which various health problems and risk factors have changed in the U.S. population over time. By identifying the important health care needs of the population, government agencies and private organizations can establish policies and plan research, education, and health promotion programs that help improve the present health status and will help prevent future health problems (37). Examples of some of the important accomplishments derived from NHANES are as follows:

- Past surveys have provided data to create the growth charts used nationally by pediatricians to evaluate children's growth and development.
- Blood data were instrumental in developing policy to eliminate lead from gasoline and food containers and soft drink cans.
- Overweight and obesity prevalence figures have led to the proliferation of programs emphasizing healthy weight management, stimulated additional research, and provided a means to track trends in overweight and obesity.
- Data have continued to indicate that undiagnosed diabetes is a significant health problem in the United States.

The continuation of the NHANES contributes to our knowledge about health and health care and provides for new initiatives, including (37):

- Determining whether there is a need to change vitamin and mineral fortification regulations for the nation's food supply
- Using hypertension and cholesterol data from the NHANES to direct education and prevention programs toward those individuals at risk and to measure success in curtailing risk factors associated with heart disease
- Identifying how use of e-cigarettes will affect lung function in an effort to further understand respiratory disease in the United States
- Understanding how food insecurity influences eating behaviors and health
- Examining how weight loss attempts influence the rate of overweight and obesity

Clinical Trials

Clinical trials are also a very useful way for researchers and public health officials to learn more about the role that specifically prescribed lifestyle changes, nutritional interventions, and drug treatments play in health and disease. Both healthy and diseased patients can be enrolled in clinical trials, with the purpose of the trial determining the specific population to be studied. Clinical trials are used to test the efficacy or effectiveness of interventions with a primary outcome being to determine how well the intervention works. While most clinical trials are used to test the effectiveness of new treatments using drugs or nutritional interventions, many clinical trials examine the role of physical activity or exercise in both healthy and diseased individuals.

Prominent examples of clinical trials from exercise science disciplines can be found in the Molecular Transducers of Physical Activity Consortium (MoTrPAC). This consortium established numerous interconnected clinical trials to identify and describe the assortment of molecular transducers that underlie the effects of physical activity in humans (38). An exercise-related molecular transducer is a factor or substance that converts an exercise signal into a molecular signal that alters the physiology of the human body. The goal of the MoTrPAC is to examine the molecular changes that occur during and after exercise and then use that information to advance the understanding of how physical activity improves and maintains good health (39). MoTrPAC began as a 6-year program supported by the National Institutes of Health. Clinical sites have been established throughout the continental United States, with coordination for the study provided by exercise science and allied health care professionals (38). The National Institutes of Health provides information about all registered clinical trials at the following Web site: https://clinicaltrials.gov/ct2/home.

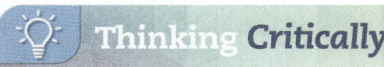

Thinking Critically

In what ways will some of the more significant longitudinal epidemiologic research studies contribute to identifying the future trends in exercise science?

USING PAST INFORMATION TO IMPROVE FUTURE HEALTH

Information that is collected from epidemiologic studies and clinical trials can be used to support subsequent research, develop health care plans and public policies to guide future decision making, and allow for recommendations of programs to enhance the health and wellness of individuals. For example, the endorsement of the health benefits obtained from physical activity and exercise by the U.S. Surgeon General was a significant landmark in the promotion of physical activity and exercise for enhancing health and reducing diseased risk within individuals (40). The Surgeon General serves as the chief health educator for the United States by providing Americans with the best scientific information available on how to improve their health and reduce the risk of illness, injury, and disease. Recommendations from the Surgeon General are developed using a structured process, whereby a panel of experts and scholars reviews past research information in a particular area that is health related and then develops a set of conclusions based on the evidence derived from that information. In 1995, the Surgeon General's Office issued a landmark report that highlighted the positive effects of physical activity on the health of the musculoskeletal, cardiovascular, respiratory, and endocrine systems, including a reduced risk of premature mortality and reduced risks of coronary heart disease, hypertension, colon cancer, and diabetes mellitus. The report also suggested that regular participation in physical activity appears to reduce depression and anxiety, improve mood, and enhance the ability to perform daily tasks throughout the lifespan. Recommendations for the appropriate amount of physical activity and exercise helped establish the standards for using exercise to assist in the treatment of diseased individuals (40). The publication of this landmark report is a classic example of how past research can be used to assist in making new policy recommendations and decisions. The U.S. Surgeon General continues to use research to make recommendations for improving the health and well-being of all Americans. The 2015 recommendation by the U.S. Surgeon General for the "Step It Up!" program as a key initiative for Promoting Walking and Walkable Communities is an excellent example of how research is used to provide leadership and direction in promoting health and fitness for the entire nation (41). The more recent 2020 U.S. Surgeon General's Call to Action to Control Hypertension seeks to minimize the adverse health effects of hypertension by utilizing three overarching goals (42):

- **Goal 1.** *Make hypertension control a national priority.*
- **Goal 2.** *Ensure that the places where people live, learn, work, and play support hypertension control.*
- **Goal 3.** *Optimize patient care for hypertension control.*

By organizing stakeholder groups such as federal, state, and local governments, public health and health care professionals, academic institutions and researchers, employers and health plan purchasers, as well as other key alliances to support the use of proven strategies in every community and for every population group, effective interventions and treatments for hypertension can be identified and implemented.

Another example of using prior research to support health promotion initiatives is the Federal Government's **Healthy People Program**, which has been in existence since the latter part of the 20th century. The Healthy People 2030 initiative consists of a set of health objectives for the United States to achieve during the third decade of the 21st century. It is intended to be used by individuals, states, communities, nonprofit and professional organizations, and others to help develop programs to improve health and well-being. Healthy People 2030 builds on the public health initiatives pursued over the previous decades of work. The 1979 Surgeon General's Report, *Healthy People* (43), and *Healthy People 2000: National Health Promotion and Disease Prevention Objectives* (44), established national health objectives and served as the basis for the development of state and local community plans. Like its predecessors, *Healthy People 2030* (http://www.healthypeople.gov/) was developed through a broad consultation process, built on the best scientific knowledge, and designed to measure goals and outcomes over time (45,46). The *Healthy People 2010 and Healthy People 2020 Final Reviews* provide a comprehensive view of the health of the United States and also include a report on each of the focus areas (47).

Healthy People 2030 was released in August 2020 and included an ambitious agenda for improving the health of Americans. The overarching goals for *Healthy People 2030* include the following (47):

Healthy People Program A program administered by the United States Department of Health and Human Services to improve the health of the nation's population.

- *Attain healthy, thriving lives and well-being free of preventable disease, disability, injury, and premature death.*
- *Eliminate health disparities, achieve health equity, and attain health literacy to improve the health and well-being of all.*
- *Create social, physical, and economic environments that promote attaining the full potential for health and well-being for all.*
- *Promote healthy development, healthy behaviors, and well-being across all life stages.*
- *Engage leadership, key constituents, and the public across multiple sectors to take action and design policies that improve the health and well-being of all.*

The health promotion initiatives, including objectives and the action plan for *Healthy People 2030*, can be found at https://health.gov/healthypeople.

WHAT WILL THE FUTURE BRING?

The health of the people of the United States is a critically important issue. It is clear that regular and sufficient amounts of physical activity and exercise, along with proper nutrition, play an important role in promoting good overall individual health. It is important, however, not to forget the broader impact of health and disease conditions. Individuals who are not in good health have an increased risk of premature morbidity and mortality, a poor quality of life, increased medical and health care expenses, and lost time from work. As a society, poor health creates an economic burden not only on the individual citizens of the United States, but also on businesses to provide health insurance, and on the health care systems of the state and federal governments to provide quality health care to everyone. For example, in 2018 the total national health care expenditure for the United States was $3.6 trillion, which translates to $11,172 per individual, and was 17.7% of the **gross domestic product** (42). Given the current trajectory, health care expenditures are expected to rise to $6.2 trillion and 19.7% of the gross domestic product by 2028 (42). Diseases associated with physical inactivity and poor nutrition not only contribute to high total health care costs but are also burdensome to individuals. For example, in hospitalizations of children a primary medical diagnosis of obesity is associated with a shorter hospital length of stay but higher hospital charges and costs when compared to overall hospitalizations. When obesity is a secondary medical diagnosis, the hospital stay is longer and there are higher charges and costs compared to overall hospitalizations (48). These high costs continue to create a considerable economic burden on our nation's health care system and cannot be endured by individuals, business, and state and federal governments. As we project into the future, individuals will be expected to assume greater responsibility for personal health and health care costs. It is clear that a cooperative effort must be made among all stakeholders to address the health issues of our nation. It is without question that exercise science and allied health professionals can and must play an important role in this effort to solve these problems. Past and current

Gross domestic product The total market value of all the goods and services produced within the borders of the United States during a specified period, usually one calendar year.

information must be used to guide future research programs, program development, and policy initiatives and advocacy. The scope of importance ranges from developing and implementing the most effective physical activity and nutrition programs for individuals across the lifespan to being an integral part of the comprehensive health care team for treating and rehabilitating individuals from disease and illness, especially those who are most vulnerable and at risk.

Future Research

Epidemiological data indicate that many nations are not meeting recommendations for physical activity and healthy behaviors. Consequently, the rates of various disease conditions remain high. In the United States, federal and state governments and private organizations and foundations play a crucial role in supporting future research on health and health care issues. Exercise science professionals will continue to play an active role in research as greater insight into how improvements in individual and population health and how effective and cost-efficient health care is delivered. Interdisciplinary research into broad program initiatives will be central to enhancing the understanding of how to develop the most cost-effective and efficacious treatments. Additional research at the cellular and molecular levels will need to continue to occur as we attempt to further understand the role of genetics and molecular responses to physical activity, physical inactivity, and exercise in both healthy and disease conditions. A better understanding of what genes are affected by lifestyle behaviors such as nutritional intake, physical activity, and regular exercise will provide key knowledge for the development of safe, effective, and cost-efficient prevention and treatment programs. Large-scale clinical trials and smaller efficacy studies will be essential for creating a knowledge base for advancing both the individual and public health efforts of government and public and private organizations. Some prospective areas of principal future research are highlighted in Table 13.3. Each of these areas will require a coordinated effort among exercise science and allied health professionals to develop sound research studies that answer our most important questions about health and health care.

Program Development

As exercise science and allied health professionals work to enhance the health and reduce the disease risk of the clients and patients they work with, it will be important to expand effective programs and develop new and innovative intervention programs and examine the efficacy of those interventions. Exercise science professionals must and will play a major role in the implementation and advancement of these programs. Effective program development and execution must occur at the national, state, and local levels. As a result, exercise science professionals will be at the forefront of program delivery, working with clients and patients of all ages in diverse health care settings, including private practice, hospitals, fitness centers, employee health and wellness programs, outpatient clinics, and residential care facilities. The delivery of health and wellness programs to individuals and groups is critical for reducing disease risk and improving health. Numerous private and public organizations and businesses must also continue to develop and implement programs at the local level for enhancing health and promoting chronic disease risk reduction. For example, employee health promotion and well-being programs are being effectively used to improve the health of working individuals and reducing health care costs for both individuals and employers (Figure 13.6). Governmental agencies such as the Centers for Disease Control

Table 13.3 Future Research in Health and Disease Risk Reduction

EXERCISE SCIENCE AREA	POTENTIAL RESEARCH AREAS
Exercise physiology	• Genetic influence on disease risk and health promotion across all population levels • Mechanisms involved with improvements in physiologic function • Role of physical inactivity in physiological function
Clinical exercise physiology	• Role of exercise and reduction of disease risk, including CVD, cancer, diabetes, hypertension, obesity, and metabolic syndrome • Promotion of effective and cost-efficient rehabilitation programs for diseased individuals
Athletic training and sports medicine	• Development of effective treatment and rehabilitation strategies for individuals injured during participation in physical activity and exercise • Further development of medical interventions that enhance physical activity in health and disease conditions
Exercise and sport nutrition	• Role of nutrition in health promotion • Efficacy of specialty diets in weight management and health promotion programs
Exercise and sport psychology	• Development of effective intervention strategies for promoting behavioral change and enhancing exercise adherence • Role of behavior in promoting physical activity and exercise
Motor behavior	• Understanding how individuals with differing levels of ability or development can effectively participate in exercise • How does neurophysiologic functioning affect health and/or prevent disease
Clinical and sport biomechanics	• Developing effective movement patterns for individuals with biomechanical limitations • Prevention of injuries during physical activity and exercise

FIGURE 13.6 Employee fitness and wellness programs can be used to reduce health care costs. (Photo courtesy of Shutterstock.)

and Prevention (www.cdc.gov) and the President's Council on Fitness, Sports and Nutrition (www.fitness.gov), as well as state and local health departments will continue to work with professional organizations such as the American Heart Association and the American College of Sports Medicine to develop and implement programs to improve the health and well-being of individuals. Exercise is Medicine™ is a national health promotion program that is a combined initiative of the American Medical Association and the American College of Sports Medicine. The overarching vision of this program is to make physical activity and exercise a standard part of a disease prevention and medical treatment paradigm in the United States (49). As a primary goal, Exercise is Medicine™ desires that "physical activity be considered by all health care providers as a vital sign in every patient visit, and that patients are effectively counseled and referred as to their physical activity and health needs, thus leading to overall improvement in the public's health and long-term reduction in health care cost" (49). Exercise is Medicine™ provides physicians and other health care providers, the public, health and fitness professionals, and the media with important and useful knowledge and information about the role of physical activity and exercise in the promotion of good health. Additional information about Exercise is Medicine™ can be found at http://exerciseismedicine.org/. Some examples of other current programs for the promotion of good health and disease risk reduction are listed in Table 13.4. For additional information about program initiatives, you are encouraged to visit the Web sites of the various governmental, private, and professional organizations whose mission is to promote good health. This will be a good way to gauge some of the future directions of exercise science.

Table 13.4 Programs Designed to Enhance Health and Reduce Disease Risk

AGENCY/ORGANIZATION	PROGRAM INITIATIVE
Centers for Disease Control and Prevention	• Coordinated School Health Program • School Health Index • State-based Nutrition and Physical Activity Program to Prevent Obesity and Other Chronic Diseases
American Heart Association	• Racial and Ethnic Approaches to Community Health (REACH) • The Heart of Diabetes • Million Hearts 2022 • WISEWOMAN
American College of Sports Medicine	• Exercise is Medicine
Robert Wood Johnson Foundation	• Healthy Disparities • Disease Prevention and Health Promotion • Social Determinants of Health
Directors of Health Promotion and Education	• Plan4Health
American Alliance for Health, Physical Education, Recreation and Dance	• Physical Best Program • Every Student Succeeds Act
American Diabetes Association	• Project Power
Academy of Nutrition and Dietetics	• National Nutrition Month

Table 13.5 Policy Initiatives and Advocacy Efforts in Exercise Science

AGENCY/ORGANIZATION	POLICY/ADVOCACY INITIATIVE
American Heart Association	• Heart disease and stroke: you're the cure
American College of Sports Medicine	• Science Partner of the President's Council on Physical Fitness, Sports, and Nutrition
National Association of State Boards of Education	• Fit, Healthy, and Ready to Learn: A School Health Policy Guide
Academy of Nutrition and Dietetics	• Kids Eat Right
American Diabetes Association	• Diabetes Advocacy Leadership Program
United States Department of Agriculture Food and Nutrition Division	• National School Lunch Program
National Academy of Medicine	• Culture of Health
Robert Wood Johnson Foundation	• Healthy Children and Families

Policy Initiatives and Advocacy

Federal and state governments as well as private and professional organizations promote and support policy initiatives designed to enhance health and reduce disease risk for individuals and populations. The development of sound policies includes having an infrastructure that provides sophisticated data collection and analysis, professional guidance, funding for the initiation of the policies, and an evaluation to assess the effectiveness of the policies. Professional and private organizations also use advocacy efforts to support the policy initiatives. Examples of some policy initiatives and advocacy efforts in exercise science are contained in Table 13.5.

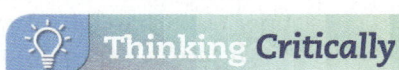

Thinking Critically

How will increased knowledge about the human genome and molecular biology contribute to developing new and effective strategies for treating lifestyle-related diseases?

EXERCISE SCIENCE AND SPORT AND ATHLETIC COMPETITION

Performance enhancement is critical for helping individuals participating in sport and athletic competition reach their highest performance potential. Individuals from beginners to elite athletes are continuously looking for ways to improve training and conditioning, practice strategies, and, ultimately, competition and game performance. Performance enhancement can take the form of improving individual movement skills and team strategy, as well as conducting a comprehensive assessment of physical and psychological abilities and developing an individualized training prescription and nutritional program. Performance enhancement also includes equipment development, aspects of mental preparation and behavioral training, the reduction of injury risk, and the recovery from illness and injury.

Exercise science professionals have long played an important role in improving sport and athletic performance. From the early work of Coleman R. Griffith, who in 1925 established the first Athletic Research Laboratory at the University of Illinois (50) to the recent work examining the efficacy of live-high and train-low altitude training for elite endurance athletes (51) to the use of **metabolomics** for understanding physiological function, exercise science professionals have been working to enhance individual and team-sport and athletic performance. Professional organizations such as the American College of Sports Medicine, the National Athletic Trainers Association, the National Strength and Conditioning Association, the Association for Applied Sport Psychology, and others have been instrumental in advancing the knowledge base of physiological, psychological, nutritional, and rehabilitative aspects of performance enhancement. Although it is difficult to predict where the future of exercise science will move in sport and athletic performance enhancement, it is safe to say that exercise science professionals will play an important role in that future.

Thinking Critically

Why should athletes and coaches rely on the knowledge of exercise science professionals for enhancing individual and team-sport and athletic performance?

WHAT WILL THE FUTURE BRING?

Many of the areas of study comprising exercise science have a component focused on improving performance in sport and athletic competition and improving the medical care and safety of athletes. Most importantly, it has been recognized that individualized attention is critical to improving performance. It is without question that the professionals working in exercise physiology, athletic training and sports medicine, sport nutrition, sport psychology, and motor behavior and biomechanics will use the knowledge gained from prior research to formulate new rehabilitation programs, novel nutritional interventions, heightened mental coaching, new training strategies, and new equipment for competition. As these areas are implemented, future trends of study will be identified. Sporting events and athletic competitions require athletes of all ages to perform activities with different types of movements, rigorous training and conditioning, and advanced mental preparation. From a single weightlifting event (Figure 13.7), lasting a few seconds, to the

FIGURE 13.7 Weightlifting movements only last a few seconds. (Photo courtesy of Shutterstock.)

Metabolomics The study of metabolites that are produced and released through physiological processes at both the cellular and systemic levels.

FIGURE 13.8 A two-hundred-meter sprint lasts about 20 seconds. (Photo courtesy of Shutterstock.)

two-hundred-meter sprint (Figure 13.8), lasting approximately 20 seconds, to the Iron Man Triathlon (Figure 13.9), lasting between 7 and 17 hours, athletes are required to coordinate the various systems and components of the mind and body into a sustained effort that results in their best individual performance at that moment in time. Whether it is playing competitive tennis (Figure 13.10), driving a race car (Figure 13.11), or climbing a mountain (Figure 13.12), the body must function optimally so that individual performance is maximized. Furthermore, team sports (Figure 13.13) have the added challenge of needing each individual athlete performing at their physical and psychological best to have a successful outcome during competition.

To examine athletic performance during the sporting event or individual competition is to provide only part of the picture. Athletes participate in regular training, conditioning, and practice sessions for numerous hours each week often for years at a time to maximize their performance during a competition that sometimes lasts only a few seconds. During training, athletes refine their movements through repetitive systematic training and adjustments to their biomechanical movement patterns. To be best prepared for rigorous practice and difficult competition it is imperative that athletes have optimal nutritional intake.

FIGURE 13.9 The Iron Man Triathlon can last for 7–17 hours. (Photo courtesy of Shutterstock.)

FIGURE 13.10 Competitive tennis. (Photo courtesy of Shutterstock.)

FIGURE 13.11 Race car driving. (Photo courtesy of Shutterstock.)

FIGURE 13.12 Mountain climbing. (Photo courtesy of Shutterstock.)

FIGURE 13.13 Team-sport performance requires top-level performance by all players. (Photo courtesy of Shutterstock.)

Appropriate macronutrient and micronutrient consumption allows athletes to meet the energy demands of training and recovery, as well as enhance the functioning of the various systems of the body that are important for successful performance in competition. When injured, athletes rely on athletic trainers and sports medicine personnel to make an accurate diagnosis, develop an effective medical intervention, and implement a successful rehabilitation strategy so that they may fully return to competition quickly.

High-level participation in sporting events and athletic competition will require knowledge from the exercise science disciplines to play an important role in assisting athletes in maximizing their performance. For example, over the years the various exercise science professionals have been instrumental in advancing our understanding of performance and developing new strategies for training, enhancing equipment and techniques for competition, and providing insights into the best nutritional intake for optimal performance. As we look to the future, we can examine several areas of study — training for performance; equipment and aids for performance; health and safety for performance — that will play instrumental roles in improving sport and athletic performance. The selection of these areas of study is not meant to minimize the importance of study in other areas of sport and athletic competition but to provide you with some examples of future areas that will be examined.

Training for Performance

Advances in the understanding of muscle contraction and relaxation have resulted in exercise science professionals advocating different types of training for improving sport performance. For example, the use of **plyometric training** (Figure 13.14) for improving muscle power output (52) during training and competition has increased in popularity. Plyometric training results in an increase in muscle force development, which has the potential to increased sports performance. Improvements in both eccentric and concentric muscle actions (53) and change of direction movements (54) have been shown to occur following plyometric training. Enhanced sport performance has been demonstrated in women's soccer (55), men's soccer (56), volleyball (57), and other team-sport athletes (58). Continued

Plyometric training A type of exercise training designed to produce fast, powerful movements and improve the functions of the neuromuscular system for the purposes of improving sport performance.

FIGURE 13.14 Plyometric training enhances muscle strength and power. (Shutterstock.)

advancements in the understanding of how plyometric training improves performance will result in a progressive expansion of its use for training athletes in various types of sports.

Another area for future study and advancement is focused on the potential for **high-altitude training** to enhance performance (59). Training at high altitude induces hypoxia because the individual breathes air with a lower oxygen content. Training under a hypoxic condition results in, among several physiological changes, an increase in red blood cell number and blood volume. The increase in red blood cells provides for greater oxygen transport to the working muscles and tissues of the body, improving performance in endurance events. Hypoxic training appears to be a worthwhile strategy for improvement in high-intensity running performance in team-sport athletes (60), with enriched performance lasting for at least 4 weeks post training (61); however, not all experts agree that performance benefits can be derived from engaging in this practice (62). Regardless, the potential for hypoxic training to improve athletic performance has led to conducting training programs (63) at high altitudes (Figure 13.15), the use of elevation training masks (64,65), and the development of simulated altitude tents (Figure 13.16) for those individuals who do not have access to a geographically high elevation. Additionally, as a result of a better understanding of how changes in altitude affect physiological and psychological function there are now preparticipation examination guidelines for individuals who are involved in climbing sports at high altitude (66,67) where hypoxic conditions exist.

Through the collection and the use of **big data**, exercise science professionals and others have been influential in searching for ways to enhance sport and athletic performance (68). The recent expansion of wearable technology, mobile phone applications, and related devices for monitoring physiological function and responses during rest, training, and competition provides large data sets of extensive and detailed measurements (69).

High-altitude training The practice of athletes living and performing exercise at high altitudes with the expectation of an improved sport performance at lower altitudes.

Big data The use of extremely large sets of data to inform the decision-making process.

FIGURE 13.15 High-altitude training improves low-altitude performance. (Photo courtesy of Shutterstock.)

Novel and innovative approaches to utilizing these large data sets more fully could provide a basis for more objective assessment of training programs and plans, coaching strategies, and new research methods (69). The proper use of big data technology has the potential to empower athletes to train more effectively for improving performance and reducing injury by adjusting appropriate training intensity, duration, frequency, and mode of activity (70). The use of big data is not limited to monitoring physiological function, as a particularly interesting area of big data focuses on the behavior of athletes (71) and how they perform during competition (72).

Equipment and Aids for Performance

The study of how athletes use nutritional (73) and pharmacologic supplements (74) to enhance performance and the associated side effects from using these supplements (75) will continue to be at the forefront of much exercise science research (76). Metabolomics is a specific area of study that involves the measurement of selected metabolites, transcription factors, and proteins in an effort to understand the overall metabolism of the human body. As a field of scientific study, metabolomics can be used to provide greater insight into how the body responds to acute bouts of exercise and chronic exercise training (77). The intent of metabolomics, as related to sport and athletic competition, is to better understand how metabolic processes are altered within individuals performing regular high-level training

FIGURE 13.16 Simulated high-altitude tent. (Photo courtesy of Hypoxico, Inc.)

and competition (78,79). As athletes and coaches better understand the changes in metabolism with acute exercise and chronic training, there can be a focus on what types and amounts of nutritional and pharmacological supplements can best be utilized to enhance performance. The continued search for methods of performance enhancement will involve exercise science professionals in determining whether any benefit can be derived from these methods and what the associated risks, if any, might be (75,80).

The evaluation of the effectiveness and efficacy of psychological strategies for performance enhancement at both the individual (81) and group level (82) is also a critical area of focus for advancement and future research. Specific types of psychological or social interventions, including mental practice, goal setting, team building, self-talk, and stress management, can lead to improved sport and athletic performance (83). Examples of questions to consider for future exploration include whether the psychological or social interventions are effective for both sexes and across a range of competition levels, whether qualities of the intervention provider influence treatment effectiveness and efficacy, whether effects vary based on the type of intervention and if single types of interventions are more or less effective than multicomponent interventions, and whether interventions have lasting effects on sport and athletic performance (83).

Athletes, coaches, and the general public have also become more interested in the role that molecular biology can play in identifying the different genes that are responsible for predicting athletic performance (84). This increased public interest in genetic testing for predicting sport performance and talent identification has led to the development of consensus statements by various experts in the discipline of molecular genetics (85,86). Successful performance in sport and athletic competition is complex, resulting from a combination of an extensive assortment of physiological and psychological traits and attributes (87). It is clear that there are specific components associated with high-level performance across a range of sports; however, what remains to be determined is whether a specific gene can be identified and associated with successful performance. Factors such as biological phenotypes, environmental influences, and behavioral characteristics play important roles in whether an individual is successful as an athlete. While the effectiveness of genetic testing for predicting sports performance is currently being debated (88), numerous companies are offering services to the general public. Exercise science professionals will play a critical role in helping athletes and coaches better understand the opportunities and limitations associated with using genetic testing for talent identification in sport and athletic performance.

The use of banned or illegal substances for enhancing sport and athletic performance has been common among athletes for decades. There have been numerous instances of high-profile amateur and professional athletes who have been stripped of awards and banned from competition in their respective sports. Examples of substances and methods that are banned for use by athletes include anabolic agents, peptide hormones and growth factors, beta-2 agonists, diuretics and masking agents, stimulants and narcotics, and beta-blockers (89). Traditional methods for detecting the use of a banned substance involved the direct detection of doping by analytical doping controls. In 2009, the World Anti-Doping Agency (WADA) issued the Athlete Biological Passport Operating Guidelines. According to WADA, "The fundamental principle of the Athlete Biological Passport (ABP) is to monitor selected biological variables over time that indirectly reveal the effects of doping rather than attempting to detect the doping substance or method itself" (89). In 2014, the WADA introduced the Steroidal Module, which can be used to establish longitudinal profiles of an individual athlete's steroid variables measured in urine samples. As new substances and methods have been developed to enhance performance, the WADA Athlete Biological Passport Guidelines have been refined to improve the efficacy of detecting anti-doping rule violations. The WADA continues to develop the Athlete Biological Passport in consultation with stakeholders, by refining the present modules as well as adding new ones as they are developed (89).

Health and Safety for Performance

Years of epidemiological and prospective research have provided a greater understanding of the complex issues surrounding the **female athlete triad** (90); however, there is still much to learn and comprehend (91). Our current understanding has resulted in better nutritional practices, improved diagnostic assessments, and more effective interventions for females with low bone mineral density (92,93). Educational programs are critical for ensuring that female athletes recognize the signs and symptoms of the female athlete triad and initiate action for minimizing the potentially harmful effects of the condition (94,95). Future research will continue to be needed to safeguard females at risk for the female athlete triad when participating in various sports and athletic competition (96).

The systematic evaluation of strategies for evaluating and enhancing equipment (97), as well as reducing the risk of injuries during training and competition (98), will continue to be prime areas of interest and responsibility for exercise science researchers. This is especially true in the area of developing equipment and assessments for reducing the risk level and severity of traumatic brain injury and concussion during sport and athletic competition (99). While much attention has been focused on how helmets impact the severity (100) and characteristics (101,102) of concussion during participation in American football, there are other sports such as ice hockey (103) and equestrian (104) where appropriate helmet use can play a pivotal role in concussion risk and severity. Important discussions about the required equipment for reducing concussion risk and severity will continue to occur (105).

The evaluation (106) and rehabilitation (107) of injured athletes by exercise science professionals is important for ensuring the health and safety of individuals who participate in all types of sports and athletic competition. This is especially true as more female athletes train longer and more intensely and become involved in contact sports such as rugby or ice hockey (Figure 13.17). The development of gender-specific evaluations and treatments for recovery from concussion will be critical for returning athletes to participate in sport (108). In the past, the primary focus of sports injury research has been on physical factors associated with the injury or the rehabilitation. This is despite the understanding that when an athlete sustains an injury there are also psychosocial factors that have a potential influence on the outcomes of the rehabilitation program (109). Exercise science professionals will continue to investigate the most appropriate training and recovery strategies in an effort to reduce the relative risk of injury during training and competition.

Female athlete triad A combination of three different disorders that affect female athletes: osteoporosis, eating disorders, and amenorrhea.

FIGURE 13.17 Females are increasingly involved in contact sports.

Finally, the COVID-19 pandemic had an astonishing impact on sport and athletic competition, from the way athletes were allowed to train for competition to the cancellation of high school and collegiate sports and professional leagues to the postponement of the Olympics. The pandemic has had more of an impact on sport than any of the cataclysmic events of our lifetime. Acute responses to the pandemic resulted in an awareness of how COVID-19 could spread more easily from one individual to another in high-contact sports such as football and soccer but less easily in low-contact sports such as baseball and softball. As a result of this knowledge, diagnostic testing will become more important in high-contact sports (110). In the foreseeable future, the way exercise science and sports medicine professionals evaluate and treat athletes with potentially infectious diseases will be altered to ensure the health and safety of participants and accessory personnel (110,111). The interaction of physical and psychological stressors that accompany high-intensity training and competition can alter immune system function, making athletes potentially more susceptible to infectious diseases (110). Future research will certainly need to consider how to not only ensure but also maximize the health and safety of athletes engaged in training and participation in sports and athletics (110).

Table 13.6 provides examples of possible future trends in exercise science that are centered on the enhancement of sport and athletic performance. Advancement of these trends will require a coordinated effort of exercise science professionals, allied health professionals, athletes, and coaches to determine what is effective and safe for use during training and competition. It is certain that all exercise science professionals will play an important role in the understanding and advancement of sport and athletic performance for many years to come.

Thinking Critically

How can exercise science professionals continue to contribute to enhancing individual and team performance in sport and athletic competition?

Table 13.6 Potential Trends of Future Sport and Athletic Performance Enhancement in Exercise Science

EXERCISE SCIENCE AREA	FUTURE TRENDS
Exercise physiology	• Identifying the mechanisms of legal and illegal human growth–promoting agents • Using technology to refine optimal training programs for individuals of different genetic profiles, different sports, and different ages
Athletic training and sports medicine	• Identification of techniques and signals to improve the detection of injuries and the prevention of potential injuries • Improvement of treatment modalities, including enhanced medical and surgical techniques, that will improve individual outcomes following an injury
Exercise and sport nutrition	• Refinement of appropriate macronutrient and micronutrient intakes for performance enhancement of different sports and different age groups • Clarification of the role of nutritional supplements in performance enhancement
Exercise and sport psychology	• Identification of effective psychosocial performance enhancement techniques for individuals of both sexes and various ages • Increased utilization by athletes of all sports
Motor behavior	• Further refinement of effective motor learning and control strategies to enhance skill acquisition • Continued clarification of optimal practice strategies
Clinical and sport biomechanics	• Improvement of movement patterns for sport techniques • Development of equipment that will enhance performance and reduce the risk of injury

Interview

Kathryn Schmitz, PhD, MPH, FACSM, FTOS

Professor, Department of Public Health Sciences and Department of Physical Medicine and Rehabilitation, Penn State University College of Medicine

Brief Introduction – I grew up in Maryland and attended the University of North Carolina-Greensboro and the North Carolina School of the Arts as an undergraduate. I was a modern dancer from childhood through my late twenties. At some point during college, I decided that a degree in economics would serve me better in the long run than a degree in dance, so I graduated from UNC-Greensboro in 1984 with a BA in Economics. I was third generation at UNC-G, both my mother and grandmother graduated college there as well, when it was called the "Women's College of North Carolina." I spent 8 years dancing professionally and working as a fitness professional in New York City before returning to academic pursuits to get an MS Ed in Exercise Physiology from Queens College CUNY. I then moved from the Big Apple to Minneapolis for doctoral training in Kinesiology at the University of Minnesota. I stayed on at UMN for a postdoctoral fellowship (during which I completed an MPH in Epidemiology) and then continued on as a faculty member in the Division of Epidemiology at the School of Public Health. In 2005, I was recruited to the University of Pennsylvania, where I worked until 2016. In 2016, I was recruited to the Penn State College of Medicine to lead the Oncology Nutrition and Exercise program. My research program focuses on exercise oncology.

Q: What are two or three of your most significant career experiences?

Up to when I was a postdoctoral fellow, the focus of my research had been on exercise and metabolic and cardiovascular diseases. I was interested in finding another path that would allow me to take larger leaps forward with each study, a "frontier" area that had very little research. In 2000, I read a paper by Anne McTiernan that was a call to action for researchers like me (focused on exercise and metabolic or cardiovascular diseases) to consider changing focus to exercise and cancer. I reached out to Anne, who became a mentor to me for years, and I've focused on exercise oncology since that time. In 2009, and again in 2018, I was part of the leadership team for the ACSM Roundtables on Exercise and Cancer. These roundtables have both contributed meaningfully to forward progress in the field of exercise oncology, setting guidelines for exercise for people living with and beyond cancer, and a call to action for clinicians to make assessment, advice, and referral to exercise standard practice in oncology. In 2017, I became president elect of the American College of Sports Medicine. The executive leadership of ACSM includes three people: President Elect, President, and Immediate Past President. I entered the 3-year period of executive leadership duties in June 2017 and finished in June 2020, a period of tremendous change in the college. I am humbled and honored to have served with Walt Thompson, Liz Joy, Bill Kraus, and NiCole Keith on the executive committee and to have had the opportunity to start the Moving Through Cancer initiative (www.exerciseismedicine.org/movingthroughcancer) as my presidential initiative. The bold goal of the Moving Through Cancer initiative is that exercise assessment, advice, and referral will be standard practice in oncology by 2029.

Q: Why did you choose to become an "exercise scientist"?

When I was a fitness professional in New York City, I had clients who were very well educated. They would ask me questions I could not answer. I went to get my Master's at Queens College, with Bill McArdle, John Magel, Mike Toner, and Paul Fardy because I wanted to be able to answer my well-educated clients' questions. What I did not anticipate was that in the process of getting that degree, I would fall in love with exercise science research.

Q: What are some of the key issues that will be at the forefront/horizon of exercise science for the next 20 years?

The field of exercise oncology has grown exponentially in the past 20-plus years. The first randomized trials in humans were completed in 1988, and the first review of trials, in 1996, included four trials. In 2005, I published a meta-analysis of the field that included 22 trials. In 2010, I updated that meta-analysis, and there were 82 trials. According to PubMed, there was a 281% increase in exercise and cancer research between 2010 and 2018. That said, we are still a long way from having exercise included as standard of care for people living with and beyond cancer. The leadership of the Moving Through Cancer initiative has set an agenda that outlines the steps we think are needed toward the bold goal of making exercise standard practice in oncology: awareness, workforce enhancement (training of exercise professionals and oncology clinicians), program development, policy changes, and further research on the efficacy of exercise for outcomes related to cancer. I implore students to consider exercise oncology as an area of focus, regardless of whether you seek to be an exercise professional, a rehabilitation clinician, or an exercise scientist.

Q: What advice would you have for a student exploring a career in any exercise science profession?

Raise the standard. Ensure that anything you tell someone (anyone) about the benefits of exercise is backed by the highest quality science. Find an area where you can really make a difference. Feeling needed is really motivating! Keep in touch with your mentors. They invest a lot in you and want to know how you are doing.

Interview

Jonathan Finnoff, DO, FACSM, FAMSSM

Chief Medical Officer, United States Olympic and Paralympic Committee; Professor, Department of Physical Medicine and Rehabilitation, Mayo Clinic College of Medicine and Science, Rochester, MN

Brief Introduction – I was born and raised in Boulder, Colorado, and attended the University of Colorado in Boulder, where I received my BA in Molecular, Cellular, and Developmental Biology. I completed my medical studies at the University of New England. My postgraduate education included an internship at Delaware Valley Medical Center, a physical medicine and rehabilitation residency at the University of Utah, and a sports medicine fellowship at Mayo Clinic in Rochester, MN. My research interests include advanced diagnostic and interventional ultrasound, regenerative medicine, and injury and illness prevention.

Q: What are two or three of your most significant career experiences?

One year following graduation from my sports medicine fellowship, I had the opportunity to begin serving as the head team physician for Utah State University. Directing and delivering medical care to the athletes at this NCAA Division 1 school provided me with the foundational knowledge and experience required to achieve my career goals. In 2002, I was selected by the Salt Lake Organizing Committee to direct the Athlete Medical Clinic at the Soldier Hollow Venue during the Winter Olympic Games and to serve as the Venue Medical Officer for the Soldier Hollow Venue during the Winter Paralympic Games. I was inspired by the spirit and passion of the Olympic and Paralympic movement, which led me to volunteer as a team physician for the United States Ski and Snowboard Team for the next two decades, serve as a team physician at two additional Winter Olympic Games, and eventually be selected to serve as the Chief Medical Officer for the United States Olympic and Paralympic Committee.

Q: Why did you choose to become a sports medicine physician?

As a child, I dreamed of being a professional athlete. I was fortunate enough to have the natural gifts and determination to make that dream a reality as a professional mountain bike racer. When I entered college, it was to obtain a business degree in preparation for an eventual transition into the cycling industry after my athletic career ended. However, my life changed when I came upon a person who committed suicide. When I discovered them, they were still alive. I called 911, but they died before the ambulance arrived. I decided to enroll in an emergency medical technician class so I would have some medical skills if I ever came upon an emergency in the future. I loved it and realized that medicine was my calling. I changed my major, applied to medical school, and the rest is history.

Q: What are some of the key issues that will be at the forefront/horizon of sports medicine for the next 20 years?

Right now, the topic in the front of my mind is infectious disease. The vaccines for COVID-19 likely won't be 100% effective, many people will choose not to get vaccinated, and vaccine availability will be variable around the world. In addition, there are other infectious diseases that can negatively impact sport. Therefore, I think it will be important for the sports medicine community to learn from our current situation and adopt strong infection prevention strategies that are carried into the future. I also believe early sports specialization is a significant problem facing the sports medicine profession. Early sports specialization leads to increased injury, illness, and burnout. At the opposite end of the spectrum is the obesity pandemic. Addressing both of these problems is critical to ensuring the health of our society.

Q: What advice would you have for a student exploring a career in any exercise science profession?

As a sports medicine physician, I'm asked everyday by patients and athletes what they can do to reduce their stress; improve their sleep; enhance their athletic performance; and reduce their risk of injury, illness, and disease. I tell them that the magic ingredient for all of these is exercise. I can't think of more impactful professions than sports medicine or exercise science. You get to meet interesting and motivated people and help them solve problems that improve their quality of life. If you choose to pursue this career path, I believe you will find it to be extremely rewarding.

SUMMARY

- Much of the information used to predict future health promotion and disease risk reduction trends in exercise science comes from longitudinal epidemiologic research.
- Future trends in health promotion and disease risk reduction will be in the areas of research, program development, and policy initiatives.
- The enhancement of sport and athletic performance requires all areas of exercise science to provide an integrated and coordinated effort to improve the psychological and physiologic components that affect performance.
- Future trends in sport and athletic competition will most certainly be centered on the molecular and genetic factors that influence performance.

FOR REVIEW

1. Define the following terms:
    a. Epidemiological
    b. Longitudinal
    c. Clinical trial
2. What was the primary purpose of the Framingham Heart Study?
3. List five important outcomes of the Framingham Heart Study that help guide the current recommendations for reducing disease risk.
4. What is the primary purpose of the continuous NHANES?
5. What was the primary rationale for conducting the HERITAGE Family Study?
6. Describe the primary conclusion of the 1995 United States Surgeon General's report on Health Promotion and Disease Prevention.
7. What are the primary purposes of the Healthy People 2010, 2020, and 2030 programs?
8. Describe the relationships among research, program development, and policy initiatives as they relate to the future of exercise science.
9. Explain how each of the following areas in exercise science might contribute to enhancing individual and team-sport and athletic performance:
    a. Exercise physiology
    b. Athletic training and sports medicine
    c. Sport nutrition
    d. Sport psychology
    e. Motor behavior
    f. Sport biomechanics

Project-Based Learning

1. Identify five articles published within the last 5 years that have a focus on physical activity or exercise. Within the discussion section of each article, identify the recommendations for future research made by the authors. Prepare a presentation that includes at least five key future research recommendations that will impact physical activity and exercise and ultimately improve health and reduce disease risk. In your presentation, include why the five key areas are critical to study?
2. Identify five articles published within the last 5 years with a focus on improving sport or athletic competition. Prepare a presentation that includes at least five key future research recommendations that will impact sport or athletic performance with the purpose of enhancing performance. In your presentation, include why the five key areas are critical to study?

REFERENCES

1. Ashley FW, Kannel WB. Relation of weight change to changes in atherogenic traits: the Framingham Study. *J Chronic Dis*. 1974;27:103–14.
2. Blair SN, Kohl HW, Paffenbarger RS, Clark DG, Cooper KH, Gibbons LW. Physical fitness and all-cause mortality. *JAMA*. 1989;262(17):2395–401.
3. Sesso HD, Paffenbarger RS, Lee IM. Physical activity and coronary heart disease in men: the Harvard Alumni Health Study. *Circulation*. 2000;102:975–80.
4. American Heart Association Web Site [Internet; cited 2020]. Available from: www.heart.org.
5. Paffenbarger RS, Hale WE. Work activity and coronary heart mortality. *N Engl J Med*. 1975;292:545–50.
6. Paffenbarger RS, Hyde RT, Wing AL, Hsieh CC. Physical activity, all-cause mortality, and longevity of college alumni. *N Engl J Med*. 1986;314(10):605–13.
7. Powell KE, Thompson PD, Caspersen CJ, Kendrick JS. Physical activity and the incidence of coronary heart disease. *Annu Rev Public Health*. 1987;8(1):253–87.
8. Siscovick DS, Weiss NS, Fletcher RH, Schoenbach VJ, Wagner EH. Habitual vigorous exercise and primary cardiac arrest: effect of other risk factors on the relationship. *J Chronic Dis*. 1984;37:625–31.
9. Booth FW, Roberts CK, Laye MJ. Lack of exercise is a major cause of chronic disease. *Comp Physiol*. 2012;2(2):1143–211.
10. LaHue SC, Comella CL, Tanner CM. The best medicine? The influence of physical activity and inactivity on Parkinson's disease: physical activity, inactivity, and PD. *Mov Disord*. 2016;31(10):1444–54.
11. Pinto AJ, Roschel H, de Sá Pinto AL, et al. Physical inactivity and sedentary behavior: overlooked risk factors in autoimmune rheumatic diseases? *Autoimmun Rev*. 2017;16(7):667–74.
12. Jia H, Zack MM, Gottesman II, Thompson WW. Associations of smoking, physical inactivity, heavy drinking, and obesity with quality-adjusted life expectancy among US adults with depression. *Value Health*. 2018;21(3):364–71.
13. Blond K, Brinkløv CF, Ried-Larsen M, Crippa A, Grøntved A. Association of high amounts of physical activity with mortality risk: a systematic review and meta-analysis. *Br J Sports Med*. 2020;54(20):1195–201.
14. Myers A, Gibbons C, Finlayson G, Blundell J. Associations among sedentary and active behaviours, body fat and appetite dysregulation: investigating the myth of physical inactivity and obesity. *Br J Sports Med*. 2017;51(21):1540–4.
15. Booth FW, Hargreaves M. Understanding multi-organ pathology from insufficient exercise. *J Appl Physiol*. 2011;111(4):1199–200.
16. Booth FW, Laye MJ, Roberts MD. Lifetime sedentary living accelerates some aspects of secondary aging. *J Appl Physiol*. 2011;111(5):1497–504.

17. Chodzko-Zajio W, Proctor DN, Fiatarone-Singh MA, et al. Exercise and physical activity for older adults. *Med Sci Sports Exerc.* 2009;41(7):1510–30.
18. Donnelly JE, Blair SN, Jakicic JM, Manore MM, Rankin JW, Smith BK. Appropriate physical activity intervention strategies for weight loss and prevention of weight regain for adults. *Med Sci Sports Exerc.* 2009;41(2):459–71.
19. Garber CE, Blissmer B, Deschenes MR, et al. Quantity and quality of exercise for developing and maintaining cardiorespiratory, musculoskeletal, and neuromotor fitness in apparently healthy adults: guidance for prescribing exercise. *Med Sci Sports Exerc.* 2011;43(7):1334–59.
20. Dunn AL, Trivedi MH, O'Neal HA. Physical activity dose-response effects on outcomes of depression and anxiety. *Med Sci Sports Exerc.* 2001;33(6):S587–97.
21. Thune I, Furberg AS. Physical activity and cancer risk: dose-response and cancer, all sites and site-specific. *Med Sci Sports Exerc.* 2001;33(6):S530–50.
22. U.S. Department of Health and Human Services, Centers For Disease Control and Prevention, National Center for Chronic Disease Prevention and Health Promotion, President's Council on Physical Fitness and Sports. *Physical Activity and Health: A Report of the Surgeon General.* 1995.
23. Centers for Disease Control and Prevention Web site [Internet; cited 2021]. Available from: http://www.cdc.gov.
24. The HERITAGE Family Study Web site [Internet; cited 2020]. Available from: www.pbrc.edu/heritage/.
25. Miller KE, Martz DC, Stoner C, et al. Efficacy of a telephone-based medical nutrition program on blood lipid and lipoprotein metabolism: results of our healthy heart: telephone-based medical nutrition and blood lipids. *Nutr Diet.* 2018;75(1):73–8.
26. Rogers KM, Littlefield LA, Taylor JK, Papadakis Z, Grandjean PW, Moncada-Jiménez J. The effect of exercise intensity and excess postexercise oxygen consumption on postprandial blood lipids in physically inactive men. *Appl Physiol Nutr Metab.* 2017;42(9):986–93.
27. Ray KK, Corral P, Morales E, Nicholls SJ. Pharmacological lipid-modification therapies for prevention of ischaemic heart disease: current and future options. *Lancet.* 2019;394(10199):697–708.
28. Raghavan S, Porneala B, McKeown N, Fox CS, Dupuis J, Meigs JB. Metabolic factors and genetic risk mediate familial type 2 diabetes risk in the Framingham Heart Study. *Diabetologia.* 2015;58(5):988–96.
29. Bouchard C, Leon AS, Rao DC, Skinner JS, Wilmore JH, Gagnon J. The HERITAGE family study. Aims, design, and measurement protocol. *Med Sci Sports Exerc.* 1995;27(5):721–9.
30. Bouchard C, An P, Rice T, et al. Familial aggregation of VO_2max response to exercise training: results from the HERITAGE family study. *J Appl Physiol.* 1999;87(3):1003–8.
31. Rankinen T, Perusse L, Borecki IB, et al. The Na(+)-K(+)-ATPase alpha 2 gene and trainability of cardiorespiratory endurance: the HERITAGE family study. *J Appl Physiol.* 2000;88(1):346–51.
32. Rankinen T, Sung YJ, Sarzynski MA, Rice TK, Rao DC, Bouchard C. Heritability of submaximal exercise heart rate response to exercise training is accounted for by nine SNPs. *J Appl Physiol.* 2012;112(5):892–7.
33. Vellers HL, Verhein KC, Burkholder AB, et al. Association between mitochondrial DNA sequence variants and $\dot{V}O_2$ max trainability. *Med Sci Sports Exerc.* 2020;52(11):2303–9.
34. Perusse L, Rice T, Province ME, et al. Familial aggregation of amount and distribution of subcutaneous fat and their responses to exercise training in the HERITAGE family study. *Obesity.* 2000;8(2):140–50.
35. Horton R, Crawford G, Freeman L, Fenwick A, Wright CF, Lucassen A. Direct-to-consumer genetic testing. *BMJ.* 2019;367:l5688.
36. Delaney S, Christman M. Direct-to-consumer genetic testing: perspectives on its value in healthcare. *Clin Pharmacol Ther.* 2016;99(2):146–8.
37. Centers for Disease Control and Prevention Web site [Internet cited; 2020]. Available from: http://www.cdc.gov.
38. Molecular Transducers of Physical Activity Consortium. MoTrPAC 2021 [Internet cited 2021]. Available from: www.motrpac.org/index.cfm.
39. Sanford JA, Nogiec CD, Lindholm ME, et al. Molecular transducers of physical activity consortium (MoTrPAC): mapping the dynamic responses to exercise. *Cell.* 2020;181(7):1464–74.
40. Pate RR, Pratt M, Blair SN, et al. Physical activity and public health. *JAMA.* 1995;273(5):402–7.
41. United States Department of Health and Human Services. *Step It Up! The Surgeon General's Call to Action to Promote Walking and Walkable Communities.* Washington (DC): United States Department of Health and Human Services; 2015.
42. Center for Medicade and Medicare Services [Internet cited; 2021]. Available from: https://www.cms.gov/.

43. U.S. Department of Health Education, and Welfare. *Healthy People: The Surgeon General's Report on Health Promotion and Disease Prevention.* Report No.: 79-55071. 1979.
44. National Center for Health Statistics. *Healthy People 2000 Final Review.* Report No.: 76-641469. Hyattsville (MD): National Center for Health Statistics; 2001.
45. Healthy People 2030 [Internet cited; 2020]. Available from: https://health.gov/healthypeople.
46. Healthy People 2020 [Internet cited; 2011]. Available from: www.healthypeople.gov.
47. Office of Public Health Scientific Services Web site [Internet cited; 2020]. Available from: www.cdc.gov/ophss/.
48. Kompaniyets L, Lundeen EA, Belay B, Goodman AB, Tangka F, Blanck HM. Hospital length of stay, charges, and costs associated with a diagnosis of obesity in US children and youth, 2006–2016. *Med Care.* 2020;58(8):722–6.
49. Exercise is Medicine [Internet cited; 2016]. Available from: www.exerciseismedicine.org.
50. Vealey RS. Smocks and jocks outside the box: the paradigmatic evolution of sport and exercise psychology. *Quest.* 2006:128–59.
51. Millet GP, Chapman RF, Girard O, Brocherie F. Is live high–train low altitude training relevant for elite athletes? Flawed analysis from inaccurate data. *Br J Sports Med.* 2019;53(15):923–6.
52. Potteiger JA, Lockwood RH, Haub MD, et al. Muscle power and fiber characteristics following 8 weeks of plyometric training. *J Strength Cond Res.* 1999;13(3):275–9.
53. Behrens M, Mau-Moeller A, Mueller K, et al. Plyometric training improves voluntary activation and strength during isometric, concentric and eccentric contractions. *J Sci Med Sport.* 2016;19(2):170–6.
54. Asadi A, Arazi H, Young WB, de Villarreal ES. The effects of plyometric training on change-of-direction ability: a meta-analysis. *Int J Sports Physiol Perform.* 2016;11(5):563–73.
55. Nonnato A, Hulton AT, Brownlee TE, Beato M. The effect of a single session of plyometric training per week on fitness parameters in professional female soccer players: a randomized controlled trial. *J Strength Cond Res.* 2020. doi: 10.1519/JSC.0000000000003591.
56. Yanci J, Los Arcos A, Camara J, Castillo D, García A, Castagna C. Effects of horizontal plyometric training volume on soccer players' performance. *Res Sports Med.* 2016;24(4):308–19.
57. Wang M-H, Chen K-C, Hung M-H, et al. Effects of plyometric training on surface electromyographic activity and performance during blocking jumps in college Division I men's volleyball athletes. *Appl Sci.* 2020;10(13):4535.
58. Slimani M, Chamari K, Miarka B, Del Vecchio FB, Chéour F. Effects of plyometric training on physical fitness in team sport athletes: a systematic review. *J Hum Kinet.* 2016;53(1):231–47.
59. Beidleman BA, Muza SR, Fulco CS, et al. Seven intermittent exposures to altitude improves exercise performance at 4300 m. *Med Sci Sports Exerc.* 2008;40(1):141–8.
60. Faiss R, Girard O, Millet GP. Advancing hypoxic training in team sports: from intermittent hypoxic training to repeated sprint training in hypoxia. *Br J Sports Med.* 2013;47(Suppl 1):i45–i50.
61. Hamlin MJ, Lizamore CA, Hopkins WG. The effect of natural or simulated altitude training on high-intensity intermittent running performance in team-sport athletes: a meta-analysis. *Sports Med.* 2018;48(2):431–46.
62. Bejder J, Nordsborg NB. Specificity of "live high–train low" altitude training on exercise performance. *Exerc Sport Sci Rev.* 2018;46(2):129–36.
63. Brocherie F, Millet GP, Hauser A, et al. "Live high–train low and high" hypoxic training improves team-sport performance. *Med Sci Sports Exerc.* 2015;47(10):2140–9.
64. Porcari JP, Probst L, Forrester K, et al. Effect of wearing the elevation training mask on aerobic capacity, lung function, and hematological variables. *J Sports Sci Med.* 2016;15(2):379–86.
65. Jung HC, Lee NH, John SD, Lee S. The elevation training mask induces modest hypoxaemia but does not affect heart rate variability during cycling in healthy adults. *Biol Sport.* 2019;36(2):105–12.
66. Campbell AD, Davis C, Paterson R, et al. Preparticipation evaluation for climbing sports. *Clin J Sport Med.* 2015;25(5):412–7.
67. Campbell AD, McIntosh SE, Nyberg A, Powell AP, Schoene RB, Hackett P. Risk stratification for athletes and adventurers in high-altitude environments: recommendations for preparticipation evaluation. *Clin J Sport Med.* 2015;25(5):404–11.
68. Rein R, Memmert D. Big data and tactical analysis in elite soccer: future challenges and opportunities for sports science. *Springerplus.* 2016;5(1):1–13.
69. Passfield L, Hopker JG. A Mine of information: can sports analytics provide wisdom from your data? *Int J Sports Physiol Perform.* 2017;12(7):851–5.

70. Guan H, Zhong T, He H, et al. A self-powered wearable sweat-evaporation-biosensing analyzer for building sports big data. *Nano Energy.* 2019;59:754–61.
71. Morgulev E, Azar OH, Lidor R. Sports analytics and the big-data era. *Int J Data Sci Anal.* 2018;5(4):213–22.
72. Sarlis V, Tjortjis C. Sports analytics — evaluation of basketball players and team performance. *Inf Syst.* 2020;93:101562.
73. Rosenbloom C, Murray B. Risky business: dietary supplement use by athletes. *Nutr Today.* 2015;50(5):240–6.
74. Yesalis CE. *Anabolic Steroids in Sport and Exercise.* 2nd ed. Champaign (IL): Human Kinetics; 2000.
75. Maughan RJ, Burke LM, Dvorak J, et al. IOC consensus statement: dietary supplements and the high-performance athlete. *Br J Sports Med.* 2018;52(7):439–55.
76. Weihrauch M, Handschin C. Pharmacological targeting of exercise adaptations in skeletal muscle: benefits and pitfalls. *Biochem Pharmacol.* 2018;147:211–20.
77. Heaney LM, Deighton K, Suzuki T. Non-targeted metabolomics in sport and exercise science. *J Sports Sci.* 2017;37(9):959–67.
78. Bongiovanni T, Pintus R, Dessi A, et al. Sportomics: metabolomics applied to sports. The new revolution? *Eur Rev Med Pharmacol Sci.* 2019;23(24):11011–9.
79. Bragazzi NL, Khoramipour K, Chaouachi A, Chamari K. Toward sportomics: shifting from sport genomics to sport postgenomics and metabolomics specialties. promises, challenges, and future perspectives. *Int J Sports Physiol Perform.* 2020;15(9):1201–2.
80. Hoffman JR, Kraemer WJ, Bhasin S, et al. Position stand on androgen and human growth hormone use. *J Strength Cond Res.* 2009;23(5):S1–59.
81. Barwood MJ, Thelwell RC, Tipton MJ. Psychological skills training improves exercise performance in the heat. *Med Sci Sports Exerc.* 2008;40(2):387–96.
82. Bruner MW, Eys MA, Wilson KS, Cote J. Group cohesion and positive youth development in team sport athletes. *Sport Exerc Perform Psychol.* 2014;3(4):219–27.
83. Brown DJ, Brown DJ, Fletcher D, Fletcher D. Effects of psychological and psychosocial interventions on sport performance: a meta-analysis. *Sports Med.* 2017;47(1):77–99.
84. Yang N, MacArthur DG, Gulbin JP, et al. ACTN3 genotype is associated with human elite athletic performance. *Am J Hum Genet.* 2003;73(3):627–31.
85. Webborn N, Williams A, McNamee M, Bouchard C, Pitsiladis YP, Ahmetov I, et al. Direct-to-consumer genetic testing for predicting sports performance and talent identification: Consensus statement. *Br J Sports Med.* 2015;49(23):1486–91.
86. Vlahovich N, Fricker PA, Brown MA, Hughes D. Ethics of genetic testing and research in sport: a position statement from the Australian Institute of Sport. *Br J Sports Med.* 2017;51(1):5–11.
87. Mattsson CM, Wheeler MT, Waggott D, Caleshu C, Ashley EA. Sports genetics moving forward: lessons learned from medical research. *Physiol Genomics.* 2016;48(3):175–82.
88. Pickering C, Kiely J, Grgic J, Lucia A, Del Coso J. Can genetic testing identify talent for sport? *Genes.* 2019;10(12):972.
89. World Anti-Doping Agency [Internet cited; 2020]. Available from: https://www.wada-ama.org/.
90. Nattiv A, Loucks AB, Manore MM, Sanborn CF, Sundgot-Borgen J, Warren MP. The female athlete triad. *Med Sci Sports Exerc.* 2007;29(5):1867–82.
91. Williams N, Koltun KJ, Strock NCA, De Souza MJ. Female athlete triad and relative energy deficiency in sport: a focus on scientific rigor. *Exerc Sport Sci Rev.* 2019;47(4):197–205.
92. Fletcher JA. Canadian Academy of Sport and Exercise Medicine position statement: osteoporosis and exercise. *Clin J Sport Med.* 2013;23(5):333–8.
93. Joy EA, De Souza MJ, Nattiv A, et al. 2014 female athlete triad coalition consensus statement on treatment and return to play of the female athlete triad. *Curr Sports Med Rep.* 2014;13(4):219–32.
94. Krick R, Brown K, Ramsay S, Brown AF. Changes in knowledge of the female athlete triad among female high school athletes following a brief nutrition education intervention. *J Acad Nutr Diet.* 2017;117(10):A142.
95. Koltun KJ, Strock NCA, Southmayd EA, Oneglia AP, Williams NI, De Souza MJ. Comparison of female athlete triad coalition and RED-S risk assessment tools. *J Sports Sci.* 2019;37(21):2433–42.
96. Koltun KJ, Williams NI, Souza MJD. Female athlete triad coalition cumulative risk assessment tool: proposed alternative scoring strategies. *Appl Physiol Nutr Metab.* 2020;45(12):1324–31.
97. Luttrell MD, Potteiger JA. Effects of powercranks training on cardiovascular fitness and cycling efficiency. *J Strength Cond Res.* 2003;17(4):785–91.

98. Curtis CK, Laudner KG, McLoda TA, McCaw ST. The role of shoe design in ankle sprain rates among collegiate basketball players. *J Athl Train*. 2008;43(3):230–3.
99. Sone JY, Kondziolka D, Huang JH, Samadani U. Helmet efficacy against concussion and traumatic brain injury: a review. *J Neurosurg*. 2017;126(3):768–81.
100. Greenhill DA, Navo P, Zhao H, Torg J, Comstock RD, Boden BP. Inadequate helmet fit increases concussion severity in American high school football players. *Sports Health*. 2016;8(3):238–43.
101. Collins CL, McKenzie LB, Ferketich AK, Andridge R, Xiang H, Comstock RD. Concussion characteristics in High School Football by helmet age/recondition status, manufacturer, and model: 2008–2009 through 2012–2013 academic years in the United States. *Am J Sports Med*. 2016;44(6):1382–90.
102. Bailey AM, McMurry TL, Cormier JM, et al. Comparison of laboratory and on-field performance of American football helmets. *Ann Biomed Eng*. 2020;48(11):2531–41.
103. Clark JM, Taylor K, Post A, Hoshizaki TB, Gilchrist MD. Comparison of ice hockey goaltender helmets for concussion type impacts. *Ann Biomed Eng*. 2018;46(7):986–1000.
104. Connor TA, Clark JM, Jayamohan J, et al. Do equestrian helmets prevent concussion? A retrospective analysis of head injuries and helmet damage from real-world equestrian accidents. *Sports Med Open*. 2019;5(1):1–8.
105. Comstock RD, Arakkal AT, Pierpoint LA, Fields SK. Are high school girls' lacrosse players at increased risk of concussion because they are not allowed to wear the same helmet boys' lacrosse players are required to wear? *Inj Epidemiol*. 2020;7(1):18.
106. Broglio SP, Cantu RC, Gioia GA, et al. National Athletic Trainer's Association position statement: management of sport concussion. *J Athl Train*. 2014;49(2):245–65.
107. Kuster MS, Spalinger E, Blanksby BA, Gachter A. Endurance sports after total knee replacement: a biomechanical investigation. *Med Sci Sports Exerc*. 2000;32(4):721–4.
108. Chiang Colvin A, Mullen J, Lovell MR, Vereeke West R, Collins MW, Groh M. The role of concussion history and gender in recovery from soccer-related concussion. *Am J Sports Med*. 2017;37(9):1699–704.
109. Forsdyke D, Smith A, Jones M, Gledhill A. Psychosocial factors associated with outcomes of sports injury rehabilitation in competitive athletes: a mixed studies systematic review. *Br J Sports Med*. 2016;50(9):537–44. doi:10.1136/bjsports-2015-094850.
110. Denay KL, Breslow RG, Turner MN, Nieman DC, Roberts WO, Best TM. ACSM call to action statement: COVID-19 considerations for sports and physical activity. *Curr Sports Med Rep*. 2020;19(8):326–8.
111. Diamond AB, Narducci DM, Roberts WO, et al. Interim guidance on the preparticipation physical examination for athletes during the SARS-CoV-2 pandemic. *Curr Sports Med Rep*. 2020;19(11):498–503.

Index

Note: Page numbers in *italics* denote figures; those followed by 't' denote tables.

A

Academic discipline, defined, 7
Academic preparation in exercise science, 22–23, *23*
    careers in health care, 24–25, *25*
    chiropractic school, 27
    dental school, 26
    graduate school, 30–31, 31t
    medical school, 25–26
    physical and occupational therapy, 29–30, *29*
    physician assistant programs, 28–29, *28*
    podiatric medical school, 27–28
    undergraduate coursework, 23–24, 24t
Academy of Nutrition and Dietetics (AND), 253–254, 468t
Accelerometers, 144, 144t, 437, 438, *439*
Acquired immune deficiency syndrome (AIDS), 99
Acquired immunity, 98, 98t
ACSM (*see* American College of Sports Medicine)
Acute load, 394
Acute responses, 121
    to physical activity and exercise, 121–122, 122t
    research in, 122t
Ad libitum, 279
Adenosine diphosphate (ADP), 101
Adenosine triphosphate (ATP), 101
Adipose tissue, 259
Adolescence, 344, *345*
Adrenaline, 95
Adult obesity–attributable medical expenditures, 140–141, *140*
Adulthood, 345
Aerobic metabolism, 133
Aerobics and Fitness Association of America, 493t
Afferent (sensory) neurons, 72
AIDS (*see* Acquired immune deficiency syndrome)
Air displacement plethysmography, 440
Allergenic, 186
Allied health care, 23
    professionals, *25*
Allied Health Professions Aptitude Test, 30
Allopathic medicine, 25
Alzheimer disease, 353
Amenorrhea, 138
American Alliance for Health, Physical Education, Recreation, and Dance (AAHPERD), 339
American Association of Cardiovascular and Pulmonary Rehabilitation (AACVPR), 480, 481t
American College of Sports Medicine (ACSM), 16–17, 212, 467t, 480, 481t, 482, 514
    early development of, 17–21
    key exercise science-related publications from, 20–21t
    position stands from, 18–19t
    role of, 21–22
American Council on Exercise (ACE), 467t, 480, 481t
American Heart Association, 514
American Institute of Nutrition (AIN), 252
American Medical Association, 210
American Medical Society for Sports Medicine (AMSSM), 212
American Occupational Therapy Association (AOTA), 30
American Orthopaedic Society for Sports Medicine (AOSSM), 212
American Physical Therapy Association (APTA), 29
American Physiological Society (APS), 482
American Society of Biomechanics (ASB), 481t, 482
American Society of Exercise Physiologists (ASEP), 119, 481t, 482–483
Anabolic state, 96, 142, 435
Anabolic steroids, 77
Anabolism, 261
Anatomic reference axes, 386t
Anatomic reference planes, 386t
Anatomic reference position, 386t
Androgenic effects, 96
Anemia, 445
    sports, 276
Angina pectoris, 161, 180, 182
Angular motion, 384–385
Animal spirits, 339
Animal Welfare Act, 52
Anorexia nervosa, 138
Anterior cruciate ligament (ACL)
    injuries, 234–236, *235*
    reconstruction, 229–231, *230*
Anthropometric measurements, 442
Anthropometry, 13
Anticarcinogenic action, 92
Antioxidant vitamins, 275
Anxiety, 312, 313–314
    disorders, 192–193

Applied research, 43–44, 44–45t
Aristotle, 379
Arousal, 308–309
   in athletic competition, *308*
   and performance using drive theory, 309–310, *310*
   and performance using inverted U hypothesis, 310–311, *311*
Arrhythmia, 182
Arthritis, 189–190
Arthroscopy, 228–229, *230*
Association for Applied Sport Psychology (AASP), 298, 472, 481t, 483
American Association of Cardiovascular and Pulmonary Rehabilitation (AACVPR), 465–466
Asthma, 186
Asthma attack, 186
Atherogenic dyslipidemia, 188
Atherosclerosis, 82, 182
Athletic trainers, 214–215, 215t, 469–470, *470* (*see also* Athletic training)
   examination, assessment, and diagnosis, 216, *216*, *217*
      course of action, 219
      primary survey, 217
      record keeping, 219, 219t
      secondary survey, 217–219, 218t
   health care administration and professional responsibility, 222
      legal and insurance issues, 224–225
      professional development and responsibility, 223
      providing coverage, 223–224
      working with team physician, 225–226, *226*
   immediate and emergency care, 219–220
   injury/illness prevention and wellness protection, 215–216
   therapeutic intervention, 220
      exercise activities, 221–222
      therapeutic modalities, 222, *223*, 224t
Athletic training, 206–208
   areas of study
      anterior cruciate ligament injuries in females, 234–236, *235*
      concussion management, 232–233, *233*
      mental health issues, 233–234
   athletic trainers, *206*
   early influences, 209
   history of, 209–211
   injuries treated at emergency rooms, *208*
   injury rates, 207–*208*
   professionals, primary responsibility areas (*see* Athletic trainers)
   twentieth-century influences, 209–210
   twenty-first-century influences, 210–211, 211t
Atrial fibrillation, 182
Attention, 311–312, 313t
Attribution theory, 306–307, *307*

Author guidelines, 56
Autologous chondrocyte implantation (ACI), 231–232
Autonomic nervous system, 72
Autonomic neurons, 72
Axial forces, 392

B

Basal ganglia, 361
Basal metabolic rate, 251
Basic research, 43, 44–45t
Behavior change therapy, 269
Behavioral and mental health disease
   anxiety disorders, 192–193
   depression, 193
Behavioral psychology, 301 (*see also* Sport psychology)
   qualitative assessments, 453–454, 454t
   quantitative assessments, 453
"Benchtop-to-bedside" research, 44
Big data, 520
Bioelectric impedance, 441, *441*
Bioelectricity, concept of, 339, *339*
Biomechanics, 378–379
   areas of study, 396–397
      clinical biomechanics, 397–399, *398*
      ergonomics, 399–400, 400t
      sport biomechanics (*see* Sport biomechanics)
   complex movement, 394
      kinetic link principle, 395–396
      projectiles, 395
   early influences, 379–381, *380–381*
   historic events in development of, 383t
   history of, 379–383
   movement, 390–391
      acute *vs.* chronic loads, 394
      concepts and definitions related to kinetics, 392t
      loading, effects of, 393–394
      mechanical loads on human body, 391, 392t
      tension and combined loads, 392–393, *393*
   study of, 383–384
      body motion, 384–385, *384–385*
      joint movement terminology, 386, *387*, 387t
      mechanical systems, 385–386
      qualitative analysis of human movement, 388–390, *391*
      spatial reference systems, 387–388, *388–389*
      standard reference terminology, 386, 386t
   twentieth-century influences, 382
   twenty-first-century influences, 382–383
Biomechanist, 470
Biomedical engineer, 470
Biopsy, 231 (*see also specific biopsies*)
Blood collection
   blood measures, 445–446
   general equipment, 442, *444*, 445

Blood pressure, 173, 174t
    assessment, 426–427, *427*
    measurements of, 11, *12*
Blood pressure cuff, 427, *427*
Board of Certification in Professional Ergonomics, 493t
Board of Certification, Inc. (BOC), 210
Body composition, 102
    assessment
        anthropometric measurements, 442
        bioelectric impedance, 441, *441*
        densitometry, 439–440, *440*
        dual energy X-ray absorptiometry, 441
        skinfold assessments, 442, *443*
Body mass index (BMI), 141
Body motion, types of, 384–385, *384–385*
Body stabilization, 349–350
Bone densitometer, 433
Bone metabolism
    physical activity and exercise, implications for, 136–137
    sport and athletic performance, implications for, 138, *138*
Bone mineral density, 79, *80*, 138, *138*
Borg scales for perceived exertion, 174, *175*
Bradykinesia, 367
Brain activity measurement, 451–452, *452*

C

Caffeine ingestion, 280
Calcium intake, 276
Cancer, 193
    exercise effect on, 99
Carbohydrates, 258–259, *258* (*see also specific types*)
    consumption, 130, *130*
    ingestion of, 92
    intake for sport, 270–271, 271t
    loading and supplementation, 278
Carcinogens, 92
Cardiac arrhythmia, 182
Cardiac cycle, 173
Cardiac muscle, 76
Cardiac output, 74, 182, 426
Cardiac rehabilitation programs, 82–83, 161
Cardiopulmonary system, 82, 85
Cardiorespiratory system, 82, 85
Cardiovascular disease (CVD), 179–180, *179*, 501
    angina pectoris, 182
    cardiac arrhythmia, 182
    chronic heart failure, 183
    coronary artery disease, 181–182
    Framingham Heart Study, 503–506
    HERITAGE Family Study, 506–507
    hypertension, 184
    myocardial infarction, 180–181, *181*
    peripheral vascular disease, 183, *183*
    potential benefits of exercise for, 180t
    risk factors for, 501t
    valvular heart disease, 182–183
Cardiovascular system, *83*
    and exercise science, 82–84, *84*, 85t
    functions of, 82, 85t
    primary components of, 82
    and pulmonary function assessment
        blood pressure assessment, 426–427, *427*
        electrocardiograph equipment, 424, *424–425*
        ergometers, 417–420, *418–419*
        metabolic measurement equipment, 420–421, *421*
        pulmonary function equipment, 421–424, *422–423*
        pulse oximeter, 425–426, *426*
        treadmills, 417–420
Careers and professional issues, 24–25, *25*, 462–463, *462*
    additional organizations and agencies, 493, 493t
    career employment and professional opportunities, 468–469
        athletic trainer, 469–470, *470*
        biomechanist, 470
        biomedical engineer, 470
        chiropractor, 471
        clinical exercise physiologist, 471
        dietitian/sport dietician, 471–472, *472*
        ergonomist, 472
        exercise and sport psychologist, 472
        graduate school and researcher, 473–474, 473t
        medical doctor, 474
        occupational therapist, 474
        personal trainer, 475
        physical therapist, 475–476, *475*
        physician assistant, 476, *476*
        podiatrist, 476
        public and private school teacher, 477
        recreational therapist, 477
        strength and conditioning coach, 477, *478*
        wellness and fitness industry professional, 478–479, 479t
    certification, 463–464
        defined, 464–466, 467t
    licensure, 463–464
        defined, 466, 467t
    professional organizations, 467–468t, 479–480 (*see also* Professional organizations)
    registration, 463–464
        defined, 466, 467t
    U.S. government agencies
        Centers for Disease Control and Prevention, 491–492, *491*
        Department of Health and Human Services, 487–489, *488*, 489t
        Deputy Director of Public Health Science and Surveillance, 492, 492t
        National Institutes of Health, 489–491, 490t

Cartesian coordinate system, 388
Catabolism, 261
Catastrophic injury insurance, 225
Causal attributions, 306
Centers for Disease Control and Prevention (CDC), 491–492, *491*, 512, 514
Cerebellum, 360–361
Cerebral concussion, 233
Cerebral palsy, 73, 192
Certification, 463–464
  defined, 464–466, 467t
Certified athletic trainers, 23, *23*
Challenge point, 358
Childhood, 343–344, *344*
Chiropractic medicine, 27
Chiropractic school, 27
Chiropractor, 471
Cholesterol, 259
Chondrocytes, 231
ChooseMyPlate, 267, *267*
Chronic adaptations, 123
  to physical activity and exercise, 123–124, 123t
  research in, 124t
Chronic bronchitis, 184
Chronic heart failure, 183
Chronic injury, 394
Chronic load, 394
Chronic obstructive pulmonary disease (COPD), 85
Chronic traumatic encephalopathy (CTE), 233
Chronometric method, 349
Circumplex model, 311
Clinical biomechanics, 378, 397–399, *398*
Clinical depression (*see* Depression)
Clinical exercise physiologist (CEP), 471
Clinical exercise physiology
  areas of study in, 193–194
    exercise location and modality, 194–195
    individualized exercise, 195
    technology, use of, 194
  defined, 160
  early influences, 161
  history of, 160–164, 165t
  late twentieth-century influences, 161–163, 163t
  physiologists' duties and responsibilities (*see* Physiologists, clinical exercise)
  significant events in history of, 165t
  specific disease conditions (*see specific diseases*)
  twenty-first-century influences, 163–164
Clinical studies, 501
Clinical trials, 44, 501, 509
Closed kinetic chain, 221
Closed-loop system, 363–364, *363*
Coauthors, 56
Cognition, 295, 311–312, 313t
Cognitive function, exercise and, 316–317

Cognitive psychology, 348
Cold environmental conditions, 148–149
Commission on Accreditation of Allied Health Education Programs (CAAHEP), 465
Commission on Accreditation of Athletic Training Education (CAATE), 210, 469
Committee on the Accreditation for the Exercise Sciences (CoAES), 465
Competitive tennis, 517, *518*
Complex carbohydrates, 258, *258*
  intake for sport, 271
Compressive force, 392–393, 392t
Computed tomography, 433, *433*
Concentric muscle actions, 221
Concussion management, 232–233, *233*
Conduction, 148
Conflicts of interest, 58
Constructionist approach, 453
Contextual interference, 354–355, *354*
Continuing education units (CEUs), 232, 465
Continuing medical education (CME), 232
Control of walking, 368–369
Controllability, 306
Convection, 148
Conventional approach, 401
Cooper Institute, 493t
Copyright agreement, 58
Coronary artery disease (CAD), 82, 181–182
Course of action, athletic trainer, 219
COVID-19 pandemic, 500, 524
Creatine phosphate, 101, 281
Credentialing, 463
Credentialing organization, 463
Cross-sectional research, 47
Cross-sectional studies, 342
Crossover point, 127, *128*
Cryotherapy, 446
Crystal intelligence, 346–347
Curvilinear translation, 384
CVD (*see* Cardiovascular disease)
Cystic fibrosis, 186

D

DASH Eating Plan, 265, 267
Data analysis, 54
Data collection, 52–53, 53t
De moti animalium (Borelli), 379
Decoding, 352
Deconditioned state, 479
Degenerative joint disease (*see* Osteoarthritis)
Dehydration, 279
Delayed onset muscle soreness, 76–77
Densitometry, 439–440, *440*
Dental Admissions Test (DAT), 26

Dental school, 26
Deoxyhemoglobin, 426
Department of Health and Human Services (DHHS), 487–489, *488*, 489t
Dependent variable, 51
Depression, 193, 312, 314
Deputy Director of Public Health Science and Surveillance (DDPHSS), 492, 492t
Descriptive research, 47
Diabetes mellitus, 132, 186–188, 259
Diagnostic testing, 165, *166*, 414
Diaphoresis, 182
Diastolic blood pressure, 173
Diet, 251
    and nutrition, 248
Dietary Guidelines Advisory Committee (DGAC), 266
Dietary recall method, 262t, 263–265
Dietary record method, 262t, 263–265, *263*
Dietary reference intake (DRI), 141, 257
Dietitian/sport dietician, 471–472, *472*
Diffuse brain injury, 233
Digestive system
    and exercise science, 92–93, 93t
    functions of, 90–91, 93t
    primary components of, 91, *91*
Direct calorimetry, 251
Directional terms, 386t
Discobolus statue, 10, *10*
Discus Thrower, 10, *10*
Dispositional approach, 302–304, *303*
Dissemination of findings, 54–55
Distraction hypothesis, 315–316
Doctoral degree in physical therapy (DPT), 29, 30
Doctorate of occupational therapy (OTD), 30
Dopamine, 316
Doubly labeled water method, 144, 144t, 437
Drive theory, 309–310, *310*
Dual energy X-ray absorptiometry (DXA/DEXA)
    body composition assessment, 441
    musculoskeletal system assessment, 433–434, *434*
Duration, 123
Dynamic systems theory, 362–363
Dynamics, 378
Dyspnea, 177
Dystrophin, 191

E

Early childhood, 343–344, *344*
Eccentric force, 392
Eccentric muscle actions, 77, 221
Echocardiography, 175–176
Ecological models, 299
Educational Assistance Act of 1972, 234
Efferent (motor) neurons, 72

Efficacy, 145
Electrocardiograph (ECG), 168, 175, *176*
    equipment, 424, *424–425*
Electroencephalography (EEG), 451, *452*
Electrolytes, 90
Electromyography (EMG), 381
    motor performance assessment, 450
    musculoskeletal assessment, 428–429, *428*
Electronic technology in cardiac rehabilitation programs, 194
Embryonic myotubes, 134
Emergency care, 219–220
Emotion and performance, 308–309, *308*
    circumplex model, 311
    drive theory, 309–310, *310*
    inverted U hypothesis, 310, *311*
Emotional well-being, 314–315, *315*
    exercise and psychological well-being, 315–316, 316t
Employee fitness and wellness programs, 512, *513*
Encoding, 352
Endocrine system, *71*
    clustering of metabolic syndrome risk factors, relationship among, 95
    endocrine glands and selected secreted hormones, 94t
    and exercise science, 95–96, 97t
    primary components of, 94
    primary hormones and functions of, 93–94, 97t
Endorphins, 316
    hypothesis, 316
Energy assessment methods, 144t
Energy balance
    assessment
        energy expenditure (*see* Energy expenditure, measuring)
        energy intake, 434–435
    and weight control, 139, *139*
        physical activity and exercise, implications for, 140–141, *140*, *142*
        sport and athletic performance, implications for, 141–142
Energy expenditure
    measuring, 435–436, *435*
        accelerometers, 437, 438, *439*
        doubly labeled water, 437
        heart rate monitors, 437, *437*
        pedometers, 437, 438, *438*
        physical activity questionnaires, 437, 438
        whole room indirect calorimeter, 436–437, *436*
    and physical activity, 143–144
        energy assessment methods, 144t
        invasive muscle biopsy, 143, *143*
        pedometers, 143, *143*
        physical activity and exercise, implications for, 144–145
        sport and athletic performance, implications for, 145

Energy intake assessment, 434–435
Energy systems
    components of, 101, 104t
    energy production systems in body, 102t
    and exercise science, 102–103, *103*, 104t
    functions of, 100–101, 104t
Enriched water, 437
Environment, 319
Environmental exercise, 145–146
    cold environmental conditions, 148–149
    heat-related illness, guidelines for, 147t
    physical activity and exercise, implications for, 146–147, *146*
    sport and athletic performance, implications for, 147
Epidemiologic studies, 500
Epidemiological studies, 252
Epidemiology, 95–96, 99, 127, 128, 502
    and health promotion, 502–503
        clinical trials, 509
        Framingham Heart Study, 503–504, *504*, 505–506t
        HERITAGE Family Study, 506–507, *507*
        National Health And Nutrition Examination Survey, 507–508, *508*
Equipment, 414
    and aids for performance, 521–523
    behavioral and psychological assessment
        qualitative assessments, 453–454, 454t
        quantitative assessments, 453
    blood collection
        blood measures, 445–446
        general equipment, 442, *444*, 445
    body composition assessment
        anthropometric measurements, 442
        bioelectric impedance, 441, *441*
        densitometry, 439–440, *440*
        dual energy X-ray absorptiometry, 441
        skinfold assessments, 442, *443*
    cardiovascular and pulmonary function assessment
        blood pressure assessment, 426–427, *427*
        electrocardiograph equipment, 424, *424–425*
        ergometers, 417–420, *418–419*
        metabolic measurement equipment, 420–421, *421*
        pulmonary function equipment, 421–424, *422–423*
        pulse oximeter, 425–426, *426*
        treadmills, 417–420
    energy balance assessment
        energy expenditure (*see* Energy expenditure, measuring)
        energy intake, 434–435
    improvement, sport biomechanics, 402–403, *402*, *404*
    motor performance assessment
        brain activity measurement, 451–452, *452*
        electromyography, 450
        eye-tracking instruments, 450–451
        goniometers, 449, *449*
        information processing measurement, 451, *451*
        motion capture systems, 450, *450*
        potentiometers, 449
    musculoskeletal assessment
        computed tomography, 433, *433*
        dual energy X-ray absorptiometry, 433–434, *434*
        electromyography, 428–429, *428*
        force platforms, 429, *429*
        handgrip dynamometer, 430
        isokinetic dynamometers, 430, *430*
        magnetic resonance imaging, 430–431, *431*
        magnetic resonance spectroscopy, 431
        muscle biopsy equipment, 432–433, *432*
        pressure-sensitive insoles, 429
    pretesting guidelines and procedures, 414–416
        individual issues to control, 416
        laboratory and field testing, 416–417
    rehabilitation
        resistance devices and exercise equipment, 446
        therapeutic ultrasound, 447, *448*
        thermotherapy, 446
        transcutaneous electric nerve stimulation, 447–448, *448*
        whirlpool tubs, 447, *447*
Ergogenic aids, 101, 280–281, 282t
Ergometers, 417–420, *418–419*
Ergonomics, 399–400, 400t
Ergonomist, 472
Erythropoietin, 80
Essential amino acids, 261
Euhydration, 277
Evaporation, 148
Evidence-based medicine, 503
Evidence-based practice
    defined, 60–61
    model, *60*
    working of, 61–62, 61t
Exclusion criteria, 52
Exercise
    activities, 221–222
    adherence, 318
        determinants of, 321–322, *321*, 322t
    behavior, 317–318
        theories and models, 318–320, 319t, *320*
    defined, 6
    induced asthma, 86, 186
    location and modality, 194–195
    prescription, 178–179
    testing and evaluation, 165–166, *166*
Exercise and sport psychologist, 472
Exercise is Medicine (EIM), 22, 163, 164, 514
Exercise physiology, 117–118
    areas of study, 126
        bone metabolism, 136–138, *137–138*
        carbohydrate consumption during exercise, *130*

energy balance and weight control, 139–142, *139–141*
energy expenditure and physical activity, 143–145, *143*, 144t
environmental exercise, 145–149, *146*, 147t, *148*
factors controlling substrate metabolism, 127–130, *128*, 128t
muscle control of glucose uptake, 130–134, *131*
skeletal muscle physiology, 134–136, *135*
basis of study
acute responses to physical activity and exercise, 121–122, 122t
chronic adaptations to physical activity and exercise, 123–124, 123–124t
training and conditioning programs, 124–126
defined, 117
history of, 117–121
other areas of, 149
twentieth-century influences
early, 117–118, *118*
late, 118–119
twenty-first century influences, 119–120, 120–121t
Exercise Preparticipation Health Screening Questionnaire, 168, *169*
Exercise psychology, 294–296, 312
anxiety, 313–314
areas of study
determinants of exercise adherence, 321–322, *321*, 322t
imagery and performance enhancement, 323–324
injury and mental health, 322–323
conceptual framework of, *294*
defined, 294
depression, 314
early influences, 296
emotional well-being, 314–315, *315*
exercise and psychological well-being, 315–316, 316t
exercise and cognitive function, 316–317
exercise behavior, 317–318
theories and models, 318–320, 319t, *320*
history of, 296–297
mind and body (*see* Mind and body)
significant events in development of, 300t
twentieth-century influences, 297–299, *297*
twenty-first century influences, 299
Exercise science, 2–7
academic preparation in, 22–23, *23*
careers in health care, 24–25, *25*
chiropractic school, 27
dental school, 26
graduate school, 30–31, 31t
medical school, 25–26
physical and occupational therapy, 29–30, *29*
physician assistant programs, 28–29, *28*
podiatric medical school, 27–28
undergraduate coursework, 23–24, 24t
age-adjusted prevalence of overweight and obesity, 3, *3*

and American College of Sports Medicine, 16–17
early development of, 17–21
key exercise science-related publications from, 20–21t
position stands from, 18–19t
role of, 21–22
areas of study in, 7–8, 8t
causes of mortality, 2, *2*
definition of, 6–7
early influences, 10–11, *10–11*
as field of study, 8–9
high school athletics, 4, *5*
history of, 9–16
intercollegiate athletics, 4, *5*
nineteenth-century influences, 11–12
physical activity guidelines, adults aged 18 years and older, 2, *3*
relationships of disciplines and subdisciplines of, 7–8, *7*
twentieth-century influences
early, 12–14, *14*
late, 14–15, *15*
twenty-first-century influences, 15–16
Exogenous hormone supplementation, 96
Exogenous insulin, 132
Expanded Food and Nutrition Education Program (EFNEP), 266
Experimental research, 47
study, designing, 49–50
Experimental variable, 51
Extrinsic motivation, 304
Eye-tracking instruments, 450–451
Eysenck's model, 302

F

Fading knowledge of results, 356
Family Educational Rights and Privacy Act (FERPA), 225
Fat, 259–260
as energy substrate, 128–129
intake for sport, 274
oxidation, 129
-soluble vitamins, 261
Female athlete triad, 138, *138*, 276, 523
Festination, 367
Fibrosis, 186
Field of study, Exercise Science as, 8–9
Field test, 416–417
Financial support, 58
Fitness and functional capacity testing, 414
Fitness industry, 478
Fluid intake and fluid replacement, 278–280, 279t
Fluid intelligence, 346–347
Fluid replacement
fluid intake and, 278–280, 279t
hydration status and, 276–277, *277*

Fluorometer, 445
Food deserts, 269–270
Food frequency questionnaire, 263
Food swamps, 269–270
Force platforms, 429, *429*
Fosbury flop, 401
Framingham Heart Study, 503–504, *504*, 505–506t
Free fatty acids, 259
Frequency, 123
Functional benefits, 160
Functional capacity, 176
    testing, 166
Functional magnetic resonance imaging, 452
Future research, 511–512 (*see also* Twenty-first century)

G

Gait/gait pattern, 368, 397, *398*
Galley proof, 59
Gas chromatograph/mass spectrometer, 445
Gastric emptying, 92–93
Gastrointestinal system, 92
Gatorade Sports Science Institute, 255
Gene polymorphism and total fat mass, 507, *507*
General motion, 385
General motor control systems, 362–363
    closed-loop system, 363–364, *363*
    open-loop system, 364–365, *364*
Genetic epidemiology, 506
Genetics, in health and fitness, 119
Genotype, 506
Gestational diabetes, 187
Glucagon, 128
Glucose polymer solutions, 130
Glucose transport protein 4 (GLUT 4), 131, *131*, 259
Glucose uptake, muscle control of (*see* Muscle control of glucose uptake)
Glycogen, 258
Glycogenolysis, 101
Glycolysis, 101
Goniometers, 449, *449*
Graded exercise tests (GXTs), 166
    maximal graded exercise testing, 177
    submaximal graded exercise tests, 176–177
Graduate Record Examination (GRE), 30
Graduate school, 30–31, 31t
Graduate school and researcher, 473–474, 473t
Graduates of programs, 462 (*see also* Careers and professional issues)
Gross domestic product, 511
Ground reaction forces, 429, *429*

H

Habit, arousal and, 309
Habit Theory, 318
Halter monitors, 424

Handgrip dynamometer, 430
Health, 2
    equity, 16
    exercise science and, 500–502, *502–503*
    risk, pretest screening for, 167
    status, 167–168, *169*
Health Behavior Model, 318–319
Health care administration and professional responsibility, 222
    legal and insurance issues, 224–225
    professional development and responsibility, 223
    providing coverage, 223–224
    working with team physician, 225–226, *226*
Health Insurance Portability and Accountability Act (HIPAA), 225
Health nutrition, 248–250, *248–249*, 250t, 265–266
    areas of study, 268–270
        effective diets for weight loss, 268–269
        food deserts and food swamps, 269–270
        strategies for promoting healthy weight loss, 269
    dietary changes and risk for common disease conditions, 265t
    dietary guidelines for health, 266–268, *267*
    early influences, 250
    history of, 250–256
    twentieth-century influences, 250–252, *251*, 253t
    twenty-first-century influences, 253–254
Health promotion, 502
    epidemiology and
        clinical trials, 509
        Framingham Heart study, 503–504, *504*, 505–506t
        HERITAGE Family Study, 506–507, *507*
        National Health And Nutrition Examination Survey, 507–508, *508*
Health-related physical fitness testing and interpretation, 177–178, 178t
Healthy People 2030, 141, 318, 487–489, 510
Healthy People Program, 510
Heart, 82
Heart attack (*see* Myocardial infarction)
Heart failure, chronic, 183
Heart rate, 169, 173
    –monitoring method, 144
    monitors, 437, *437*
Heat loss, 148
Heat-related illness, 146–147
    guidelines for, 147t
Hematology, 445
Hematopoiesis, 79
Hemoglobin, 276
HERITAGE Family Study, 119, 506–507, *507*
Heterogeneous group, 134
High-altitude training, 520, *521*
High school athletics, 4, *5*
Historical research, 49
Homeostasis, 94, 121

Hormones, effect of, 95
HUMAN calorimeter, 251, *251*
Human Movement Science, 341
Hydration status and fluid replacement, 276–277, *277*
Hypercholesterolemia, 141, 260
Hyperinsulinemia, 96, 140
Hyperlipidemia, 140, 188, 260
Hypertension, 173, 184
Hypnotic catalepsy, 296
Hypoestrogenemia, 138
Hypoglycemia, 132, 187, 259, 271
Hypoglycemic condition, 129
Hyponatremia, 147, 277
Hypothalamic amenorrhea, 138
Hypothermia, 148
Hypothesis, 51

I

Imagery and performance enhancement, 323–324
Immediate and emergency care, 219–220
Immediate energy sources, 101
Immune system
    components of innate and acquired immunity, 98, 98t
    exercise and upper respiratory illness relationship, 99
    and exercise science, 99–100, 100t
    functions of, 97–98, 100t
Inactivity physiology, 116
Inclusion criteria, 52
Independent variable, 51
Indirect calorimetry, 251, 420
Individualized exercise, 195
Infancy, 343
Information processing, 348–351, *350*
    measurement, 451, *451*
Informed consent, 168, *170*
Injury, 394
    prevention, sport biomechanics, 405
Innate immunity, 98, 98t
Institutional Animal Care and Use Committee (IACUC), 51–52
Institutional Review Board (IRB), 51–52
Insulin, 95–96
    shock, 132
Intensity, 123
Interactive components, 465
Intercollegiate athletics, 4, *5*
International Federation of Sports Medicine (FIMS), 212
International Journal of Sport Nutrition, 255
International Journal of Sport Nutrition and Exercise Metabolism, 255
International Journal of Sport Psychology, 298
International Journal of Sports Biomechanics, 382
International Society of Biomechanics (ISB), 481t, 483
International Society of Biomechanics in Sport (ISBS), 481t, 483
International Society of Motor Control (ISMC), 481t, 484
International Society of Sport Psychology, 298

Intrinsic motivation, 304
Inverted U hypothesis, 310, *311*
Iron, 276
Iron Man Triathlon, 517, *517*
Ischemia, 180
Isokinetic dynamometers, 221, 430, *430*
Isokinetic muscle actions, 221
Isometric muscle action, 221
Isotonic muscle actions, 221
Isotope, 437

J

Joint movement terminology, 386, *387*, 387t
Journal of biomechanics, 382
Journal of Human Movement Studies, 341
Journal of Motor Behavior, 341

K

K-rations, 251
Ketoacidosis, 187
Kidneys, 88
Kinematic information, 450
Kinematics, 378
Kinesiology, 7
Kinetic link principle, 395–396
Kinetics, 378
    concepts and definitions related to, 392t
    defined, 390
Knee replacements, 398–399
Knowledge
    for qualitative analysis, 389–390
    of results, motor learning, 355–356
        fading, 356
        summary, 356
    use of, motor development, 347–348, *347*
Kraus–Weber physical fitness test, 14, *15*

L

Laboratory and field testing, 416–417
Lactic acid, 101
    formation of, 133
    removal of, 104
Later childhood, 343–344, *344*
Le Mouvement (Woodworth), 340
Lead author, 55, 56
Learning difficult motor skills, 357–359, *357–358*
Licensure, 463–464
    defined, 466, 467t
Lifespan stages, 342–343
    adolescence, 344, *345*
    adulthood, 345
    early childhood and later childhood, 343–344, *344*
    infancy, 343
    older adulthood, 345–346, *346*, 346t
    prenatal, 343

Linear motion, 384
Lipases, 128
Lipoprotein, 188
Locus of causality, 306
Long-term memory, 351–353
Longitudinal research, 47
Longitudinal studies, 342
Loss of bone mass, 190, *191*
Lumen, 182

M

Macronutrients, 90, 257
Magnetic resonance imaging (MRI), 430–431, *431*
Magnetic resonance spectroscopy (MRS), 431
Magnetoencephalography (MEG), 452
Major depressive disorder (*see* Depression)
Master of science in occupational therapy (MSOT), 30
Maturational theorists, 343
Maximal graded exercise testing, 177
Maximal lactate steady state, 103–104
Maximal oxygen consumption, 83–84, 168
Mechanical fatigue, 399
Mechanical loads, 391
    acute *vs.* chronic, 394
    effects of, 393–394
    on human body, 391, 392t
    tension and combined loads, 392–393, *393*
Mechanical systems, 385–386
Medical College Aptitude Test (MCAT), 26
Medical contraindications, 167
Medical doctor, 474
Medical school, 25–26
Mediterranean diet, 251
Memory, 351–353, *352–353*
"Memory drum" theory, 340
Menopause, 138
Mental health
    defined, 233–234
    injury and, 322–323
    issues, 233–234
Mental preparation for performance, 297
Meta-analysis, 268
    research, 48
Metabolic disease
    diabetes mellitus, 186–188
    hyperlipidemia, 188
    metabolic syndrome, 188–189
    obesity, 188
    potential benefits of exercise for, 187t
    prevalence of, *187*
Metabolic measurement cart, 420
Metabolic measurement equipment, 420–421, *421*
Metabolic syndrome, 95, 188–189
    risk factors, *95*

Metabolism (*see specific metabolisms*)
Metabolites, 431
Metabolomics, 516
Metatheoretical approaches, 299
Micronutrients, 90, 257
Million Hearts 2022, 164
Mind and body
    attention and cognition, 311–312, 313t
    emotion and performance, 308–309, *308*
        circumplex model, 311
        drive theory, 309–310, *310*
        inverted U hypothesis, 310, *311*
    motivation
        attribution theory, 306–307, *307*
        social cognitive theory, 304–306, *305*
    personality, 301–302, *301*
        dispositional approach, 302–304, *303*
    physical activity, exercise, sport, and athletic competition, 300–301
    self-determination theory, 307–308
Minerals, 262
    intake for sport, 274–276, 275t
Mixed methods research, 46
Monoamine hypothesis, 316
Monoglycerides, 259
Morbidity, 2
Mortality, 2, *2*
Motion capture systems, 450, *450*
Motivation
    attribution theory, 306–307, *307*
    Extrinsic, 304
    Intrinsic, 304
    social cognitive theory, 304–306, *305*
Motor behavior, 336–337
    areas of study
        control of walking, 368–369
        equipment, games, and playing environment modification, 365–366, *366*
        Parkinson disease, 367–368, *367*
    at early age, *338*
    history of, 337–341
    motor control (*see* Motor control)
    motor development (*see* Motor development)
    motor learning (*see* Motor learning)
    organizational structure of motor, *337*
Motor blocks, 367
Motor control, 336
    early influences, 339–340, *339–340*
    general motor control systems, 362–363
        closed-loop system, 363–364, *363*
        open-loop system, 364–365, *364*
    neuroanatomy, 359, *360*
        basal ganglia, 361
        cerebellum, 360–361
        motor cortex, 361

peripheral motor system, 362
premotor cortex, 362
primary motor cortex, 361
supplementary motor cortex, 361-362
recent influences, 340-341, 342t
Motor cortex, 361
Motor development, 336, 341-342
early influences, 337-338, *338*
lifespan stages, 342-343
adolescence, 344, *345*
adulthood, 345
early childhood and later childhood, 343-344, *344*
infancy, 343
older adulthood, 345-346, *346*, 346t
prenatal, 343
recent influences, 338-339
use of knowledge, 347-348, *347*
Motor learning, 336
early influences, 339-340, *339-340*
information processing, 348-351, *350*
knowledge of results, 355-356
fading, 356
summary, 356
memory, 351-353, *352-353*
part-whole practice, 356-357
learning difficult skills, 357-359, *357-358*
practice organization and contextual interference, 354-355, *354*
variability of practice, 355
recent influences, 340-341, 342t
Motor performance
assessment
brain activity measurement, 451-452, *452*
electromyography, 450
eye-tracking instruments, 450-451
goniometers, 449, *449*
information processing measurement, 451, *451*
motion capture systems, 450, *450*
potentiometers, 449
Mountain climbing, 517, *518*
Movement, 390-391
acute *vs.* chronic loads, 394
complex, 394
kinetic link principle, 395-396
projectiles, 395
concepts and definitions related to kinetics, 392t
loading, effects of, 393-394
mechanical loads on human body, 391, 392t
qualitative analysis of human movement, 388-390, *391*
tension and combined loads, 392-393, *393*
Multiple sclerosis, 191-192
Multistore memory model, 351
Muscle biopsy, 135, 143, *143*, 144t, 145
equipment, 432-433, *432*
history, 254

Muscle control of glucose uptake, 130-132, *131*
physical activity and exercise, implications for, 132-133
sport and athletic performance, implications for, 133-134
Muscle fiber hyperplasia, 77
Muscle fiber hypertrophy, 77, *78*
Muscle glycogen, 270
loading, 278
Muscular dystrophy, 191
Muscular strength in older adults, 135
Muscular system, *71*
and exercise science, 76-78, *78*, 78t
functions of, 75-76, 78t
nomenclature, and specific contractile and metabolic characteristics of skeletal muscle fibers, 76t
primary components of, 76
Musculoskeletal system
assessment
computed tomography, 433, *433*
dual energy X-ray absorptiometry, 433-434, *434*
electromyography, 428-429, *428*
force platforms, 429, *429*
handgrip dynamometer, 430
isokinetic dynamometers, 430, *430*
magnetic resonance imaging, 430-431, *431*
magnetic resonance spectroscopy, 431
muscle biopsy equipment, 432-433, *432*
pressure-sensitive insoles, 429
MyActivity Pyramid, 502, *503*
Myocardial infarction, 180-181, *181*

N

National Athletic Trainers' Association (NATA), 209-211, 468t, 481t, 484
Research and Education Foundation, 211
National Collegiate Athletic Association (NCAA), 4, *5*
National Health And Nutrition Examination Survey (NHANES), 3, *3*, 507-508, *508*
nutrition and health, 252, 253
National Heart, Lung, and Blood Institute, 503
National Institutes of Health (NIH), 489-491, 490t
National School Lunch Program (NSLP), 266
National Strength and Conditioning Association (NSCA), 467t, 481t, 484
Naturalistic research, 46
Nervous system
components of, 74-75
disorders and benefits of exercise, 74t
and exercise science, 73-75, 75t
functions of, 72-73, 75t
organizational structure of, *73*
Neuroanatomy, 359, *360*
basal ganglia, 361
cerebellum, 360-361

Neuroanatomy (*continued*)
    motor cortex, 361
    peripheral motor system, 362
    premotor cortex, 362
    primary motor cortex, 361
    supplementary motor cortex, 361–362
Neurotransmitters, 314, 316–317
Nonaxial force (*see* Eccentric force)
Nonessential amino acids, 261
Noradrenaline, 95
Norepinephrine, 95–96, 128, 314, 316
Normative data, 178
North American Society for the Psychology of Sport and Physical Activity (NASPSPA), 298, 341, 481t, 485
Nutrition (*see also* Health nutrition; Sport nutrition)
    defined, 248
    diet and, 248
    intake
        dietary recall and dietary record, 262t, 263–265, *263*
        disease conditions and, 248, *249*
        measurement, 262–265
    nutrients, 257–258, 257t
        carbohydrate, 258–259, *258*
        fat, 259–260
        minerals, 262
        protein, 260–261, *260*
        vitamins, 261
        water, 262
    strategies for enhancing sport and athletic performance, 250t

O

Obesity, 140, 188 (*see also* Energy balance and weight control)
Objective changes, 295
Obstructive lung diseases, 423
Obstructive pulmonary disease (COPD), 184–185, *185*
Occupational therapist, *29*, 474
Occupational therapy (OT), 29–30, *29*
Occupational therapy assistant (OTA), 29
Older adulthood, 345–346, *346*, 346t
Older adults
    cardiovascular, musculoskeletal, and psychomotor function, primary changes in, 346t
    exercise and motor skill development, 345–346, *346*
    improvements in cognitive function, 352, *352*
    muscular strength in, 135
Olympic Games, 4, *4*
Omni scale, 174
Open access journals, 59
Open kinetic chain, 221
Open-loop system, 364–365, *364*
Oral presentations, at professional conference, 55
Orthopedic and neuromuscular disease
    arthritis, 189–190
    cerebral palsy, 192
    multiple sclerosis, 191–192
    muscular dystrophy, 191
    osteoporosis, 190–191, *191*
    potential benefits of exercise for, 190t
    prevalence of, *189*
Orthopedic medicine, 212
Oscilloscope, 173
Osmolarity, 88
Osteoarthritis, 189–190, 397–398
Osteopathic medicine, 25
Osteopenia, 190, 433
Osteoporosis, 79, 136, *136*, 190–191, *191*, 276, 433
Outcome-dependent effect, 306, *307*
Overload, 125
Overtraining syndrome, 99–100
Oxidative metabolism, 101
Oxygen consumption, 176
    at maximal effort exercise, 420
Oxygen free radicals, 275
Oxyhemoglobin, 426

P

Palpation, 218
Paralympic Games Olympic Games, 4, *4*
Parasympathetic nervous system, 72
Parenchymal lung disease, 185
Parkinson disease, 367–368, *367*
Part-whole practice method, 356–357
    learning difficult skills, 357–359, *357–358*
Pathophysiology, 160
Pedometers, 143, *143*, 144, 144t, 437, 438, *438*
Peer-reviewed journals, 54, 56–59, *57*, 58t
Peptides, 260
Perceptual changes, 295
Performance
    defined, 354
    enhancement, 323–324, 515
    equipment and aids for, 521–523
    health and safety for, 523–524, *524*, 525t
    psychology, 295
    training for, 519–521, *520–521*
Peripheral artery disease, 183, *183*
Peripheral motor system, 362
Peripheral vascular disease, 183, *183*
Personal trainer, 475
Personality, 301–302, *301*
    dispositional approach, 302–304, *303*
Phenotypes, 506
Physical activity, 6
    energy expenditure and (*see* Energy expenditure and physical activity)
    questionnaires, 144, 144t, 437, 438
Physical Activity Guidelines Advisory Committee Scientific Report, 119

Physical activity questionnaires, 144, 144t, 437, 438
Physical education, 8, 12
Physical environment, 319–320
Physical examination, 167
Physical therapist, *29*, 475–476, *475*
Physical therapy (PT), 29–30, *29*
Physical therapy assistant (PTA), 29
Physician assistant, 476, *476*
    programs, 28–29, *28*
Physicians (*see* Medical doctor)
Physiologists, clinical exercise, 160, 164
    exercise prescription, 178–179
    exercise testing and evaluation, 165–166, *166*
    health-related physical fitness testing and interpretation, 177–178, 178t
    performing test, 168, *171–172*
        blood pressure, 173, 174t
        echocardiography, 175–176
        electrocardiogram, 175, *176*
        heart rate, 169, 173
        maximal graded exercise testing, 177
        oxygen consumption and functional capacity, 176
        Rating of Perceived Exertion, 174, *175*
        submaximal graded exercise tests, 176–177
    pretesting procedures
        health status, 167–168, *169*
        informed consent, 168, *170*
        physical examination, 167
        pretest screening for health risk, 167
Plaque, 181–182
    buildup of, 82, *84*
Plasma, 88
Plyometric exercise training, 405
Plyometric training, 519, *520*
Podiatric medical school, 27–28
Podiatric medicine, 27
Podiatrist, 476
Policy initiatives and advocacy, 515, 515t
Polyphenols, 281
Position stand, 18, 18–19t
Positron emission tomography, 452
Postabsorptive state, 127
Poster presentations, at professional conference, 55
Potentiometers, 449
Practice organization and learning
    contextual interference, 354–355, *354*
    variability of practice, 355
Prediabetes, 259
Premotor cortex, 362
Prenatal, 343
Preprint servers, 59
Prescription, exercise, 178–179
President's Council on Fitness, Sports and Nutrition, 514
Pressure liquid chromatograph, 445
Pressure-sensitive insoles, 429
Primary motor cortex, 361

Primary sampling units, 507
Primary sources, 49
Primary survey of injury assessment, 217
Procarcinogens, 92
Professional conferences, 55–56
Professional Education in Athletic Training, 211
Professional liability insurance, 225
Professional organizations
    in exercise science, 467–468t, 479–480
        American Association of Cardiovascular and Pulmonary Rehabilitation, 480, 481t
        American College of Sports Medicine, 480, 481t, 482
        American Council on Exercise, 480, 481t
        American Physiological Society, 482
        American Society of Biomechanics, 481t, 482
        American Society of Exercise Physiologists, 481t, 482–483
        Association for Applied Sport Psychology, 481t, 483
        International Society of Biomechanics, 481t, 483
        International Society of Biomechanics in Sport, 481t, 483
        International Society of Motor Control, 481t, 484
        National Athletic Trainers' Association, 481t, 484
        National Strength and Conditioning Association, 481t, 484
        North American Society for the Psychology of Sport and Physical Activity, 481t, 485
        Society for Health and Physical Educators, 481t, 485
        Sports, Cardiovascular, and Wellness Nutrition Dietetics Practice Group, 485
    related to exercise science, 486, 486t
Program accreditation, 465
Progressive-part method of practice, 357
Projectiles, 395
Proprioceptors, 362
Prospective studies, 500
Protein, 260–261, *260*
    intake for sport, 271–274, 272t, *273*
Prothrombotic state, 188
Protocol, 52
Psychological core, 302
Psychological phenomena, 295
psychological well-being, exercise and, 315–316, 316t
Psychology and Athletics (Griffith), 297
Psychology of Coaching (Griffith), 297
Psychomotor function, 346
Public and private school teacher, 477
Pulmonary function equipment, 421–424, *422–423*
Pulmonary rehabilitation, 162, 163t
Pulmonary system
    and exercise science, 85–87, 87t
    functions, 84–85, 87t
    primary components of, 85, *86*
Pulse oximeter, 425–426, *426*
Pyruvic acid, 101

Q

Quadriceps (Q angle), 235
Qualitative analysis of human movement, 388–390, *391*
Qualitative assessments, 453–454, 454t
Qualitative research, 46
Quantitative assessments, 453
Quantitative research, 45–46

R

Race car driving, 517, *518*
Radiation, 148
Randomized controlled trial, 48
Range of motion (ROM), 218, 220
Rating of Perceived Exertion (RPE), 168, 174, *175*
Reaction time, 349, 451, *451*
Recombinant cycle ergometers, 417
Recombinant human erythropoietin (rEPO), 81
Recommended dietary allowances (RDA), 257
Record keeping, athletic trainer, 219, 219t
Recreational therapist, 477
Rectilinear translation, 384
Registered dietitians, 23, *23*
Registration, 463–464
    defined, 466, 467t
Regression equations, 440
Rehabilitation
    resistance devices and exercise equipment, 446
    therapeutic ultrasound, 447, *448*
    thermotherapy, 446
    transcutaneous electric nerve stimulation, 447–448, *448*
    whirlpool tubs, 447, *447*
Relapse Prevention Model, 318
Relative task difficulty, 358
Relaxation phase of heart (diastole), 173
Relaxation set-method, 297
Reliability, 53
Reliable assessment, 415
Renal insufficiency, 183
Repetitive load (*see* Chronic load)
Research
    abstract, 55
    in acute responses, 122t
    additional descriptions, 47–49 (*see also specific entries*)
    animal model, 40, *41*
    approval, obtaining, 51–52
    computer modeling, 40, *42*
    defined, 40
    evidence-based practice
        defined, 60–61
        model, *60*
        working of, 61–62, 61t
    human model, 40, *41*
    practices, 40–42, *41–42*, 43t
    process, *49*
    analysis, performing, 54
    data collection, 52–53, 53t
    dissemination of findings, 54–55
    experimental research study, designing, 49–50
    interpreting and presenting results, 54
    peer-reviewed journals, 56–59, *57*, 58t
    professional conferences, 55–56
    research approval, obtaining, 51–52
    study methodology, 51
    subjects, identifying, 52
    technical reports, preprint servers, and webinars, 59–60
    variables, defining, 51
    student, 62–63, 63t
    subjects, identifying, 52
    tissue samples and cell cultures, 40, *42*
    types of, 43–47, 44–45t (*see also specific types*)
Research Quarterly for Exercise and Sport, 341
Researchers, attitudes and characteristics of, 43t
Residual lung volume, 423
Resistance devices and exercise equipment, 446
Respiratory disease
    asthma, 186
    cystic fibrosis, 186
    obstructive pulmonary disease, 184–185, *185*
    potential benefits of exercise for, 184t
    restrictive pulmonary disease, 185
Respiratory exchange ratio (RER), 102, 435
Respiratory gas analysis, 144, 144t
Response programming, 349, 351
Response selection, 349, 351
Restrictive lung disease, 422
Restrictive pulmonary disease, 185
Reversibility, 125
Rhabdomyolysis, 279
Rheumatoid arthritis, 190
Road cycling, 403, *404*
Role-related behaviors, 302
Rotary motion/rotation (*see* Angular motion)

S

Satiety, 259
Scaling process, 365
Scientists, attitudes and characteristics of, 43t
Secondary hypertension, 184
Secondary sources, 49
Secondary survey of injury assessment, 217–219, 218t
Selective attention, 352
Self-concept, 298
Self-determination theory, 307–308
Self-efficacy, 298, 305–306
    model of, *305*
Self-esteem, 298
Self-reported physical inactivity, Prevalence of, 501, *502*

Sequential kinetic link principle, 395–396
Sequential motion (*see* Sequential kinetic link principle)
Sequential research, 48
Serotonin, 314, 316
Serum, 445
Shear force, 392t
Short-term memory, 351–353
Short-term sensory store, 351–353
Shutterstock, *78*
Sign and symptom-limited stress test, 177
Silent ischemia, 182
Simple carbohydrates, 258
Simple sugars (*see* Simple carbohydrates)
Simulated high-altitude tent, 520, *521*
Simultaneous kinetic link principle, 396
Single load (*see* Acute load)
Situation, 320
16PF questionnaire, 302, *303*
Skeletal muscle, 76
    physiology
        physical activity and exercise, implications for, 134–135, *135*
        sport and athletic performance, implications for, 135–136
Skeletal system
    and exercise science, 79–81, *80*, 81t
    functions of, 79, 81t
Skinfold assessments, 442, *443*
Skinned fiber technique, 135
Smooth muscle, 76
Soccer style, 401
Social cognitive theory, 304–306, *305*, 318, 319–320
Social Ecological Model, 318
Social environment, 319
Society for Health and Physical Educators (SHAPE), 481t, 485
Society for Sport, Exercise, and Performance Psychology, 299
Society of Health and Physical Educators (SHAPE), 339
Soft-tissue endoscopy, 213
Somatic neurons, 72
Spatial reference systems, 387–388, *388–389*
Specificity, 125
Spectrophotometer, 445
Sphygmomanometer, 173, 427
Sport and athletic competition, 6, 515–516
    equipment and aids for performance, 521–523
    future research, 516–519, *516–519*
    health and safety for performance, 523–524, *524*, 525t
    training for performance, 519–521, *520–521*
Sport biomechanics, 378, 400
    equipment improvement, 402–403, *402*, *404*
    injury prevention, 405
    technique improvement, 401, *401*, 402t
    training improvement, 403, 405

Sport nutrition
    areas of study, 277–278
        carbohydrate loading and supplementation, 278
        ergogenic aids, 280–281, 282t
        fluid intake and replacement, 278–280, 279t
    carbohydrate intake, 270–271, 271t
    early history, 254
    fat intake, 274
    hydration status and fluid replacement, 276–277, *277*
    protein intake, 271–274, 272t, *273*
    twentieth-century influences, 254–255
    twenty-first-century influences, 255–256, 256t
    vitamin and mineral intake, 274–276, 275t
Sport psychology, 294–296
    areas of study
        determinants of exercise adherence, 321–322, *321*, 322t
        imagery and performance enhancement, 323–324
        injury and mental health, 322–323
    conceptual framework of, *294*
    defined, 294
    early influences, 296
    history of, 296–297
    mind and body (*see* Mind and body)
    significant events in development of, 300t
    twentieth-century influences, 297–299, *297*
    twenty-first-century influences, 299
Sports anemia, 276
Sports, Cardiovascular, and Wellness Nutrition Dietetics Practice Group (SCAN), 485
Sports, Cardiovascular, and Wellness Nutritionists (SCAN), 255
    Dietetic Practice Group, 255
Sports medicine, 206–208
    advances in sports-related injuries treatment, 227
        anterior cruciate ligament reconstruction, 229–231, *230*
        arthroscopy, 228–229, *230*
        autologous chondrocyte implantation, 231–232
        ulnar collateral ligament reconstruction, 231
    areas of study
        anterior cruciate ligament injuries in female, 234–236, *235*
        concussion management, 232–233, *233*
        mental health issues, 233–234
    defined, 17
    early influences, 211–212
    history of, 211–214
    injuries treated at emergency rooms, *208*
    injury rates, *207–208*
    physicians, guidelines for, 226–227
    sports-related injuries, reasons for increased, 227t
    team physician consensus statements, 228t
    twentieth-century influences, 212–213
    twenty-first-century influences, 213–214, 214t

Sports-related injuries
    reasons for increased, 227t
    treatment, 227
        anterior cruciate ligament reconstruction, 229–231, *230*
        arthroscopy, 228–229, *230*
        autologous chondrocyte implantation, 231–232
        ulnar collateral ligament reconstruction, 231
Stability, 306
Standard reference terminology, 386, 386t
Starches (*see* Complex carbohydrates)
Statics, 378
Stethoscope, 173
Stimulus recognition, 349, 351
Stimulus–response alternatives, 350
Stimulus–response compatibility, 350
Strength and conditioning coach, 477, *478*
Stress echocardiogram, 176
Stress injury, 394
Stroke volume, 74
Student research activities, 62–63, 63t
Study methodology, 51
Subjective, Objective, Assessment, and Plan (SOAP), 219, 219t
Submaximal graded exercise tests, 176–177
Substrate metabolism, factors controlling, 127–128, *128*
    factors affecting fuel utilization, 128t
    physical activity and exercise, implications for, 129
    sport and athletic performance, implications for, 129–130, *130*
Substrates, 127
    utilization, 436
Summary knowledge of results, 356
Summer Olympic Games, 4, *4*
Supplemental Nutrition Assistance Program Education (SNAP-Ed), 266
Supplementary motor cortex, 361–362
Survey, interview, and observational research, 48
Sweating, 277, *277*
Symmetric weakness, 191
Sympathetic nervous system, 72
Symptomatic ischemia, 182
Systematic review, 48
Systems approach, 70 (*see also specific systems*)
Systems of body, *71*
Systolic blood pressure, 173

T

Task complexity, 356
Task difficulty
    defined, 357
    learning and, *358*
    and practice performance, 357
    relative, 358

Team Physician Consensus Statements, 213, 227, 228t
Team physician, working with, 225–226, *226*
Team-sport performance, 517, *519*
Technical reports, 59
Technique improvement, sport biomechanics, 401, *401*, 402t
Telehealth, 194
Tensile force, 392–393, 392t
Thallium testing, 424
Theory of Planned Behavior/Reasoned Action, 318, 319
Therapeutic benefits, 160
Therapeutic modalities, athletic trainers, 222, *223*, 224t
Therapeutic ultrasound, 447, *448*
Thermogenic hypothesis, 316
Thermotherapy, 446
Third-party payment for services, 225
Tommy John surgery, 231
Torque, 430
Torsion, 393
Total fat mass, gene polymorphism and, 507, *507*
Total knee replacements, 398–399
Total lung capacity, 423
Total vascular peripheral resistance, 427
Trace minerals, 261
Training and conditioning programs, 124–126
Training effect, 394
Training for performance, 519–521, *520–521*
Training improvement, sport biomechanics, 403, 405
Traits, 302
Transamination, 260–261
Transcutaneous electric nerve stimulator (TENS), 447–448, *448*
Translational research, 44, 44–45t
Transtheoretical Model of Behavior, 318, 320
Treadmills, 417–420
Treatment variable, 51
Tryptophan, 314
Twenty-first century, 500
    epidemiology and health promotion, 502–503
        clinical trials, 509
        Framingham Heart study, 503–504, *504*, 505–506t
        HERITAGE Family Study, 506–507, *507*
        National Health And Nutrition Examination Survey, 507–508, *508*
    exercise science and health, 500–502, 501t, *502–503*
    exercise science and sport and athletic competition, 515–516
        equipment and aids for performance, 521–523
        future research, 516–519, *516–519*
        health and safety for performance, 523–524, *524*, 525t
        training for performance, 519–521, *520–521*
    past information for future health, 509–511
        future research, 511–512, 513t
        policy initiatives and advocacy, 515, 515t
        program development, 512, *513*, 514, 514t

24-h recall, 263
Two-hundred-meter sprint, 517, *517*
Type 1 diabetes mellitus, 132, 187
    carbohydrates and, 259
Type 2 diabetes mellitus, 132, 187
    carbohydrates and, 259

U

Ulnar collateral ligament reconstruction, 231
Ultrasound treatment, *223*
Undergraduate coursework, 23–24, 24t
Undifferentiated satellite cells, 77
Upper respiratory illness, exercise and, 99, *99*
Ureter, 88
Urinary system
    and exercise science, 88–90
    functions of, 87–88, 90t
    male and female, primary components of, 88, *89*
Urine, 88
U.S. government agencies
    Centers for Disease Control and Prevention, 491–492, *491*
    Department of Health and Human Services, 487–489, *488*, 489t
    Deputy Director of Public Health Science and Surveillance, 492, 492t
    National Institutes of Health, 489–491, 490t
USDA Food Guide, 267

V

Valid assessment, 415
Validity, 53
Valvular heart disease, 182–183
Variability of practice, 355
Variables, defining, 51
Vasodilation, 183
Vegetarian, 435
Vitamin, 261
    fat-soluble vitamins, 261
    intake for sport, 274–276, 275t
    water-soluble vitamins, 261

W

Walking, control of, 368–369
Water, 262
    -soluble vitamins, 261
Webinars, 60
Weight control, energy balance and (*see* Energy balance and weight control)
Weight loss
    effective diets for, 268–269
    intervention programs, 141, *142*
    strategies for promoting healthy, 269
Wellness and fitness industry professional, 478–479, 479t
Western roll high-jump technique, 401, *401*
Whirlpool tubs, 447, *447*
Whole body plethysmograph, 422, *422*
Whole room indirect calorimeter, 436–437, *436*
Winter Olympic Games, 4, *4*
Working memory, 352
World Anti-Doping Agency (WADA), 213
World Health Organization, 493t

Y

Youth Sport Safety Alliance (YSSA), 210–211